# Proceedings of the Fourth International Meeting of Anthracology

Brussels, 8–13 September 2008
Royal Belgian Institute of Natural Sciences

Edited by

Freddy Damblon

BAR International Series 2486
2013

Published in 2016 by
BAR Publishing, Oxford

BAR International Series 2486

*Proceedings of the Fourth International Meeting of Anthracology*

ISBN 978 1 4073 1100 5



COVER IMAGE Hunters in Snow *(1565) Pieter Bruegel the Elder © Kunsthistorisches Museum Wien (Austria)*



BAR Publishing is the trading name of British Archaeological Reports (Oxford) Ltd. British Archaeological Reports was first incorporated in 1974 to publish the BAR Series, International and British. In 1992 Hadrian Books Ltd became part of the BAR group. This volume was originally published by Archaeopress in conjunction with British Archaeological Reports (Oxford) Ltd / Hadrian Books Ltd, the Series principal publisher, in 2013. This present volume is published by BAR Publishing, 2016.

Printed in England

BAR
PUBLISHING

BAR titles are available from:

BAR Publishing
122 Banbury Rd, Oxford, OX2 7BP, UK
EMAIL info@barpublishing.com
PHONE +44 (0)1865 310431
FAX +44 (0)1865 316916
www.barpublishing.com

## Table of contents

## List of Illustrations

**E. Allevato *et al.*: Evidence from charcoal analysis for the extensive exploitation of cork-oak (*Quercus suber*) forest in the Roman imperial period: the vicus of Thamusida (NW Morocco)**



**F. Antolín *et al.*: Changes in the perception of and the interaction with the environment from the Mesolithic (10300-8500 cal BC) to the Early Neolithic (c. 5400 cal BC) in Can Sadurní Cave (Begues, Barcelona province, Spain). A view from the archaeobotanical data**



**Y. Carrión *et al.*: Neolithic woodland in the North Mediterranean Basin: the case of *Olea europaea* L**



**L. Caruso Fermé: Forest resources exploitation and management by Selknam hunter-gatherer societies: Results of the archaeobotanical analysis of Ewan site (Tierra del Fuego, Argentina)**

**D.J. Goldstein & I. Shimada: An experimental and ecological approach to modeling ancient fuel use**



**C. Lancelotti *et al.*: Vegetation and wood exploitation at Harappa, Punjab (Pakistan). First results of the charcoal analysis**



**M. Martin Seijo & R. Piqué i Huerta: New data about Wood Use in the Northwest of the Iberian Peninsula**

**M. Moskal-Del Hoyo *et al.*: Anthracological analysis from a Bronze Age necropolis at Kokotów (Poland)**



**O. Nelle *et al.*: Charcoals from iron smelting furnaces – fuel supply and environment of a medieval iron smelting site near Peppange/Luxembourg**

## Addresses of the authors and co-authors

ALLÉE, Philippe. Géolab UMR-6042-CNRS, 39E Rue Camille Guérin 87036 Limoges, France. philippe.allee@unilim.fr

ALLEVATO, Emilia. Dipartimento di Archeologia e Storia delle Arti, Facoltà di Lettere e Filosofia, Università degli Studi di Siena, via Roma 56-53100 Siena, Italy. eallevat@unina.it

ALLUÉ MARTÍ, Ethel. Institut Català de Paleoecología Humana i Evolució Social (Unitat asociada al CSIC), University Rovira i Virgili, Tarragona, Spain. eallue@prehistoria.urv.cat

ANDERSON, Darwin. W. Department of Soil Science, University of Saskatchewan, Saskatoon, Saskatchewan, S7N 5A8 Canada. darwin.anderson@usask.ca

ANTOLÍN, Ferran. Universitat Autònoma de Barcelona, Campus de Bellaterra, Edifici B, Facultat de Lletres, Departament de Prehistòria, 08193 Bellaterra, Barcelona, Spain. ferranantolin@hotmail.com, fantolin@csic.imf.es

ARRIZABALAGA, Alvaro. Dept. Geografía, Prehistoria y Arqueología (University of the Basque Country UPV/EHU). Tomás y Valiente s/n, 01006 Vitoria-Gasteiz, Spain.

BADAL, Ernestina. Universitat de València, Av. Blasco Ibáñez 28, 46010 Valencia, Spain. Ernestina.badal@uv.es

BALLESTEROS, Anna. IPHES, Institut Català de Paleoecologia Humana i Evolució Social, Plaça Imperial Tarraco, 1, 43005,Tarragona, Spain. anna.ballesteros@estudiants.urv.cat

BELLAVIA, Valentina. Dipartimento di Arboricoltura, Botanica e Patologia Vegetale, Facoltà di Agraria, Università Federico II di Napoli, via Università 100 I-80055 Portici, Italy.

BURJACHS, Francesc. 1) ICREA. Barcelona, Catalonia, Spain. francesc.burjachs@urv.cat; 2) IPHES, Institut Català de Paleoecologia Humana i Evolució Social, Plaça Imperial Tarraco, 1. Universitat Rovira i Virgili, 43005 Tarragona,.Catalonia, Spain. francesc.burjachs@urv.cat

BUXÓ, Ramon. Museu d'Arqueologia de Catalunya. C/Pedret 95, 17007 Girona, Spain. rbuxo@gencat.cat

CARACUTA, Valentina. Laboratory of Archaeobotany and Palaeoecology, University of Salento, Italy. girolamo.fiorentino@ateneo.unile.it

CARRIÓN, Yolanda. Centro de Investigaciones sobre Desertificación (CIDE), CSIC, Carretera Moncada-Náquera Km 4,5. 46113 Moncada (Valencia), Spain. Yolanda.carrion@uv.es

CARUSO FERMÉ, Laura. Laboratori d'Arqueobotanica. Facultat de Lletres. Departament de Prehistoria. Universitat Autonoma de Barcelona (08193) Bellaterra, Cerdanyola del Vallès. Catalunya, Spain. lcarusoferme@gmail.com

CHABAL, Lucie. Centre de Bio-Archéologie et d'Ecologie, (USTL/CNRS/EPHE), Institut de Botanique, 163 rue Auguste Broussonet, 34090 Montpellier, France. Lucie.Chabal@univ-montp2.fr

COLAIANNI, Giampiero. Università del Salento, Lecce, Italy. quercus74@hormail.com

CROSSLAND, Donna. Kejimkujik National Park and National Historic Site, Maitland Bridge, Annapolis County, NS, B0T 1B0 Canada, Donna.Crossland@pc.gc.ca

DAMBLON, Freddy. Royal Belgian Institute of Natural Sciences, Department of Palaeontology, Section of Palaeobotany. Rue Vautier 29, B-1000 Brussels, Belgium. Freddy.damblon@naturalsciences.be

DI PASQUALE, Gaetano. 1) Dipartimento di Archeologia e Storia delle Arti, Facoltà di Lettere e Filosofia, Università degli Studi di Siena, via Roma 56 - 53100 Siena, Italy. 2) Dipartimento di Arboricoltura, Botanica e Patologia Vegetale, Facoltà di Agraria, Università Federico II di Napoli, via Università 100 I-80055 Portici, Italy. gaetano.dipasquale@unina.it

D'ORONZO, Cosimo. Università Cattolica del Sacro Cuore, Milano, Italy. girolamo.fiorentino@ateneo.unile.it

DOUCET, Jean-Louis. University of Liège, Gembloux, Agro-Bio Tech, Unit of Forest and Nature Management, Laboratory of Tropical and Subtropical Forestry, B-5030 Gembloux, Belgium. doucet.jl@fsagx.ac.be

DUFRAISSE, Alexa. CR CNRS, Archéozoologie, archéobotanique: sociétés, pratiques et environnements. Muséum national d'Histoire Naturelle, Département EGB, UMS 303-UMR 7209, CP 56, 55 rue Buffon, 75005 Paris, France. dufraisse@mnhn.fr

DURAND, Aline. Laboratoire d'Archéologie Médiévale Méditerranéenne (UMR 6572) et Université d'Aix-Marseille I, Maison Méditerranéenne des Sciences de l'Homme, 5, rue du Château de l'Horloge, BP 647, 13 094 Aix-en-Provence cédex 1, France. adurand@mmsh.univ-aix.fr

EDO, Manel. CIPAG. Facultat de Prehistòria, Història Antiga i Arqueologia. Universitat de Barcelona. C/ Onze de setembre, 32, 08859, Begues, Barcelona, Spain. medo@telefonica.net

EICHHORN, Barbara. Institute of Archaeological Sciences, Department of Pre- and Protohistory, Archaeology and Archaeobotany of Africa, Goethe University, Grüneburgplatz 1, D-60323 Frankfurt, Germany. barbara_eichhorn@yahoo.de b.eichhorn@em.uni-frankfurt.de

EUBA, Itxaso. ICAC Institut Català d'Arqueología Clàssica, Plaça Rovellat s/n; 43003, Tarragona, Spain. ieuba@icac.net

FABRE, Laurent. 1) INRAP, DIR Méditerranée, 561, rue Etienne Lenoir, Km Delta, 30900 Nîmes. 2) Centre de Bio-Archéologie et d'Ecologie, (USTL/CNRS/EPHE), Institut de Botanique, 163 rue Auguste Broussonet, 34090 Montpellier, France. l.fabre@thau-agglo.fr

FIGUEIRAL, Isabel. 1) INRAP, DIR Méditerranée, 561 rue Etienne Lenoir, Km delta, 30900 Nîmes. 2) Centre de Bioarchéologie et d'Ecologie (UMR 5059), Institut de Botanique, 163, rue Auguste Broussonnet, 34 090 Montpellier, France. figueir@univ-montp2.fr

FIORENTINO, Girolamo. Laboratory of Archaeobotany and Palaeoecology. University of Salento, Italy. girolamo.fiorentino@ateneo.unile.it

GARCÍA MARTÍNEZ, María Soledad. Dpto. de Biología Vegetal (Botánica), Facultad de Biología, Universidad de Murcia, Campus de Espinardo 30100- Espinardo, Murcia, Spain. msgarmar@um.es

GILLET, Jean-François. University of Liège, Gembloux Agro-Bio Tech, Unit of Forest and Nature Management, Laboratory of Tropical and Subtropical Forestry, Nature plus asbl, B-5030 Gembloux, Belgium, jf.gillet@natureplus.be

GOLDSTEIN, David John. United States National Park Service, Christiansted National Historic Site, Christiansted, Virgin Islands, 00820, USA, David_Goldstein@nps.gov

GRASSO, Anna Maria. Università degli Studi di Siena, Siena, Italy. grasso_annamaria@libero.it

GRAU ALMERO, Elena. Departament de Prehistòria i Arqueologia, Universitat de València, Av.Blasco Ibañez, 28, 46010 Valencia, Spain. Elena.grau@uv.es

JANSEN, Doris. Graduate School, "Human Development in Landscapes", Christian-Albrechts-University Kiel, Olshausenstr. 40, D-24098 Kiel, Germany. djansen@ecology.uni-kiel.de

KRAUSE, Rüdiger. Institute of Archaeology, University of Frankfurt, Grüneburgplatz 1, 60629 Frankfurt am Main, Germany, R.Krause@em.uni-frankfurt.de

LANCELOTTI, Carla. IMF-CSIC c/Egipciaques, 15; 08001 Barcelona, Spain, carla.lancelotti@imf.csic.es

LOO, Judy. Atlantic Forestry Centre, 1350 Regent Street South, Fredericton, New Brunswick, Canada. Judy.Loo@NRCan-RNCan.gc.ca

MARTELLA, D. Laboratory of Archaeobotany and Palaeoecology, Salento University, Italy.

MARTÍN SEIJO, María. Grupo de Estudos para a Prehistoria do Noroeste-Dep Historia I, Universidade de Santiago de Compostela, 15782 Santiago de Compostela, A Coruña, Spain, maria.martin.seijo@gmail.com

MARZIANO, Mario. Dipartimento di Archeologia e Storia delle Arti, Facoltà di Lettere e Filosofia, Università degli Studi di Siena, via Roma 56, 53100 Siena, Italy.

MATOGA, Andrzej. Archaeological Museum of Kraków, ul. Senacka 3, 31-002 Kraków, Poland. Andrzej.Matoga@ma.krakow.pl

MENSUA, Carmen. Laboratori d'Arqueobotànica, Departament de Prehistòria, Universitat Autònoma de Barcelona, Campus de Bellaterra, Edifici B, Facultat de Lletres, Departament de Prehistòria, 08193 Bellaterra, Barcelona, Spain, carmenmensua@gmail.com

MOSKAL-DEL HOYO, Magdalena. Department of Prehistory and Archaeology, University of Valencia, Av. Blasco Ibáñez 28, 46010 Valencia, Spain; magdalena.moskal@uv.es

NELLE, Oliver. Institute for Ecosystem Research, Christian-Albrechts-University Kiel, Olshausenstr. 40, D-24098 Kiel, Germany. onelle@ecology.uni-kiel.de

NTINOU, Maria. Agg. Sikelianou 2, 45221 Ioannina, Greece. Maria.Ntinou@uv.es

OEGGL, Klaus. Institute of Botany, University of Innsbruck, Sternwartestrasse 15, 6020 Innsbruck, Austria. Klaus.Oeggl@uibk.ac.at

OVERBECK, Michael. Department for Prehistoric Archaeology, Westfälische Wilhelms-University Münster, Robert-Koch-Str. 29, D-48149 Münster, Germany. michaeloverbeck@gmx.de

PALET, Josep Maria. ICAC Institut Català d'Arqueología Clàssica, Plaça Rovellat s/n; 43003, Tarragona, Spain. jpalet@icac.net

PAPI, Emanuele. Dipartimento di Arboricoltura, Botanica e Patologia Vegetale, Facoltà di Agraria, Università Federico II di Napoli, via Università 100 I-80055 Portici, Italy.

PARADIS, Sandrine. GEOLAB, UMR-CNRS 6042, 39E rue Camille Guérin, 87036 Limoges, France. sandrine.paradis@unilim.fr

PECCI, Alessandra. Dipartimento di Arboricoltura, Botanica e Patologia Vegetale, Facoltà di Agraria, Università Federico II di Napoli, via Università 100, I-80055 Portici, Italy.

PERRET, Sébastien. Earth Sciences – Archaeometry, Department of Geosciences, University of Fribourg, Chemin du Musée 6, 1700 Fribourg, Switzerland. mailto:sebastien.perret@unifr.ch

PICORNELL GELABERT, Llorenç. Seminary of Prehistoric Studies and Research, University of Barcelona (Spain); Arqueobalear Research Team, University of the Balearic Islands, Spain. tokelau24@hotmail.com

PIQUÉ i HUERTA, Raquel. Laboratori d'Arqueobotànica, Departament de Prehistòria, Universitat Autònoma de Barcelona, Edifici B, Campus de la UAB, 08193 Bellaterra (Cerdanyola del Vallès), Despatx: B9/119, Spain. raquel.pique@uab.cat

PONOMARENKO, Elena. Ecosystem Archaeology Services, 1139 Agincourt Rd., Ottawa ON, K2C 2H8 Canada, eponomarenko@yahoo.com

RIERA MORA, Santiago. Seminary of Prehistoric Studies and Research, University of Barcelona, Spain. mailto:rieram@ub.edu

ROBION-BRUNNER, Caroline. Department of Anthropology, University of Geneva, Rue Gustace Revilliod 12, Case postale CH-1211 Genève 4, Switzerland. caroline.robion@unige.ch

Maria Oliva. Centro Andaluz de Arqueología Ibérica, Edif. B-1 Universidad de Jaén, 23071 Jaén, Spain. moliva@ujaen.es

ROS SALA, Maria Milagrosa. Dpto. Prehistoria, Arqueología, Historia Antigua, Historia Medieval y CC.TT. Historiográficas, Facultad de Letras, Universidad de Murcia. C/ Santo Cristo, 1. 30001-Murcia, Spain.

ROUAUD, Romain. Géolab UMR-6042-CNRS, 39E Rue Camille Guérin 87036 Limoges, France. romain.rouaud@free.fr

RUIZ-ALONSO, Mónica. Grupo de Investigación en Arqueobiología. Instituto de Historia Centro de Ciencias Humanas y Sociales (CCHS). Consejo Superior de Investigaciones Científicas (CSIC). c/ Albasanz, 26-28; 28037 Madrid, Spain. monica.ruiz@cchs.csic.es

SCHWARZ, Anton Stefan. Institute of Botany, University of Innsbruck, Sternwartestrasse 15, 6020 Innsbruck, Austria. Stefan.Schwarz@uibk.ac.at

SERNEELS, Vincent. Earth Sciences – Archaeometry, Department of Geosciences, University of Fribourg, Chemin du Musée 6, 1700 Fribourg, Switzerland. vincent.serneels@unifr.ch

SERVERA VIVES, Gabriel. Seminary of Prehistoric Studies and Research, University of Barcelona (Spain); Arqueobalear Research Team, University of the Balearic Islands (Spain); GEOLAB Limoges, UMR 6042 CNRS, 39E rue Camille Guérin, 87000 Limoges, France. serveravives@gmail.com

SHIMADA, Izumi. Southern Illinois University at Carbondale, Department of Anthropology, Faner Hall, Carbondale, IL, 62901-4208, U.S.A., ishimada@siu.edu

STELLATI, Angela. Università degli Studi di Bari, Bari, Italy angela.stellati@libero.it

TENGBERG, Margareta, UMR 7209 – Archéozoologie, archéobotanique: Sociétés, pratiques et environnements. Muséum national d'Histoire naturelle, CP 56, 55 rue Buffon, 75005 Paris, France. tengberg@mnhn.fr

THIÉBAULT, Stéphanie. UMR 7209 – Archéozoologie, archéobotanique : Sociétés, pratiques et environnements. Muséum national d'Histoire naturelle, CP 56, 55 rue Buffon, 75005 Paris, France. thiebault@mnhn.fr

THIRIOT, Jacques. Laboratoire d'Archéologie Médiévale Méditerranéenne (UMR 6572) et Université d'Aix-Marseille I, Maison Méditerranéenne des Sciences de l'Homme, 5, rue du Château de l'Horloge, BP 647, 13 094 Aix-en-Provence cédex 1, France. thiriot@mmsh.univ-aix.fr

VASCHALDE, Christophe. Laboratoire d'Archéologie Médiévale Méditerranéenne (UMR6572) et Université d'Aix-Marseille I, Maison Méditerranéenne des Sciences de l'Homme, 5, rue du Château de l'Horloge, BP 647, 13 094 Aix-en-Provence cédex 1, France. vaschalde@mmsh.univ-aix.fr; christophe_vaschalde@yahoo.fr

ZAPATA, Lydia. Dept. Geografía, Prehistoria y Arqueología (University of the Basque Country UPV/EHU). Tomás y Valiente s/n. 01006 Vitoria-Gasteiz, Spain. lydia.zapata@ehu.es

ZYSNARSKA, Małgorzata. Archaeological Museum of Kraków, ul. Senacka 3, 31-002 Kraków, Poland. Andrzej.Matoga@ma.krakow.pl

# PROCEEDINGS OF THE FOURTH INTERNATIONAL MEETING OF ANTHRACOLOGY
## Royal Belgian Institute of Natural Sciences, Brussels, 8 – 13 September 2008

# INTRODUCTION TO THE VOLUME

The present volume is made of 25 papers as the result of communications to the fourth International Meeting of Anthracology held in Brussels at the Royal Belgian Institute of Natural Sciences (RBINS) between the 8 and 13 September 2008.

During the meeting 57 oral communications and 57 posters were presented with the participation of some 236 authors and co-authors from 25 countries in Europe, Africa, North and South America, Australia and New Zealand.

Taking the main interests of research into account, the meeting was subdivided in five sections:

Section 1. Methods, taphonomy, dating (11 oral communications; 9 posters). This section was devoted to several themes mainly focused on combustion, taphonomy, identification, measurements and $^{14}C$ dating on charcoal.

Section 2. Pre-Quaternary charcoal (5 oral communications; 1 poster). Charcoal was put in evidence in various Devonian to Cenozoic deposits. Palaeoecological and climatological approaches were attempted successfully.

Section 3. Pedo-anthracology (6 oral communications; 8 posters). This relatively new discipline seems having produced nice results on forest history allowing an easier approach of the human impact with regard to climate influence on the landscape.

Section 4. Archaeo-ethno-anthracology (28 oral communications; 35 posters). This theme clearly appears as the most attractive for charcoal analysts because it is directly related to archaeological investigations asking for thorough information on the use of wood and the transformation of the environment under human pressure. Here too taphonomy appears as a major tool for understanding all steps of wood use and transformation, including final combustion and preservation.

Section 5. Climato-anthracology (7 oral communications; 4 posters). The role of climate variability and fire regime becomes one of the most relevant topic for research notably because it may provide essential information to understand present day fire-climate interactions and their role for nature conservation.

Moreover, five keynotes were presented in relation with the theme of each section by:

1. Isabelle THÉRY-PARISOT and Lucie CHABAL: Anthracology and taphonomy, from wood gathering to charcoal analysis. A review of the taphonomical processes modifying anthracological assemblages.
2. Andrew C. SCOTT: Charcoal in deep time and the role of fire in earth system processes.
3. Paul HAESAERTS: Long loess sequences with charcoal in Eurasia: their implication in the chronology of the Upper Pleistocene climate sequences.
4. Stéphanie THIÉBAULT: Archaeological charcoal records: experience and perspectives.
5. Mitchell POWER: Climate driven changes in fire regimes during the Late Quaternary.

### WORKSHOPS

In the same time, three workshops were organized with the goal of informing non specialists and widen the debate on the following field of work:

1. Andrew C. SCOTT: Charcoal recognition / taphonomy. This workshop had a great success due to a savant combination of theory and practice with demonstration in microscopy aiming to recognize and identify charred material derived from natural fires in various ecosystems. This was a very useful complement to the keynote of the same speaker on the role of fire in earth system processes.
2. Mitchell POWER: The importance of databases in anthracology and possible connection with other databases in Palaeobotany. Some 25 participants were debating about various systems of databases on wood and charcoal remains and their use in reconstructing the history of fire on earth at different times of the Late Quaternary as well as the relation between fire regime and climate changes.
3. Freddy DAMBLON: International Association of Charcoal Research. The aim of this workshop was prospecting the possibility to coordinate and promote charcoal research internationally amongst a broad range of scientific disciplines. The hope was also to

form an organizing committee able to establish the basis for setting up an International Association for Charcoal Research. A first document for discussion was prepared by A.C. Scott and F. Damblon which led the assembly to propose a first executive committee with G. Di Pasquale, I. Stuijts and F. Damblon, following the model for the Irish Wodan database. Unfortunately, up today, the tokens of interest seem rather limited for such a kind of organization due probably to a lack of time by the participants.

## VISITING THE AFRICA MUSEUM TERVUREN AND THE NATIONAL BOTANIC GARDEN OF BELGIUM

The main interest in visiting the Africa Museum Tervuren was to understand the organization and functionning of the xylarium, probably the most important for Africa and other tropical or temperate countries in the world. The visitors were guided by Dr Hans BEECKMAN in charge of the xylarium and wood laboratory with some 14,000 species and around 60,000 specimens of wood from worldwide. The field of research concerns technology, ecology, palaeontology, art history or archaeology. Three databases constitute the information accompanying the collection: (1) the Tervuren Xylarium Taxa, (2) Tervuren Xylarium Specimens and (3) Tervuren Xylarium Images.

Another interest was the opportunity to guided visit of the exhibition "Knock the wood" that showed all forms of the use of wood in Africa, especially in art.

The National Botanic Garden of Belgium was also visited under the lead of Dr Gert AUSLOOS. The visitors were especially interested in the Plant Palace with its 13 interconnected glasshouses among which the Victoria House as the most humid of all houses and covering the world's biggest water lilies along with marshloving plants, carnivorous plants and papyrus. The north side of the Plant Palace contain plants from the world's richest ecosystem, the tropical rainforest. The House containing the Montane Rainforest comprises many plants adapted to the foggy clouds high up in tropical mountains including fuchsias and orchids. On the contrary, the Dry House is filled with various champions of drought, cacti and succulent plants. Here too the visitors had some problems to leave such fascinating plant world.

## FIELD TRIPS (13 September 2008) TO HAUTRAGE (Lower Cretaceous) AND TO SCLAYN (Scladina cave, Upper Pleistocene)

One full day was devoted to the visit of two important sites that provided nice palaeobotanical records giving information on climate and plant environment. This was an opportunity for 20 authors to present an up to date synthesis on geology, paleontology, sedimentology, palynology, plant mega- and mesofossils, and wood macroremains found on the Wealden site of the Hautrage quarry (Haine Basin) as well as new data on geology, charcoal and pollen from the Scladina cave in the Meuse Basin. The visit ended with a general discussion with the archaeologists and palaeoenvironmentalists in the research station (Dr Dominique BONJEAN) linked to the Scladina cave.

## ACTS OF THE MEETING

A first Special Issue edited by A.C. SCOTT and F. DAMBLON "Charcoal and its use in palaeoenvironmental analysis. Selected papers from: 4th International meeting of Anthracology, Brussels, 8-13 September 2008" has been published in the journal Palaeogeography, Palaeoclimatology, Palaeoecology, 291 (2010): 1-165. This volume was more especially devoted to methodological topics including taphonomy, stratigraphy and $^{14}C$ dating, pre-Quaternary deposits, Pleistocene records and to microcharcoal analysis.

Now, the present volume of BAR International Series as "Proceedings of the Fourth International Meeting of Anthracology, Brussels, 8-13 September 2008" edited by F. DAMBLON is more specifically focused on charcoal analysis in archaeological context with a number of important case studies on diverse prehistoric and historic sites in Europe, the Near and the Middle East, North, West and Central Africa, North and South America. Some authors also develop particular methodological aspects to interpret charcoal analysis. We hope that the present series of articles will be a source of fruitful debates.

**Acknowledgements**

The organizing committee is happy to acknowledge all contributors to the success of the 4th International Meeting of Anthracology, Brussels, in particular:

Dr Camille PISANI, Director of the Royal Belgian Institute of Natural Sciences who allowed the meeting to occur in excellent conditions at the institute.

The members of the Scientific Committee who spent time for reading, reviewing and selecting the papers for publication:

Prof. Ernestina BADAL GARCIA, University of Valencia, Spain
Dr Hans BEECKMAN, Royal Museum of Central Africa, Belgium
Dr Freddy DAMBLON, Royal Belgian Institute of Natural Sciences, Brussels, Belgium
Prof. Barbara EICHHORN, University of Frankfurt, Germany
Dr Isabel FIGUEIRAL, University of Montpellier, France
Prof. Girolamo FIORENTINO, University of Lecce, Italy
Prof. Philippe GERRIENNE, University of Liège, Belgium
Dr Dominique MARGUERIE, University of Rennes, France
Prof. Oliver NELLE, University of Kiel, Germany

Prof. Mitchell POWER, University of Salt Lake City, USA

Prof. Andrew C. SCOTT, Royal Holloway Univ., London, U.K.

Dr Stéphanie THIÉBAULT, University of Paris I, Nanterre, France

Prof. Marcel OTTE, University of Liège, Belgium

Prof. Jean-Louis VERNET, University of Montpellier, France

Prof. Jacques VERNIERS, University of Gent, Belgium

The organizers of two main workshops, Prof. Andrew C. SCOTT (on charcoal recognition / taphonomy) and Prof. Mitchell POWER (on databases).

The contributors to the Field Guidebook of the excursion:

Dr Jean-Marc BAELE (Faculté Polytechnique de Mons)

Dr Dominique BONJEAN (Archéologie Andennaise, Sclayn)

Dr Freddy DAMBLON (RBINS), co-editor

Dr Véronique DAVIERO-GOMEZ (Université de Lyon 1, Claude Bernard)

Dr Jean DEJAX (Muséum National d'Histoire Naturelle, Paris)

Prof. Christian DUPUIS (Faculté Polytechnique de Mons)

Prof. Philippe GERRIENNE (Université de Liège), co-editor

Thomas GERARDS (Université de Liège)

Dr Thomas GILLOT (Université de Lyon 1, Claude Bernard)

Dr Pascal GODEFROIT (RBINS)

Dr Bernard GOMEZ (Université de Lyon 1, Claude Bernard)

Prof. Paul HAESAERTS (RBINS)

Prof. Denise PONS (Université Paris VI)

Dr Stéphane PIRSON (RBINS, Région wallonne), co-editor

Dr Paul SPAGNA (RBINS)

Dr Yves VANBRABANT (RBINS, Geological Survey)

Prof. Johan YANS (Facultés Notre-Dame-de-la-Paix, Namur)

The editorial board of the Geological Survey (Léon DEJONGHE and Michel DUSAR) which allowed us to print the program of the meeting and the field trip guidebook.

The whole support team of the RBINS: Françoise ANTONUTTI, Gilbert CLAES, Gael DE DEYNE, Eric DERMIENCE, Christelle DEVITS, Eric DE WEER, Alain DRÈZE, Kareen GOLDFEDER, Thierry HUBIN, Christine KERWYN, Guy LONCKE, Wilfried MISEUR, Olivier NINANE, Carole PALECO, Olivier RETOUT, Anouk SCHOETERS, Adriano VANDERSYPEN, Ann VENMANS, Jacqueline VERHEYEN, Ingrid VERCAMBRE and all collaborators who helped in the success of the meeting organization.

The team of the Africa Museum Tervuren:

Dr Hans BEECKMAN, Wannes HUBAU

The team of the National Botanic Garden of Belgium:

Dr Gert AUSLOOS, Patrick BOCKSTAEL, Chris KOSOLOSKY

The different sponsors of the meeting:

- The Direction of the RBINS (Royal Belgian Institute of Natural Sciences) (Dr Camille PISANI)
- The Geological Survey of the RBINS (Dr Cecile BAETEMAN, Dr Michiel DUSAR)
- The Department of Palaeontology of the RBINS (Prof. Etienne STEURBAUT)
- FNRS (Fonds de la Recherche Scientifique)
- FWO (Fonds voor Wetenshappelijk Onderzoek)
- BRUSSELS INTERNATIONAL (BIP)
- LEICA Microsystems Belgium B.V.B.A.
- FEI Europe B.V. – Belgium branch
- JEOL Europe B.V.
- Kunsthistorisches Museum Wien

Freddy DAMBLON
and members of the organizing committee
Mona COURT-PICON
Aurélie SALAVERT
Cecile BAETEMAN
Royal Belgian Institute of Natural Sciences
29, Rue Vautier, B-1000 Brussels, Belgium

# GROUP AND SOUVENIR PHOTOS

Participants group photo: 1. Tescari F.; 2. Doutrelepont H.; 3. Bellavia V.; 4. Hubau W.; 5. Impagliazzo S.; 6. Di Pasquale G.; 7. Pessin H.; 8. Allevato E.; 9. Out W.; 10. Socratous M.; 11. Stuijts I.; 12. Ponomarenko E.; 13. Allué E.; 14. Delhon C.; 15. Euba I.; 16. Chrzavzez J.; 17. Henry A.; 18. Deryck P.; 19. Power M.; 20. Court-Picon M.; 21. Damblon F.; 22. Touflan P.; 23. Picornell L.; 24. Kazmer M.; 25. Garcia Martinez M.; 26. Grau Almero E.; 27. Oliva M.; 28. Caracuta V.; 29. Uzquiano P.; 30. Ruiz M.; 31. Amrani S.; 32. Eichhorn B.; 33. Höhn A.; 34. Kuypers S.; 35. Schoch W.; 36. Marguerie D.; 37. Salavert A.; 38. Dotte Sarout E.; 39. Scott A.; 40. Maames K.; 41. Cywa K.; 42. Ludemann T.; 43. Paysen A.; 44. Botta L.; 45. Magnan G.; 46. Petr L.; 47. Chester P.; 48. Coubray S.; 49. Dufraisse A.; 50. Marcoux N.; 51. Ntinou M.; 52. Thery-Parisot I.; 53. Figueiral I.; 54. Badal E.; 55. Jansen D.; 56. Moskal M.; 57. Bouchaud C.; 58. Kocar P.; 59. Allee P.; 60. Heiss A.; 61. Kocarova R.; 62. Lancelotti C.; 63. Masi A.; 64. Paradis S.; 65. Rouaud R.; 66. Carrion Y.; 67. Benes J.; 68. Deforce K.; 69. Robin V.; 70. Schwarz A. S.; 71. McParland L.; 72. Hudspith V.; 73. Novak J.; 74. Hall G.

## Communications

1. Paul Haesaerts

1 bis: Workshop Andrew Scott

## Poster session

2. Marie-José Gaillard; Aline Durant; Rita Scheel-Ybert

3. Alexa Dufraisse on microscopy demonstration with P. Deryck

4. Poster session

## Coffee break atmosphere

5. Coffee break

6. Elena Grau Almero; Ernestina Badal; Maria Ntinou; Yolanda Carrion

7. Girolamo Fiorentino; Rita Scheel-Ibert; Freddy Damblon

8. Isabel Figueiral; Maria Socratous

## Dinner

9. Andrew Scott; Mitchell Power; Ingelise Stuijts

10. Dominique Marguerie; Mona Court-Picon; Michel Thinon; Pamela Chester

11. Emilie Dotte; Sylvie Coubray;
Aurélie Salavert; Claire Delhon

13. Trio Trad

12. Werner Schoch; Thomas Ludemann;
Oliver Nelle

## Visit Africa Museum

14. David Goldstein; Alexa Höhn; Barbara Eichhorn;
Wannes Hubau; Hans Beeckman

## Excursion

15. Paul Spagna; Elena Ponomarenko; Johan Yans; Thomas Gerards; Mona Court-Picon. (Quarry of Hautrage)

16. Philippe Gerrienne; Cecile Baeteman; Andrew Scott; Mona Court-Picon; Jean Dejax; Johan Yans; Thomas Gerards. (Quarry of Hautrage)

17. Dominique Bonjean. (Scladina cave)

18. Stephane Pirson. (Scladina cave)

# PREFACE

The reader is going to discover in this volume a major part of the scientific contributions presented during the fourth international meeting of anthracology held in Brussels in 2008. This fourth meeting gathered more than 236 authors coming from 25 countries, which presented 57 oral communications and 57 posters.

Since the first meeting, in Montpelier in 1991, the success and the participations are always increasing. It is important to underline, as this volume testifies of it, the important part taken by young researchers in Brussels, so showing the development of the second generation of anthracologists.

In the name of all the participants, it is particularly pleasant to me to thank the steering committee and more particularly Dr Freddy Damblon, as well for the organization of these days at the Royal Belgian Institute of Natural Sciences in Brussels, as for his remarkable scientific work.

The presented articles can be grouped around common themes.

The theme methodology especially concerned the taphonomy, the post-deposit processes, the systems of identification and the quantitative eco-anatomy. It is interesting to notice that this field of work becomes attached more to the indicators of current or very ancient fires, as shown in the articles of J.-F. Gillet and J.-L. Doucet, or of E. Ponomarenko *et al.* The techniques of tree rings analysis are clarified by the works of S. Paradis *et al.*

The experiments and the ethnographical observations are more and more used to obtain current references with the aim of testing their results to those of the archaeological observations. So are observed the Selknam hunter-gatherer societies of Tierra del Fuego (Argentina) by L. Caruso and the Dogon of Mali by B. Eichhorn *et al.* while experiments to model the ancient fuel uses are developed by D.J. Goldstein and I. Shimada.

It is important to underline that the majority of the articles presented here attempt to study the dynamics of past vegetations, but always in connection with the exploitation, the management or the use of wood fuel by the societies of the past. The studies understanding the relation of humans with their woody environment are dominating. We can enumerate the contributions of M. Ruiz-Alonzo *et al.* for the Palaeolithic in the Basque Country; of F. Antolin *et al.* for the Mesolithic and Neolithic in Catalonia; of M. Martin-Seijo and R. Piqué in the North-Western Spain during Neolithic; of C. D'Oronzo *et al.* in the South-Eastern Italy during the Holocene. Or, for more recent periods, in Syria (G. Fiorentino and V. Caracuta), in the South of Spain (M.S. Garcia *et al.*) and South-Eastern Spain (M.O. Rodriguez-Ariza), in Balearic Islands (L. Picornell *et al.*), in Eastern Alps (A.S. Schwarz *et al.*), in Pakistan (C. Lancelotti *et al.*), in Poland (M. Moskal-Del-Hoyo *et al.*), in Luxembourg (O. Nelle *et al.*); and in France, during historic times in Dordogne (R. Rouaud and P. Allée) and Southern France, in Provence (C. Vaschalde *et al.*) and Languedoc (I. Figueiral *et al.*) or since Neolithic until Modern Age in the mountain of Eastern Pyrenees (I. Euba and J.M. Palet Martinez).

Finally, two species of trees are more particularly exposed: the cork oak during the Roman period (E. Allevato *et al.*) and the olive tree during the Neolithic and its place in the North of the Mediterranean Basin (Y. Carrion *et al.*).

The readers of these articles will be certainly richer in knowledge in botany and in long term vegetation history. These data allow to better understand the evolution of the biodiversity and the implementation of the landscapes, but also to understand indeed the uses, the transformations and the management of the woody resources. They enlighten us on the impacts of the human interventions, in very diverse geographical regions, in the implementation of the current biodiversity.

We cannot finish this introduction without thanking all the authors of theses articles patiently collected by Freddy Damblon, which are the support of fruitful researches for the future.

Stéphanie THIEBAULT
Directeur de recherche au CNRS
MNHN-CNRS-UMR7209
Archéozoologie-Archéobotanique: Sociétés, Pratiques et Environnements
55, rue BUFFON 75005 Paris

# EVIDENCE FROM CHARCOAL ANALYSIS FOR THE EXTENSIVE EXPLOITATION OF CORK-OAK (*QUERCUS SUBER*) FOREST IN THE ROMAN IMPERIAL PERIOD: THE VICUS OF THAMUSIDA (NW MOROCCO)

Emilia ALLEVATO
Dipartimento di Arboricoltura, Botanica e Patologia Vegetale, Facoltà di Agraria,
Università Federico II di Napoli, via Università 100 I-80055 Portici, Italy
eallevat@unina.it

Alessandra PECCI
Equip de Recerca Arqueològica i Arqueomètrica, Universitat de Barcelona (ERAAUB), Spain

Emanuele PAPI
Dipartimento di Arboricoltura, Botanica e Patologia Vegetale, Facoltà di Agraria,
Università Federico II di Napoli, via Università 100 I-80055 Portici, Italy

Gaetano Di PASQUALE
Dipartimento di Arboricoltura, Botanica e Patologia Vegetale, Facoltà di Agraria,
Università Federico II di Napoli, via Università 100 I-80055 Portici, Italy.
Dipartimento di Archeologia e Storia delle Arti, Facoltà di Lettere e Filosofia,
Università degli Studi di Siena, via Roma 56 – 53100 Siena, Italy
gaetano.dipasquale@unina.it

**Abstract**: *This work shows the preliminary results of the study of charcoal collected in the* vicus *of* Thamusida – *Sidi Ali ben Ahmed (Morocco), in the Roman province of Mauretania Tingitana. The charcoal assemblage gives information on the woody flora and the wood usage during a period from the second half of the 1st cent. AD to the 3rd cent. AD, during the Roman occupation of the site. Data suggest the presence of a* Quercus suber *forest in the close surrounding of the site, with larger extension with respect to the present-day Mamora forest. The presence of* Vitis vinifera *and* Olea europaea *in the agrarian landscape was also detected. The presence within the charcoal assemblages of* Castanea sativa *it is worth to note since its presence in the wild vegetation in the surrounding of the site is rather improbable.*

**Key words:** *Charcoal analysis, Roman imperial period, Mamora forest, Firewood, Timber*

## INTRODUCTION

The Roman *vicus* of *Thamusida*, situated along the Sebou River, about 50 km kilometres away far from the modern town of Rabat (Figure 1), was firstly excavated in the first half of 19th cent. by French archaeologists; more recent excavation was performed by the Università di Siena in collaboration with the Institut National de Sciences de l'Archéologie et du Patrimoine de Rabat (INSAP) between 1999 and 2006. The archaeological surveys showed the presence of a military camp built in the 1st cent. AD, on the Roman Empire edge, following the establishment of *Mauretania Tingitana* as a Roman province. The existence of a settlement surrounding the camp, namely the *vicus*, with all the principal infrastructures was detected too: houses, thermal baths, temples, taverns, shops for the manufacture of the weapons, barns and factories for the production of *garum* (fish sauce), ovens to produce amphorae, bricks and ceramics. After the withdrawal of the Roman army, at the end of the 3rd cent. AD, *Thamusida* continued to be inhabited until the Arabian conquest (7th – 8th cent. AD).

In this paper we show the results obtained by means of charcoal analysis on samples collected in the stratigraphic units dated between the second half of the 1st cent. AD and the 3rd cent. AD. This work aims to provide information about the vegetation landscape during the Roman occupation and about the usage of wood resources.

## THE ENVIRONMENT OF THAMUSIDA STUDY SITE

The coastal region of Kenitra-Rabat is under the thermomediterranean sub-humid bioclimate; the annual mean precipitation is 600 mm/yr and the rainiest periods are in winter (up to 100 mm in December) and sea breezes in summer occur by the entering wet air masses from Atlantic Ocean. The average yearly temperature is

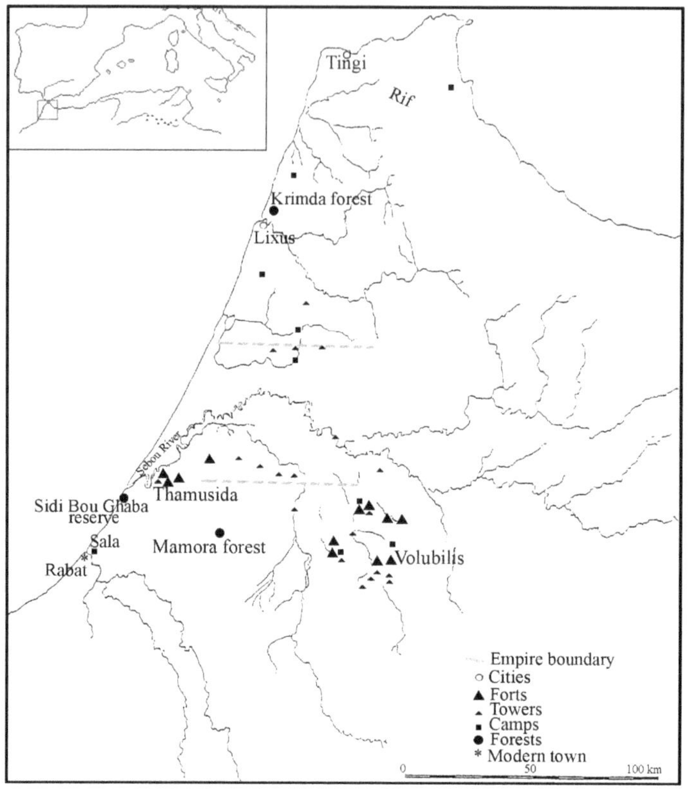


Figure 1. Modified from Akeraz *et al.* 1993. Location of the sites cited in the text

17.2°C, with extreme seasonal values of 13°C in winter and 27°C in summer (Ben Kabbour *et al.* 2006).

The current vegetation in the close proximity of the site is represented by prevailing *Chamaerops humilis* and scattered shrubs of *Ziziphus lotus* and *Olea europaea.* Alongside the Sebou River, riparian vegetation is represented by scattered plants of *Vitex agnus castus* and *Crataegus monogyna.* Along the coast, 20 kilometres away from the site, the "Biological Reserve of Sidi Bou Ghaba" (Fig. 1) includes shrubby formations with *Juniperus phoenicea, Phillyrea angustifolia, Pistacia lentiscus, Rhamnus oleoides* and *Retama monosperma.* Riparian vegetation with prevailing *Populus alba* characterises the surrounding area of the Merja (lagoon) of Sidi Bou Ghaba. The Mamora forest (Figure 1) is located a few kilometres away from the site; it is at present one of the most degraded forests in Morocco with few woody taxa (Rejdali, 2004). *Quercus suber* is the only tree taxon and is accompanied by few shrubby species, such as the endemic *Pyrus mamorensis* and sporadically *Olea europaea, Phillyrea latifolia* and *Teline linifolia.*

## MATERIALS AND METHODS

Sediment samples dated between the second half of the 1st cent. AD and the 3rd cent AD were collected by the archaeologists. Each sample ranged between 5 and 10 kg according to the stratigraphical unit (SU) thickness and extension. The sediments from 50 SU in 8 excavation areas (Figure 2) were floated and then dry sieved on a sieve column with mesh size of 4.0 and 2.0 mm. Two trenches close to a hill defined productive area were excavated and sampled in each layer. Charcoal fraction up to 4 mm was naked eye sorted while fragments >2 mm were sorted with the help of a dissecting stereomicroscope with a low magnification (10x – 40x). All the sorted charcoal fragments were analysed with an incident light microscope (100x – 1000x) and identified referring to wood anatomy atlases (Schweingruber, 1990; Neumann *et al.* 2001; Vernet *et al.* 2001), as well as to our reference collection. Additional samples for the reference collection has been collected by mean of a botanical survey within a 10 km radius from the site; to create charcoal from the wood samples, the air-dried samples were covered by sand to restrict the oxygen supply to avoid total combustion, and heated in a muffle furnace at 350°C for 20 min (Machado Yanes, 1992). Nomenclature follows the Flore pratique du Maroc (Fennane *et al.* 1999; Fennane *et al.* 2007).

## RESULTS

2610 charcoal fragments were analysed and 21 taxa were identified. The results are represented in Figure 3, where the charcoal fragment counts and the percentages are shown; count refers to the sum of fragments from the SU in each area. The number of SU in which each taxon is present, namely ubiquity, is shown too. The number of taxa in each area ranges between 3 and 14.

The more common charcoal fragments belong to the *Quercus* genus; they were present in all areas and

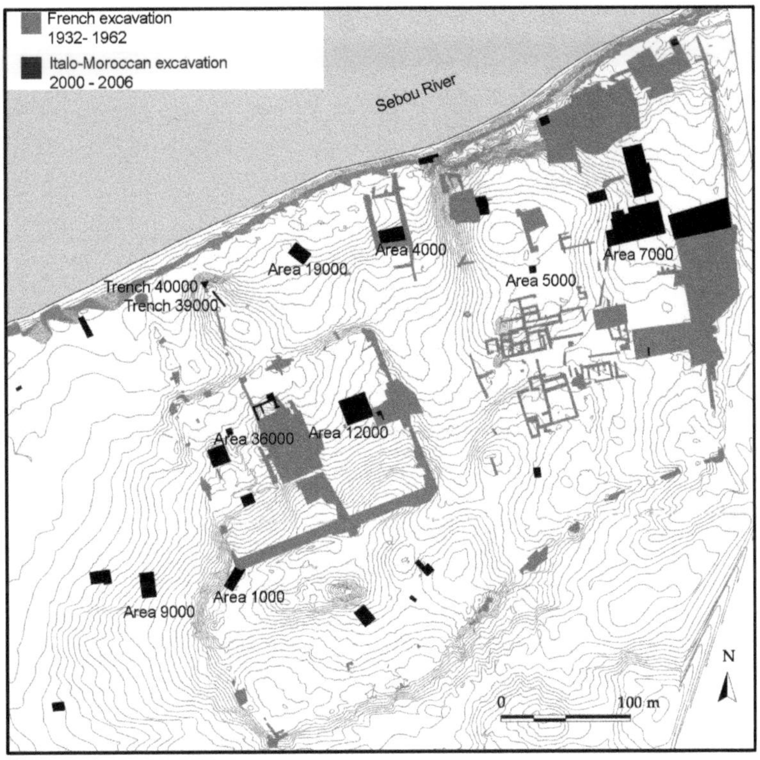


Figure 2. Extension, functional type and location of sampled areas. Area 1000: Huge fortified wall, Area 4000: Granary for the storage of the supply coming from Rome., Area 5000: Roman crossroads, Area 7000 Buildings – living floors, Area 9000: Suburbium house – fireplace, Area 12000: Barracks with storage rooms, Area 19000: Industrial furnaces Area 36000: Oven for bread and food cooking

represented the unique materials in four out of nine areas; *Quercus suber* always prevails followed by *Q. ilex* type. *Q. suber/ilex* and *Quercus* are also present among the identification ranks because identification to the species rank for the genus *Quercus* is not ever feasible. More precisely, charcoal fragments were attributed to *Q. suber* following Schweingruber (1990) only when we were able to observe the transition from early to late wood in large and complete growth rings. Additional testing was also performed by measuring the vessel diameter following Vernet *et al.* (2001).

The analysis of the 2 trenches revealed the presence of 5 layers entirely constituted by charred cork. Charred cork was also found in all areas, apart from the fire place and the constant presence of *Q. suber* charcoal in all the contexts is also remarkable. The remaining taxa follow at a far distance, with values often < 1%. The larger amount of taxa has been found in the granary and the barracks with 14 taxa (Figure 3).

## DISCUSSION

### The ancient Mamora forest

From our results we propose that the Marmora forest, at that time probably bigger than today, was used for wood and cork supply. Its composition had already been influenced by human impact.

The high amount of charcoal as well of charred cork testifies for the intense and continuous use of this wood for fires, and also for structures as shown by the high percentages of this wood among the remains coming from the collapse layers of the fortified walls (Area 1000). The layers with charred cork in the trenches are a clear evidence of the extensive use of cork, which was related to its high availability in the close surrounding of the site; unfortunately the interpretation of the function of cork in these contexts is hard since no archaeological comparison exist.

Two interesting comparisons of a huge use of *Q. suber* come from Sicily (Italy); at Segesta (S-W Sicily) this wood was largely used for building purposes in the Middle Age (Castiglioni and Rottoli, 1997), in the necropolis of the Greek colony of Himera (648 – 409 BC, Northen Sicily) cork oak was the main fuel for the funerary pires (Di Pasquale unpubished).

The almost constant presence of cork fragments in the different typological contexts testifies for its habitual harvesting. It seems that the cork oak forests represented multi-functional forests for cork harvesting and wood exploitation, just like the present ones (Quézel and Médail, 2003). The Mamora forest (Figure 1) seems to have been the most probable supply area for wood and cork. Sauvage (1961) attested the presence of isolated individuals of *Q. suber* in proximity of *Thamusida*, on the opposite shore of the Sebou river. In Morocco, cork oak

| | Area 1000 Fortified walls 2 SU | | | Area 4000 Granary 4 SU | | | Area 5000 Roads 2 SU | | | Area 7000 Buildings 9 SU | | | Area 9000 Fireplace 1 SU | | Area 12000 Barracks 11 SU | | | Area 19000 Furnaces 16 SU | | | Area 36000 Bread oven 5 SU | | | Total 50 SU | | |
|---|---|---|---|---|---|---|---|---|---|---|---|---|---|---|---|---|---|---|---|---|---|---|---|---|---|---|
| **Taxa** | Count | % | Ubiquity | Count | % | Ubiquity | Count | % | Ubiquity | Count | % | Ubiquity | Count | % | Count | % | Ubiquity | Count | % | Ubiquity | Count | % | Ubiquity | Count | % | Ubiquity |
| *Quercus suber* | 61 | 61,0 | 2 | 147 | 56,8 | 4,0 | 34 | 48,6 | 2 | 252 | 77,5 | 8 | 227 | 87,6 | 102 | 23,0 | 11 | 843 | 84,4 | 15 | 134 | 87,0 | 4 | 1800 | 68,97 | 47 |
| *Quercus suber/ilex* | 13 | 13,0 | 2 | 30 | 11,6 | 4,0 | 10 | 14,3 | 2 | 34 | 10,5 | 5 | 30 | 11,6 | 84 | 18,9 | 6 | 133 | 13,3 | 5 | 6 | 3,9 | 2 | 340 | 13,03 | 27 |
| *Quercus* | 8 | 8,0 | 2 | 21 | 8,1 | 4,0 | 10 | 14,3 | 2 | 20 | 6,2 | 3 | 1 | 0,4 | 80 | 18,0 | 9 | 2 | 0,2 | 1 | 2 | 1,3 | 2 | 144 | 5,52 | 24 |
| *Castanea sativa* | – | – | – | 1 | 0,4 | 1,0 | – | – | – | – | – | – | – | – | – | – | – | – | – | – | – | – | – | 1 | 0,04 | 1 |
| *Populus* | – | – | – | 2 | 0,8 | 2,0 | – | – | – | – | – | – | – | – | – | – | – | – | – | – | – | – | – | 2 | 0,08 | 2 |
| Cf *Prunus* | – | – | – | – | – | – | – | – | – | – | – | – | – | – | 9 | – | 3 | – | – | – | – | – | – | 9 | 0,34 | 3 |
| Rosaceae | – | – | – | 4 | 1,5 | 1,0 | – | – | – | – | – | – | – | – | – | – | – | – | – | – | – | – | – | 4 | 0,15 | 1 |
| *Frankenia* | – | – | – | 3 | 1,2 | 2,0 | – | – | – | – | – | – | – | – | 8 | 1,8 | 2 | – | – | – | – | – | – | 11 | 0,42 | 4 |
| Gymnospermae | 1 | 1,0 | 1 | 1 | 0,4 | 1,0 | – | – | – | 1 | 0,3 | 1 | – | – | 4 | 0,9 | 3 | – | – | – | – | – | – | 7 | 0,27 | 6 |
| *Tetraclinis articulata* | – | – | – | 1 | 0,4 | 1,0 | – | – | – | 1 | 0,3 | 1 | – | – | 20 | 4,5 | 4 | – | – | – | – | – | – | 22 | 0,84 | 6 |
| *Pinus pinaster* | – | – | – | 1 | 0,4 | 1,0 | – | – | – | – | – | – | – | – | – | – | – | – | – | – | – | – | – | 1 | 0,04 | 1 |
| *Daphne-Thymelaea* | – | – | – | 1 | 0,4 | 1,0 | – | – | – | – | – | – | – | – | 1 | 0,2 | 1 | – | – | – | – | – | – | 2 | 0,08 | 2 |
| *Cistus* | – | – | – | 2 | 0,8 | 1,0 | – | – | – | – | – | – | – | – | – | – | – | – | – | – | – | – | – | 2 | 0,08 | 1 |
| *Erica* cf *scoparia* | – | – | – | – | – | – | – | – | – | – | – | – | – | – | 2 | 0,5 | 2 | – | – | – | – | – | – | 2 | 0,08 | 2 |
| *Ceratonia siliqua* | 3 | 3,0 | 1 | – | – | – | – | – | – | – | – | – | – | – | – | – | – | – | – | – | – | – | – | 3 | 0,11 | 1 |
| Cf *Crataegus monogyna* | – | – | – | – | – | – | – | – | – | – | – | – | – | – | 2 | 0,5 | 2 | – | – | – | – | – | – | 2 | 0,08 | 2 |
| *Arbutus unedo* | – | – | – | – | – | – | 1 | 1,4 | 1 | – | – | – | – | – | 1 | 0,2 | 1 | – | – | – | – | – | – | 2 | 0,08 | 2 |
| *Pistacia lentiscus* | – | – | – | – | – | – | – | – | – | – | – | – | – | – | 1 | 0,2 | 1 | – | – | – | – | – | – | 1 | 0,04 | 1 |
| Fabaceae | – | – | – | 1 | 0,4 | 1,0 | – | – | – | – | – | – | – | – | – | – | – | – | – | – | – | – | – | 1 | 0,04 | 1 |
| *Olea europaea* | – | – | – | – | – | – | – | – | – | 1 | 0,3 | 1 | – | – | 4 | 0,9 | 2 | – | – | – | 2 | 1,3 | 1 | 7 | 0,27 | 4 |
| *Vitis vinifera* | – | – | – | 1 | 0,4 | 1,0 | – | – | – | – | – | – | – | – | 1 | 0,2 | 1 | – | – | – | – | – | – | 2 | 0,08 | 2 |
| Undetermined | 14 | 14,0 | | 43 | 16,6 | | 15 | 21,4 | | 16 | 4,9 | | 1 | 0,4 | 125 | 28,2 | | 21 | 2,1 | | 10 | 6,5 | | 245 | 9,39 | |
| **Total** | **100** | | | **259** | | | **70** | | | **325** | | | **259** | | **444** | | | **999** | | | **154** | | | **2610** | | |
| Cork remains | x | | | x | | | x | | | x | | | – | | x | | | x | | | x | | | | | |

Figure 3. Charcoal fragments count and percentage; count is the sum of identified fragments from the SU in each area. Ubiquity represents the number of SU in which the taxon is present; Ubiquity has been omitted for Area 9000 since a single US was analysed. *Taxa* are arranged in broad ecological groups

forest is judged very old (Emberger, 1928a; Sauvage 1961) and probably extended on the Atlantic plane at least from the Holocene optimum (Ballouche and Damblon, 1988). Floristic impoverishment, recent cover and density decrease due to human impact have been revealed by ecological studies (Rejdali, 2004; Sauvage, 1961; Natividadae, 1950) who considered the Mamora as one of the most degraded forests in Morocco and suggested that just the intense anthropic exploitation caused its drastic reduction. Also Emberger (1928b, 1934) hypothesised a past greater extension of *Q. suber* forest in northern Morocco. A major extension of the Mamora forest is also supported by pollen analysis carried out in Morocco where isolated stands of *Q. suber* yet exist. These data suggest a greater extension of *Q. suber* forests (Ballouche and Damblon 1988).

Our data testify for a strong cork oak-man interaction during the Roman occupation of the site but we are unable to state the beginning of this forest exploitation. Palynological data from the close Lake Sid Bou Ghaba (Figure 1; Reille, 1979) attested the forest cover reduction and ascribed it to man action from 6500 BP onward. The scarcity of other tree taxa in the charcoal assemblages induces us to believe that this forest was already largely anthropised during the investigated period. Carrion *et al.* (2000) consider, for the Iberian Peninsula, the cork oak mono-specific forests as the result of human influence. The charcoal analysis does not allow the presence of other sclerophyllous and deciduous *Quercus* to be excluded even if it is also conceivable that part of the charcoal identified as *Quercus* and *Quercus suber ilex* could belong at the same to *Quercus suber*.

Taxa as *Cistus, Pistacia lentiscus, Erica* cf *scoparia, Daphne-Thymelaea* and *Arbutus unedo* could represent the understorey vegetation of the cork oak forest and could be present in the assemblages as kindling cuttings used for fuel. Metro and Sauvage (1955) report the scattered presence of these species until the second half of the last century; the disappearance of *Erica arborea* and a strong reduction of *Erica scoparia* was attested in recent time (Sauvage, 1961).

*Erica arborea* still persist in the Krimda forest (Damblon, 1991), pointing out a probable wider extension of this shrub in the past when the anthropic pressure was less heavy.

To sum up, we hypothesise that the strong anthropisation and the heavy use of this forest did not affected, at least until Roman time, its extension and composition. A similar situation can be traced for the coastal cork-oak forest of Krimda (Figure 1), where pollen analysis shows that human impact did not strongly affect the extension and the composition of the forest until the 11th cent. AD (Damblon, 1991).

**Other wild plants**

*Tetraclinis articulata* was found in the granary. It is widely present in northern Africa being diffuse on the Atlas slopes, on the Plateau Central, from the Rif to the Anti-Atlas (Fennane *et al.* 1984). The presence of this species in the surrounding of the site cannot be ruled out for its ecology (Quézel and Médail, 2003). It is considered as a precious tree because its wood is very appreciated for its natural beauty and homogeneity and is used for marquetry and cabinet work. Wood of *T. articulata* identified from furniture in the Vesuvius area (Herculaneum, Italy) in the 1st cent. AD (Mols, 2002) testifies for the knowledge of the good quality and beauty of the wood and its wide circulation around the Mediterranean area during the Roman times.

Small amount of *Pinus pinaster* was found in the granary; the presence of maritime pine in the surrounding of the site seems very improbable, since this tree lives at present on the Rif and on the Middle Atlas between 1200 and 2000 asl. Thus, we suppose that the recovered charcoal could belong to an imported woodwork coming from one of these north African mountais or at the most from Iberia.

*Populus* was found only in the granary (<1%) and it could be referred to *Populus alba*, today present around the coastal lake of Sidi Bou Ghaba. Probably, it was common in the riparian vegetation along the Sebou river. *Pistacia lentiscus* and *Frankenia* testify for wood supply from the coastal dunes, too. Only one fragment of *Castanea sativa* was identified in the granary of the town, but the hypothesis of chestnut cultivation in this region ***is*** suggestive but rather improbable. Just one station of chestnut tree is present in Morocco, on the Jbel Mahfoura in the Rif, in a sub-humid context. The presence of this species is relative to only one fragment recovered in the civil granary in a level dated to the 2nd – 3rd cent. AD. This recovery is extremely interesting because the beginning of chestnut cultivation in western Mediterranean is still object of debate: the beginning of cultivation is attested in Italy and in the Iberian Peninsula at the beginning of the 1st cent. AD (Conedera *et al.* 2004) but its wide cultivation only started from the 5th – 6th cent. AD. The presence of chestnut wood/timber is rarely attested in western Mediterranean before Medieval times. Sadori and Susanna (2005) attested structural use of chestnut timber in the 4th cent. AD in central Italy while recent studies in Naples region (southern Italy) predate the extensive use of its timber to the 1st cent. BC and suggest the presence of a restricted refuge area on Vesuvius' foothills (Di Pasquale *et al.* 2010). Our hypotesis is that this wood fragment probably comes from a wooden work coming from Iberia and is not related with the vegetation around the site.

**Olea and Vitis**

The presence of charcoals of *Olea europaea* together with *Vitis vinifera* strongly suggests their cultivation in the surrounding of the study site. These taxa were present in areas related to food storage and consumption (granary and barracks).

The wild oleaster is present in Morocco while *Vitis vinifera* is considered sub-spontaneous in the Moroccan flora (Fennane, 2007). According to Zohary and Hopf (2000) wild grape is spontaneous in north-west Africa. Archaeological data show the presence of equipment for olive oil production between the 2nd and the 3rd cent. AD in the close Roman site of *Volubilis* (Figure 1) (Akeraz and Lenoir, 1982). Archaeobotanical data from *Lixus,* in contexts spanning between the 8th and the 1st cent. BC, show the presence of charcoals and seeds of olive and wine grape (Grau Almero, 2005; Grau Almero *et al.* 2001). Brun (2003) attests the spread of oliviculture in northern Africa, probably after the Phoenician expansion. Actually, our data testify, at least for the Roman period, *Olea* and *Vitis* cultivation, but it was not necessarily linked to oil and wine production. In fact, chemical analysis of the contents of some amphorae recovered in the warehouse of *Thamusida* suggests the import of oil and wine. Oil was identified in three amphorae (type *Dressel* 20) coming from the Iberian peninsula (Salvini *et al.* 2007), while wine was probably contained in two amphorae (types Dre*ssel* 30 and *Africana* II.1) coming from Algeria and Tunisia respectively (Pecci, unpublished).

**CONCLUSIONS**

The data here presented show the strong and multifunctional use of *Quercus suber* coming from the Mamora forest during the Roman occupation of the site, between the 1st and the 3rd cent. AD. The Mamora forest had a slightly higher biodiversity with respect to the present-day, with a shrubby understorey; a sporadic presence of other tree taxa, such as deciduous and sclerophyllous oaks cannot be excluded. Thus, we hypothesise that although strongly exploited, the Mamora forest was not heavy degraded.

Evidence of olive and grape cultivation seems not intended for oil and wine production.

Further analysis at the site in the pre-Roman and Islamic layers will give a diachronic frame of the vegetation changes clarifying the onset of human impact on this forest.

**Acknowledgements**

The authors express gratitude to two anonymous referees and the editor for their helpful comments.

**References**

AKERAZ, A. and LENOIR, E. 1982. Les huileries de Volubilis. *British Archaeological Reports* 14, 69-120.

AKERRAZ, A., BROUQUIER-REDDE, V., LENOIR, E. 1993. L'occupation antique de la plaine du Gharb, *Colloque International sur l'archéologie et l'histoire de l'Afrique du nord,* 25-29 octobre 1993, Pau, 311.

BALLOUCHE, A. and DAMBLON, F. 1988. Nouvelles données palynologiques sur la végétation Holocène du Maroc. In *Actes du Dixième Symposium de Association des Palynologues de Langue Française* 83-90, Bordeaux, Institut Français Pondichéry.

BEN KABBOUR, B., ZOUHRI L., MANIA J., COLBEAUX J.P. 2006. Assessing groundwater contamination risk using the DASTI/IDRISI GIS method: coastal system of western Mamora, Morocco. *Bulletin of Engineering Geology and the Environment* 65, 463-470.

BRUN, J.P. 2003. *Le vin et l'huile dans la Méditerranée antique: viticulture, oléiculture et procédés de transformation.* Paris, Errance.

CARRION, J.S., PARRA, I., NAVARRO, C. and MUNUERA, M. 2000. Past distribution and ecology of the cork oak (*Quercus suber*) in the Iberian Peninsula: a pollen-analytical approach. *Diversity and distributions* 6, 29-44.

CASTIGLIONI, E. and ROTTOLI, M. 1997. I macroresti vegetali. In Molinari A. (ed.) *Segesta II Il castello e la Moschea. Scavi 1989-1995*, Palermo, 235-257.

CONEDERA, M., KREBS, P., TINNER, W. and TORRIONI, D. 2004. The cultivation of *Castanea sativa* (Mill.) in Europe, from its origin to its diffusion on a continental scale. *Vegetation History and Archaeobotany* 13, 161-179.

DAMBLON, F. 1991. Contribution pollenanalytique à l'histoire des forêts de chêne-liège au Maroc: la subéraie de Krimda. In Heine K. (ed.), *Palaeoecology of Africa and the surrounding islands* (Ballouche A., Maley J. (guest eds.) Proceeding of the First Symposium on African Palynology, Rabat 1989, A.A. Balkema, Rotterdam, Brookfield 22, 171-183.

Di PASQUALE, G., ALLEVATO, E., RUSSO ERMOLLI, E., LUBRITTO, C., YONEDA, M., TAKEUCHI, K., KANO, Y., De SIMONE, F. 2010 Reworking the idea of chestnut (*Castanea sativa* Mill.) cultivation in Roman times: new data from ancient Campania. *Plant Biosystem* 144, 896-904.

EMBERGER, L. 1928a. L'importance du chêne liège dans le paysage marocain avant la destruction des forêts. *Revue géographique marocaine* 7, 121-124.

EMBERGER, L., 1928b. Les limites naturelles de la forêt de la Mamora. *Bulletin Société Scientifique Naturelle Marocaine.* 8, 220-222.

EMBERGER, L. 1934. *Les plantes fossiles dans leurs rapports avec les végétaux vivants.* Paris, Masson.

FENNANE, M., BARBERO, M., QUEZEL, P. 1984. La thuya de Berbérie au Maroc: aperçu phytogéographique et écologique. *Bulletin Institut Scientifique de Rabat* 8: 115-134.

FENNANE, M., IBN TATTOU, M., MATHEZ, J., OUYAHA, A. and El OUALIDI, J. 1999. *Flore pratique du Maroc*, Vol. 1. *Travaux Institut Scientifique*, série botanique, Rabat.

FENNANE, M., IBN TATTOU, M., MATHEZ, J., OUYAHA, A. and El OUALIDI, J. 2007. *Flore pratique du Maroc*, Vol. 2. *Travaux Institut Scientifique*, série botanique, Rabat.

GRAU ALMERO, E., PEREZ JORDA, G., IBORRA ERES, M.P. 2001. Gestion de recursos y economia. Lixus-colonia fenicia y ciudad punica-mauritana. Anotaciones sobre sus ocupacion medieval. *Saguntum extra* 4, 191-199.

GRAU ALMERO, E. 2005. Estudio antracologico. In: Aranegui, C. (Ed.) Lixus 2 Ladera Sur. *Excavaciones arqueológicas Marroco-Españolas en la colonia fenicia. Campañas 2000-2003. Saguntum Extra* 6, 219-220.

MACHADO YANES, M.d.C. 1992. Introducción al analisis antracológico en la isla de Tenerife. *Bulletin de la Société Botanique de France* 139, 495-506.

MARAÑÓN, T., AJBILOU, R., OJEDA, F. and ARROYO, J. 1999. Biodiversity of woody species in oak woodlands of southern Spain and northern Morocco. *Forest Ecology and Management* 115, 147-156.

METRO, A. and SAUVAGE, C. 1955. *Flore des végétaux ligneux de la Mamora.* Collection la nature au Maroc Société des sciences naturelles et physique du Maroc, Rabat.

MOLS, S.T.A.M. 2002. Identification of the wood used in the furniture at Herculaneum In Jashemski, W.F. and Meyer, F.G. (eds): *The Natural History of Pompeii.* 225-234. Cambridge University Press, Cambridge.

NATIVIDADE, J.V. 1950. *Subericultura.* Ministerio da Economia, Pescas e Alimetacao, Direccao Geral das Florestas e Aquicolas. Lisboa.

NEUMANN, K., SCHOCH, W., DÉTIENNE, P., SCHWEINGRUBER, F.H. 2001. *Woods of the Sahara and the Sahel / Bois du Sahara et du Sahel / Hölzer der Sahara und des Sahel.* Birmensdorf, Eidg.

Forschungsanstalt WSL. Bern, Stuttgart, Wien, Haupt.

QUÉZEL, P. and MÉDAIL, F. 2003. *Ecologie et Biogéographie des Forêts du Bassin Méditerranéen.* Paris, Elsevier, Collection Environnement.

REILLE, M. 1979. Analyse pollinique du lac Sidi Bou Ghaba. *Ecologia Mediterranea,* 61-65.

REJDALI, M. 2004. Forest cover changes in the Maghreb countries with special reference to Morocco. In Mazzoleni, S., Di Martino, G. and Rego, F. (Eds.) *Recent dynamics of the Mediterranean vegetation and landscape.* 13–20.Chichester, Wiley & Sons Ltd.

SADORI, L. and SUSANNA, F. 2005. Hints of economic change during the late Roman Empire period in central Italy: a study of charred plant remains from "La Fontanaccia", near Rome. *Vegetation History and Archaeobotany,* 14, 386–393.

SALVINI, L., PECCI, A., PAPI, E. and CERRI, L. 2007. Analisi dei residui organici assorbiti in due anfore romane (Thamusida, Marocco): GC-MS ed ESI-MS, in D'Amico C. (Ed.) *Atti del IV Congresso nazionale dell'Associazione Italiana di Acheometria,* Pisa 2006, 741-749, Bologna, Pàtron Editore.

SAUVAGE, C. 1961. *Recherche géobotaniques sur les subéraies marocaines.* Travaux de l'Institut Scientifique Chérifien, Séries botanique, Rabat.

SCHWEINGRUBER, F. 1990. *Anatomy of European woods: an atlas for the identification of European trees, shrubs, and dwarf shrubs,* Bern.

VERNET, J.L., OGEREAU, P., FIGUEIRAL, I., MACHADO YANES, C. and UZQUIANO, P. 2001. *Guide d'identification des charbons de bois préhistoriques du sud-ouest de l'Europe.* CNRS, Paris.

ZOHARY, D., HOPF, M. 1994. *Domestication of plants in the Old world.* Oxford, Clarendon Press.

# CHANGES IN THE INTERACTION BETWEEN SOCIETY AND THE ENVIRONMENT FROM THE MESOLITHIC (10300-8500 CAL BC) TO THE EARLY NEOLITHIC (C. 5400 CAL BC) IN CAN SADURNÍ CAVE (BEGUES, BARCELONA PROVINCE, SPAIN) A VIEW FROM THE ARCHAEOBOTANICAL DATA

Ferran ANTOLÍN
Laboratori d'Arqueobotànica, Departament de Prehistòria, Universitat Autònoma de Barcelona.
CIPAG. Predoctoral JAE Scholarship, GASA – Unidad Asociada al CSIC-IMF
fantolin@csic.imf.es

Raquel PIQUÉ
Laboratori d'Arqueobotànica, Departament de Prehistòria, Universitat Autònoma de Barcelona
raquel.pique@uab.cat

Anna BALLESTEROS
IPHES, Institut Català de Paleoecologia Humana i Evolució Social, Tarragona, Catalonia, Spain;
URV. Universitat Rovira i Virgili, Tarragona, Catalonia, Spain
anna.ballesteros@estudiants.urv.cat

Francesc BURJACHS
IPHES, Institut Català de Paleoecologia Humana i Evolució Social, Tarragona, Catalonia, Spain;
URV. Universitat Rovira i Virgili, Tarragona, Catalonia, Spain;
ICREA. Barcelona, Catalonia, Spain
francesc.burjachs@urv.cat

Ramon BUXÓ
Museu d'Arqueologia de Catalunya. C/Pedret 95, 17007 Girona
rbuxo@gencat.cat

Carmen MENSUA
Laboratori d'Arqueobotànica, Departament de Prehistòria, Universitat Autònoma de Barcelona.
CIPAG. Predoctoral JAE Scholarship, GASA – Unidad Asociada al CSIC-IMF
carmenmensua@gmail.com

Manel EDO
CIPAG. Facultat de Prehistòria, Història Antiga i Arqueologia. Universitat de Barcelona.
medo@telefonica.net

***Abstract**: Societies build a determinate perception of and certain ways of interaction with the environment in order to maintain their systems of production and reproduction. Perception in Prehistory can be studied from the material evidences left by the interaction between different groups and their environment. In this paper, we will be approaching interaction mainly from the archaeobotanical data (both charcoal and seed/fruit remains) of the late Mesolithic and early Neolithic layers of Can Sadurní Cave (Barcelona, Spain). Our results show clear differences in resource exploitation between the last hunters and gatherers and the first farmers of the area, which are not only related to the changes in the environment but also to the use of the space in the cave (the interaction with the environment) and the social perception of the environment.*

***Keywords**: archaeobotany, Iberian Peninsula, agriculture, gathering, fuel*

## THE PERCEPTION OF THE ENVIRONMENT BY PREHISTORIC GROUPS AND THEIR INTERACTION – Theoretical background

As several authors have already stated (Devara Group, 2006; Piqué, 1999; Terradas, 2001; Zurro, 2006), productive and reproductive activities are organized by each society in a particular physical space. We will use the word environment to refer to this space. According to the above mentioned authors, the way people interact with the environment is labour.

The perception of the environment is developed during the interaction of individuals with it through their own

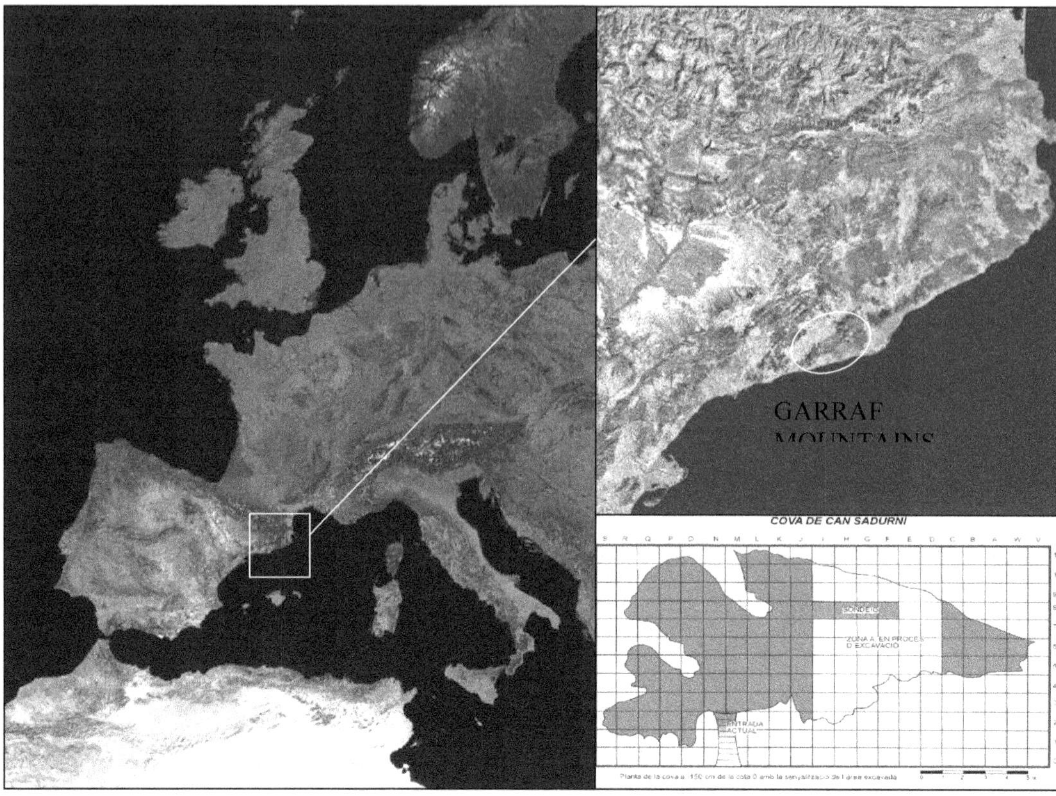


Figure 1. Location and plan of the site

experience and the interaction with other individuals from their own group (through education) or other groups (Ingold, 2000). Societies control the perception of the environment of each type of individual. Adults and children or men and women can have different perceptions of their environment because the systems of production and reproduction depend on it. Studying processes of labour, then, can bring us to answer the question of how people perceived their environment. Archaeobotanical remains are residues of processes of production and, thus, are indispensable to study the relationship between human beings and their environment (even though we need a holistic archaeological approach if we want to have a complete view of it).

## MATERIALS AND METHODS

### The site

Can Sadurní Cave is located on the southern slope of a low hill that overlooks the fertile plain of Begues (Barcelona, Spain), at 421 m.a.s.l. (Figure 1). Some 27 different layers have been identified, dating from 11,000 cal BC to the past century (Edo *et al.* 2011). Layers 18 (Early Neolithic, 5476-5309 cal BC) and 21 (Mesolithic, with two distinct subphases: 21E6, dated on 10840-10290 cal BC, and 21, dated on 8730-8540 cal BC) are analysed here (Edo *et al.* 2011). The results available for these layers come from a 1 x 4 m trial trenching (squares F8, G8, H8 and I8, see Figure 1) that is still ongoing. The materials here presented show the state of research in 2008, when this paper was written. Layer 18 is a unique case of Early Neolithic burial deposit. Remains of between 6 and 11 individuals (according to DNA studies, see Gamba *et al.* 2011) have been recovered so far. These remains were associated with apparel: several pottery vessels full of charred seeds (found broken but still with seeds attached to the vessels' walls), intentionally deposited goat extremities, shell plaques and other lithic and macrolithic pieces (Blasco *et al.* 2005). The archaeobotanical material of this layer could have been produced in one single action, as it seems to be the result of a burial ritual that would have involved the burning of part of the human bones (cranial fragments) and the apparel (Antolín *et al.* 2011). Layer 21, on the other side, was formed in two distinct phases (see dates above), as a result of an incalculable number of occupations of the cave by hunters and gatherers. Combustion structures are found and clear evidences of meat processing activities, especially of wild rabbit (*Oryctolagus cunniculus*) (Edo *et al.* 2011) took place inside the cavity.

The totality of the layer as been treated by flotation, except 30 litres of sediment from layer 21, which were water sieved, in all cases the sieves were of 1 mm and 0.5 mm mesh size. Seed and charcoal have been identified for both layers, but the absolute number of recovered remains is significantly different between each layer (Figure 2). All of the 348 seed remains have been studied for layer 21 and nearly 75% for layer 18 (41,908). Nevertheless, layer 21 has a very important problem of soil contamination due to the accidental falling of sediment from the walls of the trial trenching, which results in a very high percentage of domestic cereal remains (over 80% of the total) in the assemblage. Since

| **Taxa** | **Seed/charcoal** | **Represented part** | **Layer 18** | **Layer 21** |
|---|---|---|---|---|
| | | | **Neol.** | **Mesol.** |
| *Hordeum vulgare* | seed | grain | 120 | 2 |
| | | postc fragment | 215 | 3 |
| | | prec fragment | 6 | |
| | | **Total remains** | 341 | 5 |
| | | **NMI** | 306 | |
| *Hordeum vulgare* var. *nudum* | seed | grain | 29 | 4 |
| | | postc fragment | 28 | 1 |
| | | **Total remains** | 57 | 5 |
| | | **NMI** | 46 | |
| *Triticum aestivum/durum/turgidum* | seed | grain | 1884 | 51 |
| | | postc fragment | 357 | 5 |
| | | prec fragment | 6 | |
| | | **Total remains** | 2247 | 56 |
| | | **NMI** | 2107 | |
| *Triticum dicoccum* | seed | grain | 2634 | 23 |
| | | postc fragment | 2361 | 23 |
| | | prec fragment | 223 | |
| | | spikelet fork | 49 | |
| | | glume base | 14 | |
| | | **Total remains** | 5281 | 46 |
| | | **NMI** | 4012 | |
| *Triticum monococcum* | seed | grain | 616 | 29 |
| | | postc fragment | 366 | 36 |
| | | prec fragment | 225 | |
| | | spikelet fork | 8 | |
| | | glume base | 9 | |
| | | **Total remains** | 1224 | 65 |
| | | **NMI** | 908 | |
| *Triticum dicoccum/monococcum* | seed | spikelet fork | 16 | 2 |
| | | glume base | 59 | 1 |
| | | **Total remains** | 75 | 3 |
| *Triticum* sp. | seed | grain | | 10 |
| | | postc fragment | 31561 | 45 |
| | | prec fragment | 667 | |
| | | embryo | 88 | |
| | | glume base | 1 | 1 |
| | | **Total remains** | 32317 | 56 |
| *Hordeum* sp. | seed | postc fragment | 31 | 5 |
| *Hordeum/Triticum* | seed | postc fragment | 136 | 39 |
| cf. *Avena* sp. | seed | fragment | 1 | |
| *Chenopodium album* | seed | complete seed | 1 | |
| *Chenopodium hybridum* | seed | complete seed | 1 | |
| *Lathyrus* sp. | seed | fragment | 1 | |
| Papilionaceae | seed | fragment | 4 | |
| *Polygonum* sp. | seed | fragment | 1 | |
| cf. Umbelliferae | seed | complete seed | 1 | |

| Taxa | Seed/charcoal | Represented part | Layer 18 | Layer 21 |
|---|---|---|---|---|
| | | | Neol. | Mesol. |
| *Acer* sp. | charcoal | | 2 | 23 |
| *Arbutus unedo* | seed | fragment | 5 | |
| | charcoal | | 14 | 2 |
| Cistaceae | charcoal | | | 1 |
| Ericaceae | charcoal | | 6 | |
| *Juniperus* sp. | charcoal | | 7 | 3 |
| Leguminosae | charcoal | | | 1 |
| *Olea europaea* | charcoal | | 8 | |
| *Pinus halepensis* | charcoal | | 6 | |
| *Pinus* sp. | seed | fragment | 3 | 12 |
| *Pistacia lentiscus* | charcoal | | 2 | |
| *Pistacia* sp. | charcoal | | 5 | |
| *Prunus* sp. | seed | fruit | | 1 |
| | charcoal | | 30 | 25 |
| *Quercus* subg. *Quercus* | charcoal | | 162 | |
| *Quercus ilex* | charcoal | | 109 | |
| *Quercus* sp. | seed | acorn fragment | 1 | |
| | charcoal | | 37 | |
| *Rhamnus* sp. | charcoal | | 1 | |
| Rosaceae | charcoal | | 6 | |
| cf. Rosaceae type 1 | seed | fruit | | 1 |
| | | fragment | | 4 |
| cf. Rosaceae type 2 | seed | fragment | | 2 |
| *Rubus idaeus* | seed | complete seed | 1 | |
| Unidentified fruit | seed | fragment | 2 | 11 |
| Unidentified fruit type 1 | seed | fragment | | 19 |
| Unidentified fruit type 2 | seed | fragment | | 3 |
| Unidentified fruit type 3 | seed | fragment | | 9 |
| Unidentified fruit type 4 | seed | fragment | | 1 |
| Unidentified | charcoal | node | 8 | |
| Unidentified | charcoal | bark | 15 | |
| Unidentified | charcoal | | 57 | |
| | seed | complete seed | 17 | |
| | | fragment | 160 | 5 |
| | | **Total of seed remains** | 41908 | 348 |
| | | **Total of charcoal remains** | 475 | 67 |

Figure 2. Results of the archaeobotanical analysis of layers 18 and 21
(prec: fragment produced prior to charring; postc: fragment produced after charring)

we have not dated any of these remains and we believe that in our archaeological context one cannot consider them other than contaminations we are not going to deal with them in the present analysis. On the other hand, the rest of the taxa have not been treated as contaminations, since, except in the case of pine cone fragments, none of them are documented in the upper layers (Antolín, 2008). In the case of pine cone scale fragments, only 6 of these have been identified among over 50,000 analyzed items from 14 layers. The fact that more remains have been recovered in the Mesolithic layer is consistent enough. Radiocarbon dating, which should be carried out in the future, will certainly solve this issue. A representative number of the total of charcoal remains has been studied for layer 18 (475 fragments) and only coordinated (extracted during the excavation and, for this reason,

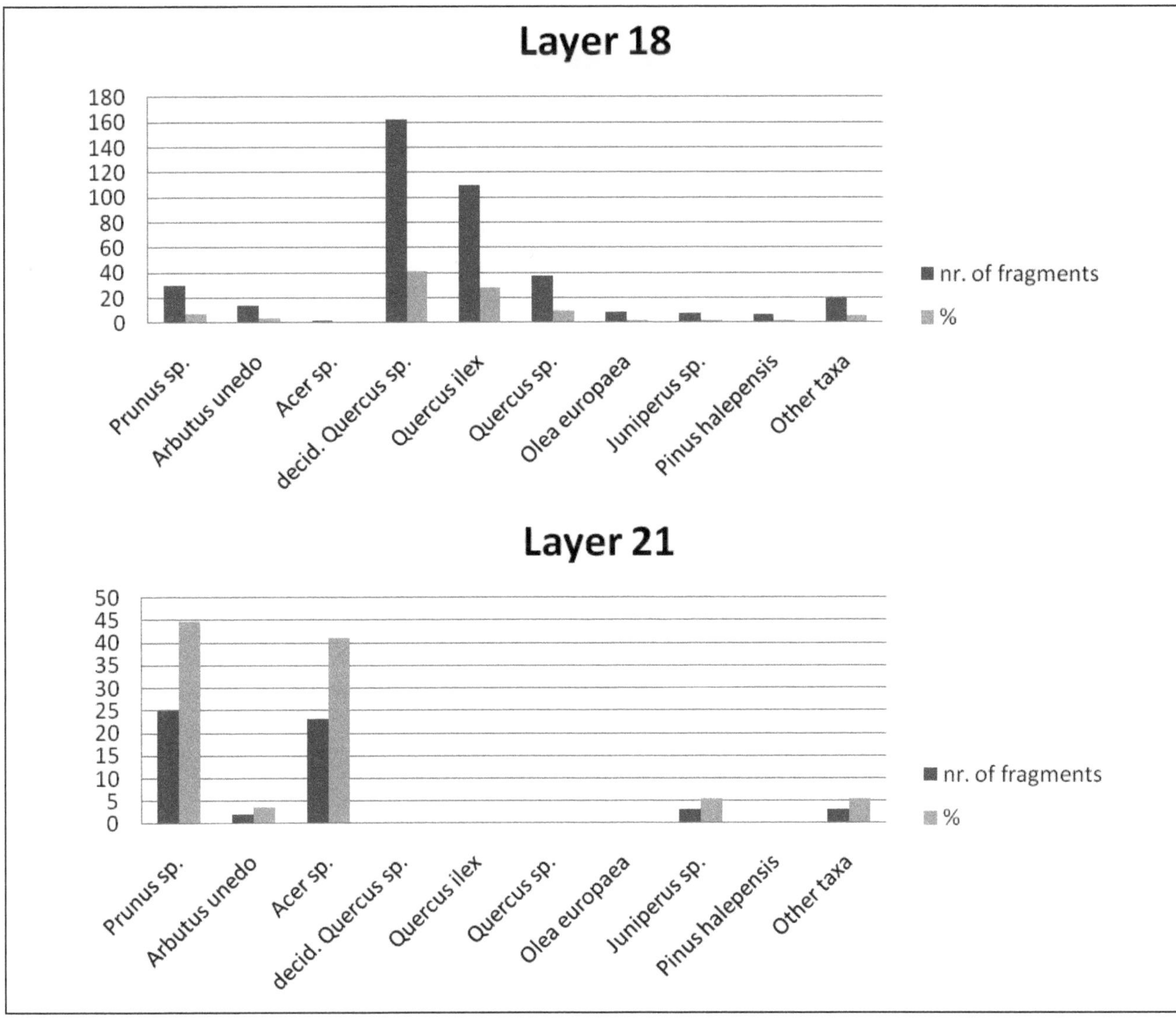


Figure 3. Number of fragments and relative percentages of the main taxa found in the anthracological analysis of layers 21 (top) and 18 (bottom)

clearly belonging to the layer) charcoal remains have been studied for layer 21 (67) so as to avoid the contamination problems observed in the samples treated by flotation.

**Charcoal analysis**

Charcoal analysis has mainly been oriented to taxonomic determination (using the authors' reference collection and literature such as Schweingruber, 1990) but the presence of cracks in the fragments and post-depositional effects have also been noted for layer 18. The combination of this technique with the results of the soil micromorphology analysis (Bergadà and Cervelló, 2011) has allowed us to approach to the taphonomic processes that could have caused the assemblage.

**Charred seeds and fruit analysis**

The seed and fruit identification has been carried out with the help of the authors' reference collection, also using specialized atlas (e.g. Cappers *et al.* 2006). Besides, any consequences of the processing of the harvest on the grains themselves were also registered. For this reason fragments produced prior to charring, probably produced during the dehusking of the grains, were quantified (following Bouby *et al.* 2005; Valamoti, 2002; 2011). The minimum number of individuals (MNI) has been calculated for cereal grains according to previous works (Antolín, 2008; Antolín and Buxó, 2011).

**RESULTS AND DISCUSSION**

For the Mesolithic period, 13 taxa have been identified considering both charcoal and seed and fruit analyses (see Figure 2). All the carpological remains are fragmented and in a bad state of preservation except one fruit of *Prunus* sp. They appear in a very low density (0-0.4 remains per litre of sediment aproximately). The low density of seed/fruit items in contexts of hunter and gatherer groups is very common (e.g. Aura *et al.* 2005) and, as other authors have stated (Hather and Mason, 2002), the number of remains is deeply affected by

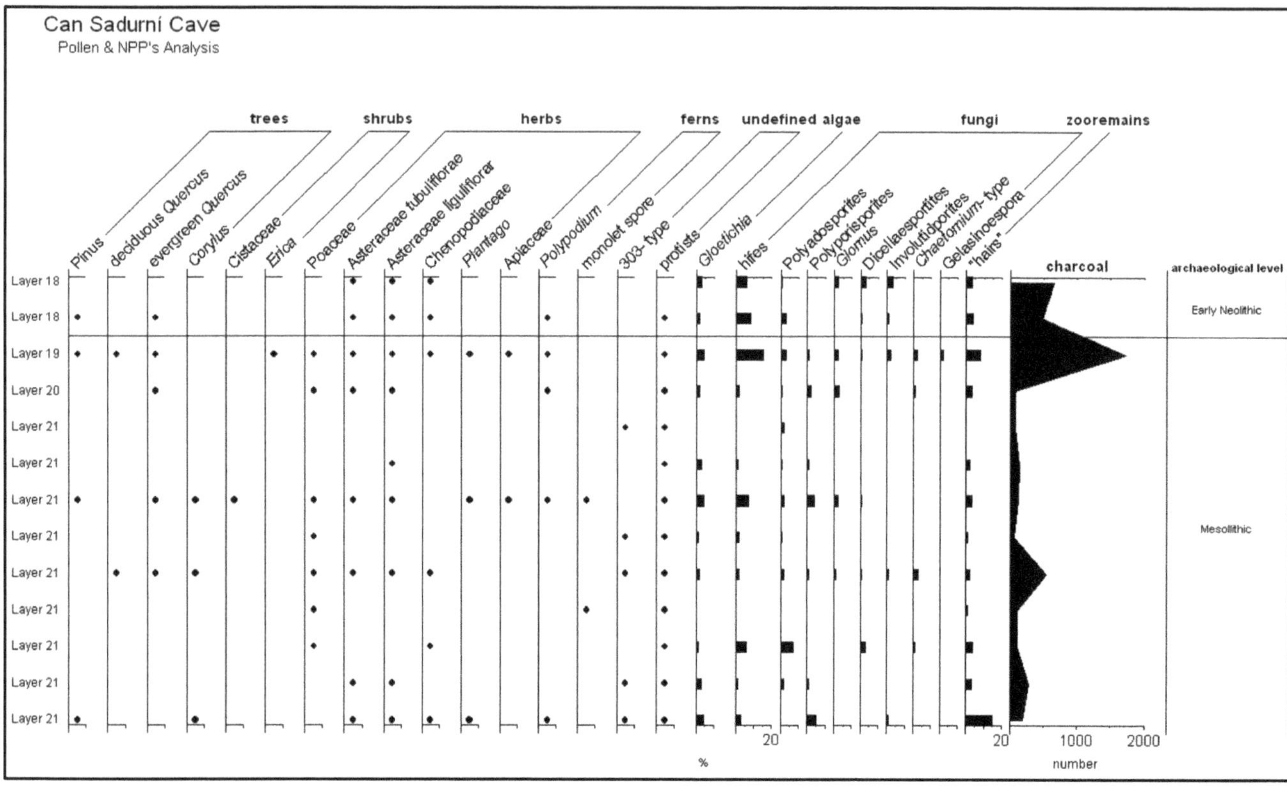


Figure 4. Diagram with the results of the pollen analysis

taphonomic agents and it is very likely that it does not represent past economies. The sample available for anthracological analysis of layer 21 is clearly scarce and, maybe for this reason, the results show a low taxonomic diversity (6 taxa, Figure 3).

Primarily supramediterranean taxa (*Acer* sp., *Prunus* sp.) have been identified, but bushes are also present (*Arbutus unedo, Juniperus* sp. and Cistaceae). The carpological analysis shows that evergreen taxa (*Pinus* sp.) also lived in the plain. *Acer* and *Prunus* would grow in the most humid valleys and only occasionally between deciduous oaks, while the bushes would grow in evergreen oak woods, in open tree formations or places where the soil wouldn't allow well-developed woods. The palynological analysis carried out in the northern profile of the trench (A. Ballesteros and F. Burjachs, unpublished) documents, in very low numbers, the presence of *Corylus* sp., *Pinus* sp. and evergreen and deciduous *Quercus* sp (Figure 4). These results are of relevant significance, since several taxa that were available near the cave are not documented in the analysis of archaeobotanical macroremains.

In the Early Neolithic layer, 22 taxa have been recovered (Figure 2). Anthracological remains are very abundant. A considerable number of them have radial-oriented cracks (some of the fragments weren't identifiable to species level because of the abundance of such cracks) and in a few cases some type of calcification has occurred (Figure 5). The presence of radial-oriented cracks is generally associated to the use of green timber, although other causes could produce such effects on the wood. (Théry-Parisot, 2001).

Most of the taxa identified for the Neolithic period belong to oak woods. Both deciduous and evergreen *Quercus* are dominant in this phase (Figure 3). They were probably the most available taxa around the site. Open tree formations are also present, since we find *Juniperus* sp., *Pistacia* sp. and *Olea europaea*, and more ruderalized areas where *Avena* sp., *Chenopodium album* or *C. hybridum* could grow, which probably indicate the first signs of forest degradation in our region. Pollen analysis (Riera *et al.* 2007) indicate a predominance of deciduous species but a more important presence of evergreen trees, like *Pinus* and *Buxus,* than what could be observed for the previous period in the region. Pollen analysis carried out in the cave confirms these trends (Figure 4).

Seed and fruit remains are mainly composed by domestic cereal crops (over 99% of the assemblage), among which glume wheats, especially emmer (*Triticum dicoccum*) dominate (Figure 6). The few numbers of wild fruits associated with the grains are autumn fruits. As mentioned above, they were associated to cardial ware decorated vessels (partly stuck to the pottery fragments but mostly spread all over the layer). More than 80% of the grains were fragmented and only 2% of this fragmentation occurred before carbonization. This high percentage of fragmentation seems due to recent postdepositional effects: excavation techniques, trampling, etc. On the other hand, fragmentation produced prior to charring would have been produced during the processing of the grain. Most of the fragments produced prior to charring belong to glume wheats. Experimental works have demonstrated that some dehusking techniques can produce a low proportion of

Figure 5. Charcoal fragments from layer 18. Top left: Fragment of charcoal of a deciduous *Quercus* with radial-oriented cracks in the transversal plane; top right: Fragment of charcoal with evidences of calcification; bottom: Unidentifiable fragment of charcoal with very pronounced cracks

grain fragments (Meurers-Balke and Luning, 1992). Around 25% of MNI of einkorn were fragmented prior to charring, while emmer shows less than 5% of these cases (Figure 7), which indicates that they could have been produced in different activities, like a separate dehusking process (Antolín, 2008; Antolín and Buxó 2011). Barely any cases of fragments produced prior to charring belong to naked wheat, probably because it was processed independently.

## THE INTERACTION BETWEEN SOCIAL GROUPS AND THE ENVIRONMENT IN CAN SADURNÍ CAVE

In order to approach the processes of production that took place in the cave, we will approach the four different phases that constitute these processes: obtaining, processing, storage and consumption.

### Obtaining processes

The obtaining processes during the Mesolithic show repeated patterns. The fact that submediterranean taxa were exploited for food and fuel purposes along this period prooves that they were part of the system of production of those societies, even though their wood is hard and less suitable as fuel than other available taxa like pine or oak wood.

On the other hand, Neolithic groups had a distinct interaction with the environment. At least two vegetal communities were exploited for fuel: one with deciduous taxa, probably mainly constituted by oaks (mixed with

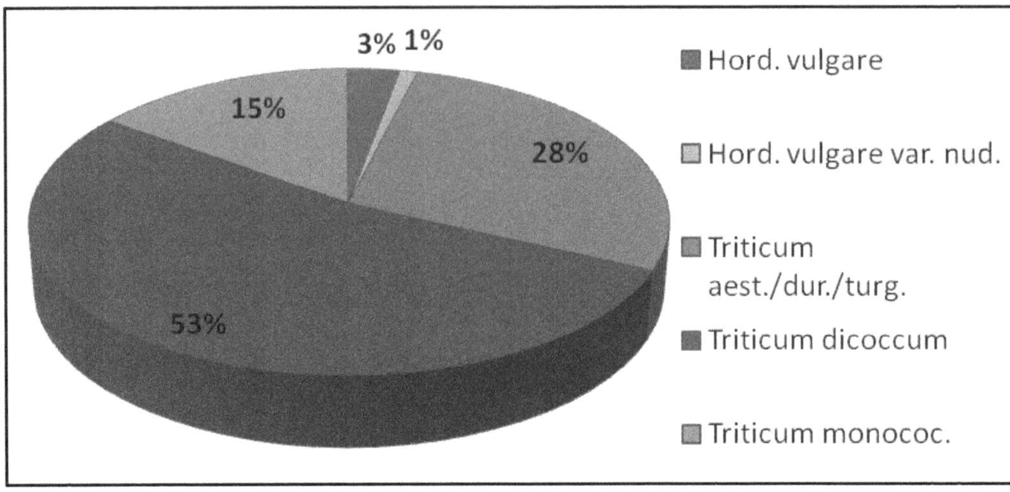


Figure 6. Relative percentages of domestic cereal crops of layer 18

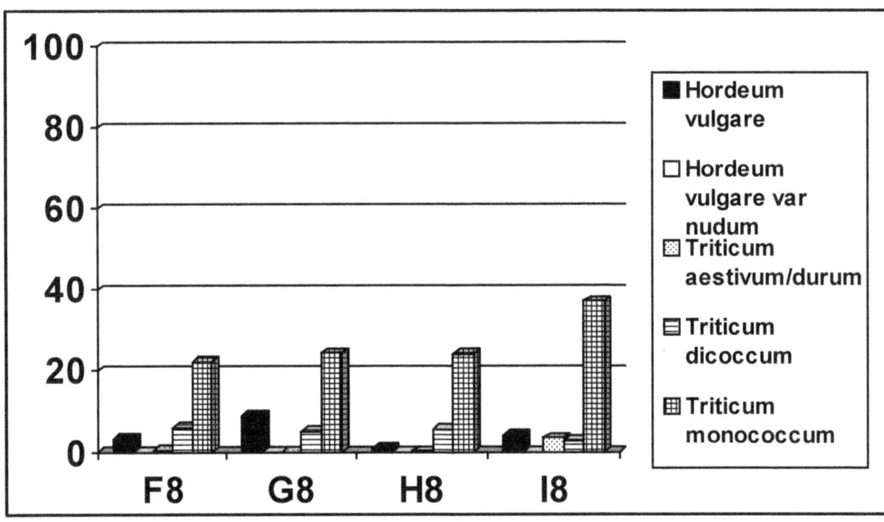


Figure 7. Percentage of fragments of cereal grains produced prior to charring of layer 18

*Prunus* and *Acer*), and one with evergreen taxa. Fruit remains are very scarce but they all are autumn fruits. Cracked oak wood could indicate the use of green wood as fuel, which means that its obtention would have required specific tools (such as axes) of which we do not have any archaeological evidence. Another possibility to be considered is that wet dead wood was used, for which no such tools would have been needed.

The first domesticated cereal remains appear in this period but we can't calculate the importance they signified in the diet of the first neolithic communities. We think that some species have been grown separately, as we explain further on, because particular processing activities have been detected on some taxa. One piece of flint was found to have been used as a sickle (Gibaja *et al.* 2011) which helps to demonstrate that the production of the cereal could have been local.

**Processing**

There are no archaeological evidences of this phase of the process of production during the Mesolithic period. Transportation had to take place at some point, but we cannot calculate the investment of labour that was needed.

More details are available for this phase during the neolithic occupation. Bark was identified during the anthracological analysis but one cannot know whether it is a wood processing residue that was (deliberately or not) thrown to the fire or a non-processed-wood combustion residue. Further research on the origin of this charred bark fragments could lead us to a better knowledge of the form in which wood was obtained, processed and consumed.

More can be said about the cereals' processing tasks. Both emmer and einkorn were dehusked before charring and it is very likely that during the dehusking process some fragmentation of the grains occurred. The heavier proportion of fragmented einkorn points to a possible independent process for each glume wheat. The data presented suggest that at least the glume wheats, which received an specific type of processing, were cultivated and processed separately from the free-threshing wheat.

### Storing

No archaeological evidences of storing activities have been found during the Mesolithic occupation.

We can state that at least part of the wood consumed during the neolithic occupation was not stored for very long, as it was possibly green when burned. On the other hand, there are no evidences of storing of the cereals besides the pottery vessels that contained them. If, according to the ripening time of the recovered wild fruits associated with the charred cereal, the burial ritual happened in autumn (of which we are not sure), cereals might have been stored for some months before their consumption. No evidences of insects or sprouting were found, which indicates that cereal had been kept in optimal conditions (Antolín, 2008).

### Consuming

As mentioned above, the use of the cave is different in both layers. Layer 21 shows a repeated number of occupations and the cave seems to be used as shelter for productive and consuming activities (Edo *et al.* 2011). All charcoal taxa identified for the Mesolithic period could have been consumed as fuel. On the other hand, all the fragments of fruit could have had an alimentary value. The preservation of these fruit remains is probably accidental, as experimental works show (Sievers and Wadley, 2008). Despite this, we consider that their constant appearance in Mesolithic sites in the region (e.g. Allué *et al.* 2007) is a proof of some intentional gathering.

Layer 18, on the other hand, is a funerary context. The taphonomic study of the materials from this layer shows that pottery vessels full of grain, some human cranial bones, animal bones, lithic and macrolithic artefacts were exposed to fire action (Antolín *et al.* 2011). Low temperatures were achieved during the fire: soil micromorphological analysis has detected temperatures below 300ºC (Bergadà and Cervelló, 2011), while human bone analysis (Antolín *et al.* 2011) has determined termoalteration produced between 230 and 300ºC for some cases and between 300-370ºC for others. These low temperatures could be due to the use of green wood as fuel, which would not allow higher temperatures and at the same time explains the good preservation of the grains (Antolín, 2008) and the low presence of ashes in layer 18 (Bergadà and Cervelló, 2011). Thus plant remains in this Neolithic level could have been ritually consumed by depositing apparel to the buried members of the community.

## CONCLUSIONS: CHANGES IN THE PERCEPTION OF AND THE INTERACTION WITH THE ENVIRONMENT IN CAN SADURNÍ CAVE

The archaeobotanical material recovered from Mesolithic occupations in Can Sadurní seems to be the result of a particular social pattern, as it is repeated over the centuries. Not only the cave is occupied with the same economical purposes, but also the same types of sources are chosen. This pattern is very much concerned on *what* is obtainable and obtained from the environment, which makes the *where* a secondary variable. The pollen analysis shows availability of other more suitable taxa but for some reasons, the focus is on *Acer* and *Prunus*. It is important to notice that other sites of the same period located in different ecosystems like Balma del Gai and Balma de la Griera (Allué *et al.* 2007; Nadal, 2000) show very similar patterns, which could be evidencing that it is a supraregional model of subsistence.

Neolithic societies' data seem to show a special significance on *where* the materials are obtained. Since the labour investment before the extraction of the material is much higher than in Mesolithic societies, the place becomes a central aspect in the obtaining process. The perception of the environment, in this case, is built with the association of a concrete place and a certain activity. This is very clear with cereal growing, as it seems that the three types of wheat would have been grown separately. The funerary ritual celebrated in the cave is another evidence of the significance of spaces in Neolithic societies. Instead of focusing in a low number of taxa, the taxonomic diversity increases and reflects vegetal communities more diversely.

A larger surface of the site must be excavated, together with a thorough radiocarbon control to discriminate contaminations, and more sites must be analyzed in the region in order to have a complete view of all the activities that took place in the plain and, eventually, of how its inhabitants perceived and interacted with their environment.

### Acknowledgements

F. Antolín and R. Piqué belong to the consolidated research project AGREST (2009 SGR 734, AGAUR) and this research would not have been possible without the archaeology team working in Can Sadurní Cave, which at the time when this paper was written (December 2008) was leaded by Anna Blasco, Manel Edo and Pepa Villalba (CIPAG), to whom we are very grateful. The authors would like to acknowledge S. Thiébault and F. Damblon for their comments on an earlier version of this paper.

### References

ALLUÉ, E., NADAL, J., ESTRADA, A. and GARCÍA-ARGÜELLES, P. 2007. Los datos antracológicos de la Balma del Gai (Bages, Barcelona): una aportación al conocimiento de la vegetación y la explotación de los recursos forestales durante el Tardiglaciar en el NE Peninsular. *Trabajos de Prehistoria* 64, n. 1, 87-97.

ANTOLÍN, F. 2008. *Aproximació a l'estudi de la percepció i la interacció amb l'entorn vegetal en societats caçadores recol·lectores i agricultores ramaderes (10,000-4000 cal ANE). Resultats de l'estudi arqueobotànic del jaciment arqueològic de la Cova de Can Sadurní (Begues, Baix Llobregat).* Unpublished Master Thesis, Universitat Autònoma de Barcelona. http://hdl.handle.net/2072/40656

ANTOLÍN, F., ACHE, M., BERGADÀ, M., BLASCO, A., BUXÓ, R., EDO, M., GIBAJA, J.F., MENSUA, C., PALOMO, A., PIQUÉ, R., RUIZ, J., SAÑA, M., VERDÚN, E., VILLALBA, M.J. 2011. Aproximació interdisciplinària a l'acció del foc en les inhumacions i aixovars del Neolític antic cardial de Can Sadurní (Begues, Baix Llobregat). In A. Blasco, M. Edo, & M.J. Villalba, (coord.), *La cova de Can Sadurní i la Prehistòria de Garraf. Recull de 30 anys d'investigació*, 151-157. Milano, EDAR-Hugony.

ANTOLÍN, F. and BUXÓ, R. 2011. Proposal for the systematic description and taphonomic study of carbonized cereal grain assemblages: a case study of an early Neolithic funerary context in the Cave of Can Sadurní. *Vegetation History and Archaeobotany* 20, 53-66.

AURA, J.E., CARRIÓN, Y., ESTRELLES, E., PÉREZ-JORDÀ, G. 2005. Plant economy of hunter-gatherer groups at the end of the last Ice Age: plant macroremains from the cave of Santa Maira (Alacant, Spain) ca. 12000-9000 BP, *Vegetation History and Archaeobotany* 14, 542-550.

BERGADÀ, M. and CERVELLÓ, J.M. 2011. Estratigrafia, micromorfologia i paleoambient de la Cova de Can Sadurní (Begues, Baix Llobregat) des dels *c.*11,000 fins els 5,000 anys BP. In A. Blasco, M. Edo, & M.J. Villalba, (coord.), *La cova de Can Sadurní i la Prehistòria de Garraf. Recull de 30 anys d'investigació*, 97-108. Milano, EDAR-Hugony.

BLASCO, A., EDO, M., VILLALBA, M.J., BUXÓ, R., JUAN-TRESSERRAS, J., SAÑA, M. 2005. Primeros datos sobre la utilización sepulcral de la Cueva de Can Sadurní (Begues, Baix Llobregat) en el Neolítico Cardial. In P. Arias, R. Ontañón and C. García-Moncó (eds.). *III Congreso del Neolítico en la Península Ibérica. Santander 5 a 8 de octubre de 2003*, 625-633.

BOUBY, L., FAGES, G. and TREFFORT, J.M. 2005. Food storage in two Late Bronze Age caves of Southern France: palaeoethnobotanical and social implications. *Vegetation History and Archaeobotany* 14, 313-328.

CAPPERS, R.T.J., BEKKER, R.M. and JANS, J.E.A. 2006. *Digitale Zadenatlas van Nederland.* Groningen, Barkhuis publishing & Groningen University Library.

EDO, M., BLASCO, A., VILLALBA, M.J. 2011. La cova de Can Sadurní, guió sintètic de la prehistòria recent de Garraf. In A. Blasco, M. Edo, & M.J. Villalba, (coord.), *La cova de Can Sadurní i la Prehistòria de Garraf. Recull de 30 anys d'investigació*, 13-95. Milano, EDAR-Hugony.

GAMBA, C., FERNÁNDEZ, E. and ARROYO-PARDO, E. 2011. Análisis de AND antiguo: los restos humanos de Can Sadurní. In A. Blasco, M. Edo, & M.J. Villalba, (coord.), *La cova de Can Sadurní i la Prehistòria de Garraf. Recull de 30 anys d'investigació*, 141-149. Milano, EDAR-Hugony.

Devara Group 2006. Análisis etnoarqueológico del valor social del producto en sociedades cazadorasreco-lectoras. In *Etnoarqueología de la Prehistoria: más allá de la analogía*, 189-208.Barcelona, CSIC (Treballs d'Etnoarqueologia, 6).

GIBAJA, J.F., PALOMO, A., BLASCO, A., VILLALBA, M.J. and EDO, M. 2011. El utillaje lítico del neolítico antiguo de Can Sadurní: primeros datos. In A. Blasco, M. Edo, & M.J. Villalba, (coord.), *La cova de Can Sadurní i la Prehistòria de Garraf. Recull de 30 anys d'investigació*, 121-128. Milano, EDAR-Hugony.

HATHER, J.G. and MASON, S.L.R. 2002. *Hunter-Gàtherer Archaeobotany. Perspectives from the northern temperate zone.* London, Institute of Archaeology, UCL.

INGOLD, T. 2000. *The Perception of the Environment: Essays on Livelihood, Dwelling and Skill.* Routledge, London/New York.

MEURERS-BALKE, J. and LÜNING, J. 1992. Some aspects and experiments concerning the processing of glume wheats. In P.C. Anderson (ed.), *Préhistoire de l'agriculture: nouvelles approches expérimentales et ethnographiques*, 341-362. Paris, CNRS (Monographie du CRA, 86).

NADAL, J. 2000. La fauna de mamífers al Garraf i els seus voltants a través del registre arqueològic. *III Trobada d'estudiosos del Garraf*, Monografies, 30, 165-170.

PIQUÉ, R. 1999. *Producción y uso del combustible vegetal: una evaluación arqueológica.* Treballs d'Etnoarqueologia, 3, CSIC, Madrid.

RIERA, S., ESTEVE, X. and NADAL, J. 2007. Systèmes d'exploitation et anthropisation du paysage méditerranéen du Néolithique Ancien au premier âge du Fer: le cas de la dépression de Penedès (nord-est de la péninsule ibérique). In H. Richard, M. Magny and C. Mordant (eds.), *Environements et cultures à l'âge du Bronze en Europe Occidentale. Actes des congres nationaux des Sociétés historiques et scientifiques 129e, Besançon, 2004*, 121-141. Paris, Editions du CTHS (Documents préhistoriques, 21).

SIEVERS, C. and WADLEY, L. 2008. Going underground: experimental carbonization of fruiting structures under hearths. *Journal of Archaeological Science* 35, 2909-2917.

SCHWEINGRUBER, F. 1990. *Anatomy of European Woods.* Bern/Stuttgart, Paul Haupt.

TERRADAS, X. 2001. *La gestión de los recursos minerales en las sociedades cazadoras-recolectoras.* Madrid, CSIC. Treballs d'Etnoarqueologia, 4.

THERY-PARISOT, I. 2001. *Économie des combustibles au Paléolithique.* Paris, CNRS. Dossier de Documentation Archéologique, 20.

VALAMOTI, S.M. 2002. Food remains from Bronze Age Archondiko and Mesimeriani Toumba in northern Greece. *Vegetation History and Archaeobotany* 11, 17-22.

VALAMOTI, S.M. 2011. Ground cereal food preparations from Greece: the prehistory and modern survival of traditional Mediterranean "fast foods". *Archaeological and Anthropological Sciences* 3, 19-39.

ZURRO, D. 2006. El análisis de fitolitos y su papel en el estudio del consumo de recursos vegetales en la prehistoria: bases para una propuesta metodológica materialista. *Trabajos de Prehistoria*, 63, nº2, 35-54.

# NEOLITHIC WOODLAND IN THE NORTH MEDITERRANEAN BASIN: A REVIEW ON *OLEA EUROPAEA* L

Yolanda CARRIÓN
[1]Centro de Investigaciones sobre Desertificación (CIDE), CSIC, Carretera Moncada-Náquera Km 4,5.
46113 Moncada (Valencia), Spain
Yolanda.carrion@uv.es

Maria NTINOU
Agg. Sikelianou 2, 45221 Ioannina, Greece
Maria.Ntinou@uv.es

Ernestina BADAL
Universitat de València, Av. Blasco Ibáñez 28. 46010 Valencia, Spain
Ernestina.badal@uv.es

**Abstract**: *The aim of this paper is to specify the natural distribution of* Olea europaea *L. during the Early Holocene in the Northern Mediterranean by means of the identification of wood charcoal remains of this species at prehistoric sites. For this purpose, we have reviewed the relevant literature and extracted the data in which* Olea *charcoal has been identified. We have taken into consideration the biogeographical and chrono-cultural contexts in which the species is present, its frequency of occurrence at different locations and the associated plant taxa with the aim of tracking the Holocene history of the oleaster. Based on this information we suggest that the species started expanding during the Preboreal from Pleistocene thermophilous tree* refugia *located in the Levant, Cyprus, Sicily and the southern parts of the Iberian Peninsula. Its presence was confined to the thermomediterraenan bioclimatic level. The expansion dynamics of* Olea *after the Boreal are better understood in the western Mediterranean. There the species becomes very abundant or dominant in the thermophilous plant formations of the Atlantic period and expands to favorable enclaves outside the limits of the thermomediterraenan bioclimatic level.*

**Key words**: Olea *wood charcoal, Pleistocene refugia, Holocene, thermomediterranean level*

## INTRODUCTION

At present, *Olea europaea* L. var. *sylvestris* is an important element of the vegetation in the circummediterranean area. The species is considered an accurate thermal bioindicator for the definition of the thermomediterranean bioclimatic level while its cultivated variety is an emblematic and genuine plant of the Mediterranean cultures from protohistoric times. In the archaeological record the presence of *Olea* remains, is recorded since prehistoric times in context associated to the Epipalaeolithic, the Neolithic and the Bronze Age periods while great part of the discussion concerning this plant has been dedicated to its domestication and cultivation (Besnard *et al.* 2002; Breton *et al.* 2006; Contento *et al.* 2002; Liphschitz *et al.* 1991; Rodríguez-Ariza and Montes 2005; Zohary and Hopf 2000).

In this paper our aim is to discuss the presence of *Olea europaea* in the thermophilous vegetation since early post-glacial times in order to define the early Holocene distribution and consolidation of this species in the Mediterranean basin. For this purpose we use the information that originates from charcoal analyses carried out at various prehistoric sites of the northern Mediterranean basin. We believe that wood charcoal is probably the most adequate category of archaeobotanical remains for approaching the above-mentioned issues; if *Olea* had existed in a given area, its wood would have been used as fuel in domestic fires much before the management of the tree for its edible fruits and/or its cultivation. *Olea* wood charcoals at a site most probably reflect its presence in a certain area while by way of charcoal analysis the species becomes an integral part of the charcoal assemblages that reflect the characteristics of the local vegetation. Therefore, by using charcoal sequences from prehistoric sites we may document the vegetation types in which *Olea* was established in early Holocene times, its frequencies in formations of the Atlantic *optimum* (c. 8800-5600 cal. BP) and its behaviour in relation to human activity.

The chronological framework in this paper is delimited by the beginning of the Preboreal and the end of the Atlantic (c. 11500-5600 cal. BP), approximately coinciding with the Mesolithic and Neolithic period. The origins of olive domestication and the expansion of techniques and/or cultivars during the Bronze Age in the eastern Mediterranean (Terral *et al.* 2004) and during the Iron Age in the western Mediterranean are important topics that lay beyond the scope of this article.

## SETTING

In the present-day, the olive is a prominent feature of the whole Mediterranean basin and constitutes a complex of wild forms, cultivated varieties and secondary feral

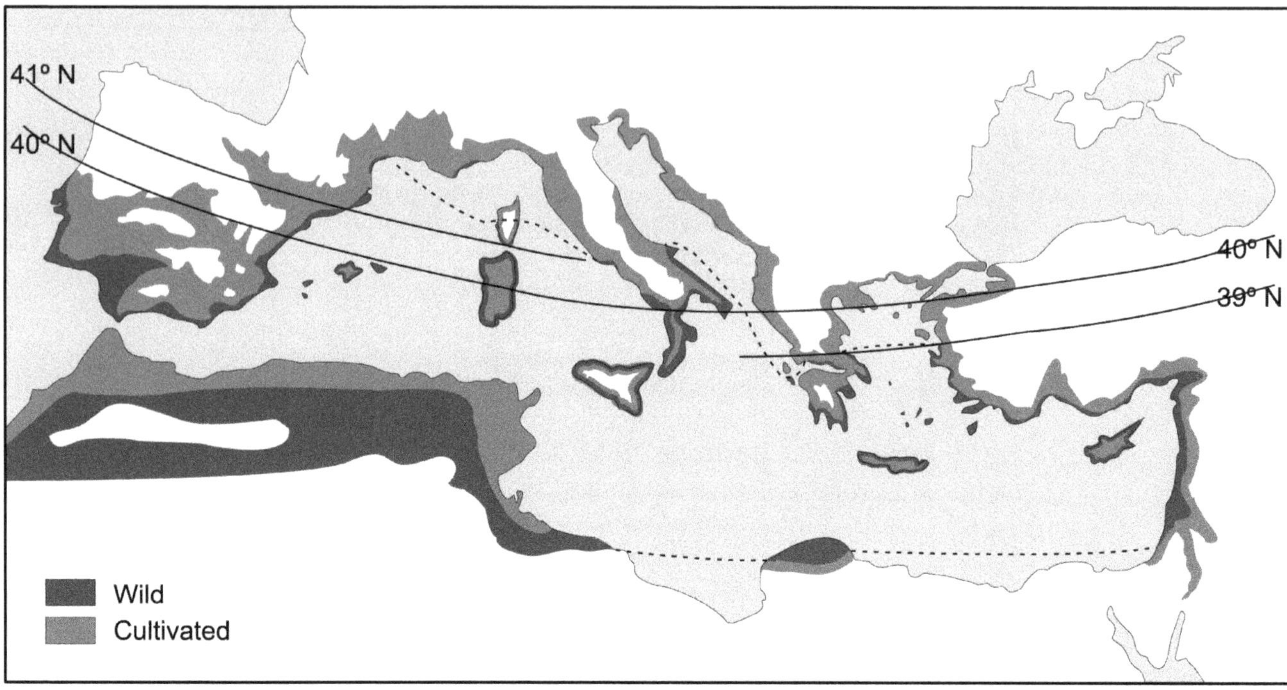


Figure 1. Current distribution of wild and cultivated olive tree (*Olea euroapaea* L.) in the Mediterranean basin

forms. The oleaster occupies the warmest areas of the Mediterranean, approximately coinciding with the thermomediterranean bioclimatic level or with the lower mesomediterranean (Ozenda 1975; Rivas-Martínez 1987) (Figure 1). In the western Mediterranean, it extends over low, warm lands with a mean annual temperature of between 17-19ºC. A limiting factor for its development is the mean temperature of the coldest month that should not be below 6ºC (Rubio *et al.* 2002, 343, figure 1). In terms of altitude, its limit in the Iberian Peninsula is around 500 m. In the eastern Mediterranean, the oleaster thrives in the *Ceratonion-Pistacion lentisci* plant association (Browics 1983; Zohary and Orshan 1959). In the plant associations of both the western and the eastern Mediterranean, thermophilous elements like *Pistacia lentiscus* L., *Quercus* sp. evergreen, *Myrtus communis* L.; *Rhamnus lycioides* L., *Rosmarinus officinalis* L., etc., accompany *Olea*. In areas that have suffered little human activity, these species may form well-structured woodlands with trees up to 15 metres high, as the case of Serra d'Arrabida in Portugal shows (Costa *et al.* 2001).

Since late Prehistory, the olive has been grown for its oil-rich fruit. The cultivated variety, *Olea europaea* L. var. *europaea*, has become more flexible to climatic and environmental conditions and therefore extends beyond the previously described area (Figure 1). It penetrates towards higher, colder and more continental lands, mainly on calcareous soils, *terra rossa,* and sandy marls.

## MATERIAL AND METHOD

The paper focuses on *Olea* macroremains, particularly on wood charcoal. In the last 30 years, charcoal analyses have been carried out at archaeological sites across the Mediterranean basin and have provided ample information on the plant taxa used as fuel by human groups. Wood charcoal is particularly interesting as it concerns the local flora of each region, which responds to certain thermoclimatic parameters. Moreover, wood charcoal remains of selected taxa can be directly AMS dated, therefore offering the possibility of documenting their first appearance in a region, tracing their expansion and/or detecting taphonomic problems.

The paper is based on a thorough bibliographic review of charcoal analyses data in which *Olea* occurred (Figure 3). It is evident that not all regions and chronological periods rely on a sample of sufficient and comparable data. Despite the limitations, we use the information concerning the presence and frequency of *Olea europaea*, the accompanying flora, the associated chrono-cultural sequence, the biogeographical contexts, and the AMS dates provided by *Olea* charcoal or endocarps as tools for tracing the Holocene history of the species. For this task, we have elaborated maps for the distribution of the oleaster at present (Figure 1) and during the early Holocene (Figure 2), in order to delimit its biogeographical affinities now and in the past. These maps combined with an evaluation of the accompanying flora -in those cases that charcoal sequences were available- and with a list of AMS dates on *Olea* macroremains, constitute the bases for reassessing the issues of the postglacial presence and expansion of oleaster populations in the sclerophyllous vegetation of the Mediterranean basin. The radiocarbon dates cited in this paper are presented in calendar years BP. Dates in other formats from the cited references have been calibrated to 2 sigma using the CalPal-2007 program (Weninger *et al.* 2009) and the CalPal-2007-Hulu calibration data set (Weninger and Jöris, 2008).

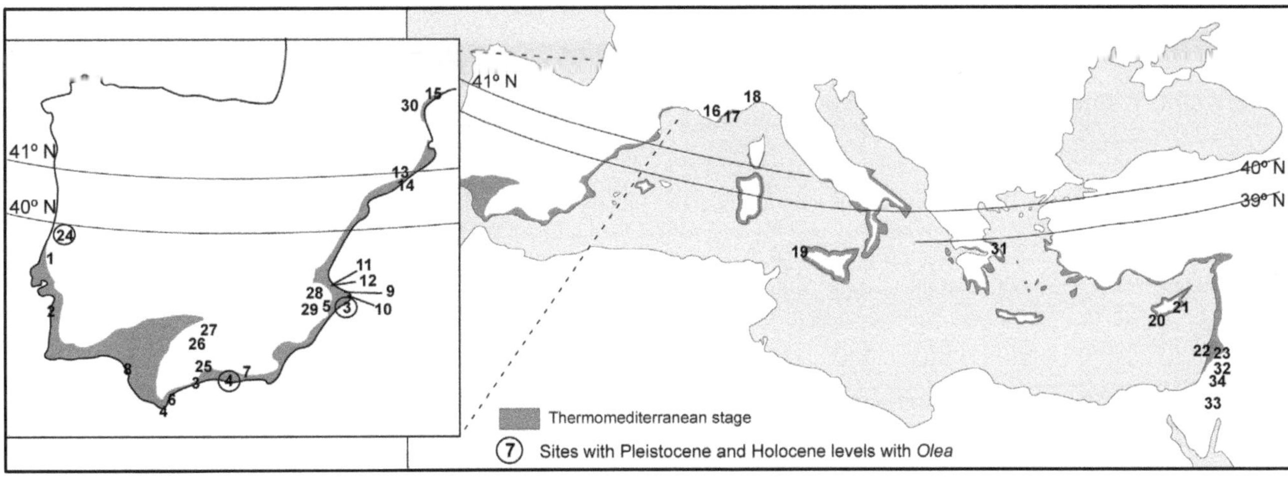


Figure 2. Holocene distribution of *Olea europaea* L. wood charcoal finds in the north Mediterranean (see information of sites in Figure 3)

Concerning the accompanying flora, we consider as thermophilous formations the association of species that usually accompanies the oleaster nowadays, which includes i.e. *Pistacia lentiscus, Rosmarinus officinalis, Rhamnus-Phillyrea, Pinus halepensis, Pinus pinea-pinaster, Quercus* sp. evergreen.

## RESULTS

*Olea* remains have been identified in 34 archaeological sites (Figure 2, Figure 3). The majority concentrates in the western Mediterranean, although this may be due to the unequal number of charcoal analyses carried out in different regions. Concerning the chronology, *Olea* is present in only 6 sequences of the Preboreal, 11 sequences of the Boreal and 31 sequences of the Atlantic. The scarce presence of *Olea* during the Preboreal might be related to the hiatuses observed in archaeological sites coinciding with the Pleistocene-Holocene transition.

The early presence of *Olea* in southern Iberia, Sicily, Cyprus and the Levant points to the existence of thermophilous *refugia* in these regions during the Pleistocene, a hypothesis that for the case of the former is supported by other lines of evidence (Carrión *et al.* 1995; Carrión *et al.* 2008; Pantaleón Cano *et al.* 1999). From the Boreal onwards, there is a progressive increase of sites in which *Olea* finds have been identified, a trend that consolidates during the Atlantic period, therefore allowing tracing the distribution of the species in more detail:

- The first expansion of *Olea* coincides with the thermomediterranean bioclimatic level.
- During the Atlantic period, *Olea* is more frequently identified and more abundant in the western Mediterranean.
- In the Iberian Peninsula, when the thermophilous woodland is established, it is then when *Olea* expands beyond the thermomediterranean level, surpassing its original distribution area.

### The Eastern Mediterranean

The earliest *Olea* finds in the eastern Mediterranean come from sites located in the thermomediterraenan level, particularly in the Levant and in Cyprus (Figure 2, Figure 3). In the Levantine area the presence of *Olea* is attested as early as the 10th millennium BP in Jericho (Western 1971) and at Abu Salem (Liphschitz *et al.* 1991). During the Pre-Pottery Neolithic (hereafter PPN) B and C (10500-8250 cal. BP), the species was present at the now submerged settlement of Atlit Yam (Galili *et al.* 1993) and also at the sites of Horvat Galil (Liphschitz 1997) and Divshon (Liphschitz *et al.* 1991). The oleaster would have formed part of the Mediterranean vegetation in the coastal areas and it would have probably grown in regions that are presently characterized by Irano-Turanian and Saharo-Arabian formations.

The presence of *Olea* on the island of Cyprus is well-documented as early as the mid-11th millennium cal. BP in Early Neolithic A contexts at Shillourokambos (Thiébault 2003) and it is also present during the 9th and 8th millennium cal. BP, at the PPN site of Khirokitia (Thiébault 2003). Therefore, the evidence from Cyprus indicates a continuous and locally abundant presence of *Olea* in the vegetation from the Preboreal and until the Atlantic period.

In Greece and the Aegean area *Olea* is infrequent during the earlier part of the Holocene. One reason for the meager identification of the species is probably the fact that the majority of the available charcoal analysis results come from northern sites (Ntinou 2002) that lay outside the area of the natural oleaster distribution, which is confined to the south of parallel 39° N (Ozenda 1975). The available data, mostly from pollen sequences, indicate the late appearance of *Olea* (towards the end of the Atlantic period) even in southerly thermomediterranean locations where the species could potentially grow (Bottema and Sarpaki 2003; Turner and Greig 1975; Jahns 1993; Moody *et al.* 1996; Wright 1972). At the site of Knossos in Crete wood charcoal

| Site | | Climatic period | | | Cultural context | References |
|---|---|---|---|---|---|---|
| | | Preboreal | Boreal | Atlantic | | |
| ***Olea* finds in the thermomediterranean stage** | | | | | | |
| 1 | Abrido da Pena d'Água | | ■ | ■ | Epipalaeolithic, Neolithic | Figueiral 1998 |
| 2 | Vale Pincel I | | | ■ | Epipalaeolithic, Neolithic | Carrión 2005 |
| 3 | Cova de les Cendres | | | ■ | Upper Palaeolithic, Neolithic | Badal *et al.* 1994 |
| 4 | Cueva de Nerja | ■ | ■ | ■ | Epipalaeolithic, Neolithic | Badal 1990 |
| 5 | Cova de Santa Maira | ■ | ■ | ■ | Magdalen., Epipaleo., Mesol. | Badal 1999; Carrión 2005 |
| 6 | Río Palmones | | ■ | ■ | Epipalaeolithic | Rodríguez-Ariza 2004 |
| 7 | Murciélagos de Albuñol | | | ■ | Neolithic | Rodríguez-Ariza, Montes 2005 |
| 8 | El Retamar | | | ■ | Neolithic | Arnanz and Uzquiano 2002 |
| 9 | Cova Bolumini | | | ■ | Neolithic | Badal 1990 |
| 10 | Cova Ampla del Montgó | | | ■ | Neolithic | Vernet *et al.* 1983 |
| 11 | Cova del Llop | | | ■ | Neolithic | Vernet *et al.* 1983 |
| 12 | Cova de la Recambra | | | ■ | Neolithic | Vernet *et al.* 1983 |
| 13 | Can Sadurní | | | ■ | Neolithic | Ros 1992 |
| 14 | Can Tintorer | | | ■ | Neolithic | Ros 1992 |
| 15 | Cova de l'Espérit | | | ■ | Mesolithic, Neolithic | Solari and Vernet 1992 |
| 16 | Caucade | | | ■ | Neolithic | Thiébault 2001 |
| 17 | Giribaldi | | | ■ | Neolithic | Thiébault 2001 |
| 18 | Arene Candide | | | ■ | Neolithic | Nisbet 1997 |
| 19 | Grotta dell'Uzzo | ■ | ■ | ■ | Mesolithic, Neolithic | Constantini 1989 |
| 20 | Shillourokambos | ■ | ■ | | Neolithic | Thiébault 2003 |
| 21 | Khirokitia | | ■ | ■ | Neolithic | Thiébault 2003 |
| 22 | Atlit Yam | | ■ | | pre-Pottery Neolithic | Galili *et al.* 1993 |
| 23 | Horvat Galil | | ■ | | pre-Pottery Neolithic B | Liphschitz 1997 |
| ***Olea* finds in the mesomediterranean stage** | | | | | | |
| 24 | Buraca Grande | | ■ | ■ | Upper Palaeol. to Neolithic | Figueiral and Terral 2002 |
| 25 | Cueva del Toro | | | ■ | Neolithic | Rodríguez-Ariza 2004 |
| 26 | Murciélagos de Zuheros | | | ■ | Neolithic | Rodríguez-Ariza 1996 |
| 27 | Polideportivo de Martos | | | ■ | Neolithic | Rodríguez-Ariza 1996 |
| 28 | Cova de l'Or | | | ■ | Neolithic | Badal *et al.* 1994 |
| 29 | La Falguera | | | ■ | Mesolithic, Neolithic | Carrión 2002; García and Aura 2006 |
| 30 | Montou | | | ■ | Middle Neolithic | Heinz *et al.* 2004. |
| 31 | Cave of The Cyclops | | | ■ | Neolithic | Ntinou in press |
| ***Olea* finds in the irano-turanian stage** | | | | | | |
| 32 | Jericho | ■ | | | pre-Pottery Neolithic A | Western 1971; Wilcox 1991, 1992 |
| ***Olea* finds in the saharo-arabic stage** | | | | | | |
| 33 | Abu Salem | ■ | | | pre-Pottery Neolithic A | Liphschitz *et al.* 1991 |
| 34 | Nahal Divshon | | ■ | | pre-Pottery Neolithic B | Liphschitz *et al.* 1991 |

Figure 3. Archaeological sites with *Olea* wood charcoal finds in the north Mediterranean

analysis (Badal and Ntinou in press) failed to detect the presence of the olive in any of the layers of the Neolithic sequence, despite the fact that typically thermomediterranean formations have been identified. In the central Aegean area, the only evidence for the presence of *Olea* derives from the Cave of the Cyclops, Youra, Northern Sporades (Figure 2, Figure 3), where few charcoal fragments have been identified in the upper part of the sequence, after *ca.* 8,500 cal. BP, coinciding with the Neolithic period (Ntinou 2011) .

## The Central Mediterranean

In the central Mediterranean, *Olea* wood charcoal finds are reported from Sicily and the Liguro-Provençal region, restricted to the thermomediterranean bioclimatic level (Figure 2, Figure 3).

At Grotta dell' Uzzo, the presence of the wild olive is confirmed by wood charcoal from the Mesolithic levels dating to between 12060-11220 and 9560-9080 cal. BP (Costantini 1989). In these levels, *Quercus* cf. *ilex* and *Phillyrea* sp. have also been identified and Costantini describes the vegetation as a xerophyll *macchia*.

In the Liguro-Provençal region, *Olea* is present at several sites, in levels corresponding to the Atlantic period, and later. These sites are strictly located in the thermomediterranean level. At Giribaldi (Thiébault 2001) and Arene Candide (Nisbet 1997), *Olea* reaches percentages of between 10% and 30%, and is accompanied by other thermophilous species, such as the lentisc and the Aleppo pine. However, *Olea* is completely absent from other sites located on the mesomediterranean level, even within a few kilometers distance from the aforementioned sites (Thiébault 2001). Only in Caucade, do *Olea* and other warm-loving taxa appear in this level, but later in the Atlantic (Thiébault 2001). The data indicate the existence in the Liguro-Provençal region of an area with warm influences constrained to the lowest parts of the thermomediterranean level.

## The Western Mediterranean

The Holocene marks the beginning of the massive expansion of *Olea* attested in most of the charcoal sequences from the warmest areas in the western Mediterranean (Figure 2, Figure 3).

In the Iberian Peninsula, the first continuous *Olea* curve is documented at Cueva de Nerja, from the end of the Pleistocene throughout the Holocene sequence (Badal 1998). At Cova de Santa Maira, *Olea* and other thermophilous taxa are continuously present with low percentages throughout the Epipalaeolithic and Mesolithic sequence (11260-11100 cal. BP and 10690-10410 cal. BP) of the cave (Aura *et al.* 2005). However, *Olea* increases substantially during the Atlantic and becomes co-dominant with *Pinus halepensis*, a fact that depicts a vegetation change (Badal 1999). During this period, the presence of *Olea* is confirmed by AMS dates (Figure 4) and there are numerous sequences on both the Mediterranean and the Atlantic façade of the Iberian Peninsula where the species constitutes more than 10% of the charcoal remains.

The largest collection of *Olea* findings is recorded to the south of the parallel 40°N. This is the case of Cova de les Cendres, Cova Bolumini, Cova Ampla, Cova de la Recambra and Cova del Llop along the eastern coast, (Badal *et al.* 1994; Vernet *et al.* 1983), Los Murciélagos de Albuñol (Rodríguez Ariza and Montes 2005) and Río Palmones (Rodríguez Ariza 2004) at the mouth of the Guadalquivir River, Vale Pincel I (Carrión 2005) and Abrigo da Pena d'Água (Figueiral 1998) along the western coast. On the Atlantic façade *Olea* is usually accompanied by *Pinus pinea* or *P. pinaster* and other taxa like *Arbutus unedo*, *Pistacia lentiscus*, evergreen *Quercus* sp., *Quercus suber*, *Rhamnus-Phillyrea*, etc.

One of the most northerly pieces of evidence for the presence of *Olea* is found at Cova de l'Espérit (Solari and Vernet 1992) in the east Pyrenees and at Can Sadurní and Can Tintorer in Catalonia (Ros 1992). However, despite the numerous anthracological analyses performed in this region, these are the only sites to the north of parallel 40°N where *Olea* is present. This latitude marks the northernmost limit of a continuous thermomediterranean belt, beyond which these bioclimatic conditions are restricted to discontinuous, isolated enclaves and it is precisely in these where *Olea* is documented, i.e. the south-eastern French sites, along the Liguro-Provençal basin.

The presence of *Olea* is also reported in the mesomediterranean level, particularly in its lower part. In this case however, the distribution and abundance of the species are quite different from those described for the thermomediterranean level. *Olea* remains seem to be strongly linked to favorable orography, such as warm and sunny slopes or valley bottoms. This is the case of Cova de l'Or, located on a south-facing slope, where *Olea* is present in percentages around 10-15% (Badal *et al.* 1994). In other more interior enclaves however, i.e. La Falguera, the presence of this species is more sporadic in the same way as other thermophilous taxa (Carrión 2002; García and Aura 2006). Many valleys tend to act as channels that let the warm coastal conditions and the thermophilous vegetation reach further inland, as deduced

| Site | Cultural context | *Olea* material | ASM B.P. | Date relevance |
|---|---|---|---|---|
| La Falguera | Epipalaeolithic | Seed | 7410 ± 70 | Relevant |
| Cova de les Cendres | Early Neolithic | Charcoal | 6650 ± 50 | Relevant |
| Vale Pincel I | Early Neolithic | Charcoal | 6350 ± 50 | Relevant |
| Cova de Santa Maira | Epipalaeolithic | Charcoal | 420 ± 40 | Irrelevant |
| Cova de les Cendres | Upper Palaeolithic | Charcoal | 6660 ± 50 | Irrelevant |

Figure 4. AMS radiocarbon dates on *Olea* macro-remains
("Date Relevance" refers to the coherence of the date with the stratigraphic and cultural context)

from the presence of *Olea* in the Guadalquivir valley. There, this species has been documented at Cueva del Toro, Murciélagos de Zuheros and Polideportivo de Martos (Rodríguez-Ariza and Montes 2005).

The most northerly presence of *Olea* in the mesomediterranean level is reported at Buraca Grande and at Montou (Figure 2). At the former, the species is present throughout the Holocene sequence, with initial dates of 9290-8890 cal. BP (Figueiral and Terral 2002), while its presence at Montou is restricted to a single level dating back to 6180-5780 cal. BP (Heinz *et al.* 2004).

## DISCUSSION

In previous sections, we have collected data concerning the presence, distribution and expansion of *Olea* in Holocene contexts. In the following paragraphs, we will consider the above-mentioned data in relation to the biogeographical situation of the sites with *Olea* remains, the accompanying flora and the available AMS dates on *Olea* wood charcoal for the purpose of specifying the establishment of the thermophilous woodland.

### The area of distribution in the early Holocene

The presence of *Olea* since the beginning of the Holocene (11700 cal. BP) has been reported in several sites in the Near East and Cyprus. In contrast, the presence of the olive in the Aegean area and Greece is not well defined. *Olea* charcoal, pollen and seed macroremains data indicate a late appearance of the olive, after the 9th millennium cal. BP, for Crete, Southern and Central Greece, and much later dates for northern Greece. In the central Mediterranean, the early presence of *Olea* is confirmed by wood charcoal from Mesolithic levels at Grotta dell' Uzzo (Costantini 1989). However, the species is systematically absent from the Italian peninsula, even later in the Holocene. In the Iberian Peninsula, the presence of *Olea* in some southern sequences (Badal 1998; Carrión 2005) is documented continuously since the end of the Pleistocene and during the early Holocene (Carrión *et al.* 2008).

The early postglacial presence of *Olea* at the south of the Iberian Peninsula, in Sicily, Cyprus and the Levant allows suggesting the location of thermophilous Pleistocene *refugia* in these areas. However, AMS dates of *Olea* remains from Full and Late Glacial contexts are needed to help clarify the distribution of the oleaster in pre-Holocene chronologies.

The Preboreal and the Boreal constitute a pioneer phase that would have implied the establishment of a biogeographical configuration similar to that of the present-day in which the wild olive distribution would have coincided with the current thermomediterranean belt. The Atlantic period would have marked the consolidation of the species, as well as the differences in its distribution and the variable dynamics between the western and eastern areas of the Mediterranenan basin.

### The expansion: biogeographical patterns

The Atlantic period marks the expansion of *Olea*, a pattern that is better observed in the western Mediterranean. This may be linked to the aforementioned presence of the species in early post-glacial contexts in this region. On the contrary, *Olea* is scarce in Italy while in mainland Greece and Crete the majority of sites with *Olea* remains date to the late Neolithic or the Bronze Age. Whether the species was introduced, as proposed for Crete (Bottema and Sarpaki 2003), or whether it expanded in the wild at a very slow rate and only became abundant with tree-tending and cultivation after the Neolithic, it remains an open question until more results are provided. In the western Mediterranean, in the Iberian Peninsula in particular, *Olea* appears during the Atlantic at virtually all the sites that are presently situated in the thermomediterranean bioclimatic level or in the lower mesomediterranean. There, the sites with *Olea* finds are distributed along the coast or at a distance of 50 km from it, generally not surpassing an altitude of 500 m. Beyond this setting, the presence of *Olea* is much scarcer, and it is only restricted to sporadic enclaves that present favorable orographic conditions.

The distribution of *Olea* finds indicates that temperature and continental conditions would have been the limiting factors for its expansion. During the Preboreal and Boreal, there are no records for the presence of *Olea* in the mesomediterranean level (except for Buraca Grande), thus the later presence of the species there should be linked to its large expansion during the Atlantic period and to the favorable orography of particular areas. In any case, the frequency of *Olea* and other thermophilous species at this level is sporadic, probably reflecting the prevailing bioclimatic conditions that would have been close to the minima for their growth.

The rapid expansion of *Olea* could have been favored by the higher temperatures during the Atlantic period. However, the human factor may have played a fundamental role in the expansion process, as well. The establishment of Neolithic groups and the farming economy are linked to evidences of deforestation and expansion of sclerophyllous woodland in many palaeobotanical sequences, probably because of woodland clearing practices or the slash-and-burn agriculture. The charcoal analysis record of the Iberian Peninsula shows that *Olea* was an intensively exploited species in most thermomediterranean and in some mesomediterranean contexts which has led several authors to consider it as evidence of an early manipulation of the species (Terral 1997; Badal 1999). In later periods, by means of cultivation, *Olea* became resistant to more continental conditions and reached the Atlantic façade of the Iberian Peninsula.

### Olea in the sclerophyllous woodland

During the Atlantic period, *Olea* forms part of the typical sclerophyllous woodland in the Mediterranean basin. If we take into account its frequency and the accompanying taxa in the available charcoal sequences, three clearly defined behaviours of the *Olea* populations can be described:

A) *Olea* dominates: *Olea* charcoal surpasses 50% of the remains and *Pistacia lentiscus, Quercus* sp. evergreen, *Phillyrea* and/or *Rhamnus* play a key role in the plant formations. This situation may be observed in coastal sites to the south of parallel 40ºN during the Boreal and the Atlantic. On the eastern coast of the Iberian Peninsula, *Pinus halepensis* is an important element of the vegetation, while *P. pinaster* and *P. pinea* are abundant on the western coast.

B) *Olea* accompanies species with a similar ecology: in some cases, *Olea* reaches significant percentages, but it is not dominant in the charcoal assemblages. The dominant taxa are evergreen *Quercus* sp., *Pinus halepensis, P. pinaster* or *P. pinea.* In these formations, *Pistacia lentiscus, Rhamnus* and/or *Phillyrea, Arbutus unedo,* Leguminosae, etc. are usually present. This would have been the case of either Cova de l'Or or the Liguro-Provençal sites.

C) *Olea* shows low frequency: the species represents less than 10% of the remains. The charcoal assemblages are dominated by evergreen *Quercus* sp., *Phillyrea* and deciduous taxa. *Pistacia lentiscus* is less represented in such contexts and together with the scant presence of *Olea,* indicates the ecological upper limits for the development of thermophilous Mediterranean woodland. Such conditions can be observed i.e. at La Falaguera and the Cave of the Cyclops, located at the mesomediterranean level, probably at the limits of what would have been the optimal oleaster distribution area.

In the two former cases, *Olea* could have formed sclerophyllous woodland as most of the maquis species may had reached a considerable height of more than 10 m. *Olea,* together with *Pistacia lentiscus, Phillyrea latifolia, Arbutus unedo, Quercus coccifera,* etc., would have formed well-structured, impenetrable thermophilous woodland with climbers and dense undergrowth.

The available data reveal the importance of *Olea* as part of the genuine Mediterranean vegetation, particularly relevant in the western part of the Mediterranenan basin. In terms of climate, the combination of *Olea* and the above mentioned thermophilous taxa indicates thermomediterranean conditions with mean annual temperatures of between 17ºC and 19ºC, and mean minimum temperatures of more than 4ºC. In terms of the rainfall pattern, this could vary from dry to subhumid since, in the present-day, the wild olive tolerates this range of precipitation regimes. We propose that, at a time when the anthropic impact on the landscape was still weak, the sclerophyllous thermomediterranean vegetation with *Olea* would have constituted woodlands that could have reached several meters in height (up to 10-15 m), as we are able to see nowadays at Serra da Arrábida in the Portuguese Alentejo (Costa *et al.* 1994). Such woodland would have been located mainly to the south of parallel 40ºN, where *Olea* is very abundant in the charcoal records. To the north of this parallel, the presence of the species is more discrete, and its expansion would have probably been linked to human deforestation processes. We therefore consider that it is necessary to provide a phytogeographical entity to the thermomediterranean formations with *Olea,* which were present from times prior to intensive anthropic landscape management.

## CONCLUSION

The available charcoal analysis data concerning *Olea* enable us to conclude the following:

A) We propose that *Olea* might have survived during the Pleniglacial in thermophilous *refugia*, located in south Iberia and Italy Peninsulas, in Cyprus and the Levant. This fact may explain the early Holocene presence of the species in these enclaves during the Preboreal and the Boreal.

B) During the Early Holocene *Olea* is more abundant in coastal sites of the Iberian Peninsula, while its presence is less conspicuous in other regions.

C) The emergence and development of the *Olea*-dominated formations took place in the Atlantic period and within the thermal, altitudinal and latitudinal limits of the thermomediterranean belt, coinciding with its present-day distribution (Figure 1). It was only during the mildest periods that this species spread further to mesomediterranean areas with a strong determining thermicity-type orography.

D) The formations in which the oleaster was an important element of the natural vegetation would have constituted the basis for the implementation of its cultivation.

### Acknowledgments

We would like to thank all the wood charcoal specialists that have identified and published *Olea* charcoals from archaeological sites. Our most sincere thanks go to Isabel Figueiral for always answering our questions and providing unpublished data.

### References

ARNANZ, A.M. and UZQUIANO, P. 2002. La evidencia arqueobotánica: los macrorrestos carbonizados del yacimiento de "El Retamar". In Lazarich González, M. and Ramos Muñoz, J. (eds.), El asentamiento de "El Retamar" (Puerto Real, Cádiz): contribución al estudio de la formación social tribal y a los inicios de la economía de producción en la Bahía de Cádiz, 205-216.

AURA, J.E., CARRION, Y., ESTRELLES, E. and PEREZ, G. 2005. Contact Information Plant economy of hunter-gatherer groups at the end of the last Ice

Age: plant macroremains from the cave of Santa Maira (Alacant, Spain) ca. 12000-9000 b.p. *Vegetation History and Archaeobotany* 14, 542-550.

BADAL, E. 1990. *Aportaciones de la antracología al estudio del paisaje vegetal y su evolución en el Cuaternario reciente, en la costa mediterránea del País Valenciano y Andalucía (18000-3000 BP)*. Ph.D. Thesis, Universitat de Valencia.

BADAL, E. 1998. El interés económico del pino piñonero para los habitantes de la Cueva de Nerja. In J.L. Sanchidrian and Mª.D. Simon (eds.) *Las culturas del Pleistoceno superior en Andalucía,* 287-300. Nerja, Patronato de la Cueva de Nerja.

BADAL, E. 1999. El potencial pecuario de la vegetación mediterránea: las cuevas redil. II Congrés del Neolític a la Península Ibèrica, *Saguntum-PLAV*, Extra-2, 69-75.

BADAL, E., BERNABEU, J., VERNET, J.L. 1994. Vegetation changes and human action from the Neolithic to the Bronze Age (7000-4000 BP) in Alicante, Spain, based on charcoal analysis. *Vegetation History and Archaeobotany* 3, 155-166.

BADAL, E., NTINOU, M. in press. Wood charcoal analysis from Neolithic Knossos: the local vegetation. In Efstratiou, N., Karetsou, A., Ntinou, M., and Banou E. (eds.), The Neolithic Settlement of Knossos in Crete: New Evidence for the Early Occupation of Crete and the Aegean Islands, INSTAP Monograph Series.

BESNARD, G., KHADARI, B., BARADAT, P. and BERVILLÉ, A. 2002. *Olea europaea* (Oleaceae) phylogeography based on choroplast DNA polymorphism. *Theoretical and Applied Genetics*. 104, 1353-1361.

BOTTEMA, S. and SARPAKI A. 2003. Environmental Change in Crete: A 9000-year Record of Holocene Vegetation History and the Effect of the Santorini Eruption. *The Holocene* 13 (5): 733-749.

BROWICS, K. 1983. *Chronology of tree and shrubs in South-West Asia and adjacent regions*. Poznan, Institute of dendrology. Polish Academy of Sciences.

BRETON, C., TERSAC, M. and BERVILLÉ, A. 2006. Genetic diversity and gene flow between the wild olive (oleaster, *Olea europaea* L.) and the olive: several Plio-Pleistocene refuge zones in the Mediterranean basin suggested by simple sequence repeats analysis. *Journal of Biogeography* 33, 1916-1928.

CARRIÓN, Y. 2002. Charcoal analysis at La Falaguera rockshelter (Alcoi, Alacant) from Mesolithic to Bronze Age: landscape and wood uses. In: Thiébault, S. (ed.) *Charcoal Analysis. Methodological Approaches, Palaeoecological Results and Word Uses.* Proceedings of the Second International Meeting of Anthracology, Paris, September 2000. British Archaeological Reports. International Series 1063, 103-108.

CARRIÓN, Y. 2005. *La vegetación mediterránea y atlántica de la Península Ibérica. Nuevas secuencias antracológicas.* Serie de Trabajos Varios del S.I.P. 104. Valencia, Diputación de Valencia.

CARRIÓN, J.S., DUPRÉ, M., FUMANAL, Mª.P. and MONTES, R. 1995. A palaeoenvironmental study in semi-arid southeastern Spain: the palynological and sedimentological sequence at Perneras Cave (Lorca, Murcia). *Journal of Archaeological Science* 22, 355-367.

CARRIÓN, J.S., FINLAYSON, C., FERNÁNDEZ, S., FINLAYSON, G., ALLUÉ, E., LÓPEZ-SÁEZ, J.A., LÓPEZ-GARCÍA, P., GIL-ROMERA, G., BAILEY, G. and GONZÁLEZ-SAMPÉRIZ, P. 2008. A coastal reservoir of biodiversity for Upper Pleistocene human populations: palaeoecological investigations in Gorham's Cave (Gibraltar) in the context of the Iberian Peninsula. *Quaternary Science Reviews* 27, 2118-2135.

COSTANTINI, L. 1989. Plant exploitation at Grotta dell'Uzzo, Sicily: new evidence for the transition from mesolithic to neolithic subsistence in southern Europe. In D.R. Harris and G.C. Hillman (eds): *Foraging and Farming. The evolution of plant exploitation.* London, Unwin Hyman Ltd, 197-206.

CONTENTO, A., CECCARELLI, M., GELATI, M.T., MAGGINI, F., BALDONI, L. and CIONINI, P.G. 2002. Diversity of *Olea* genotypes and the origin of cultivated olives. *Theoretical and Applied Genetics*. 104, 1229-1238.

COSTA, M., MORLA, C. and SAÍNZ, H. 2001. *Los bosques ibéricos. Una interpretación geobotánica.* Barcelona, Planeta.

FIGUEIRAL, I. 1998. O Abrigo da Pena d'ägua (Torres Novas): a contribuiçao da antracologia. *Revista portuguesa de Arqueología* 1(2), 73-79.

FIGUEIRAL, I and TERRAL, J.-F. 2002. Late Quaternary refugia of Mediterranean taxa in Portuguese Extremadura: Charcoal based palaeovegetation and climatic reconstruction. *Quaternary Science Reviews* 21, 549-558.

GALILI, E., HERSHKOWITZ, A., GOPHER, A., WEINSTEIN-EVRON, M., LERNAU, O., KISLEV, M. and HORWITZ, L. 1993. Atlit-Yam: A Prehistoric Site on the Sea Floor of the Israeli coast. *Journal of Field Archaeology* 20, 133-157.

GARCÍA, O. and AURA, J.E. (cood.) 2006. *El abric de la Falaguera (Alcoi, Alcant). 8000 años de ocpuación human en la cabecera del río Alcoi.* Alcoi, Diputación de Alicante, Ayuntamiento de Alcoi, Caja de Ahorros del Mediterráneo.

HEINZ, C., FIGUEIRAL, I., TERRAL, J.-F. and CLAUSTRE, F. 2004. Holocene vegetation changes in the northwestern Mediterranean: new palaeoecological data from charcoal analysis and quantitative eco-anatomy. *The Holocene* 14 (4), 621-627.

JAHNS, S. 1993. On the Holocene vegetation history of the Argive Plai (Peloponnese, southern Greece). *Vegetation History and Archaeobotany* 2, 187-203.

LIPHSCHITZ, N. 1997. Wood Remains from Two PPNB Sites: Horvat Galil and Nahal Beset. *Tel Aviv* 24, 237-239.

LIPHSCHITZ, N., GOPHNA, R., HARTMAN, M. and BIGER, G. 1991. The beginning of olive (*Olea europaea*) cultivation in the Old World: A Reassessment. *Journal of Archaeological Science* 18, 441-453.

MOODY, J., RACKHAM, O. and RAPP, G. 1996. Environmental Archaeology of Prehistoric NW Crete. *Journal of Field Archaeology* 23, 273-297.

NISBET, R. 1997. Arene Candide: Charcoal remains and prehistoric woodland use. In R. Maggi (ed.) *Arene Candide: a functional and environmental assessment of the holocehe sequence-excavations Bernabo Brea – Cardini, 1940-1950.* Vol. V, 103-118. Memoire dell'Istituto Italiano di Paleontología Umana.

NTINOU, M. 2002. *El paisaje en el norte de Grecia desde el Tardiglaciar al Atlántico. Formaciones vegetales, recursos y usos.* British Archaeological Reports, International Series 1038.

NTINOU, M. 2011. *The Cave of the Cyclops: Mesolithic and Neolithic Networks in the Northern Aegean Greece.* Volume II: Bone Tool Industries, Dietary Resources and the Paleoenvironment, and Archeometrical Studies, vol. II, Prehistory Monographs 31, INSTAP Academic Press, 297-314.

OZENDA, P. 1975. Sur les étages de végétation dans les montagnes du bassin Méditerranéen. *Documents de Cartographie Ecologique* Vol. XVI, 1-32.

PANTALEÓN, J., YLL, R. and ROURE, J.M. 1999. Evolución del paisaje vegetal en el sureste de la península Ibérica durante el Holoceno a partir del análisis polínico. II Congrés del Neolític a la Península Ibèrica, *Saguntum-PLAV*, Extra-2, 17-23.

RIVAS-MARTÍNEZ, S. 1987. *Memoria del mapa de series de vegetación de España 1:400.000.* Madrid, ICONA.

RODRÍGUEZ-ARIZA, Mª.O. 1996. Análisis antracológico de yacimientos neolíticos de Andalucía. Actas del I Congreso del neolítico en la Península Ibérica. *Rubricatum* 1, 73-83.

RODRÍGUEZ-ARIZA, M.O. 2004. Análisis antracológicos de Cerro Juré y dólmenes de Pozuelo. In F. Nocete (Coord.) *Odiel*, Arqueología Monografías 19, 35-42. Sevilla, Junta de Andalucía.

RODRÍGUEZ-ARIZA, M.O. and MONTES, E. 2005. On the origin and domestication of *Olea europaea* L. (olive) in Andalucía, Spain, based on the biogeographical distribution of its finds. *Vegetation History and Archaeobotany* 14, 551-561.

ROS, Mª.T. 1992. Contribution of charcoal analysis to the study of vegetal palaeoenvironment in Catalonia (Spain). In J.L Vernet (ed.) *Les charbons de bois, les anciens écosystèmes et le rôle de l'homme. Bulletin Societé Botanique France* 139, Actual. bot. (2/3/4), 483-494.

RUBIO, R., BALAGUER, L., MANRIQUE, E., PÉREZ, M.E. and VARGAS, P. 2002. On the historical presence of the wild olive [*Olea europaea* L. var. *sylvestris* (Miller) Lehr. (Oleaceae)] in the Eurosiberian region of the Iberian Peninsula. *Anales del Jardín Botánico de Madrid,* 59 (2), 342-344.

SOLARI, M.E. and VERNET, J.L. 1992. Late Glacial and Holocene vegetation of the Corbières based on charcoal analysis at the Cova de l'Espérit (Salses, Pyrénées orientales, France). *Review of Palaeobotany and Palynology* 71, 111-120.

TERRAL, J.F. 1997. *La domestication de l'olivier (Olea europaea L.) en Méditerranée nord-occidentale: Approche morphométrique et implications paléoclimatiques.* Unpublished PhD Thesis, Université Montpellier II.

TERRAL, J.-F., ALONSO, N., BUXO I CAPDEVILA, R., CHATTI, N., FABRE, L., FIORENTINO, G., MARINVAL, P., PEREZ JORDA, G., PRADAT, B., ROVIRA, N. and ALIBERT, P., 2004. Historical biogeography of olive domestication (*Olea europaea* L.) as revealed by geometrical morphometry applied to biological and archaeological material. *Journal of Biogeography* 31, 63-77.

THIEBAULT, S. 2001. Anthracoanalyse des établissements néolithiques de la région liguro-provençale. *Bulletin Societé Préhistorique Française* 98 (3), 399-409.

THIEBAULT, S. 2003. Les paysages végétaux de Chypre au Néolithique: Premières données anthracologiques. In J. Guilaine and A. Le Brun (eds.) *Le Néolithique de Chypre.* Actes du colloque international organisé par le département de Antiquité de Chypre et l'École Francaise d'Athènes, Nicosie 17-19 mai 2001. *Bulletin de Correspondance Hellénique*, Supplément 43, 221-230.

TURNER J. and GREIG, J.A.R. 1975. Some Holocene pollen diagrams from Greece. *Review of Palaeobotany and Palynology* 20, 171-204.

VERNET, J.L., BADAL, E. and GRAU, E. 1983. La végétation néolithique du sud-est de l'Espagne (Valencia, Alicante) d'après l'analyse anthracologique. *C.R. Academie Sciences Paris* 296, III, 669-672.

WENINGER, B. and JÖRIS, O., 2008. A $^{14}C$ age calibration curve for the last 60 ka: the Greenland-Hulu U/Th timescale and its impact on understanding the Middle to Upper Paleolithic transition in Western Eurasia. *Journal of Human Evolution* 55 (5), 772-781.

WENINGER, B., JÖRIS, O. and DANZEGLOCKE, U. 2009. CalPal-2007. Cologne Radiocarbon Calibration & Palaeoclimate Research Package. http://www.calpal.de/, accessed 2011-08-10.

WESTERN, C. 1971. The Ecological interpretation of ancient charcoals from Jericho. *Levant* 3, 31-40.

WILLCOX, G. 1991. Exploitation des espèces ligneuses au Proche-Orient: données anthracologiques. *Paléorient* 17/2, 117-126.

WILLCOX, G. 1992. Bilan des données anthracologiques du Proche-Orient. In J.L Vernet (ed.) *Les charbons de bois, les anciens écosystèmes et le rôle de l'homme. Bulletin Societé Botanique France* 139, Actual. bot. (2/3/4), 539-551.

WRIGHT, H.E., Jr. 1972. Vegetation History. In W.A. McDonald and G.R. Rapp (ed.) *The Minnesota Messenia Expedition: Reconstructing a Bronze Age regional environment*, 188-199. Minneapolis.

ZOHARY, M. and ORSHAN, G. 1959. The maquis of *Ceratonia siliqua* in Israel. *Vegetatio* 8, 285-297.

ZOHARY, D. and HOPF, M. 2000. *Domestication of plants in the Old Word.* Oxford, Oxford University Press.

# FOREST RESOURCES EXPLOITATION AND MANAGEMENT BY SELKNAM HUNTER-GATHERER SOCIETIES: RESULTS OF THE ARCHAEOBOTANICAL ANALYSIS OF EWAN SITE (TIERRA DEL FUEGO, ARGENTINA)

Laura CARUSO FERMÉ
Laboratori d'Arqueobotanica. Facultat de Lletres. Departament de Prehistoria. Universitat Autonoma de Barcelona (08193) Bellaterra, Cerdanyola del Vallès. Catalunya (Spain),
Group AGREST, funded by the Generalitat de Catalunya (AGREST-2009 SGR 734)
lcarusoferme@gmail.com

**Abstract**: *The fact that archaebotanical studies performed to date in sites attributed to Patagonian hunter-gatherer societies have been very scarce, has relatively relegated the role played by plant resources within the economy of these societies. This report is a first approach to the study of forest exploitation by the Selknam hunters-gatherers at the archaeological site Ewan, located in the central part of Isla Grande of Tierra del Fuego (Argentina) and dated in historical times (1905). The study is focused is the analysis of charcoal and wood remains of an hypothetical hut in order to characterize which taxa was used for firewood and which for building purposes as well as the way of procurement and the type of wood-processing carried out by this hunter-gatherer group. The data obtained reveal that* Nothofagus antarctica *(G. Forst) Oerst. -ñire-, was used for all purposes. The record of ring growth shape, size and marks shows that big trunks were cut or uprooted and transported to the settlement.*

**Keywords**: *Archaeobotany, wood resources, hunters-gatherers, Selknam, Tierra del Fuego*

## INTRODUCTION

Human's history and evolution are closely related to the plant environment. In all types of ecosystems there are natural materials susceptible to be converted into usable resources by members of any kind of society. Management of plant resources is directly related to the economic practices implemented by each society, which are embedded in production processes and social reproduction. At present the analysis of plant remains from archaeological sites in the Patagonian region is still in its initial stage.

In Patagonia, archaebotanical studies started in hydrological basin of the river Chubut –Argentina- (Pérez de Micou, 1979-1982; Pérez de Micou & Nacuzzi, 1983-1985).

The present research was carried out in the central area of Isla Grande of Tierra del Fuego (54° 12' 06.3'' South and 67° 16' 14.03'' West), Argentina, in the Sub-Antarctic forest area, which is integrated within Andean-Patagonic Forests (Dimitri 1972; Roig 1998). In Ewan area, forest mantles and clearings can be found combined in a wavy landscape crossed by numerous water bodies and soft elevations not higher than 200 m over sea level. Actually the most common plant species growing in the area is *Nothofagus Antarctica* (G. Forst) Oerst. –ñire-, known as *ñire*. This species can be frequently found in contact or combined with *Festuca gracillima* steppe or with *Nothofagus pumillo* forest (Tell 1997; Correa 1998).

Hunter-gatherer societies in this area, known as Selknam, have been present until the end of the XIXth century, leaving an abundant and rather complete ethnographic record (Bridges 1951; Chapman 1986; Gallardo 1910; Gusinde 1937; Lothrop 1928).

Ewan is located in the centre of Isla Grande of Tierra del Fuego (Figure 1). The site dates back towards the end of the existence of the Selknam; it is a ceremonial place were the ritual of *Hain* (an initiation ceremony) was performed (Bodganovic *et al.* 2009; Mansur *et al.* 2007a; Mansur *et al.* 2007b). It was probably occupied during weeks. Ethnographers (Gusinde 1937; Chapman 1986) mention that this ceremony could last several weeks or even months.

The site provides activity evidences in different two sectors.

- Ewan I is a conical wooden structure located in a forest clearing, its main function was ritual.
- Ewan II is located 200 m west from Ewan I; it corresponds to the place where these hunter-gatherers lived during the ceremony, several remains of habitation have been documented. One of these units - unit 1 has been totally excavated (Figure 2).

Results from dendrochronological analysis from Ewan I logs place the occupation at the beginning of XXth century (at least one of the logs was cut in 1905 spring, see Berihuete *et al.* 2009). The presence of European material in both areas (micro-flakes and tools made using European glass as well as undetermined remains of iron) reinforce the modern age of the site, as the European population arrived at Tierra del Fuego by middle XIXth century.

In this paper we are focusing on Ewan II -unit. This structure gave the possibility to study the remains of the probable hut poles used for building purposes, and therefore record the wood use–and related working

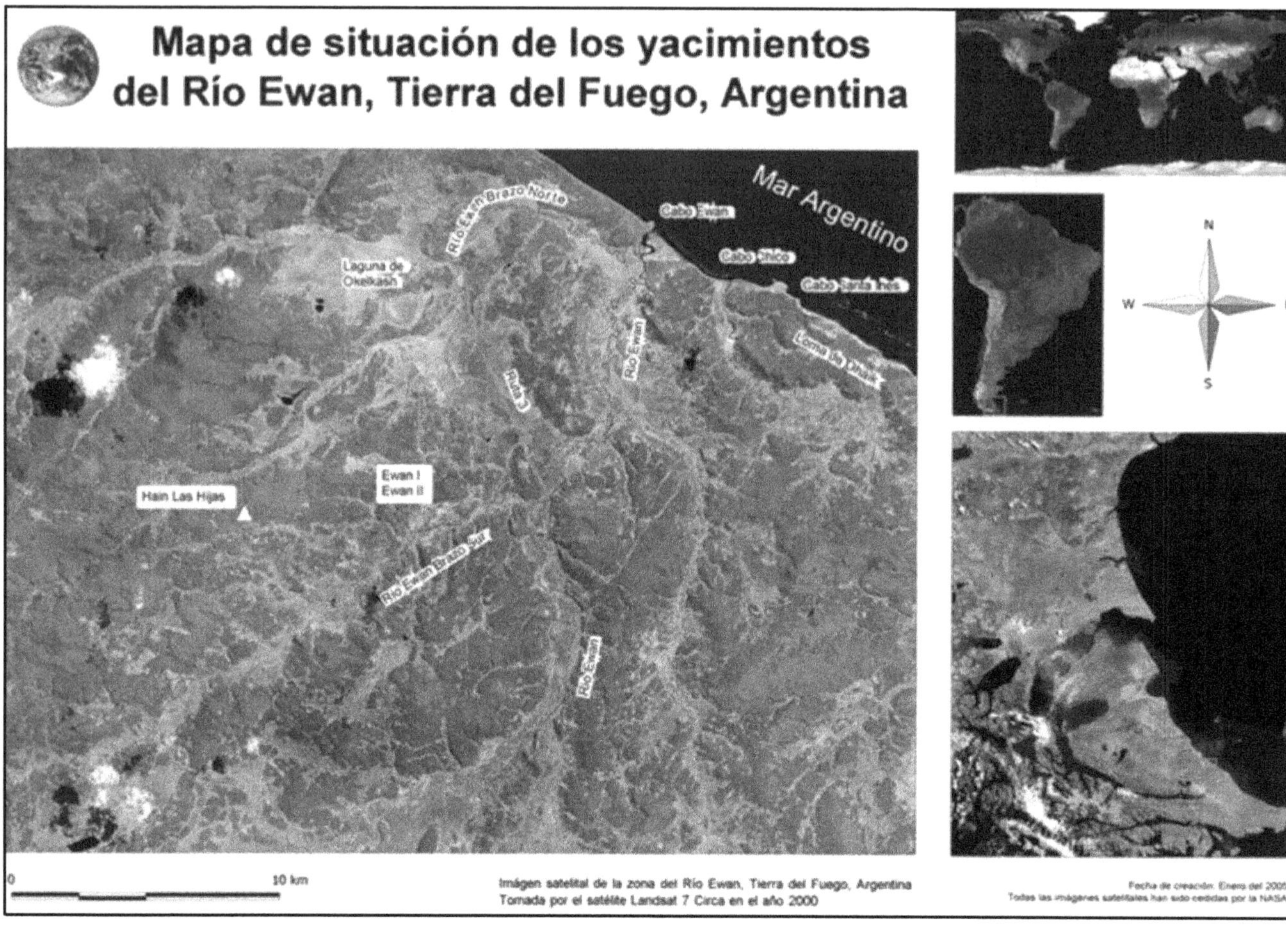


Figure 1. Ewan Map

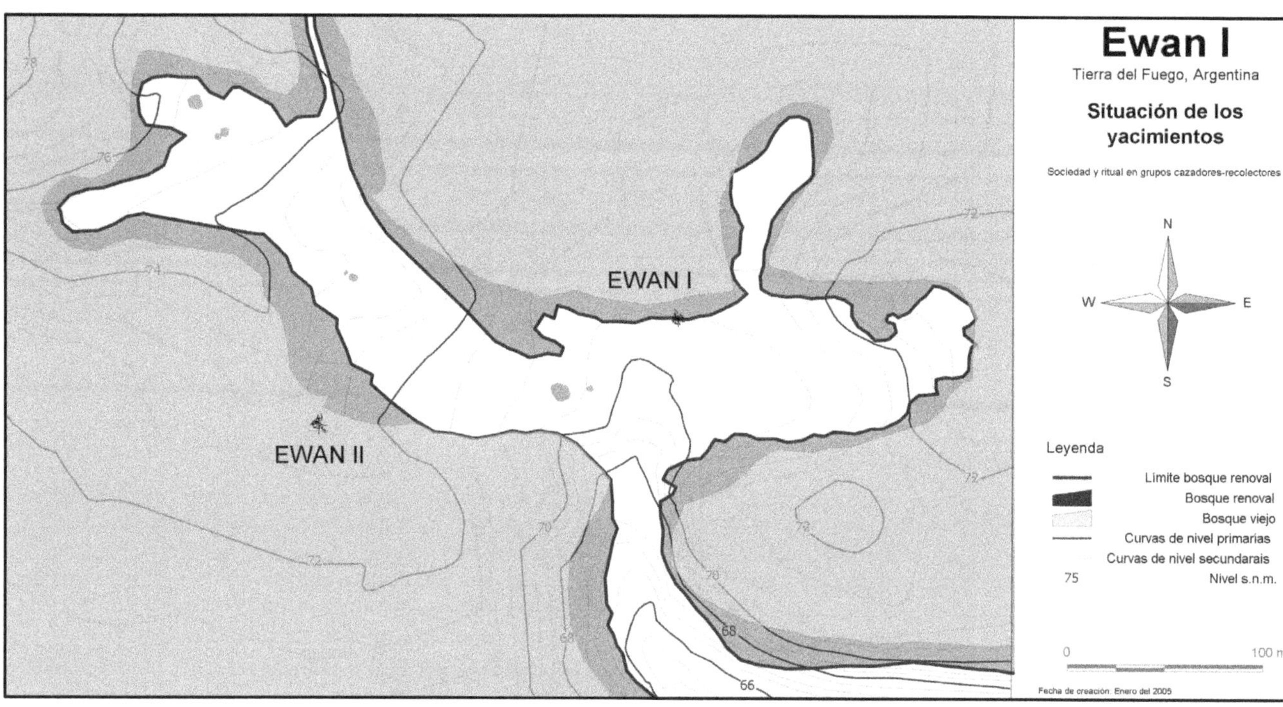


Figure 2. Area of combustion -Ewan II-unidad 1

processes. At the same time, charcoal recovered in the fireplace and its dispersal in the archaeological level gave the possibility to study exhaustively the use of wood in this Selknam settlement. On the basis of this, the main goal of this research has been to understand the ways of exploitation of woody plant resources by the Selknam, especially with reference to the dichotomy of wood for building versus wood for fuel. The purpose has also been to learn about the different exploitation and management strategies of these resources, starting from the procurement, processing/transformation and use.

The study of hunter-gatherers plant use resources is not straightforward because due to taphonomy and also because of the ways resources were used, they aren't often preserved in the archaeological record. Nevertheless, the recent chronology of Ewan site has allowed the preservation of wood remains.

Due to the recent age of the site, there is abundant ethnographic data that allow us to document many different ways of using plant resources together with different products manufactured with plant raw materials. Our research is carried out from an ethno archaeological approach. We understand ethno archaeology in a wide sense; it is used to contrast information derived from ethnographical and archaeological record of one society. Documentation allows us to generate archaeological hypothesis to interprete the archaeological record or to develop methods of research. On the other hand it also allows a better understanding of the societies because the information provided by archaeology and ethnography is often complimentary (Estevez and Vila 1995).

**Plant resources according to ethnographic writings**

Even though plant species or detailed plants activities are not mentioned, some ethnographers provide information about the usage of plant resources by this group. The most complete reports about plant resources exploitation were provided by Martín Gusinde (1937), Ángel Gallardo (1910) and Anne Chapman (1986). In synthesis, these authors describe three types of wood use: for huts or windbreaks building, as fuel and to produce tools or parts of them.

These authors mention the location, shape, characteristics and building of huts and windbreaks. According to them, the trunks used were particularly long and more o less straight. The procurement consisted of cutting trunks or directly pulling them down. Once obtained and transported, they had to be prepared before their use (Chapman 1986, 49; Gusinde 1937, 178).

In Selknam ethnographic literature, there are few references about fuel, but some general descriptions about the exploitation of wood from *Nothofagus* forests and some other aspects of fire management are described in detail. In relation to the selection and attainment of firewood, ethnography mentions that there is a clear preference to light the fire with dead wood and big trunks (Gallardo 1910: 46), which were brought to the campsite as they were needed (Gusinde 1937, 187).

**Ewan II-unit 1**

During the excavation of Ewan II-unit 1, in the surface and upper levels we found several trunks of different sizes, which had apparently fallen without a clear pattern (Figure 3). Taking into account the possibility that at least some of these trunks were part of the original wooden structure, it was decided to carry out a thorough record and a preliminary analysis of all.

Figure 3. Ewan II site, unit 1

It was also possible to identify a combustion area in which several trunks and branches partially carbonized had been preserved, as well as a large amount of charcoal (Figure 4). These trunks with combustion signs were visible on the surface of the top organic layer. The trunks used as firewood are carbonized at the extreme placed in the center of the combustion area; some of them also show cut marks at the other end. Inside the fireplace a great quantity of charcoal, heavily burnt bone fragments, shells, glass instruments, together with fragments and micro-flakes of glass were recovered.

## METHODOLOGY

The archaeobotanical methodology consisted in taxonomical and morphometric analysis of charcoal, partially carbonized firewood and the trunks. The spatial distribution of the archaeological material allowed to determine the internal and an external boundaries of Ewan II-unit1 site (Mansur *et al.* 2007a). In order to study a significant sample of charcoal remains a broad sampling strategy was carried out, taking into consideration the spatial distribution of the remains and their dimensions. All wood material from the surface of the excavated area and the upper level was analyzed, except those which were affected by strong post-depositional processes (Figure 5).

The charcoal remains were collected in sub-squares of 50 square centimeters, according to the reference grid. Big charcoal fragments and partially carbonized wood were collected by hand during excavation, but the rest were gathered through the flotation of sediments. The big amount of charcoal recovered made necessary a sub-sampling in the laboratory.

The preliminary test made in order to determine the diversity of taxa showed a homogeneous use of wood.

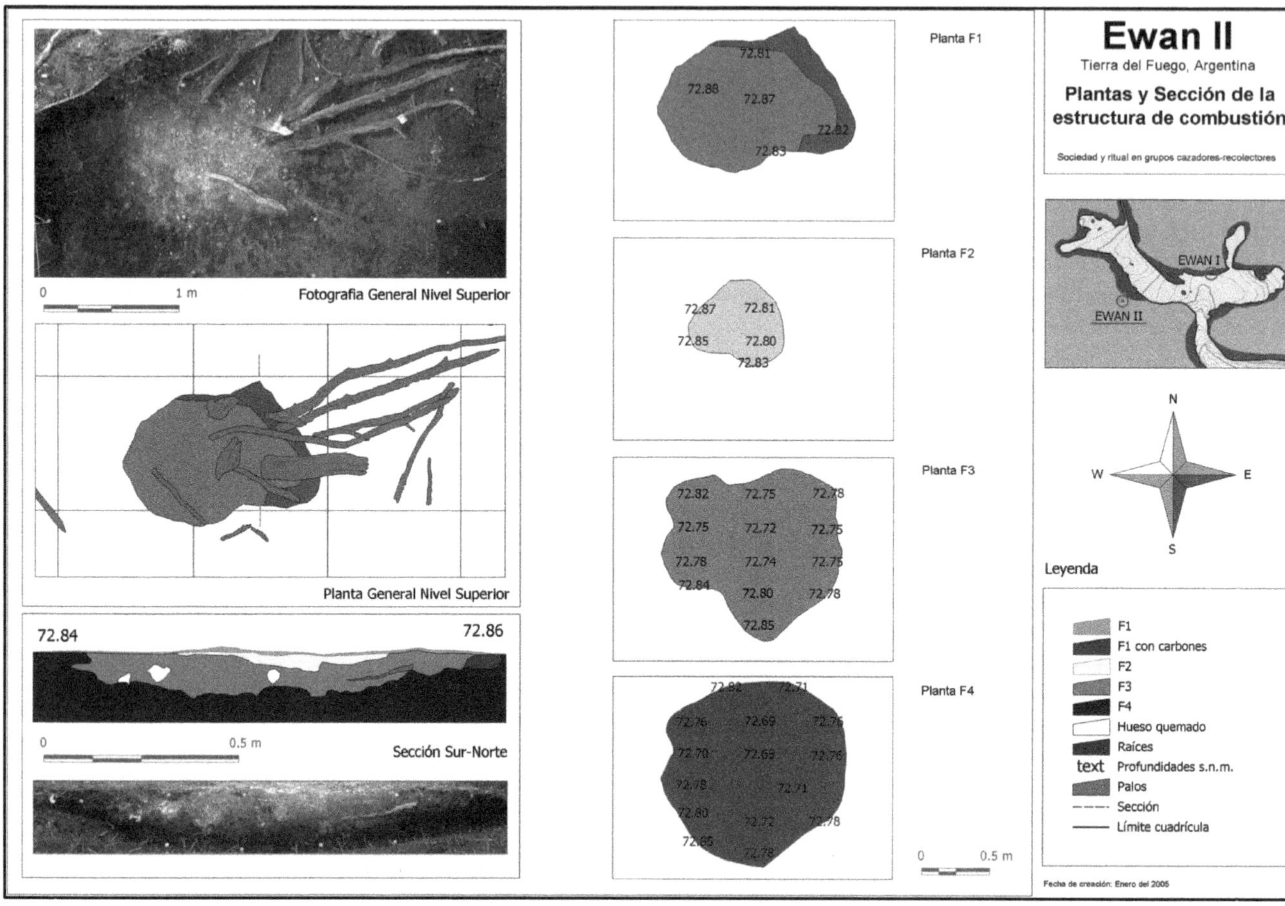


Figure 4. Combustion area -Ewan II-unidad 1

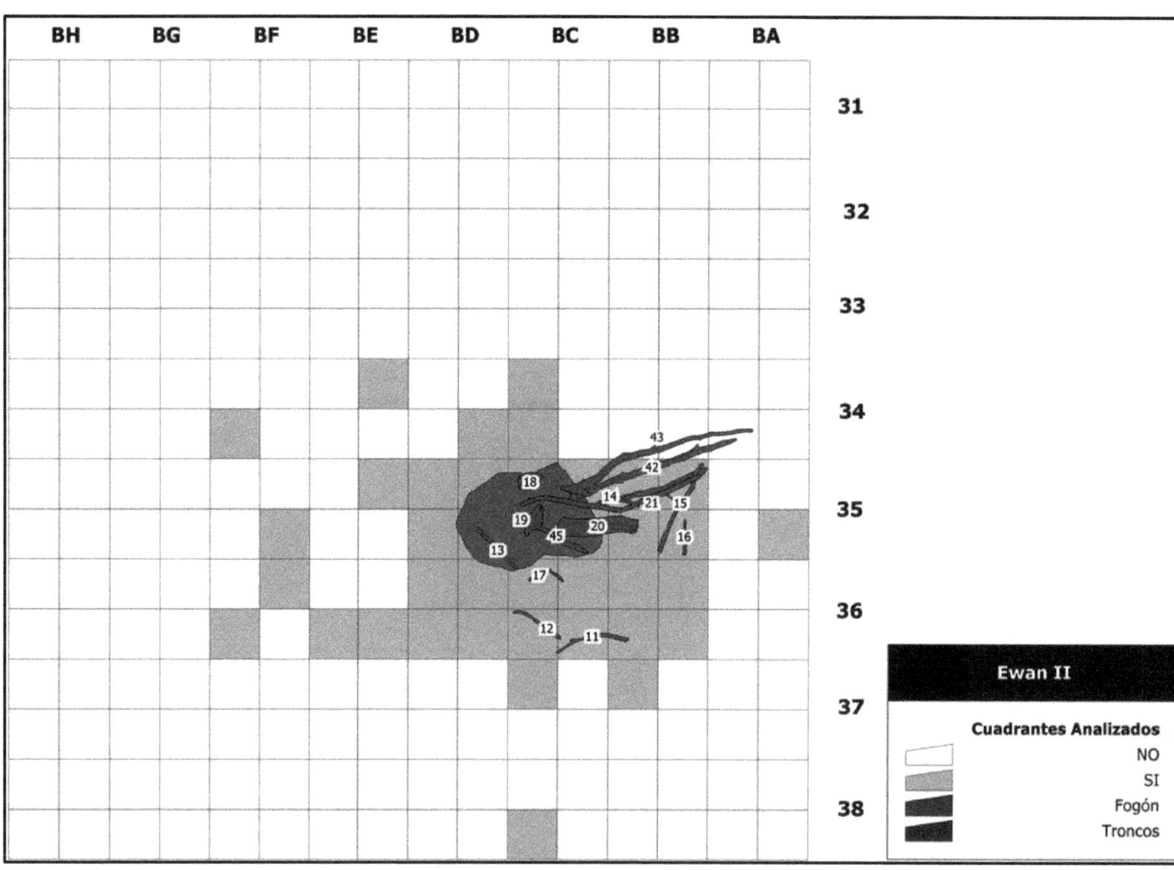


Figure 5. Analyzed quadrants of Ewan II

Given that one of the interests was the analysis of the spatial distribution of the elements we decided to analyze 25 fragments of charcoal from each 50 x 50 cm square. Because of the low taxonomic diversity we consider this to be representative for this study.

The visibility or invisibility of the curvature of the growth rings was recorded during taxonomic analysis. Two parameters were used in order to record the curvature. The observation of a significant curvature of growth rings could prove the presence of a small-sized log, belonging to a branch fragment. On the other hand, a weak curvature of the rings, which often exceeded the size of the coal, would indicate large-sized logs. The curvature of rings has been used successful in charcoals to identify the anatomical part (Ludemann and Nelle 2002; Ludemann 2006; Marguerie and Hunot 2007). In our case the curvature estimation tries to differentiates only between small branches and big branches or stems. In this sense, our approach is an adaptation of the criteria proposed by Marguerie and Hunot (2007).

Even though all trunks were analyzed following the same guidelines, before performing the morphological analysis wood trunks associated to the combustion structure were separated from the ones that were scattered in the excavation area.

At first, the morphological analysis consisted in the identification of the proximal end (the one closest to the ground) and the distal end (the farthest to the ground) of each trunk. The guiding element was the diameter, since it generally decreases from the proximal to the distal end or the presence of branches. Therefore, the smaller diameter and the split ending or tip, allows to identify the distal end. Presence of decay at one end of the trunk, is also an element that allows the identification of the proximal end. The decomposition of wood is accelerated when it is in contact with the humidity of the ground, so when the poles are standing, they start decomposing from the foundation whereas the distal part of the trunk is in a relatively better condition. Other two signs to detect the proximal end are the cut marks and the stump or root. These marks indicate that the trunk was cut with an instrument of some kind, or that it was directly uprooted from the soil. Once both ends of a pole were identified, the morphometrical analysis of all the poles was performed according to the following criteria:

1. Type/portion of the trunk used: whole trunk, half trunk and board. This allows the identification of the used part of the trunk and/or modifications carried out before using it.

2. Profile type: curved, winding or straight. The systematization of a specific profile type shall show a selective exploitation of the ligneous material used for the building of structures.

3. Dimensions of the trunks were also taken into consideration. The length and diameter allow the understanding of recurrent patterns regarding the length and thickness of the used trunks or if any kind of ligneous material has been used.

4. The extreme part of each trunk was classified according the following characteristics: cutting, uprooting, fracturing, fragmentation and decomposition, deterioration or hollowness.

5. The existence or absence of branches in the trunks was also taken into account, on the basis of the following categories:
   - Cut: there are signs of cutting
   - Torn: there are signs that show they have been torn by the roots
   - None: the trunk does not have any branches at all
   - Undetermined: due to alterations or other reasons it is not possible to determine accurately if we are dealing with torn or cut branches

To achieve our goals, it was important to differentiate between those trunks/poles that probably belonged to the possible wooden structure and those that naturally fall from forest trees and tree branches. For that purpose, following criteria were considered.

- Anthropical marks on the trunks. The presence of cut marks or the presence of the root system on the proximal end were two characteristics that made evident some kind of human activity. The evidence of wood decomposition together with the hollowing of the proximal end of the trunk –caused by the same decomposition – is a clear sign that part of the trunk was in touch with the soil (in this case it could be the result of vertical position of poles during long time).
- Existence of traces produced by environmental post-depositional agents that may help to show whether the existence of the trunks in the sites is due to natural causes –falling of a tree or belonging to a plant structure.
- The last item taken into account for the formation of the three subcategories was the morphological dimensions and characteristics of the trunks.

After this analysis we established three subcategories related with the probability to belong to a wooden structure:

– Group A: Trunks that show signs of anthropical activity; they have been cut or uprooted and they fulfill the previously established morphological characteristics (that is, trunks with a high probability to belong to a wooden structure).

– Group B: This category includes all the trunks that, in spite of fulfilling the established morphological characteristics, do not show anthropical or environmental marks that allow us to associate them to a wooden structure (that is, the trunks which show no sign of belonging to Ewan II wooden structure).

– Group C: Trunks that do not show anthropical evidence but show some decomposition or hollowing of the proximal end and also fulfill the established morphological characteristics (they have also probability to belong to a wooden structure).

## RESULTS

### Combustion structure

The anatomical analysis of the charcoal was carried out on a total sample of 664 fragments. The results show a homogenous use of the ligneous material as firewood. All the charred material belonged to *Nothofagus antarctica* (G. Forst) Oerst. -ñire-, which has the following anatomical characteristics (Wageman 1948; Solari 1993; Piqué 1999):

- Transversal section: growth rings are visible and wavy. Wood porosity is diffuse and vessels are solitary or more frequently grouped in radial lines (Figure 6).
- Tangential section: rays are mainly unicellular, although there are biseriate. Rays height ranges between 10 and 20 cells (Figure 7).

Figure 6. Transversal section of *Nothofagus antarctica* x100 (Caruso)

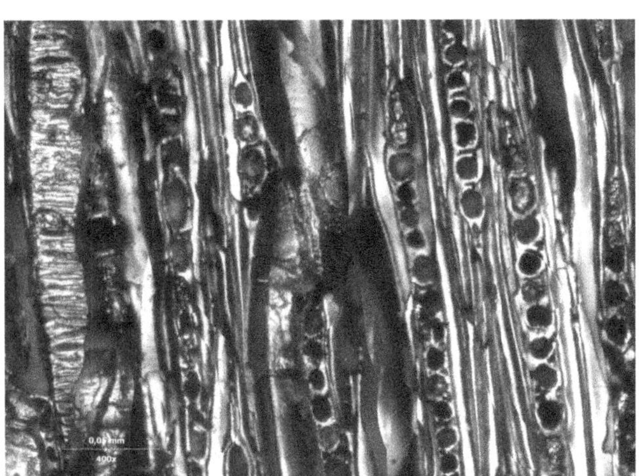

Figure 7. Tangential section of *Nothofagus antarctica* x400 (Caruso)

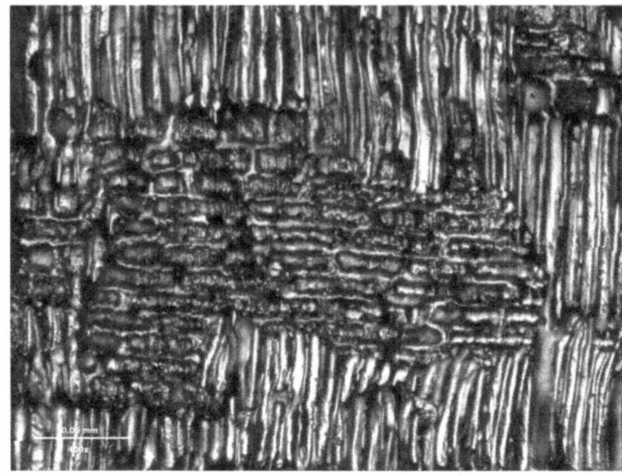

Figure 8. Radial section of *Nothofagus antarctica* x400 (Caruso)

- Radial section: rays are homogeneous. Vessels have scalariform perforations (Figure 8).

A recurrent characteristic in the charred wood studied was the low or very low curvature of the growth rings in practically all the analyzed fragments. This shows that the material employed as fuel mainly came from large-sized trunks. This evidence is also supported by the presence of eight big trunks that appear to have been burnt. The length of these trunks ranges from 190 cm to 30 cm, the proximal diameter from 6 to 19 cm and the distal diameter from 2.5 to 13 cm (Figure 9).

The profile of the trunks can be sinuous (trunks 5, 21, 43), straight (trunks 20, 42) and curved (trunk 14). Only two of them (number 19 and 43) show some cut marks in their proximal end, while the remaining five show combustion signs. The analysis of the distal end of these trunks shows that one of them has a burnt end (number 19). Four trunks (number 20, 21, 42, 43) have a fragmented end and two of them (number 5 and 14) end in a split. All the trunks still have some bark and branches.

Wood material used as fuel comes from local resources. *Nothofagus antarctica* is nowadays the most abundant and common species around the site, and it probably wasn also one hundred years ago when the settlement was occupied. The measures of the trunks also indicate a selective search of the material rather than a simple random gathering. At the same time, the selective search and collection of larger trunks implies that some effort was spent in this activity, much more than by the simple gathering of dead fallen wood.

### Wooden trunks

On the basis of the division into three groups, it was possible to distinguish that 16 of the 40 trunks recovered during the excavation may have possibly belonged to a wooden structure (12 of Group A and 4 of group B) (Figure 10).

| Trunk | Stratum | Orientation | Species | Length | Proximal Diameter | Distal Diameter |
|---|---|---|---|---|---|---|
| 5 | A1 | NW-SW | *Nothofagus antarctica* | 156 cm | 6 cm | 4.5 cm |
| 14 | A3 | E-W | *Nothofagus antarctica* | 190 cm | 6 cm | 4 cm |
| 19 | A3 | E-W | *Nothofagus antarctica* | 30 cm | 13 cm | 13 cm |
| 20 | A3 | E-W | *Nothofagus antarctica* | 83 cm | 19 cm | 13 cm |
| 21 | A3 | E-W | *Nothofagus antarctica* | 142 cm | 7 cm | 2.5 cm |
| 42 | A3 | E-W | *Nothofagus antarctica* | 175 cm | 7 cm | 5 cm |
| 43 | A3 | E-W | *Nothofagus antarctica* | 190 cm | 7 cm | 4 cm |

Figure 9. Fireplace and trunks

| Trunk | Group | Length | Proximal Diameter | Distal Diameter | Proximal End | Distal End | Orientation | Stratum |
|---|---|---|---|---|---|---|---|---|
| 1 | A | 395 cm | 9 cm | 6 cm | Cut | Fragmented | N-S | A1 |
| 2 | A | 570 cm | 11 cm | 5 cm | Cut | Fragmented | NW-SE | A1 |
| 4 | A | 128 cm | 5 cm | 4 cm | Fragmented | Cut/Fork | NE-SW | A1 |
| 8 | A | 89 cm | 12 cm | 10 cm | Cut | Fragmented | N-S | A1 |
| 10 | A | 315 cm | 13 cm | 7 cm | Fragmented | Cut/Fork | N-S | A1 |
| 28 | A | 150 cm | 6 cm | 3 cm | Cut | Cut | N-S | A1 |
| 30 | A | 280 cm | 7 cm | 4 cm | Root | Fragmented | NE-SW | A3 |
| 35 | A | 378 cm | 6 cm | 4 cm | Fragmented | Cut | E-W | A3 |
| 36 | A | 270 cm | 7 cm | 5 cm | Cut | Cut | E-W | A3 |
| 41 | A | 333 cm | 12 cm | 4 cm | Root | Horqueta | N-S | A1 |
| 44 | A | 900 cm | 8 cm | 6 cm | Cut | Fragmented | NW-SE | A3 |
| 47 | A | 295 cm | 6 cm. | 2.5 cm | Root | Fragmented | E-W | A3 |
| 22 | C | 415 cm | 11 cm | 3 cm | Hollow | Fragmented | NW-SE | A1 |
| 24 | C | 214 cm | 6 cm | 3 cm | Decomposer | Fragmented | N-S | A1 |
| 27 | C | 230 cm | 14 cm | 7 cm | Decomposer | Fork | NE-SW | A1 |
| 40 | C | 333 cm | 7 cm | 6 cm | Hollow | Fork | NE-SW | A1 |

Figure 10. Trunks structure

In general, the length of these trunks ranges from 150 to 570 cm. Their proximal ends show cut marks and has a stump – root – and range from 5 to 14 cm, prevailing those between 5 and 10 cm. Distal diameters range from 2.5 to 7 cm and in general have a split or are cut (Figure 11). In relation to the profiles, 50% are straight while the other half are winding. None of the trunks show any bark and most of them have no branches or torn except trunks 2, 4, 36 and 41, which show lateral branches. The general characteristics of these trunks show that the procurement of wood included cut or uprooting and tearing of branches. The morphology, the diameter and the length of the trunks and the presence of a split in the distal extreme show a clear selection of the raw material.

## DISCUSSION

The forest was used by the Selkman hunters-gatherer as a place for placing their dwellings and as a source for raw materials (fuel procurement and wood for inhabit

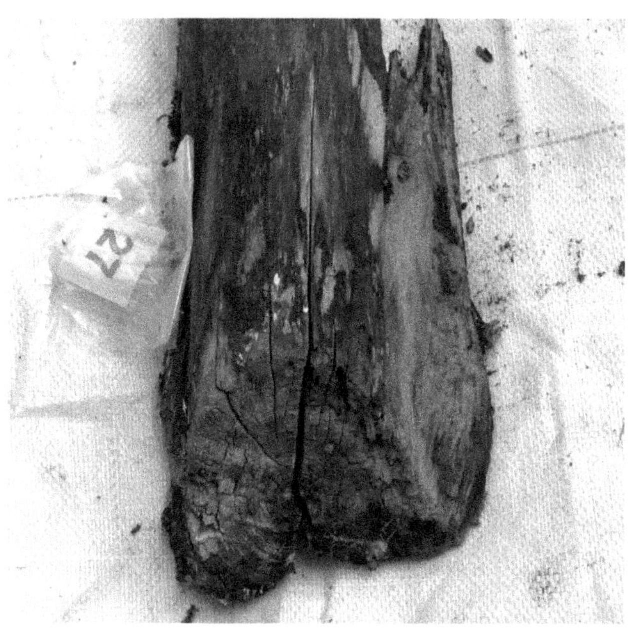


Figure 11. Cut traces

structures). The archaeobotanical results together with the information from the written sources have enabled the discussion of procurement strategies and processing as well as the use of plant resources developed by this hunter-gatherer group.

Morphological analysis of trunks belonging to Ewan II-unit 1 structure shows a selection with predetermined morphological characteristics, (longitude, diameter and profile and ending) and not to the search of a vegetal species in particular.

The characteristics of the proximal end -cut marks or uprooting- of the trunks show two different ways of procurement of the needed raw material for the building or construction. The first one –cutting of trunks- implied the use of a metal instrument and showed the growth of specific practices destined to the procurement of ligneous material. It is important to point out that at the beginning of the xx century many Selknam were working in the local ranches (*estancias*) so many of their traditional activities were modified through their contact with the European. The second way of procurement –uprooting of trunks- did not require the use of instruments, but it was still important. Once the trunks and logs were obtained, they had to be transported to the site where they were needed. The procurement –cutting or uprooting-the transportation and cutting of branches are the only activities carried out in connection with the trunks before they are used.

Ethnographically we know that trunks of this kind were used by the Selknam for constructing wooden windbreaks and huts. In this case in particular 16 trunks clearly belong to a wooden structure. Nevertheless this study does not allow us to specify which would be the type of structure. In ethnographical sources windbreaks were described as relatively conical structures, formed by 6 or 12 trunks and branches which measured 1.50m and showed a point on one end and a split in the other (Gusinde 1937; Chapman 1986). If we only take into account the ethnographic information, the poles recovered from Ewan II do not have the specific parameters of a windbreak.

To be able to determine the shape of the wooden structure, we must take into account other aspects apart from the morphological characteristics and location of the different poles. Location of the combustion structure, dispersion of the archaeological material and location of the proximal end of each one of the poles will be considered to determine the existing structure type and its dimensions. The proximal end of the trunks is placed in the peripheral areas with respect to the combustion area. The location of the different trunks would probably be surrounding those areas with the highest presence of archaeological material. It is also worth pointing out that the same falling of the poles could have altered the original position of each of them, but in no case it could have inverted the position of the extremes. The analysis of the spatial dispersion of the archaeological material -flakes and fauna remains- also suggests a close structure with activities and orientation of the logs around a central combustion structure (Maximiano 2005; Mansur *et al.* 2007b), that is to say that none of these indicators or markers suggests the existence of a relatively open structure such as a windbreak.

In relation to the procurement of wood as a fuel, the cut marks of the combustion area trunks also show that they were not the result of a simple collection, but that it was a every-day practice and that there was an organization destined to the procurement of that material. The carbonized trunks in the combustion area and the registration of the growth of slightly curved rings also confirm the search of the ligneous of great dimensions. Written sources (Gallardo 1910; Gusinde 1937; Chapman 1986) mention the use of big logs as fuel. The cut marks together with the size of the logs show that the procurement of the fuel implied a specific effort and more work compared to the one used for the collection of fallen dead wood or cutting attached wood from the trees.

The taxonomic analysis of both charred wood and trunks from the hypothetical structure show the exclusive use of a local species: *Nothofagus antarctica*. Despite this homogenous use of the wood there is a selection of the plant material. As we have already mentioned, this selection is not oriented towards the search of a determined taxon for a specific activity but to the search of trunks and logs with big dimensions. It is also worth mentioning that such taxonomic homogeneity was expected taking into account that Ewan is in an area where *Nothofagus antarctica* is the only woody taxon.

At the beginning of this work, we made reference to the little attention paid so far by archaeologists to plant resources. The results obtained from the taxonomic analysis of both wood and charred wood, together with the thorough morphological study of each one of the trunks from the Ewan II-unit 1 structure and the ethnographic information showed that the plant resources were of significant importance to the Selknam socio-economic dynamics. The morphological characteristics of the material used illustrate knowledge and mastering of specific activities for the procurement as well as for the transformation of the plant raw material in accord with its use. The forest environment was not only essential for Selknam everyday tasks, but also bears the same degree of importance as any other used resources.

## Acknowledgements

This research has been carried out thanks to a PhD grant of the Ministerio de Cultura y Educación de España given to the first author, which was carried out within the project HUM2006-01129 funded by the *Ministerio de Cultura y Educación de España (directed by Dr J.A. Barcelo)* and "Society and ritual in hunter-gatherer societies. Use of the forest as a ritual space among the Selknam of Tierra del Fuego", funded by the *Ministerio de Cultura de España* and CONICET, (directed by Dr Raquel Piqué and Dr Estela Mansur). I would also like to thank Dr Raquel Piqué and Dr Debora Zurro for their comments and discussions, Oriol Vicente for his help with some graphics and images and also to Florencia del Castillo and the rest of our fieldwork mates. Finally, thanks to the anonymous reviewers for their interesting suggestions and corrections.

**References**

BOGDANOVIC, I., CAMAROS, E., De ANGELIS, H., LASA, A., MANSUR, M.E., MAXIMIANO, A., PARMIGIANI, V., PIQUÉ, R. y VICENT, O. 2009. El paraje de Ewan, un lugar de reunión Selknam en el centro de la isla (Tierra del Fuego, Argentina) *VII Jornadas de Arqueología de la Patagonia*, 1, 941-957. CADIC-CONICET.

BERIHUETE, M., CARUSO, L., MASSACCESI, G., MANSUR, M.E., MENSUA, C. y PIQUÉ, R. 2009. El aprovechamiento de los recursos vegetales entre los Selknam de Tierra del Fuego (Argentina), una aproximación etnoarqueológica. *La alimentación en la América precolombina y colonial: una aproximación interdisciplinar*, 21-40.

BRIDGES, E.L. [1948] 2000. *El último confín de la Tierra*. Editorial Sudamericana. Buenos Aires.

CARUSO, L. 2008. *Los usos de la madera entre los cazadores-recolectores de Tierra del Fuego.* Trabajo de investigación programa de doctorado, Universidad Autónoma de Barcelona.

CARUSO, L., MANSUR, E. y PIQUÉ, R. 2008. Voces en el bosque: el uso de recursos vegetales entre cazadores-recolectores de la zona central de tierra del fuego". *Darwiniana* 46(2), 202-212. ANCEFN y CONICET.

CARUSO, L., BERIHUETE, M., MENSUA, C. 2009. Las plantas como recurso entre los cazadores-recolectores selknam: aportes del análisis arqueobotánico del sitio Ewan (Tierra del Fuego). *VII Jornadas de Arqueología de la Patagonia*, 1, 433-444. CADIC-CONICET.

CARUSO, L., MANSUR, M.E. y PIQUÉ, R. 2009. Las chozas de madera de la zona central de Tierra del Fuego. *VII Jornadas de Arqueología de la Patagonia*, 1, 445-456. CADIC-CONICET.

CORREA, M.N. 1998. *Flora patagónica.* Parte I. Colección Científica del INTA. Buenos Aires. Chapman, A. 1989. *El fin de un mundo: los Selk'nam de Tierra del Fuego.* Vazquez Mazzini Editores. Buenos Aires.

DIMITRI, M. 1972. *La región de Los Bosques Andino-Patagónicos. Sinopsis General.* Colección Científica del INTA, Buenos Aires.

ESTÉVEZ, J., VILA, A. 1995. Etnoarqueología: el nombre de las cosas. En: Estévez J. y Vila A. (comps), *Encuentros en los conchales fueguinos*, 17-23. CSIC-UAB, Barcelona.

GALLARDO, C.R. 1910. *Los Onas de Tierra del Fuego.* Zagier and Urruty Publications. Ushuaia.

GUSINDE, M. 1982. *Los indios de Tierra del Fuego. Los Selk'nam.* Centro Argentino de Etnología Americana, Buenos Aires.

LUDEMANN, T. 2006. "Anthracological Analysis of Recent Charcoal-Burning in the Black Forest, SW Germany". Charcoal Analysis: New Analytical Tools and Methods for Archaeology. A. Dufraisse, (ed). Oxford, BAR Publishing. BAR International Series 1483, 61-70.

LUDEMANN, T. and NELLE, O. 2002. Die Wälder am Schauinsland und ihre Nutzung durch Bergbau und Köhlerei. Frieburg i.Br., Forstwissenschatliche Fakultät der Universität Freiburg und Forstliche Versuchs – und Forschungsanstalt Baden-Württemberg.

MANSUR, M.E., MAXIMIANO, A., VICENTE, O. y PIQUÉ, R. 2007a. "Arqueología de Rituales en Sociedades Cazadoras-Recolectoras. Una aproximación desde el Análisis del Espacio Socialmente Producido". *Arqueología de Fuego-Patagonia. Levantando piedras, desenterrando huesos... y develando arcanos.* Punta Arenas: Fundación CEQUA, 741-754.

MANSUR, M.E., PIQUÉ, R. y VILA MITJA, A. 2007b. Etude du rituel chez les chasseurs-cueilleurs. Apport de l'ethnoarchéologie des sociétés de la Terre de Feu. *Chasseurs-cueilleurs. Comment vivaient nos ancêtres du Paléolithique supérieur*, 143-150. S. de Beaune Ed., Editions du CNRS. Paris.

MARGUERIE, D. and HUNOT, J.-Y. 2007. "Charcoal Analysis and Dendrology: Data from Archaeological Sites in North-Western France." Journal of Archaeological Science 34, 1417-1433.

MAXIMIANO CASTILLEJO, A. 2005. *Métodos geocomputacionales aplicados al análisis espacial en arqueología.* Tesis de Tercer Ciclo, Departamento de Prehistoria. Univ. Autónoma de Barcelona.

PÉREZ de MICOU, C. 1979-1982: Sitio Piedra Parada 1 (PP1), Dpto. de Languiñeo, Pcia. Chubut (Argentina). *Cuadernos del Instituto Nacional de Antropología* 9, 97-112.

PÉREZ de MICOU, C. & NACUZZI, L. 1983-1985. Los recursos vegetales de los cazadores de la cuenca del Río Chubut. *Cuadernos del Instituto Nacional de Antropologia* 10, 407-423.

PIQUÉ, R. 1999. *Producción y uso del combustible vegetal: una evaluación arqueológica.* Series Treballs d'Etnoarqueología, 3. CSIC Madrid.

ROIG, F.A. 1998. *La vegetación de la Patagonia*, 1, 48-174. Colección Científica INTA, Buenos Aires.

SOLARI, M.E. 1993. *L'homme et le bois en Patagonie et Terre de Feu au cours des six derniers millénaires:Recherches anthracologiques au Chili et en Argentine.* thèse Université Montpellier II, Montpellier. France.

TELL, G., IZAGUIRRE, I., QUINTANA D. 1997. Bosque Andino Patagónico. *Flora y Fauna Patagónicas*, 11-60. Ed Colihueche, Buenos Aires.

WAGEMANN, G. 1949. Maderas chilenas. Contribución a su anatomía e identificación. *II Congreso Sudamericano de Botánica*, 263-375. Lilloa.

# DECODING WOOD EXPLOITATION IN ARCHAEOLOGICAL SITES IN SOUTH-EASTERN ITALY


Cosimo D'ORONZO
Laboratory of Archaeobotany and Palaeoecology, Salento University, Italy
Università Cattolica del Sacro Cuore, Milano, Italy

Giampiero COLAIANNI
Laboratory of Archaeobotany and Palaeoecology, Salento University, Italy
Università del Salento, Lecce, Italy

Anna Maria GRASSO
Laboratory of Archaeobotany and Palaeoecology, Salento University, Italy
Università degli Studi di Siena, Siena, Italy

D. MARTELLA
Laboratory of Archaeobotany and Palaeoecology, Salento University, Italy

Angela STELLATI
Laboratory of Archaeobotany and Palaeoecology, Salento University, Italy
Università degli Studi di Bari, Bari, Italy

Girolamo FIORENTINO
Laboratory of Archaeobotany and Palaeoecology, Salento University, Italy
girolamo.fiorentino@ateneo.unile.it



**Abstract**: *The purpose of this work is to investigate some cases of anatomical changes in woody tissue due to biotic and abiotic factors in order to decode wood exploitation. The observation of the interaction between wood tissue and biotic factors, like fungi, insects and marine borers, and alteration products by chemical and physical agents could help to decode wood exploitation and formation processes of archaeobotanical material. Therefore a review of alterations of wood could be useful. In this work the results of taphonomical analyses carried out on wood material from archaeological contexts in south-eastern Italy (hearth structures, waterlogged material, burials) were proposed.*



**Key words**: *South-eastern Italy, taphonomy, contextual archaeology, wood exploitation*


## INTRODUCTION

The vegetational landscape of the Mediterranean area does not appear to have undergone substantial taxonomic variations attributable to natural factors during the late Holocene. It does however show ancient traces of both spatial and compositional changes due to other factors (Vernet 1993; Roberts 2000). Archaeobotanical research has demonstrated that they were the result of human actions repeated down through millennia. The environment was used as a source of plant resources, particularly wood, indispensable for the technological development of ancient civilisations. Thanks to its structural characteristics, wood was one of the key elements in the technological evolution of past societies. Its use over the centuries has undergone continuous improvements in terms of procurement and processing.

The evolving capacity for procurement and processing of wood is frequently demonstrated by special characteristics of finds recovered from archaeological contexts. The formation and subsequent conservation of the archaeological assemblage can thus be understood as the outcome of events produced by a combination of anthropic and natural agents, the reading of which entails a careful interpretation of the data in taxonomic terms. In this way, the anthracological and xylological analyses of the woody tissue can become an important source of information in the reconstruction of the behaviour of human communities in relation to their exploitation of this raw material.

## THEORETICAL BACKGROUND

All possible uses of wood by human beings are based on its procurement. As it is gathered by various means, the first choice is between living wood still on the plant and dead wood lying on the forest floor.

After gathering and possible storage, the wood can be used as:

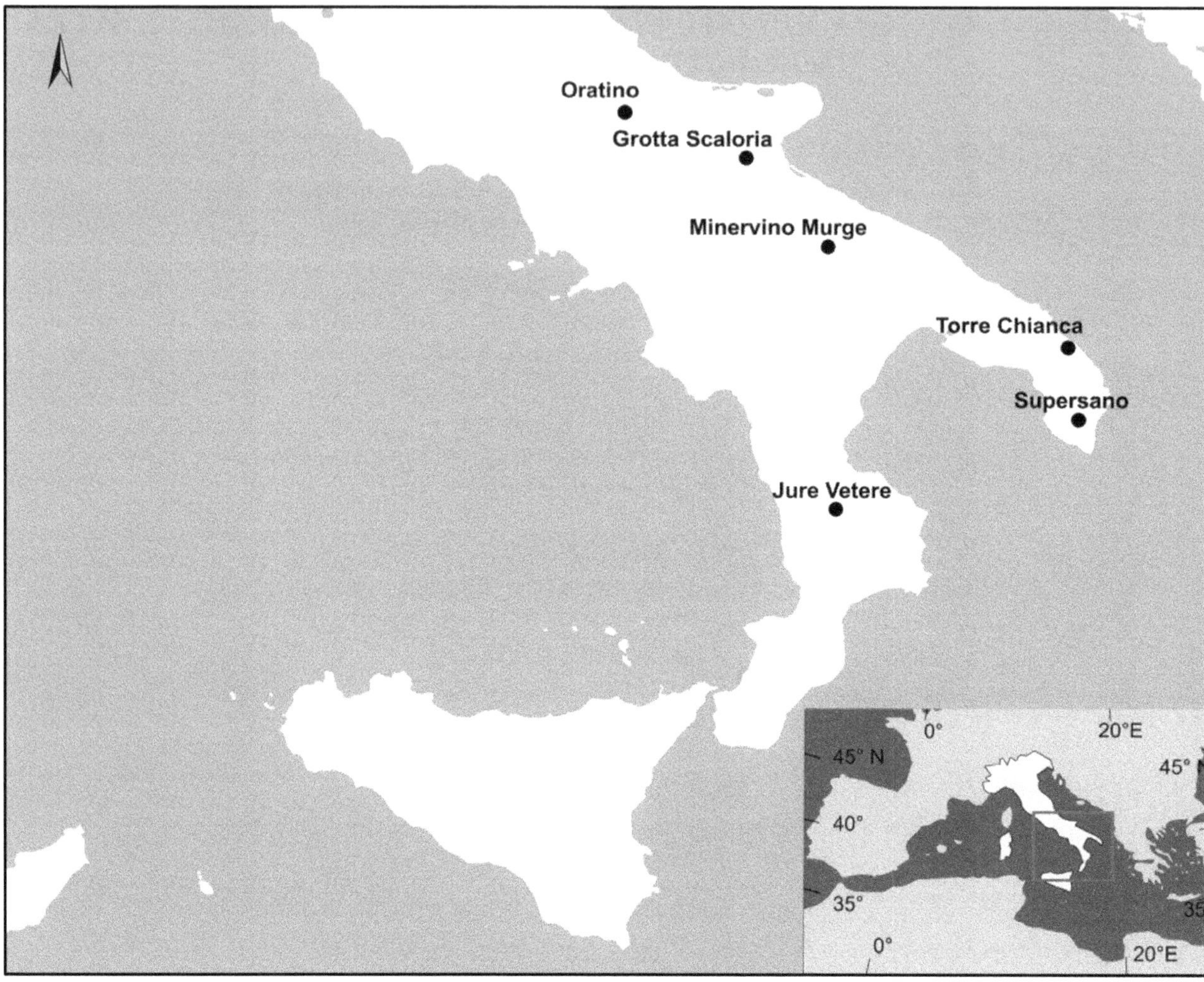


Figure 1. Investigate area with location of sites

- fuel, both in domestic hearth structures (for heating and cooking) and for craft production (energy for melting metals and firing ceramics);
- material for carpentry (building and construction of means of transport);
- material for making artefacts (handles for weapons and tools, furniture, etc.).

The ways in which the material is used can entail a series of alterations to the woody tissue. Such modifications are legible on a macro- and microscopic level and derive from biotic and abiotic factors (Loferski 1999/2001; Zabel and Morrell 1992).

Following the defunctionalisation/abandonment of a site, and in association with complex transformations of the deposit due to anthropic and natural activities, this particular category of "ecofacts" constitutes the archaeobotanical assemblage (Butzer 1984; Minnis 1981). At this point, the time period in which these transformations occurred with respect to the formation of the archaeobotanical deposit itself remains to be identified in order to recognise their presence during the pre-depositional, depositional and post-depositional dynamics.

## MATERIALS AND METHODS

The materials come from selected archaeological sites in southern Italy: Oratino (Campobasso), Grotta Scaloria (Foggia), Supersano (Lecce), Torre Chianca (Lecce), Minervino Murge (Bari) and Jure Vetere (Cosenza) (Figure 1). The sites are highly varied in terms of type (some are in the open air and some in caves) and chronology (from Prehistory to the Middle Ages). Each of them has its own specific characteristics in terms of the ways in which the plant resources were managed and exploited by the human communities that used them.

The discovery of residues of woody tissue in combustion structures (simple and complex hearths) in the Bronze Age site of Oratino raised issues concerning procurement strategies and the choice of fuel. The site is situated on a limestone outcrop a short distance from the river Biferno, in a highly distinctive bio-geographical environment that calls for reflection on some aspects of its catchment area (D'Oronzo and Fiorentino 2008). In the same way, the mid-to-late Neolithic underground religious complex of Grotta di Scaloria yielded anthracological remains from combustion structures found in both primary and secondary deposits.

The excavation of the foundations of huts discovered in the late medieval settlement of Supersano made it possible to study the use of wood as a raw material for construction. The type of context and archaeological deposit prompted a number of hypotheses regarding the use of the wood as material for carpentry, but the absence of elements that were clearly identifiable as structural, together with the limited size of the remains (in the order of a few centimetres), made it impossible to determine their original function in the place of excavation (Fiorentino 2008).

A few fragments of the wreck discovered near Torre Chianca, dated to the 3rd century AD (Martella and Fiorentino 2007) raised issues related to shipbuilding; analysis of the traces of marine organisms raised interest in the dynamics of formation of the submerged deposit and the exact period of use of the ship.

In two cases, the remains of the artefacts were found to have been replaced by metal corrosion products (Baker 1974; Graham *et al.* 1976). The recovery of residues of woody tissue belonging to the handles of two swords discovered in tombs at the Iron-Age site of Minervino Murge provided clues to the characteristics of the environment of deposition and made it possible to reconstruct the mode and time of use of the items (Fiorentino and Stellati 2008). In addition, the sampled remains of nails discovered in a tomb in the medieval monastic complex of Jure Vetere raised questions as to the methods used to make wooden coffins (Fiorentino *et al.* 2007).

All the materials described were recovered by sight or by means of systematic sampling strategies, where necessary sieving material or subjecting it to preliminary macroscopic analysis. This initial analysis, often performed on site, made it possible to conduct a preliminary assessment of the state of conservation of the remains (waterlogged, replaced by metal corrosion products, mineralised or charred).

The subsequent microscopic study was designed to determine first of all the taxa to which the remains belonged, and secondly any traces of interaction of the woody tissue with biotic and abiotic agents. By biotic agents we mean all animal and plant microorganisms that have interacted with the woody tissue before and during the use of the artefacts by human beings or subsequent to their defunctionalisation. These include fungi, whose *hyphae* interact with the anatomic elements of the wood either while it is still part of the plant (Giordano 1981; Zabel and Morrell 1992; Schwarze 2007; Osono 2007) or when it becomes part of the plant litter (Scott *et al.* 2000) (parasites of "live" and "dead" wood), and are visible (though perhaps not directly) on the residues of woody tissue. Biotic agents also include animals, such as insects, classified as either parasites (when the wood is still alive and not yet a part of an artefact) and saprophytes (when the wood is dead or a part of an artefact) (Giordano 1981; Jonsell and Weslien 2003). The taxonomic identification of the insect perforating the wood, in the absence of the individual adult (whole insect), is often difficult.

However in some cases it is possible to determine the taxonomic family on the basis of the type and orientation of the tunnels or the plant species attacked. As well as insects, in the case of waterlogged wood in a marine environment, it is possible to identify the traces of molluscs (Teredinidae *Teredo navalis*) or crustaceans (Crustacea *Balanus* sp.). In this case too, the interaction of these organisms with the wood can be diagnosed by recognising direct traces (organism) or indirect traces (tunnels, excretions, etc.) (Zabel and Morrell 1992; Giordano 1981; Oeveringa *et al.* 2001). The process and environment of conservation can prejudice the conservation of these traces, and so an experimental protocol was drawn up to assess their presence even in the ash residues of combustion structures (Figure 2) (D'Oronzo and Fiorentino 2007).

Figure 2. Experimental combustion of driftwood. Teredo sp. tunnelling in the ash (10x)

The microscopic analysis of the results of the interactions with organisms are accompanied by studies of the possible reactions of the wood to abiotic factors such as atmospheric agents, permanent loads (static and dynamic) and transport by wind or water.

## RESULTS AND DISCUSSION

The macroscopic analyses of the burnt remains from Oratino, Grotta Scaloria and Supersano show the process of transformation which, as a consequence of a complex series of chemical reactions (exsiccation, pyrolysis, carbonisation), led to distinctive variations in the colour and physical properties of the tissue. The microscopic analysis, performed in parallel, made it possible to determine whether the wood was used exclusively as fuel or whether it had been used for some other function prior to being burnt. Specifically, it was determined that the charcoals from the combustion structures of the sites of Oratino and Grotta Scaloria were used exclusively as fuel. The former had radial and tangential fractures which, after experimental analysis (D'Oronzo 2008; Fiorentino and D'Oronzo 2010), were attributed to use as semi-dry fuel (recognisable from tangential fractures for heteroxylous woods) and dry fuel (with radial fractures, whether heteroxylous or homoxylous). In the samples from the Neolithic site of Grotta Scaloria, sediment was found to have entered these cracks, probably linked to pedogenetic factors of the cave sediment. The burnt remains contained traces of fungal attacks, suggesting use of wood fuel gathered from plant litter. Such an origin is not suggested by the presence of *hyphae* in the woody tissue alone, but also by the microscopic analysis, which revealed the absence in the conducting elements (parenchymatous rays and vessels) of explosions and fractures.

A series of experimental combustion cycles performed to test the reactions of dry and semi-dry woody tissue demonstrated that dry or dead wood was less prone to produce explosions in the anatomic elements than fresh wood (Fiorentino G. and D'Oronzo C. 2010). In contrast, in the case of the remains found in Supersano the possibility that the remains were hearth residues was ruled out. The analysis performed *in situ* suggested that the combustion process, triggered directly or indirectly by human causes, concerned elements of residual carpentry used for the construction of huts and abandoned in situ. The combustion process also preserved the traces of interaction between the wood and biotic agents (parasites), which happened during the phase of use and/or abandonment of the structures. The burnt remains had clear traces of damage by borer insects (Figure 3).

Figure 3. Charcoal with insect burrow from Supersano

In the wood samples from three wrecks discovered near Torre Chianca it was observed that the use of the wood for carpentry and the subsequent interaction with seawater caused decay processes of both the biotic and abiotic type. Specifically, the macroscopic analysis of a wood fragment from Wreck B identified attack by *Balanus*, a crustacean that appeared to have anchored itself to the surface of the wood, leaving calcareous patches but not penetrating the interior. The microscopic analysis made it possible to recognise deformation of the vessels and woody tissue due to carpentry and the absorption of salts, in the form of crystals inside the vessels, resulting from the wood becoming saturated with seawater. A fragment of Wreck C had traces of activity by *Teredo navalis*, a mollusc that digs macroscopic tunnels in the wood, covering them with calcium carbonate. This mollusc can colonise the boat before it sinks, but will not attack structures protected by sediment (Beltrame 1996). Wood colonised by *Teredo* superficially seems intact, but in reality may be completely hollow inside (Figure 4). This observation enables us to advance the hypothesis that the boat was attacked while it was still in use (Zabel and Morrell 1992).

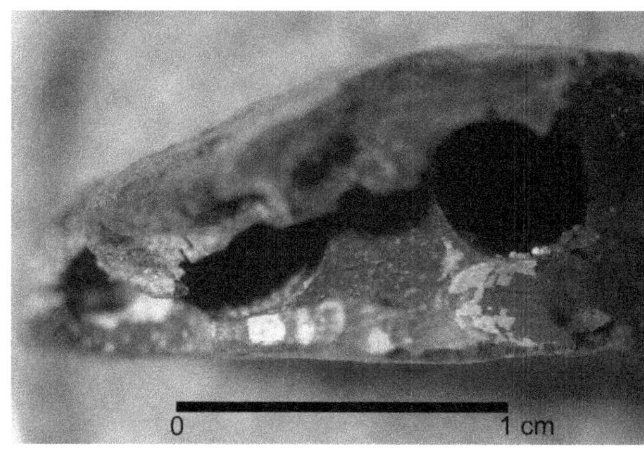


Figure 4. Fragment of waterlogged wood from shipwreck A, Torre Chianca

In the case of Minervino Murge, the study of the surface of the two swords revealed the presence of traces of wood that had been replaced by metal corrosion products. This process causes the gradual and progressive replacement of plant material with metal oxides, generating a replicate that is morphologically similar to the original but chemically different (Baker 1974; Keepax 1975).

The study of these samples has made it possible to draw some conclusions concerning the history of the artefact, from its construction to the moment of its recovery. As well as telling us about human selection of the plant species to use, it confirmed that the artefact was exposed to specific environmental conditions. Indeed, the replacement of plant tissues with metal oxides begins during the object's phase of use as a function of the wood-metal interactions, but can be accelerated or slowed down after its deposition: at this moment, in addition to variables linked to the raw materials used, such as the electro-chemical potential of the metal (Graham *et al.* 1976) and its pH, there are others that are more closely associated with the deposition environment, such as humidity, the pH of the soil, the presence of surface oxides, temperature, quantity of oxygen, the presence of salts and metallurgical factors. The samples taken, both of which belonging to the genus *Quercus*, were of different consistencies: the fragments recovered from the first sword (H4) were particularly hard and compact, whereas those of the second sword (H3) were more fragile and crumbly. Since this difference could not be attributed to variables linked to the metal from which the weapons were made, it was necessary to seek the cause in variations determined by the deposition environment (for example a slight increase in the rate of oxygen or differing degrees of humidity in the burial locations). Traces of attack by parasites, "fossilised" by the same process of replacement with metal oxides, were also recognised on both samples: tunnels excavated by insects can be seen which seem to have a mainly longitudinal orientation with exit holes in the transversal cross-section, but there is no trace of the insects themselves in either larval or adult form. These data suggest that the wood was colonised by insects of the *Lycidae* family. These coleoptera are saprophytic, i.e., attack wood that is already seasoned. The absence of insect remains suggests that the adult insects had already abandoned the artefact at the time of its deposition in the place of burial. Considering that these coleoptera have a life cycle of one year on average but depending on environmental conditions can live as long as three years, it may be assumed that the weapon had already been used before it was buried and that it did not have a purely symbolic value (Figure 5).

Figure 5. Tunnel excavated by insect in longitudinal sense from Minervino Murge

The same conservation process was seen in the fragments of woody tissue adhering to 13 iron nails discovered inside a tomb at the site of Jure Vetere (Figure 6). Analysis of the orientation of the woody tissues made it possible to identify the species used (*Abies*) and the method of assembling the board with the nails. This enabled the reconstruction of the wooden artefact, used as a coffin (Fiorentino *et al.* 2007). The replacement of woody tissues with metal oxides appears to be contemporaneous with the formation of the deposit, made up of the coffin and its contents, and thus previous to the post-depositional actions that altered its composition.

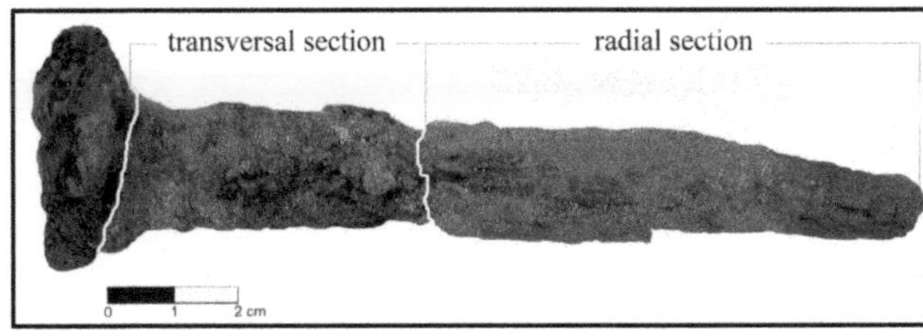


Figure 6. Nail with wood replaced by metal oxide from Jure Vetere

## CONCLUSION

The analysis of the cases treated here points to the possibility of reconstructing the uses of wood with reference to taphonomic characteristics. The anisotropic nature of wood gives it a sort of "anatomic memory" that conserves alterations caused by biotic and abiotic agents. However, this cannot constitute the only criterion in that only the context, functional and depositional, can provide information useful for generating interpretative hypotheses. The analysis of the state of preservation of the remains can provide information on the deposition environment and the preservation process, but the reconstruction of the use of the wood is possible only by comparing the traces of biotic and abiotic agents with contextual data obtained at the time of sampling and recovery of the remains.

The contextual-processual approach represents a useful tool for decoding the various phases of life of the woody tissue, including its selection, its possible use in the manufacture of objects, and the transformations brought about by use and the process by which the archaeological deposit was formed.

### References

BAKER, A.J. 1974. *Degradation of wood by products of metal corrosion.* Madison, Wisconsin, U.S. Department of Agriculture, Forest Products Laboratory.

BELTRAME, C. 1996. Archaeological evidence of the foremast on ancient sailing ships. *The International Journal of Nautical Archaeology* 25-2, 135-139.

BUTZER, K.W. 1984. *Archaeology as human ecology: method and theory for a contextual approach.* Cambridge, University Press.

D'ORONZO, C. and FIORENTINO, G. 2007. L'archeobotanica sperimentale e il combustibile legnoso spiaggiato. *L'archeologo Subacqueo* 39, 14-15.

D'ORONZO, C. and FIORENTINO, G. 2008. Le analisi archeobotaniche nel sito dell'età del Bronzo di Oratino (CB) Loc. La Rocca: implicazioni paleoeconomiche, paleoecologiche e modalità di funzionamento delle strutture pirotecniche. *Atti del "28° Convegno sulla Preistoria, Protostoria e Storia della Daunia"*, 449-473.

D'ORONZO, C. 2008. *L'Archeobotanica sperimentale per la comprensione dei processi formativi dei depositi archeologici alterati termicamente: strutture e combustibile.* Unpublished thesis, Università del Salento.

FIORENTINO, G. 2008. Analisi archeobotaniche: ricostruzione del paesaggio e modalità d'uso delle risorse vegetali. *Archeologia Medievale* XXXV, 234-236.

FIORENTINO, G., COLAIANNI, G., NOVELLIS, D. 2007. Analisi archeobotaniche al Monastero medievale di Jure Vetere: ricostruzione ambientale e uso delle risorse vegetali. In C.D. Fonseca, D. Roubis, F. Sogliani, (eds.), *Jure Vetere. Ricerche archeologiche nella prima fondazione monastica di Giacchino da Fiore (indagini 2001-2005)*, 243-267. Catanzaro, Rubbettino Editore.

FIORENTINO, G. and D'ORONZO C. 2010. Archaeobotanical and experimental approach to identify 245 fire succession in hearth structures of Apollo sanctuary at Hierapolis (Turkey). In Taphononomie des résidus organiques brûlés et des structures de combustion en milieu archéologique, Théry-Parisot I., Chabal L. & Costamagno S. (eds). Actes de la table ronde, Valbonne, 27-29 mai 2008. P@lethnologie, 2: 59-68.

FIORENTINO, G. and STELLATI, A. 2008. Compatibility of archaeobotanical analysis and restoring operations on metal objects. Workshop "Tailored strategies for conservation on the metallic artefacts in the Mediterranean Basin", Torino 16 September 2008, Unpublished.

GIORDANO, G. 1981. Tecnologia del legno I. La materia prima (2nd edition).Torino, UTET.

GRAHAM, R.D., WILSON, M.M. and OTENG-AMOAKO, A. 1976. Wood-Metal Corrosion: An Annotated Survey. *Research Bulletin* 21. Corvallis, Oregon: Forest Research Laboratory. School of Forestry, Oregon State University.

JONSELL, M., WESLIEN, J. 2003. Felled or standing retoined wood, it makes a difference for saproxylic beetles. *Forest Ecology and Management* 175, 425-435.

KEEPAX, C. 1975. Scanning Electron Microscopy of Wood Replaced by Iron Corrosion Products. *Journal of Archaeological Science* 2, 145-150.

LOFERSKI, R.J. 1999/2001. Technologies for Wood Preservation in Historic Preservation. *Archives and Museum Informatics* 13, 273-290.

MARTELLA, D. and FIORENTINO, G. 2007. Analisi archeobotanica preliminare di un carico a Torre Chianca (LE). *L'archeologo Subacqueo* 39, 16.

MINNIS, P.E. 1981. Seeds in archaeological sites; sources and some interpretative problems. *American Antiquity* 46/1, 143-152.

OEVERINGA, P., MATTHEWS, B.J., CRAGG, S.M., PITMAN, A.J. 2001. Invertebrate biodeterioration of marine timbers and sea level along the coastlines of England and Wales. *International Biodeterioration & Biodegradation* 47, 175-181.

OSONO, T. 2007. Ecology of ligninolytic fungi associated with leaf litter decomposition. *Ecological Research* 22, 955-974.

ROBERTS, N. 2000. *The Holocene: An Environmental History*. Oxford, Blackwell Publishing.

SCHWARZE, F.W.M.R., 2007. Wood decay under the microscope. *Fungal Biology Reviews* 21, 133-170.

SCOTT, A.C., CRIPPS, J.A., COLLINSON, M.E., NICHOLS, G.J. 2000. The taphonomy of charcoal following a recent heathland fire and some implications for the interpretation of fossil charcoal deposits. *Palaeogeography, Palaeoclimatology, Palaeoecology* 164, 1-31.

VERNET, J.L. 1997. *L'homme et la Forêt méditerranéenne de la Préhistoire à nos jours*. Paris, Errance.

ZABEL, R.A., MORRELL, J.J. 1992. Wood Deterioration Agents. In R.A. Zabel and J.J. Morrell (eds.), *Wood Microbiology, Decay and Its Prevention*, 21-51. San Diego, Academic Press.

# IRON METALLURGY IN THE DOGON COUNTRY (MALI, WEST AFRICA) – "DEFORESTATION" OR SUSTAINABLE USE?

Barbara EICHHORN
Institute of Archaeological Sciences, Department of Pre- and Protohistory, Archaeology and Archaeobotany of Africa, Goethe University, Grüneburgplatz 1, D-60323 Frankfurt, Germany
b.eichhorn@em.uni-frankfurt.de

Caroline ROBION-BRUNNER
Department of Anthropology, University of Geneva, Rue Gustace Revilliod 12, Case postale CH-1211 Genève 4, Switzerland

Vincent SERNEELS and Sébastien PERRET
Earth Sciences – Archaeometry, Department of Geosciences, University of Fribourg, Chemin du Musée 6, 1700 Fribourg, Switzerland

**Abstract**: *Anthracological assemblages originating from slag heaps attributed to different metallurgical traditions in the Dogon Country (Mali, West Africa) evidence the predominant use of charcoal made from wood with high burning value for iron bloomery. In spite of this, at most sites, there is no indication of exceptional selectivity for single species and the Sahelo-Sudanian woody vegetation of the area is reflected clearly in the charcoal spectra. Sites associated with the Fiko Tradition on the western margin of the Dogon Plateau are characterised by the presence of enormous metallurgical waste dumps indicating surplus iron production. Charcoal analyses at these long-term sites were carried out to reconstruct possible vegetation changes during the period of occupation. Due to the obvious mass production and changes in the charcoal record, the ecological sustainability of wood exploitation for this technology came into question, just like in other West African iron production centres before. In order to take both exploitation and natural regeneration into account, raw mass balance estimates and literature-derived data on woody biomass reproduction were combined with a timeframe based on radiocarbon dating series and oral traditions. This paper critically discusses the attempt to quantify wood use and reproduction and tries to extrapolate the results obtained in the Dogon Country to other major West African iron production sites.*

**Keywords**: *West Africa, deforestation, wood density, burning value, charcoal, metallurgy, selection, quantification, sustainability*

## INTRODUCTION

### West African metallurgy and the myth of "deforestation"

Sub-Saharan West African iron metallurgy has often been suspected to have triggered massive ecological deterioration, caused by an over-exploitation of woody resources for fuel used in the iron bloomery process (Haaland 1980, 1985; Shaw 1981; Goucher 1981, 1986; De Barros 1986; Herbert and Goucher 1987; Celis 1991; Kense and Okoro 1993; Okafor 1993; Hahn 1997; Thompson and Young 1999; Pelzer *et al.* 2003).

In this context, quantitative estimates of the metallurgical waste of several major iron production centres were established in order to draw conclusions on the amount of iron produced or, indirectly, on related wood consumption (Haaland 1980, 1985; Goucher 1981; De Barros 1986). However, as Haaland (1985) stated, due to a lack of data concerning the amount of wood available in the vegetation it was not possible to evaluate if wood off-take was above the threshold of ecological balance. For the Dapaa region in Togo, Goucher (1981) integrated data on savanna wood yield but could not sufficiently take the regenerative capacity of the woodlands into account. Despite these uncertainties, deforestation, desertification or even subsequent internal political weakening of West African empires as direct consequences of wood exploitation related to intensive metallurgy were stated, and a depletion of energy supply due to exhausted woody resources was considered one of the reasons for the decline of West African metallurgy at the beginning of the twentieth century (Haaland 1980, 1985; Goucher 1981). However, a sustainable use of woody resources is feasible, provided that yearly consumption does not exceed the annual woody biomass reproduction of a biocenosis (Von Maltitz and Scholes 1995). West African savannas are highly reproductive systems and even fields and fallows in traditional agro-forestry systems may reproduce a considerable amount of wood (Clément 1982; Ohler 1985; Breman and Kessler 1995; Nouvellet *et al.* 2003). Therefore, reasonable conclusions on the ecological consequences of wood use for iron metallurgy must take both factors – consumption and reproduction – into account (Eichhorn submitted; Eichhorn *et al.*, submitted).

### The question of metallurgy and deforestation in Central Europe and Great Britain

In the past few decades, a direct relation between prehistoric metallurgy and deforestation in Central Europe and Great Britain has been progressively doubted. Several palaeoecologists (Mighall and Chambers 1993;

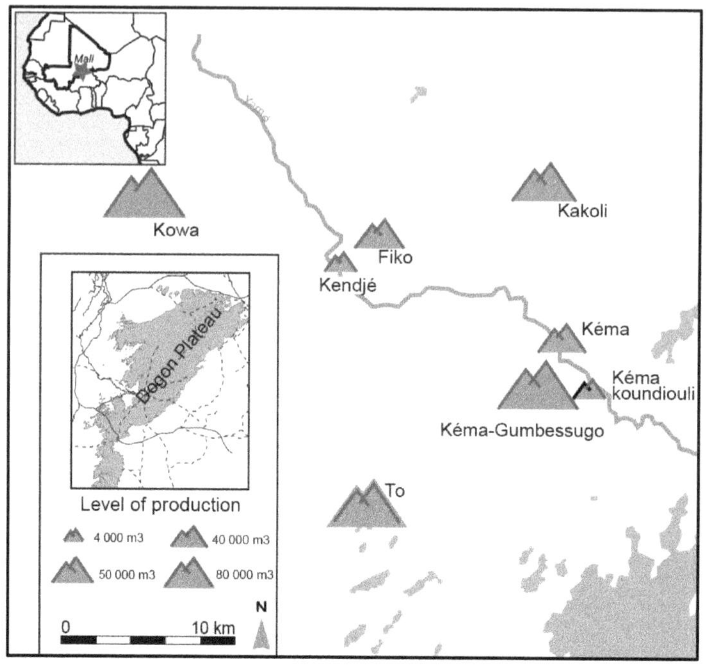


Figure 1. Map of the Fiko Tradition area, indicating the location of sites and the amount of metallurgical waste

Rösch 1994; Dörfler 1995, 2000; Dörfler and Wiethold 2000; Ludemann 2010) have put forward the opinion that the impact of wood procurement for pre-industrial metallurgy on vegetation remained negligible or at least hardly detectable. For northern Germany, a quantitative model approach combined with palaeo- and archaeobotanical methods was apt to disprove significant effects of – small scale – late Roman iron smelting activities on forest vegetation (Dörfler 1995, 2000; Dörfler and Wiethold 2000). Rösch (1994) provided, with the means of model calculations evidence that the amount of wood consumed by prehistoric societies (technical activities such as iron smelting included) in south-western Germany was below the reproduction rate, probably even during the Middle Ages when forest cover was drastically reduced due to agricultural intensification. Furthermore, Gale (2003), Cowgill (2003) and Schmidl *et al.* (2005) posited that the application of simple forest management strategies like controlled coppicing allowed for woody biomass regeneration. Even wood exploitation for intensive postmedieval charcoal production did not necessarily cause severe vegetation changes, notably in remote areas where other anthropogenic pressure on the woody vegetation was low (Ludemann 2010).

However, where direct quantitative approaches are impossible, it remains difficult to prove or disprove a possible link between metallurgical activities and detected vegetation changes, even in Europe, where pollen preserving sediments yielding palaeoecological information are markedly more common than in arid and semiarid West Africa. Once presumably anthropogenic vegetation changes are evident, it is difficult to separate the palaeobotanical signals of agricultural and metallurgical activities (Thorndycraft *et al.* 2003; Schmidl *et al.* 2005). In peat bogs, this need can partly be met by geochemical analysis based on lead isotopic geochemistry. An increase in so-called anthropogenic lead in peat cores contemporaneous to distinct changes in the pollen spectra, which point to deforestation or changes in woody species composition, may indicate direct ecological influence of palaeometallurgical activities. Anthropogenic pollen increase without any connection to lead concentration and the corresponding lead isotope composition suggests instead that agro-pastoral activities were the cause of vegetation change at a site in south-Central France at certain intervals (Jouffroy-Bapicot *et al.* 2007).

Similar systematic palaeoecological approaches which might help to prove or disprove the often assumed degradation through wood exploitation for iron metallurgy have so far not been applied to West African iron smelting sites. In a multidisciplinary approach integrating archaeological, ethnohistorical and charcoal analytical data, we have tried since 2006 to document woody species use for iron bloomery, to establish anthracological sequences and to evaluate the direct ecological consequences of wood exploitation for iron metallurgy in a large-scale West African iron production district, the Fiko Tradition on the western margins of the Dogon (= Bandiagara) Plateau in Mali (Huysecom *et al.* 2009; Robion-Brunner 2010; Robion-Brunner *et al.* submitted; Eichhorn submitted; Eichhorn and Neumann submitted; Eichhorn *et al.*, submitted). Possible ecological consequences related to atmospheric and soil pollution (cf. Jouffroy-Bapicot *et al.* 2007) have so far not been in the focus of our studies.

## ENVIRONMENTAL SETTINGS

The Fiko Tradition sites (Figure 1) are actually located at the transition between the Sahelian and the Sudanian

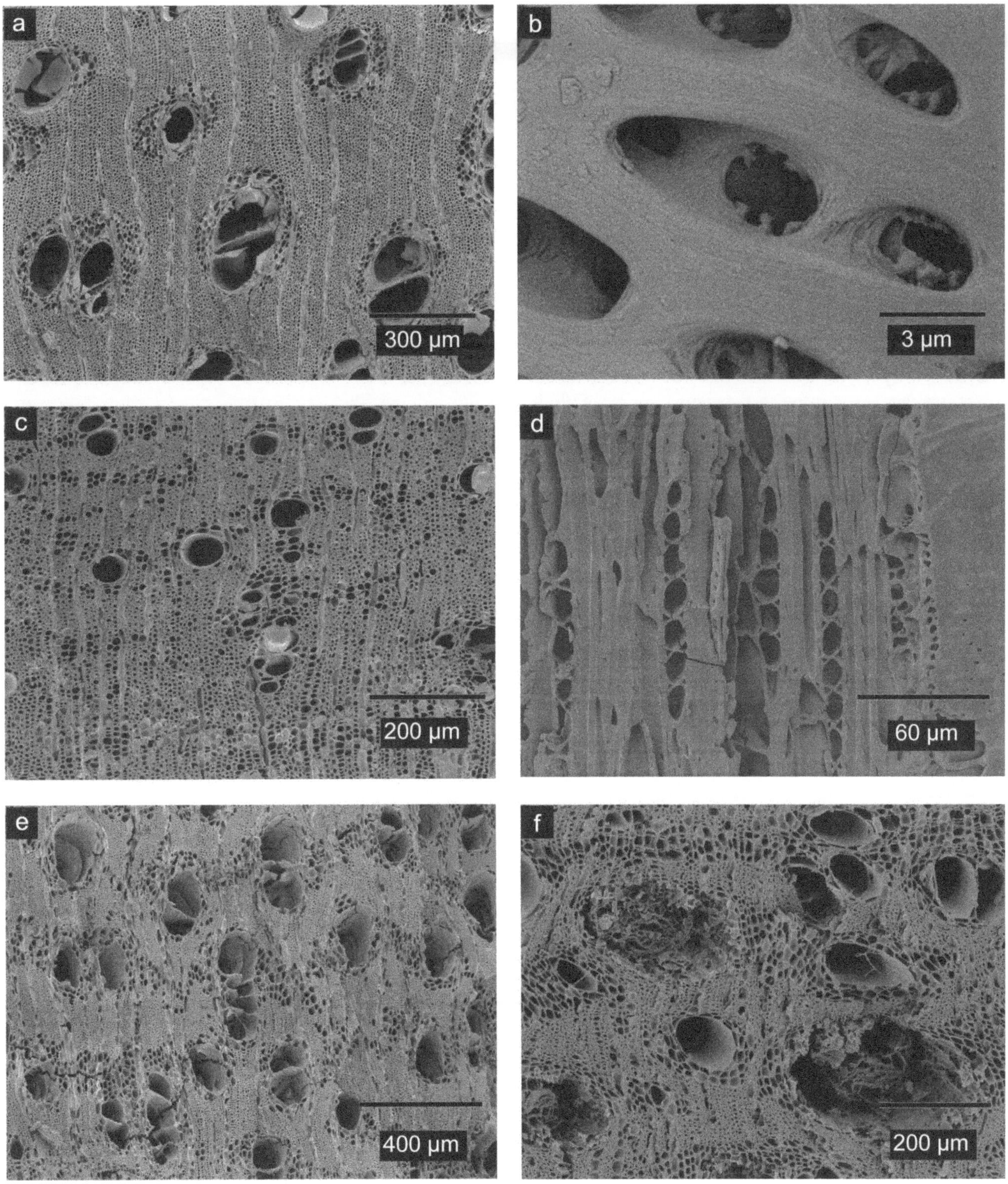


Figure 2. SEM micrographs of wood charcoal retrieved from Fiko Traditon sites (a, b *Prosopis africana*, transversal plane showing vessel groups with vasicentric parenchyma and narrow rays, tangential plane showing vestured intervessel pits; c, d *Pterocarpus* cf. *lucens*, transversal plane showing diffuse-in-aggregates and narrowly-banded parenchyma, tangential plane showing uniseriate rays, traces of calcium oxalate crystals and storied structure; e *Terminalia* sp., transversal view showing abundant paratracheal and apotracheal diffuse parenchyma; f *Combretum glutinosum*, transversal view showing included phloem, single vessels and parenchyma bands

phytogeographic zones. The climate is strictly seasonal with a rainy season of three and a half months and an average annual rainfall of about 500 mm (Nouaceur 2001). The research area lies at the limit of the Sudanian *Adansonia digitata-Bombax costatum* savannas and the Sahelian *Combretum glutinosum* savannas (Granier 2001). According to Jaeger and Winkoun (1962), the natural vegetation of favourable sites on the Dogon

Plateau is typical of a diverse Sudanian savanna characterised by *Daniellia oliveri* in association with *Vitellaria paradoxa*, *Detarium* sp., *Parkia biglobosa, Terminalia macroptera, Khaya senegalensis, Vitex doniana, Prosopis africana, Entada africana* and shrub species like *Combretum micranthum, Ozoroa insignis* and *Guiera senegalensis*. In reality, a cultural landscape covers the Dogon Plateau and the adjacent plains with extensively cultivated park savannas with few selected useful tree species. Besides these mainly fruit-bearing or soil-fertilizing trees typical fallow species able to re-sprout from suckers after cutting dominate the woody vegetation. Savanna or potential forest species like *Pterocarpus erinaceus*, *Anogeissus leiocarpus*, *Diospyros mespiliformis* or *Gyrocarpus americanus* persist mainly on steep slopes, in rock crevices, ravines and gallery forest patches. One of the species which according to the anthracological investigations was used intensively for the iron bloomery process, *Prosopis africana* (Figure 2 a, b), has almost vanished from the vicinity of the sites. Today the gallery forest of the Yamé River which runs in close proximity to the Fiko Tradition sites is floristically impoverished and fragmented due to Late and Terminal Holocene climatic deterioration and growing anthropogenic impact. However, long periods of the Holocene, as evidenced by archaeobotanical investigations at the nearby Ounjougou site complex, were characterised by the extrazonal occurrence of Sudano-Guinean species in the Yamé River's gallery forest and probably also in the surrounding savannas (Huysecom *et al.* 2004a; Ozainne *et al.* 2009; Lespez *et al.* 2011; Eichhorn and Neumann in press). Floristically diverse ravine forests with Guinean and Sudano-Guinean elements, otherwise resticted to areas of higher rainfall to the south, still persist at the escarpment of the Dogon Plateau (Jaeger and Winkoun 1962; Eichhorn and Neumann in press).

## SITE CHARACTERISTICS

Any general considerations made in the following about wood consumption and sustainability refer to the whole Fiko Tradition area, while the discussion on detailed charcoal analytical results concerns the site complexes of Fiko and Kéma. Fiko Tradition sites are distributed over an area of about 450 km$^2$ at the western margin of the Dogon Plateau and represent eight iron smelting networks on both sides of the Yamé River (Figure 1). Archaeological sites are characterised by large dumps of metallurgical waste in the form of closed craters often several metres high. On the basis of intensive archaeological prospecting and topographical mappings the total amount of metallurgical waste of all Fiko Tradition sites has been estimated at between 225,000 m$^3$ and 300,000 m$^3$ (Robion-Brunner 2010). At the site complex of Kéma-Gumbessugo (part of the Kéma complex), a number of working areas around the abandoned village Gumbessugo have been identified, each one made up of several metallurgical waste heaps close to one another. The presumably oldest preserved heaps (sector 8) are situated on the sandstone hill and nestle the closest to the abandoned village, while sectors 5, 6 and 7 are located on different terraces. These most recent heaps (sectors 1, 2 and 3) are situated on the plain, but directly in front of the sandstone hill (on its western side). This chronology scheme was initially deduced from site topography and distance to the village. According to oral traditions, Gumbessugo was raided by warriors from the Ségou Kingdom. Therefore the abandonment of the village and of its iron reduction sites has tentatively been dated to the 18th century. For the following, oral traditions convey a local history of displacement, the founding of Kéma Koundiouli on the other side of the Yamé River and, later, the present village of Kéma, both with corresponding iron reduction workshops (Huysecom *et al.* 2009). At the site of Fiko (Huysecom *et al.* 2004b, 2005, 2009) the largest and oldest slag heap (heap 1) is situated on the sandstone terraces while later iron reduction continued in the plain until the last century. In order to establish chronology firmly a high number of radiocarbon dates have been obtained at Fiko tradition sites (24 dates, Figure 3). A single date from Fiko indicates that metallurgical activities date as far back as circa 600 AD, followed by one date around 800 AD at the site of Kowa. The assemblage of all other dates indicates intensive and probably continuous production during the past millennium.

## METHODS

### Charcoal sampling

Charcoal samples were retrieved from the slag heaps during archaeological excavation respecting sediment layer limits (Fiko) or in artificial spits during sondage excavations narrowing below the surface in order to avoid the collapse of loose material (Kéma site complex). Sizes of these spits varied, according to site conditions, between 2 m x 3 m and 2 m x 5 m. All dispersed charcoal fragments of 2.5 mm and above were collected by careful dry-sieving and subsequent handpicking. Before sieving, all material was collected in buckets and large slag and furnace wall pieces removed in order to avoid unnecessary fragmentation of the charcoal during the sieving process. Wet-sieving had to be avoided as the dry material tended to break as a result of contact with water. At the Kéma sites several spongy slag blocks with enclosed wood charcoal fragments were found (fragment size 0,5-3 cm in diameter). These were carefully extracted by hammering and breaking the slag pieces and taken as separate samples. Dating samples were taken directly from the profiles after calibrating their position.

### Identification

The charcoal fragments were split into the three diagnostic planes – transverse, tangential and radial – and studied with a Laborlux S incident light microscope (range of magnifications: 50x-500x). For the analysis of minute features and photographic documentation, Scanning Electron Microscopy was applied. Identification followed the IAWA standards (Wheeler *et al.* 1989) while taxa

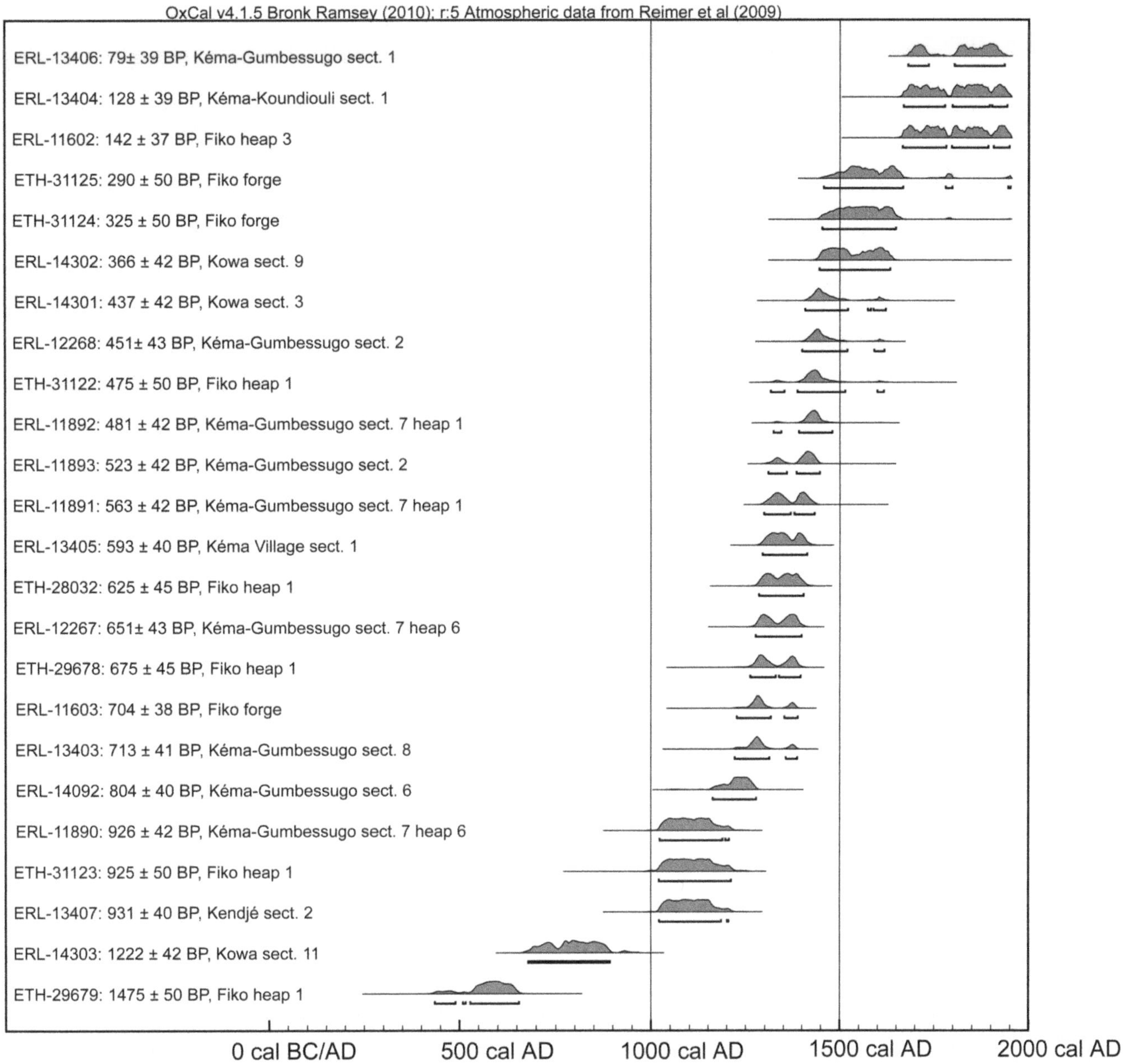


Figure 3. Radiocarbon chronology of the Fiko Tradition sites (from Eichhorn, in press)

were identified using the reference collection of modern African woods in the Frankfurt archaeobotanical laboratory, the Frankfurt DELTA/Intkey anatomical database and anatomical wood atlases and anthracological catalogues (Neumann *et al.* 2001; Eichhorn 2002; Höhn 2005).

## Detection of fuel selection

It is generally considered that the fuel value of a certain wood is correlated directly to its density or specific gravity (National Academy of Sciences 1980, 24; Raven *et al.* 2000, 711). Denser wood has a higher carbon and energy content per volume unit than lighter wood (Enquist *et al.* 1999; Svenson and Enquist 2007). In order to test if the wood used for Fiko Tradition iron bloomery was randomly collected or species were selectively chosen according to their burning quality, we drew data on wood densities of the taxa represented in our anthracological assemblages from the literature (Nygard and Elfving 2000; Armstrong *et al.* 2004; Picard *et al.* 2006). Whenever several potentially occurring species of one genus could not be differentiated anatomically during charcoal analysis (e.g. *Acacia, Terminalia*) we calculated a mean density value of the different species occurring in the area. Subsequently we developed wood density classes (in 50 kg $m^3$ steps) in which all the taxa were grouped (Figure 6). This seemed appropriate and sufficiently accurate as wood density displays considerable variability within one species (about 10% of the variations in a global wood database are attributable to intra-specific differences (Svenson and Enquist 2007)). Each charcoal fragment counted was assigned to one of the classes and their percentages calculated.

## Quantification of wood use

The only directly quantifiable factors available to us were the amount of metallurgical waste represented by the

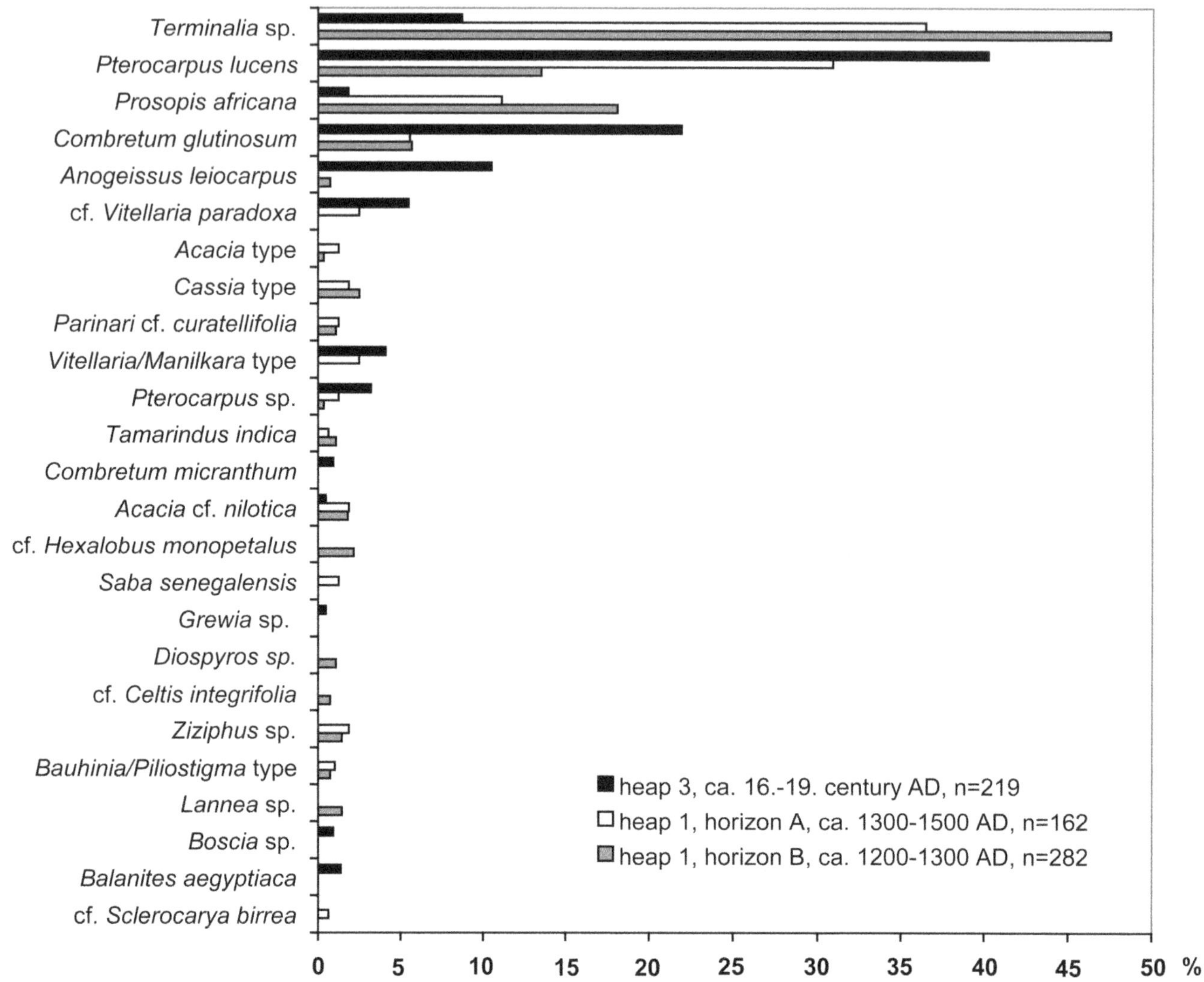


Figure 4. Charcoal analytical results from the Fiko site complex (from Eichhorn *et al.*, in press)

volume of the slag heaps, the spatial extension of the Fiko Tradition sites and a chronology of site occupation based on radiocarbon dates and oral traditions. In a model approach, not respecting spatial variations within the area, we thus tried to quantify mean annual slag output of the Fiko Tradition and to derive annual wood use and annual wood reproduction for the entire area where the Fiko Tradition sites were distributed. It is generally considered that the fuel predominantly used for West African bloomery was wood charcoal, although extremely rare exceptions involving the use of dry but un-charred wood have been observed during ethno-historical surveys (Djerma/Niger, Fouta Djalon; Celis 1991, 24). In order to quantify the woody biomass used, we first had to establish conversion rates between fresh wood with high moisture content in the vegetation, dry wood at the beginning of the charring process and the charcoal. Subsequently, the amount of slag and charcoal employed during slag formation had to be related (Dörfler 1995). Charring of tropical woods in the laboratory yielded 25-35% of the initial wood amount (Doat and Petroff 1975) while earth kiln experiments in Miombo woodland yielded a ratio slightly below 1:4 (0.23 kg charcoal per 1.0 kg wood, Chidumayo nd). We compiled an approximate mean mass ratio of 1:1.5 generalised from oven-dry and green wood from tropical tree oven-dry and green wood density values (National Academy of Sciences 1980; Wagenführ 2007). During an iron reduction process observed in Nambingué, Ivory Coast (Senufo people), Celis (1991, 25) measured with personal scales a mass relation of ore and charcoal of 1:2. This is a higher proportion of charcoal than that used by Dörfler (1995) in his model calculations but well within the range he quotes to be realistic. An ore-slag ratio of approximately 1.5:1 is adequate for the research area (Perret and Serneels 2009). This led us to estimate that the mass of charcoal used was three times the mass of slag and the mass of green wood about twenty times higher than slag mass. Data on woody biomass reproduction had to be drawn from ecological literature. The classical study of Clément (1982) relates the productivity of West African steppes, savannas and woodlands to annual precipitation and site characteristics such as soil quality and human impact. The results reveal a strict relation between annual rainfall and wood productivity. Clément subsequently developed three productivity curves – one for sites protected from fire and other severe human impact (i max), one for unprotected

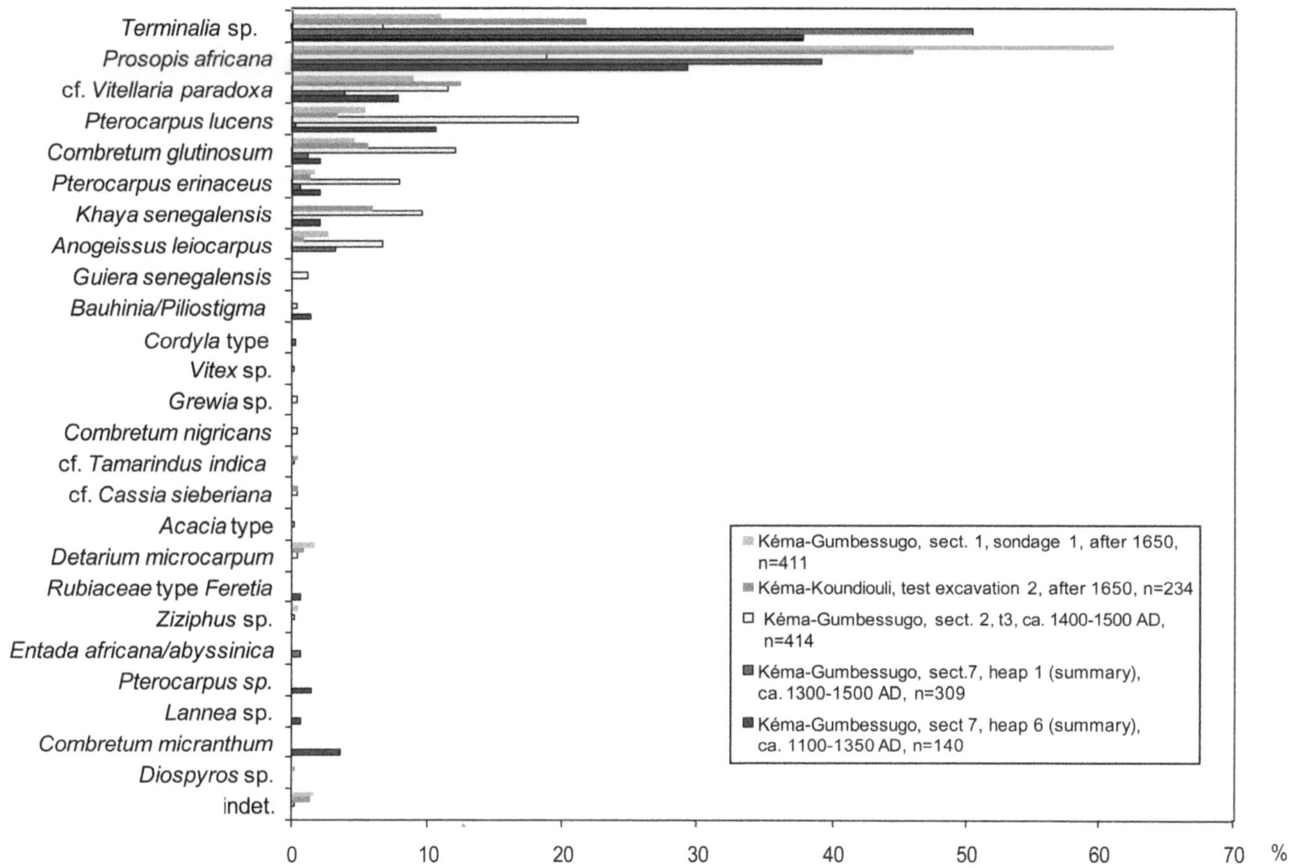


Figure 5. Charcoal analytical results from the Kéma site complex (from Eichhorn, in press)

sites and medium soil quality (i 0) and one for degraded sites on poor soils (i min, Figure 7). For the period in question, we can already presume an established agricultural system with a large area under cultivation or fallow. Shallow soils on acid sandstone surfaces covering a high percentage of the research area can be considered as slightly productive, whereas the gallery forest of the Yamé is regarded as highly productive due to its high water balance. Overall productivity for the area must thus probably have been somewhere between the values for i 0 and i min. An estimated mean basic density of 650 kg/m$^3$ compiled from ecological literature (Fearnside 1997; Nygard and Elfving 2000; Picard *et al.* 2006) was subsequently used in our calculations in order to convert wood volume into mass. Clément considers his calculations valid but only between 500 and 1600 mm annual rainfall.

In the following, we will also try to estimate the ecological impact of wood exploitation for other major Subsaharan iron production centers (Méma, Mali; Bassar, Togo; Babungo, Cameroon). For Méma we can only be sure that productivity was below the value for 500 mm and for Babungo that it was higher than the value for 1600 mm. Extrapolated values for 400 mm and 2300 mm are displayed in Figure 8.

## RESULTS

Charcoal analyses at sites belonging to different metallurgical traditions in the Dogon Country so far revealed neither extreme selectivity nor completely arbitrary fuel choice (Huysecom *et al.* 2009; Eichhorn submitted). The selection of one single species is an exception in the Dogon Country and was only observed at the Kobo site. Charcoal analyses of the site complexes of Fiko and Kéma which have already been presented in Eichhorn and Neumann (in press), Eichhorn *et al.* submitted and Eichhorn (submitted) are reproduced in Figure 4 and Figure 5.. A wide array of species has been used for fuel at the Fiko Tradition sites. Charcoal directly retrieved from slag blocks at Kéma shows, similar to the dispersed charcoal assemblages, a high diversity of taxa. In general, the charcoal analytical results reflect the typical Sudanian woody vegetation close to the limit to the Sahel. The spectra are nevertheless dominated in number by some taxa, namely *Prosopis africana* (Figure 2 a, b) and *Terminalia* species (Figure 2 e, f), *Vitellaria paradoxa* and *Pterocarpus lucens* (Figure 2 c and d) followed by members of the Combretaceae family (Figure 2 f: *Combretum glutinosum*). They are all important constituents of the surrounding savannas but several other taxa, especially many which are typical useful trees of current agroforestry parks are much lesser represented. The charcoal assemblages of both Fiko and Kéma are in all samples dominated by trees and shrubs with dense woods (Figure 6). Light-wooded species (e.g. from the families Anacardiaceae, Burseraceae, Moraceae, Verbenaceae, Sterculiaceae and Bombacaceae) are hardly represented, although quite common in the savannas and agroforestry parks of the area. In the course of occupation there are distinct changes in the Fiko sequence

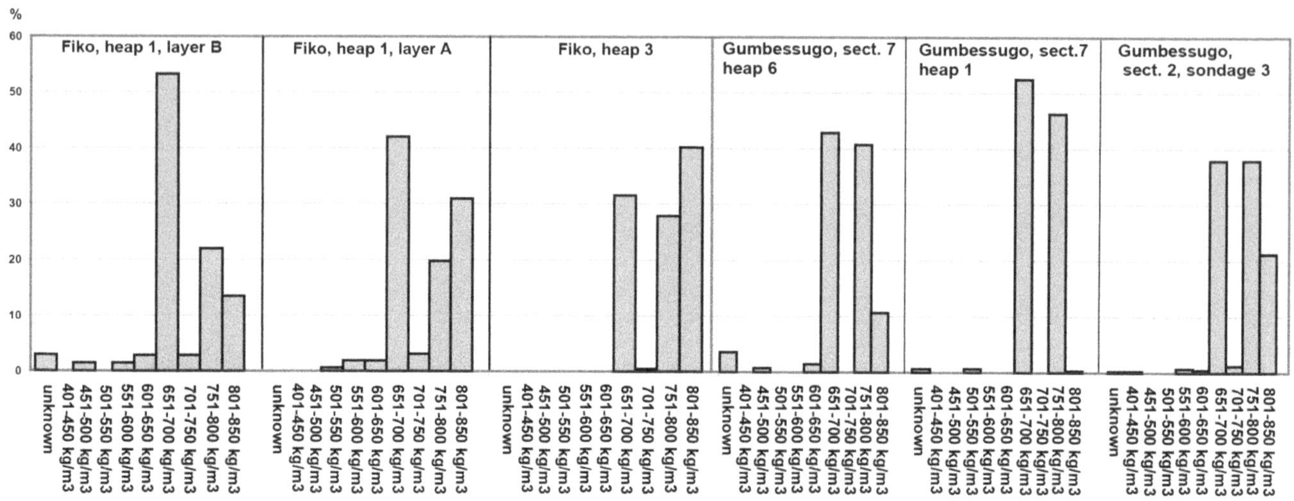


Figure 6. Percentage of charcoal fragments per wood density class for the Fiko and Kéma-Gumbessugo sites

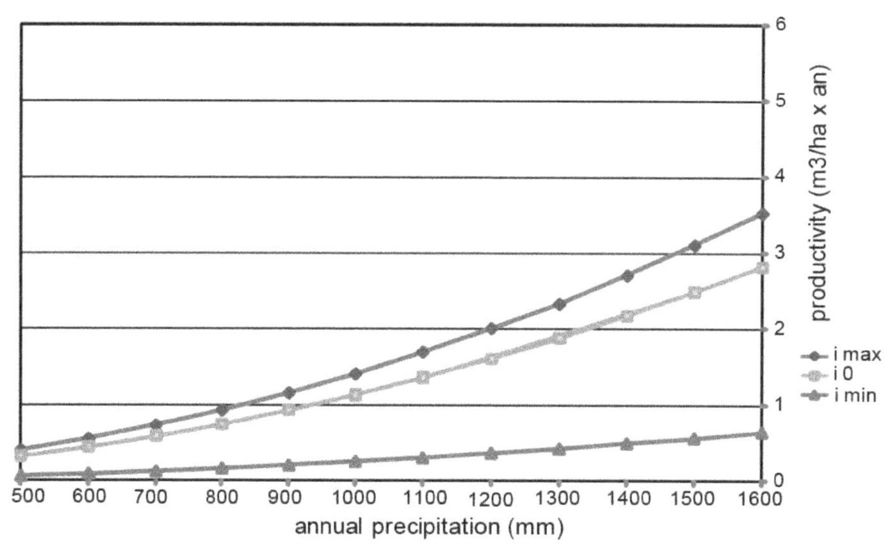


Figure 7. Annual wood reproduction in West African mixed wood-grassland vegetation according to Clément (1982)

concerning the abundance of the dominant taxa. *Terminalia* sp. and *Prosopis africana* which were preferred in the beginning decrease, whereas *Combretum glutinosum* and *Pterocarpus* cf. *lucens* increase significantly. Among the accompanying taxa, some other changes are obvious. The shea butter tree (*Vitellaria paradoxa*), a typical resident of Sudanian agoforestry parks, is increasingly used, and some Sudanian species are gradually replaced by Sahelian trees and shrubs, among them *Balanites aegyptiaca*. *Parinari* cf. *curatellifolia*, a species today only present further to the south in areas with higher precipitation, is no longer present in the assemblage. The results of charcoal analyses conducted at the Kéma complex are less unambiguous. A similar decrease of preferred taxa and the increase of species able to regenerate after cutting became evident when we compared the data available for sectors 7 and the more recent sector 2 of Kéma-Gumbessugo (Eichhorn *et al.*, submitted). However, new data from sector 1, which according to AMS-dating still operated in the past 300 years, indicate that the most preferred taxa were still sufficiently present to be highly represented in the charcoal assemblage (Eichhorn submitted). More precise dating is hampered by a radiocarbon plateau, but oral traditions indicate several movements and early displacement of iron reduction workshops to the other side of the Yamé River.

As radiocarbon dates in the second half of the first millennium AD are still rare, modelling of the land area needed to reproduce enough wood to operate reduction at the Fiko Tradition sites on a sustainable basis is shown in Figure 8 for two different lengths of occupation (1000 and 1400 years). Both smelting and bloom refining within the area would require approximately double the amount of wood-based fuel. Provided that production was more or less regular during 1400 years, without extreme peaks, and that bloom smithing took place in the area, enough wood may have been reproduced locally. Production would, albeit, have operated near the ecological threshold of affecting the standing stock if our scenario of wood production in between i 0 and i min

| site/area | Méma, Mali | Fiko Tradition | Fiko Tradition | Bassar, Togo | Babungo, Cameroon |
|---|---|---|---|---|---|
| total slag volume ($m^3$) | 45000? | 225000-300000 | 225000-300000 | 82300 | 133000 |
| production time period (years) | 150-550? | 1400 | 1000 | 575-625 | 100 |
| average annual slag output ($m^3$) | 82-300? | 160-214 | 225-300 | 132-143 | 1330 |
| mean annual precipitation (mm) | <. 400 | ca. 500 | ca. 500 | ca. 1300 | ca. 2300 |
| annual amount of wood needed for smelting (kg) | 1640-6000 | 3200-4280 | 4500-6000 | 2640-2860 | 26600 |
| annual wood productivity i min-i0-i max ($m^3$/ha) | unknown, <0,06-0,32-0,4; extrapolated: < 0,04-0,22-0,28 | 0,06-0,32-0,4 | 0,06-0,32-0,4 | 0,42-1,88-2,33 | unknown, >0,64-2,8-3,5; extrapolated: 1,3-7,5-9,4 |
| annual wood productivity i min-i0-i max (kg/ha) | unknown, < 0,04-0,21-0,26 extrapolated: < 0,026-0,14-0,18 | 0,04-0,21-0,26 | 0,04-0,21-0,26 | 0,27-1,2-1,5 | unknown, >0,42-1,75-2,3, extrapolated: 0,86-4,9-6,1 |
| surface ($km^2$) needed to yield enough mass of reproduced wood (for min value-max value slag) | imin: > 630-2307 i0: > 117-429 imax: > 91-333 | imin: 800-1070 i0: 152-204 imax: 123-165 | imin: 1125-1500 i0: 214-286 imax: 173-231 | imin: 98-106 i0: 22-24 imax: 18-19 | imin: < 415 i0: < 95 imax: < 28,3 extrapolated: imin: 309 i0: 54 imax: 44 |
| actual surface | ? | ca. 450 $km^2$ | ca. 450 $km^2$ | ca. 1600 $km^2$ | ? |

Figure 8. Table showing slag volume, the estimated number of production years, annual slag output, annual precipitation and modelled surfaces yielding sufficient wood for production on a sustainable basis (slag volume and slag annual output values retrieved from De Barros 1986, Table 5; data originally derived for Méma from Haaland 1980, for Babungo from Warnier and Fowler 1979)

according to Clément (1982) is right. We will refer to the other West African smelting centres mentioned in Figure 8 in the discussion section.

## DISCUSSION

In contrast to anthracological investigations at other West African metallurgy sites that showed the selective use of one single species, *Prosopis africana* (Neumann and Vogelsang 1996; Neumann 1999; Dupuy and Rolando 1999), we detected several hard-wooded taxa that were used by Fiko Tradition iron smelters. The diversified fuel choice corresponds well with Kiethéga (2009, 249ff.) who similarly lists a large number of species traditionally used for iron production in Burkina Faso. The same holds true for Mafa iron smelters in North Cameroon (David *et al.* 1989) and for north-eastern Nigeria (Sassoon 1964). It is evident that dense wood with a high fuel value was chosen specifically for iron smelting in the Dogon Country. Dense heavy woods emit most heat, and producing comparable heat from light-weight wood demands the use of much larger fuel volumes (Cowgill 2003). The preferred choice of hard-wooded species has also been reported by Warnier and Fowler (1979) for Cameroon and by Celis (1991, 23 ff.) for several African regions. According to Celis, the smelters stated that these species were indispensable for successful smelts. Celis did not, however, observe negative consequences for the reduction when traditional fuel was substituted by other, introduced species and concluded that the smelters' statements were driven rather by the duty to follow strong traditions than by technological reasons. It is noteworthy that *Acacia* species are hardly represented in the charcoal sequences of Fiko and Kéma even though their wood has an elevated basic density. Acacias probably played a minor role in the past vegetation of the Dogon Country, which seems also to be the case for *Balanites aegyptiaca* – only represented in the most recent slag heap at Fiko. Both taxa are today very common in the area but rare or absent in the Late and Terminal Holocene charcoal assemblages of the nearby site complex Ounjougou (Eichhorn and Neumann in press). Fluctuations in the charcoal record of Fiko indicate vegetation changes, expressed by an increase of species that are able to re-sprout from suckers and by the complete retreat of demanding species from the area. They need not inevitably be related to the ecological consequences of iron reduction. Woody vegetation may also have been altered by intensified agriculture and a desiccation trend during the period in question (Neumann *et al.* 1998; Mayor *et al.* 2005; Höhn 2007; Kahlheber and Neumann 2007). For the nearby Ounjougou reach of the Yamé River, distinct Late Holocene aridification is evident from 3800 BP onwards, broadly contemporaneous to other West African Sudano-Sahelian palaeoecological key sites (Lespez *et al.* 2011 and references therein). During the past millennium, this trend generally continued with marked oscillations (Lespez *et al.* 2011: Figure 12).

On the one hand, the disappearance of *Parinari* cf. *curatellifolia,* absent from the area today and restricted to areas with higher precipitation further to the South as well as the appearance of taxa with Sahelian affinities thriving under lower annual precipitation might be caused by the aridification trend. The increasing values of the shea butter tree *Vitellaria paradoxa*, an important agroforestry park species (Boffa 1999, Kahlheber 1999, Yaméogo *et al.* 2005), on the other hand, might simply reflect the extension of agroforestry parklands.

Our model quantification approach points to an operation of the Fiko Tradition below, but probably near the ecological threshold, especially when not only the bloomery but also bloom refining and possibly production of finished objects took place in the area. Wood for other technologies and daily uses like construction, ceramic firing or cooking has to be added. Furthermore, as hard-wooded species were preferred, it is likely that some species were more affected by wood exploitation than others and species able to re-sprout from suckers (e.g. *Combretum glutinosum*) gained importance amongst other vegetation. This is consistent with the results at Fiko where iron reduction was conducted in a small area over 1000 years, but less obvious at the Kéma site complex. It seems likely that vegetation at Fiko was transformed to a higher extent than at Kéma, where successive displacements of the reduction sites allowed for exploitation of formerly undisturbed vegetation, especially the gallery forest of the Yamé River.

Estimates of wood consumption and reproduction represent only a first model approach which does not allow for differentiating between phases of low, moderate or elevated iron production with varying amounts of wood needed for production. Furthermore, as indicated by the results at Kéma and Fiko wood for iron smelting was probably not collected all-over the areas of occupation without any differentiation. First of all, the accessibility of the fuel resource is of major importance in order to avoid unnecessary time and energy expense (Shackleton and Prins 1992; Marston 2009). Thus the radius of wood collecting most probably would have been kept as small as possible. Fruit-bearing trees in agroforestry parklands in and around the villages are generally prevented from damage through woodcutting unless they lose their productivity due to age; sacred forests are protected from wood cutting, too. In large iron production centres, the division of labour and intra-industry specialisation may occur, as in the Bassar region where a charcoal-making village existed near forested areas (De Barros 1986), thus restricting the spatial extent of wood procurement. Moreover, the attempt of quantification is related to a number of pitfalls, as due to the high number of variables that enter into the calculation and cannot be strictly calculated but only estimated, we have to admit a high cumulative error. Slag-specific gravity may have varied to a certain extent from slag heap to slag heap. At Fiko, the slag mass per volume unit has been determined at 1002 kg/m$^3$ for the early, and to 547 kg/m$^3$ for the later slag dumps (Perret and Serneels 2009). In the Cameroon Grassfields, Warnier and Fowler (1979) determined a relative value of 60% slag in a metallurgical debris heap at Bakwang, the rest being composed of different materials like furnace walls fragments, tuyères, organic material and other elements. Generally, slag amounts in our calculations are thus likely to be over-estimated. The grades of human impact on vegetation through fire, agriculture or pastoralism during the period of occupation, which according to Clément (1982) determine the reproductive capacity of each research area to a high extent, can only roughly be estimated. However, although restricted in significance due to this high cumulative error, the results indicate that generally the fuel demand for iron smelting could be covered by reproduced wood, without affecting standing stock – at least in the Bassar Region and probably at the Fiko Tradition sites. It must, however, be stressed that De Barros (1986) put forward the presence of four periods of significantly different iron production levels in the Bassar Region, and therefore periods of vegetation degradation during phases of mass production cannot be ruled out. Bloomery during the 19th century at Babungo with an extraordinarily high slag accumulation during such a short period could probably not operate based on local wood supply without affecting the standing stock, particularly as other iron producing, wood-consuming villages are present to the south at a distance of less than ten kilometres. A supply from the surrounding highlands seems possible, though. Warnier and Fowler's (1979) assignment of 133,000 m$^3$ of slag mass to the 19th century was based on oral traditions and furnace design, but this figure was not corroborated by radiocarbon dates (De Barros 1986) and might therefore contain errors. At Méma, the total slag volume values are uncertain. If an annual slag output of 300 m$^3$ is realistic, wood procurement might only be sustainable if wood was collected in a very large area. Compared to the other production centres the annual slag output is not very elevated but due to the lower annual precipitation wood regeneration is much lower than at the other sites.

## CONCLUSION

Charcoal assemblages at sites of the Fiko Tradition indicate fluctuations in woody species composition and the abundance of taxa. These vegetation changes during the past millennium can partly be assigned to aridification during that period. In addition, climatic effects were accompanied by enhanced anthropogenic impact on the vegetation, one of these factors being wood procurement for iron metallurgy. However, it seems likely that an increase of areas under cultivation affected woody vegetation more severely. The generalised view of deforestation or even desertification caused by West African iron metallurgy – as often expressed in the past in anthropogenic literature – is neither justified nor corroborated by estimates of wood use and reproduction.

## Acknowledgements

We wish to thank Katharina Neumann and Eric Huysecom for encouraging and supporting this work and thank all members of the project "Peuplement humain et paléoenvironnement en Afrique de l'Ouest" for their friendly cooperation. We are very much indebted to the Deutsche Forschungsgemeinschaft (DFG) and the Swiss National Science Foundation (FNS) for research funding. We are grateful to the "Mission Culturelle de Bandiagara" and our cooperation partners at the University of Bamako for their friendly support of this work and to the government of Mali for granting necessary visa and research permits. Many thanks are also due to Manfred Ruppel for taking the SEM photographs and to Mark Jones for language editing.

## References

ARMSTRONG, M., LELIEVRE, T., REILLY, D. and ROBERTSON, B. 2004. Evaluation of the wood quality and utilisation of plantation grown *Khaya senegalensis* (African mahagony). Conference: Prospects for high-value hardwood timber plantations in the 'dry' tropics of northern Australia, Mareeba, 19th -21st October 2004. Queensland Government, Department of Primary Industries and Fisheries, 1-16 (www.plantations2020.com.au/reports/pfng/acrobat2-3_armstrong_et_al.pdf).

BOFFA, J.-M., 1999. *Agroforestry parklands in sub-Saharan Africa.* Rome, Food and Agriculture Organization of the United Nations.

BREMAN H. and KESSLER J.-J. 1995. *Woody plants in agro-ecosystems of semi-arid regions with an emphasis on the Sahelian countries.* Berlin, Springer.

CELIS, G. 1991. *Eisenhütten in Afrika. Beschreibung eines traditionellen Handwerks in Afrika. Les fonderies Africaines du fer, un grand métier disparu.* Frankfurt, Museum für Völkerkunde.

CHIDUMAYO, E.N. nd. Inventory of wood used in charcoal production in Zambia. (http://www.worldwildlife.org/bsp/publications/africa/inventory_wood/inventory.html#Charcoal).

CLEMENT, J. 1982. Estimation des volumes et de la productivité des formations mixtes forestières et graminéennes tropicales. Données concernant les pays de l'Afrique francophone au nord de l'équateur et recommandations pour la conduite de nouvelles études. *Bois et Forêts des Tropiques* 198, 35-58.

COWGILL, J. 2003. The iron production industry and its extensive demand upon woodland resources: a case study from Creeton Quarry, Lincolnshire, in: P. Murphy and P.E.J. Wiltshire (eds.), *The Environmental Archaeology of Industry*, 48-57. Oxford, Oxbow Books.

DAVID, N., HEIMANN, R., KILLICK, D. and WAYMAN, M. 1989. Between bloomery and blast furnace: Mafa iron-smelting technology in North Cameroon. *The African Archaeological Review* 7, 183-208.

De BARROS, P. 1986. Bassar: a quantified, chronologically controlled, regional approach to a traditional iron production center in West Africa. *Africa* 56, 148-174.

DOAT, J. and PETROFF, G. 1975. La carbonisation des bois tropicaux. Essais de laboratoire et perspectives industrielles. *Bois et Forêts des Tropiques* 159, 55-72.

DÖRFLER, W. 1995. Versuch einer Modellierung des Energieflusses und des Rohstoffverbrauchs während der römisch-kaiserzeitlichen Eisenverhüttung in Joldelund, Ldkr. Nordfriesland. *Probleme der Küstenforschung im Südlichen Nordseegebiet* 23, 175-185.

DÖRFLER, W. 2000. Palynologische Untersuchungen zur Vegetations- und Landschaftsentwicklung von Joldelund, Kr. Nordfriesland, in: A. Haffner, H. Jöns and J. Reichstein (eds.), *Frühe Eisengewinnung in Joldelund, Kr. Nordfriesland. Ein Beitrag zur Siedlungs- und Technikgeschichte Schleswig-Holsteins, Teil 2,* 147-199. (Universitätsfor-schungen zur prähistorischen Archäologie 59).

DÖRFLER, W. and WIETHOLD, J. 2000. Holzkohlen aus den Herdgruben von Rennöfen und weiteren Siedlungsbefunden des spätkaiserzeitlichen Erzgewinnungs- und Siedlungsplatzes am Kammberg bei Joldelund, Kreis Nordfriesland, in: A. Haffner, H. Jöns and J. Reichstein (eds.), *Frühe Eisengewinnung in Joldelund, Kr. Nordfriesland. Ein Beitrag zur Siedlungs- und Technikgeschichte Schleswig-Holsteins, Teil 2,* 217-262. (Universitätsforschungen zur prähistorischen Archäologie 59).

DUPUY, C. and ROLANDO, C. 1999. Un four de métallurgie en stratigraphie à Koussané (Mali): Fouille, anthracologie et datations sur charbons. *Mémoires de la Société Préhistorique Française* 26, 381-385.

EICHHORN, B. 2002. *Anthrakologische Untersuchungen zur Vegetationsgeschichte des Kaokolandes, Nordwest-Namibia.* PhD thesis, University of Cologne (http://kups.ub.uni-koeln.de/volltexte/2004/1178/).

EICHHORN B. in press. Woody resource exploitation for iron metallurgy of the Fiko tradition: Implications for the environmental history of the Dogon Country, Mali/Utilisation des ressources boisées dans le cadre de la production sidérurgique des sites de Tradition Fiko: éléments nouveaux pour une reconstitution historique du paysage du pays dogon (Mali). In Proceedings of the Conference Métallurgie du Fer et Sociétés Africaines (23rd-24th April, 2010), Aix en Provence.

EICHHORN, B., ROBION-BRUNNER, C., SERNEELS, V. and PERRET, S. in press. Fuel for iron – wood exploitation for metallurgy on the Dogon Plateau, Mali. In Proceedings of the World of Iron Conference (16th – 20th February 2009: London).

EICHHORN, B. and NEUMANN, K. in press. Holocene vegetation change and land use at Ounjougou, Mali, in: D.Q Fuller and M.A. Murray (eds.), *Flora, Past Cultures and Archaeobotany in Africa.* Walnut Creek, Left Coast Press.

ENQUIST, B.J., WETS, G.B., CHARNOV, E.L. and BROWN, J.H. 1999. Allometric scaling of production and life history variation in vascular plants. *Nature* 401, 907-911.

FEARNSIDE, P.M. 1997. Wood density for estimating forest biomass in Brazilian Amazonia. *Forest Ecology and Management* 90: 59-87.

GALE, R. 2003. Wood-based industrial fuels and their environmental impact in lowland Britain, in: P. Murphy and P.E.J. Wiltshire (eds.), *The Environmental Archaeology of Industry*, 30-47. Oxford, Oxbow Books.

GOUCHER, C.L. 1981. Iron is iron until it is rust: trade and ecology in the decline of West African Iron-Smelting. *Journal of African History* 22: 179-189.

GOUCHER, C. 1986. *The iron industry of Bassar, Togo: An interdisciplinary investigation of African technological history.* Unpublished PhD-Thesis, University of Michigan, Ann Arbor.

GRANIER, C. 2001. Biogéographie. In J.C. Arnaud (ed.), *Atlas du Mali,* 22-24. Paris, Les Éditions Jeune Afrique.

HAALAND, R. 1980. Man's role in the changing habitat of Mema during the old kingdom of Ghana. *Norwegian Archaeological Review* 13, 31-46.

HAALAND, R. 1985. Iron production, its socio-cultural context and ecological implications, in: R. Haaland and P. Shinnie (eds.), *African iron working – ancient and traditional*, 50-72. Oslo, Norwegian University Press.

HAHN, H.-P. 1997. Eisentechniken in Nord-Togo: Kultur- und technikgeschichtliche Interpretationen. In R. Klein-Arendt (ed.), *Traditionelles Eisenhandwerk in Afrika*, 129-145. Köln, Heinrich-Barth-Institut (Colloquium Africanum 3).

HERBERT, E.W. and GOUCHER, C. 1987. *Resource guide for the blooms of Banjeli: Technology and gender in West African iron making.* Waterton, Documentary Educational Resources.

HÖHN, A. 2005. *Zur eisenzeitlichen Entwicklung der Kulturlandschaft im Sahel von Burkina Faso. Untersuchungen von archäologischen Holzkohlen.* PhD thesis, Goethe University, Frankfurt am Main (http://publikationen.ub.uni-frankfurt.de/volltexte/2005/2253/).

HÖHN, A. 2007. Where did all the trees go? In R. Cappers (ed.), *Fields of change. Progress in African Archaeobotany*, 35-42. Eelde, Groningen, Barkhuis Publishing.

HUYSECOM, E., OZAINNE, S., RAELI, F., BALLOUCHE, A., RASSE, M. and STOKES, S. 2004a. Ounjougou (Mali): A history of Holocene settlement at the southern edge of the Sahara. *Antiquity* 78, 579-593.

HUYSECOM, E., BALLOUCHE, A., CISSÉ, L., GALLAY, A., KONATÉ, D., MAYOR, A., NEUMANN, K., OZAINNE, S., PERRET, S., RASSE, M., ROBERT, A., ROBION, C., SANOGO, K., SERNEELS, V., SORIANO, S. and STOKES, S. 2004b. Paléoenvironnement et peuplement humain en Afrique de l'Ouest: rapport de la sixième année de recherche à Ounjougou (Mali). *Jahresbericht 2003. Fondation Suisse-Liechtenstein pour les recherches archéologogiques à l'étranger, Zürich et Vaduz,* 27-68.

HUYSECOM, E., BALLOUCHE, A., GALLAY, A., GUINDO, N., KEITA, D., KOUTI, S., LE DRÉZEN, Y., MAYOR, A., NEUMANN, K., OZAINNE, S., PERRET, S., RASSE, M., ROBION-BRUNNER, C., SCHAER, K., SERNEELS, V., SORIANO, S., STOKES, S. and TRIBOLO, C. 2005. La septième campagne de terrain et ses apports au programme interdisciplinaire Paléoenvironnement et peuplement humain en Afrique de l'Ouest. *Jahresbericht 2004. Fondation Suisse-Liechtenstein pour les recherches archéologogiques à l'étranger, Zürich et Vaduz*, 57-142.

HUYSECOM, E., OZAINNE, S., ROBION-BRUNNER, C., MAYOR, A., BALLOUCHE, A., CHAIX, L., EICHHORN, B., GUINDO, N., LE DREZEN, Y., LESPEZ, L., MEZGER, H, NEUMANN, K., PERRET, S., POUDJOUGO, M., RASSE, M., SANOGO, K, SCHNEIDER, K, SERNEELS, V., SORIANO, S., SOULIGNAC, R., TRAORÉ, B.D., and TRIBOLO, C. 2009. Nouvelles données sur le peuplement du Pays dogon: la onzième année de recherche du programme "Peuplement humain et évolution paléoclimatique en Afrique de l'Ouest". *Jahresbericht 2008. Fondation Suisse-Liechtenstein pour les recherches archéologiques à l'étranger, Zürich et Vaduz,* 71-183.

JAEGER P. and WINKOUN, D. 1962. Premier contact avec la flore et la végétation du plateau de Bandiagara. *Bulletin de l'I.F.A.N.* 24, 68-111.

JOUFFROY-BAPICOT, I., PULIDO, M., BARON, S., GALOP, D., MONNA, F., LAVOIE, M., PLOQUIN, A., PETIT, C., de BEAULIEU, J.-L. and RICHARD, H. 2007. Environmental impact of early palaeometallurgy: pollen and geochemical analysis. *Vegetation History and Archaeobotany 16*, 251-258.

NEUMANN, K., KAHLHEBER, S. and UEBEL, D. 1998. Remains of woody plants from Saouga, a medieval West African village. *Vegetation History and Archaeobotany* 7: 57-77.

KAHLHEBER, K., 1999. Indications for agroforestry. Archaeobotanical remains of crops and woody plants from medieval Saouga, Burkina Faso, in: M. Van der Veen (ed.), *The exploitation of plant resources in ancient Africa,* 89-100. New York, Kluwer Academic/Plenum.

KAHLHEBER, S. and NEUMANN, K. 2007. The development of plant cultivation in semi-arid West Africa, in T. Denham, J. Iriarte and L. Vrydaghs (eds), *Rethinking Agriculture. Archaeological and Ethnoarchaeological Perspectives*, 320-346. Walnut Creek, Left Coast Press.

KENSE, F.J. and OKORO, J.A. 1993. Changing perspectives on traditional iron production in West Africa, in T. Shaw, P. Sinclair, B. Andah and A. Okpoko (eds.), *The Archaeology of Africa: Food, Metals and Towns*, 449-458. London, Routledge.

KIETHÉGA, J.-B. 2009. *La métallurgie lourde du fer au Burkina Faso. Une technologie à l'époque précoloniale.* Paris, Éditions Karthala.

LESPEZ, L., Le DREZEN, Y., GARNIER, A., RASSE, M., EICHHORN, B., OZAINNE, S., BALLOUCHE, A., NEUMANN, K. and HUYSECOM, E. 2011. High-resolution fluvial records of Holocene environmental changes in the Sahel: the Yamé River at Ounjougou (Mali, West Africa). – *Quaternary Science Reviews,* 30: 737-756.

LUDEMANN, T. 2010. Past fuel wood exploitation and natural forest vegetation in the Black Forest, the Vosges and neighbouring regions in western Central Europe. *Palaeogeography, Palaeoclimatology, Palaeoecology* 291, 154-165.

MARSTON, J.M. 2009. Modeling wood acquisition strategies from archaeological charcoal remains. *Journal of Archaeological Science* 36, 2192-2200.

MAYOR, A., HUYSECOM, E., GALLAY, A., RASSE, M., and BALLOUCHE, A. 2005. Population dynamics and paleoclimate over the past 3000 years in the Dogon country, Mali. *Journal of Anthropological Archaeology* 24, 25-61.

MIGHALL, T.M., CHAMBERS, F.M. 1993. The environmental impact of prehistoric mining at Copa Hill, Cwmystwyth, Wales. *The Holocene* 3, 260-264.

National Academy of Sciences 1980. *Firewood crops. Shrub and tree species for energy production.* Washington, US National Research Council.

NEUMANN, K. 1999. Charcoal from West African savanna sites: Questions of identification and interpretation. In M. van der Veen (ed.), *The exploitation of plant resources in ancient Africa*, 205-219. New York, Kluwer Academic/Plenum.

NEUMANN, K. and VOGELSANG, R. 1996. Paléoenvironnement et préhistoire au Sahel du Burkina Faso. *Berichte des Sonderforschungsbereichs 268* 7, 177-186.

NEUMANN, K., KAHLHEBER, S. and UEBEL, D. 1998. Remains of woody plants from Saouga, a medieval West African village. *Vegetation History and Archaeobotany* 7: 57-77.

NEUMANN K., SCHOCH W., DÉTIENNE P. and SCHWEINGRUBER F.H.. 2001. *Hölzer der Sahara und des Sahel. Woods of the Sahara and the Sahel. Bois du Sahara et du Sahel.* Bern, Paul Haupt.

NOUACEUR, Z. 2001. Climat. In Arnaud, J.C. (ed.), *Atlas du Mali*, 16-19. Paris, Jeune Afrique.

NOUVELLET, Y., SYLLA, M.L. and KASSAMBARA, A. 2003. La production de bois d'énergie dans les jachères au Mali. *Bois et Forêts des Tropiques*, 276, 1-15.

NYGARD, R. and ELFVING, B. 2000. Stem basic density and bark proportion of 45 woody species in young savanna coppice forests in Burkina Faso. *Annals of Forest Science* 57, 143-153.

OHLER, F.M.J. 1985. The fuelwood production of wooded savanna fallows in the Sudan zone of Mali. *Agroforestry Systems* 3, 15-23.

OKAFOR, E.E. 1993. New evidence on early iron-smelting from southeastern Nigeria, in T. Shaw, P. Sinclair, B. Andah and A. Okpoko (eds.), *The Archaeology of Africa: Food, Metals and Towns*, 432–48. London, Routledge.

OZAINNE, S., LESPEZ, L., Le DREZEN, Y., EICHHORN, B., NEUMANN, K. and HUYSECOM, E. 2009. Developing a chronology integrating archaeological and environmental data from different contexts. *Radiocarbon* 51, 457-470.

PELZER, C., MÜLLER, J. and ALBERT, K.-D. 2003. Die Nomadisierung des Sahel – Siedlungsgeschichte, Klima und Vegetation in der Sahelzone von Burkina Faso in historischer Zeit, in: K.-D. Albert, D. Löhr and K. Neumann (eds.), *Mensch und Natur in West-Afrika*, 256-288. Weinheim, Wiley VCH.

PERRET, S. and SERNEELS, V. 2009. Technological characterisation and quantification of a large-scale iron smelting site in Fiko (Dogon Plateau, Mali), in: J.-F. Moreau, R. Auger, J. Chabot and A. Herzog (eds.), *Proceedings/Actes ISA 2006, 36th International Symposium on Archaeometry (Quebec City, Canada, 2-6 May 2006)*, 453-463 (Cahiers d'Archéologie du CELAT 25, Série Archéométrie 7).

PICARD, N., OUATTARA, S., DIARISSO, D., BALLO, M. and GAUTIER, D. 2006. Defining units for savanna management in Sudano-Sahelian areas. *Forest Ecology and Management* 236, 403-411.

RAVEN, P.H., EVERT, R.F. and EICHHORN, S.E. 2000. *Biologie der Pflanzen.* Berlin and New York, De Gruyter.

ROBION-BRUNNER, C. 2010. *Forgerons et sidérurgie en pays dogon. Vers une histoire de la production du fer sur le plateau de Bandiagara (Mali) durant les empires précoloniaux* (Journal of African Archaeology Monograph Series 3, Peuplement humain et paléoenvironnement en Afrique de l'Ouest, Série Monographique 1). Frankfurt, Africa Magna Verlag.

ROBION-BRUNNER C., SERNEELS, V. and PERRET, S. submitted. Variability in iron smelting practices:

confrontation of technical, cultural and economic criteria to explain the metallurgical diversity in the Dogon area (Mali). In Proceedings of the World of Iron Conference (16th – 20th February 2009: London).

RÖSCH, M. 1994. Gedanken zur Auswirkung prähistorischer Holznutzung auf Wälder und Pollendiagramme. *Dissertationes Botanicae* 234 (Festschrift G. Lang), 447-471.

SASSOON, H. 1964. Iron-smelting in the hill village of Sukur, north-eastern Nigeria. *Man* 64, 174-178.

SCHMIDL, A., KOFLER, W., OEGGL-WAHLMÜLLER, N. and OEGGL, K. 2005. Land use in the eastern Alps during the Bronze Age – an archaeobotanical case study of a hilltop settlement in the Montafon (western Austria). *Archaeometry* 47, 455-470.

SHACKLETON, C.M. and PRINS, F. 1992. Charcoal analysis and the "Principle of Least Effort" – A conceptual model. *Journal of Archaeological Science* 19, 631-637.

SHAW, T. 1981. The Nok sculptures of Nigeria. *Scientific American* 244, 154-166.

SVENSON, N.G. and ENQUIST, B.J. 2007. Ecological and evolutionary determinants of a key plant functional trait: wood density and its community-wide variation across latitude and elevation. *American Journal of Botany* 94, 451-459.

THOMPSON, G. and YOUNG, R. 1999. Fuels for the furnace. Recent and prehistoric ironworking in Uganda and beyond. In M. Van der Veen (ed.), *The exploitation of plant resources in ancient Africa*, 221-239. New York, Kluwer Academic/Plenum.

THORNDYCRAFT, V., PIRRIE, D. and BROWN, A.G. 2003: An environmental approach to the archaeology of tin mining on Dartmoor, In P. Murphy and P.E.J. Wiltshire (eds.), *The Environmental Archaeology of Industry*, 20-28. Oxbow Books, Oxford.

Von MALTITZ, G.P. and SCHOLES, R.J. 1995. The burning of fuelwood in South Africa: When is it sustainable? *Environmental Monitoring and Assessment* 38, 243-251.

WAGENFÜHR, R. 2007. *Holzatlas*. Leipzig, Fachbuchverlag.

WARNIER, J.P. and FOWLER, I. 1979. A nineteenth century Ruhr in Central Africa. *Africa* 49, 329-351.

WHEELER, E.A., BAAS, P. and GASSON, P.E. (eds.) 1989. IAWA list of microscopic features for hardwood identification by an IAWA committee. *IAWA Bulletin New Series* 10, 219-332.

YAMÉOGO, G., YÉLÉMOU, B., and TRAORÉ, D. 2005. Pratique et perception paysannes dans la création de parc agroforestier dans le terroir de Vipalogo (Burkina Faso). *Biotechnology, Agronomy, Society and Environment*, 9, 241-8.

# RECONSTRUCTION OF THE VEGETATION AND USE OF WOOD BY ANTHRACOLOGICAL ANALYSIS OF HOLOCENE MONTANE AND SUBALPINE ALTITUDE SITES OF THE EASTERN PYRENEES

Itxaso EUBA and Josep Maria PALET MARTINEZ
ICAC (Institut Català d'Arqueología Clàssica) Plaça Rovellat s/n; 43003, Tarragona, Spain
ieuba@icac.net; jpalet@icac.net

**Abstract**: *We present the results of charcoal analyses realized for 63 archaeological structures related to forest exploitation, located in two valleys at montane and subalpine altitude in the eastern Pyrenees. These structures were pastoral structures, furnaces and charcoal kilns. In the Madriu valley (Escaldes-Engordany, Andorra), 31 structures of different functionalities from 13 subalpine sites and 10 charcoal kilns situated between 1500 and 2300 m.a.s.l. near the Madriu River were analyzed. In the Vansa valley (Cadí Mountain Range, Alt Urgell, Catalonia) 22 sites from four montane and subalpine sites were analyzed. Palynological studies and radiocarbon dating have proved that these valleys were occupied since the Neolithic until Modern Age. Results of taxonomical and dendrological analysis have provided knowledge of the use of wood for different activities during the last period of the Holocene in the two valleys. Wood, mostly pine, was used as a combustible and for construction, for example for domestic fires and in artisan and industrial structures (charcoal kilns and iron first-treatment furnaces). Despite the poor taxonomical variability the results reflect high mountain vegetation with pine and Ericaceae on shadowy slopes and pine and juniper on sunny slopes. We have also identified stream bank species near the Madriu River.*

**Keywords**: *Pastoral structures, charcoal kilns, furnaces, vegetation, use of wood, High Mountain*

## INTRODUCTION

This work was integrated in a multidisciplinary and diachronic Landscape Archaeology project called 'Occupation of soil and mountain landscape forms of the eastern Pyrenees from Antiquity to the Middle Ages', directed by Dr Palet (ICAC, Tarragona) and Dr Riera (SERP-UB, Barcelona).

In this paper we present the results of charcoal analyses from different sites located in the montane and subalpine zones of two valleys in the eastern Pyrenees. The Vansa valley is situated in the Spanish pre-Pyrenees, in the Cadí Mountain Range, in a remote area with rounded peaks and a calcareous soil. The Madriu valley is located in southeast Andorra, in the municipality of Escaldes-Engordany. Andorra lies on the northern slopes of the principal axis of the eastern Pyrenees. The soil mainly consists of granitic and schistic siliceous substrates (Figure 1). In the Madriu valley we analyzed 31 structures from 13 subalpine sites (2300 – 2500 m.a.s.l.) and in the Vansa valley 22 structures from four montane and subalpine sites (1600 – 2300 m.a.s.l.) were studied. The structures identified in these sites are charcoal kilns, furnaces and pastoral structures (Palet *et al.* 2006; 2007). We also analyzed 10 charcoal kilns located between 1500 and 2300 meters of altitude that follows the GR11 of the Pyrenees. This route lies parallel to the Madriu river (Figure 2). Several radiocarbon datings were carried out which provided us with important information about the exploitation of these two valleys from the Neolithic to the Modern Period (Figure 3).

Charcoal recovered from archaeological sites is almost always of anthropic origin. Therefore anthracological analysis of archaeological sites provides us with information on the use of wood for different activities developed in the area of study and also of the vegetation of the site surroundings.

Previous charcoal-analysis studies integrated in multidisciplinary works of Landscape Archaeology and in a high mountain context have been carried out (e.g. Bonhôte and Vernet 1988; Davasse *et al.* 1997; Dubois *et al.* 1997; Durand 1998). Other anthracological studies of high mountain sites are for example Brochier *et al.* (1998), Heinz (1996), Llovera *et al.* (1997), Ros and Mora (1995), Thiébault (2005), Yáñez *et al.* (2001-2002), Zapata (2002). Studies on charcoal kiln structures are done by for example Davasse (2000), Fabre (1996) and Pèlachs (2004).

The principal aim of this study was to gather knowledge about different types of forest exploitation for different activities (herding, charcoal and resin production etc.) from the Neolithic to the Modern period. The second objective was to identify the forest vegetation in the surroundings of the different sites and in both valleys as well as to observe whether any change (such as deforestation) of these forests occurred due to these activities. As a result of the limited anthropic exploitation of these mountain landscapes, the anthracological material was very well preserved and we recovered wood (both carbonized and not carbonized) with well-conserved morphologies of utensils and structures, providing important ethnographical information.

Figure 1. Location of the Vansa Valley (Cadí Mountain Range, Alt Urgell, Spain) and the Madriu Valley (Andorra)

| Valley | Site | Structure | Type of structure | Altitude |
|---|---|---|---|---|
| Madriu valley | Sobre els orris de Setut I | 10 | Charcoal kiln | 2350 m |
| | Sobre els orris de Setut II | 15 | Charcoal kiln | 2340 m |
| | Pla de l'Ingla I | 53 | Charcoal kiln | 2185 m |
| | | 58 | Charcoal kiln | 2185 m |
| | | 59 | Enclosure | 2185 m |
| | Pla de l'Ingla II | 60 | Charcoal kiln | 2195 m |
| | Pla de l'Ingla III | 52 | Furnace | 2200 m |
| | | 102 | Charcoal kiln | 2200 m |
| | Bosc dels Estanyons | 64 | Charcoal kiln | 2250 m |
| | | 65 | Charcoal kiln | 2250 m |
| | | 66 | Charcoal kiln | 2235 m |
| | | 67 | Charcoal kiln | 2235 m |
| | | 104 | Charcoal kiln | 2180 m |
| | | 110 | Charcoal kiln | 2180 m |
| | Setut I | 1 | Charcoal kiln | 2320 m |
| | | 2 | Charcoal kiln | 2320 m |
| | Portella de Setut I | 47 | Charcoal kiln | 2335 m |
| | Carboneres de Setut | 48 | Charcoal kiln | 2290 m |
| | | 50 | Charcoal kiln | 2265 m |
| | | 129 | Charcoal kiln | 2295 m |
| | Orri del Riu dels Orris | 82 | Charcoal kiln | 2400 m |
| | Orris de Setut III | 27 | Cottage | 2300 m |
| | | 29 | *Pleta* | 2300 m |
| | | 35 | Cottage | 2300 m |
| | Pleta de les Bacives | 151 | Enclosure | 2500 m |
| | | 152 | Cottage | 2500 m |
| | | 153 | Cottage | 2500 m |

| Valley | Site | Structure | Type of structure | Altitude |
|---|---|---|---|---|
| Madriu valley | Basses de Setut | 121 | Enclosure | 2310 m |
| | | 133 | Enclosure | 2310 m |
| | | 135 | Enclosure | 2310 m |
| | | 137 | Cottage | 2310 m |
| | Altitudinal transect | C1 | Charcoal kiln | 1800 m |
| | | C2 | Charcoal kiln | 1845 m |
| | | C3 | Charcoal kiln | 1900 m |
| | | C4 | Charcoal kiln | 1900 m |
| | | C5 | Charcoal kiln | 1990 m |
| | | C6 | Charcoal kiln | 2020 m |
| | | C7 | Charcoal kiln | 2100 m |
| | | C8 | Charcoal kiln | 2100 m |
| | | C9 | Charcoal kiln | 2125 m |
| Vansa valley | Cerneres | 13 | Charcoal kiln | 1500 m |
| | | 35 | Charcoal kiln | 1605 m |
| | | 36 | Charcoal kiln | 1605 m |
| | | 37 | Charcoal kiln | 1762 m |
| | | 39 | Charcoal kiln | 1720 m |
| | | 51 | Charcoal kiln | 1775 m |
| | Goleró | 2 | Enclosure | 2020 m |
| | | 11 | Related to metallurgical activity | 2020 m |
| | | 12-1 | Related to metallurgical activity | 2020 m |
| | | 16 | Enclosure | 2020 m |
| | | 42 | Furnace | 2020 m |
| | | 43 | Furnace | 2020 m |
| | | 46 | Enclosure | 2020 m |
| | | 48 | Enclosure | 2020 m |
| | | 75 | Related to metallurgical activity | 2020 m |
| | | 79 | Related to metallurgical activity | 2020 m |
| | | 80 | Related to metallurgical activity | 2020 m |
| | Pradell | 30 | Cottage | 2025 m |
| | | 31 | Enclosure | 2025 m |
| | | 81 | Related to metallurgical activity | 2025 m |
| | Carboneres | 82 | Charcoal kiln | 2055 m |
| | | 83 | Charcoal kiln | 2055 m |
| | | 84 | Charcoal kiln | 2055 m |

Figure 2. Sites and structures with different functionalities studied in the Vansa and Madriu valleys

## METHODOLOGY

The methods used for recovering charcoal and sediment samples varied with the excavation methods and the type of structure (charcoal kilns vs other structures) (Euba 2008). From the prospected structures we collected one liter of sediment from their base. In excavated structures the sediment collection was more systematic: From charcoal kilns we gathered sediment samples every 10 cm from excavated 30 $cm^2$-surveys (two or three in each kiln) (Euba 2008). From furnaces and pastoral structures we sampled four liter from each stratigraphical unit (SU), and the remaining sediment from each SU was sieved during the excavation with 4 and 2 mm sieves. From these structures all the sediment from charcoal concentrations, such as domestic fires, was collected.

After collection, we extracted the charcoal from the samples by flotation and subsequent water sieving over a column of 4 and 2 mm sieves. This was done with

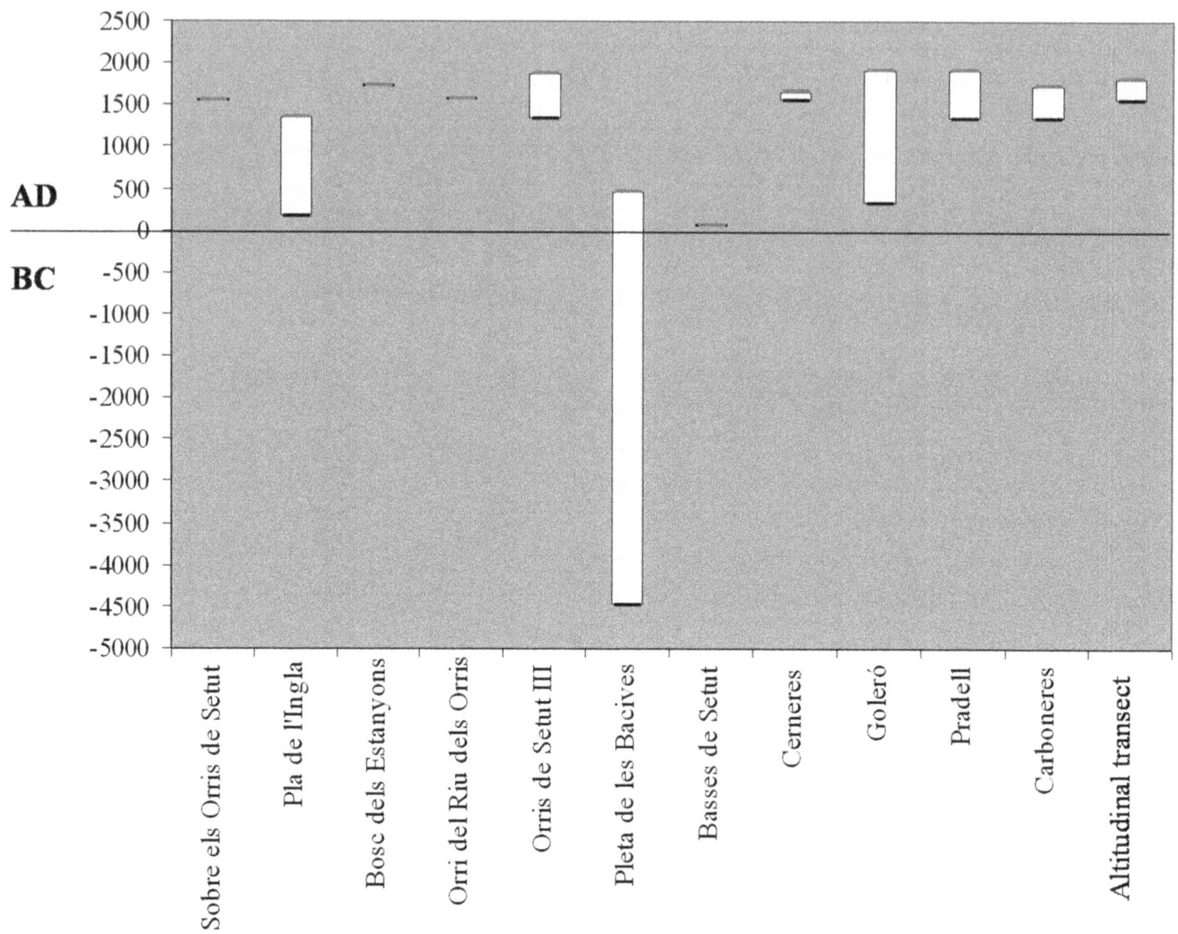


Figure 3. Chronology of the structures based on radiocarbon dating

the sediments from all structures except for sediment samples from charcoal kilns which were only sieved, not floated.

Afterwards, in the laboratory, we examined each charcoal fragment using a metallographic microscope (Olympus BX-51) with reflected light and 50x to 500x magnifications. We performed taxonomical as well as dendrological and taphonomical analyses.

Dendrology focuses on the analysis of growth rings, whose development depends on diverse external and internal factors such as climate, animal activities, natural stand dynamics, competition between trees, anthropic activities, etc. (Carrión 2007; Dufraisse 2006; Euba 2008; Schweingruber 1996). Charcoal taphonomy is the study of wood cell alteration produced during combustion and post-depositional processes. This provides information on anthropic and/or natural formation processes of charcoal deposits (Badal and Carrión 2004; Di Lernia and Fiorentino 1995; Euba 2008; Théry-Parisot 2001). This type of study is becoming more frequent since it can provide a better understanding of the archaeological record.

The accuracy and level of charcoal identification depends on the size of the charcoal fragments, the alterations that they present and the taxonomical variability within the species/genus/family (Allué 2002; Schweingruber 1990). We identified up to species level (*Buxus sempervirens*), genus level (*Betula*, *Juniperus*, *Salix* or *Pinus* sp.) and family level (Ericaceae or Monocotiledonea).

Within the genus *Pinus* the Mediterranean type pines can be distinguished from the mountain type pines. Mountain type pines have large fenestriform pits in the cross-section from the parenchyma cells to the tracheids (Schweingruber 1990). Two mountain type pines are present in the Pyrenees: *Pinus sylvestris* (Scots pine) and *Pinus uncinata* (Mountain pine). These are very difficult to distinguish anatomically, but *Pinus uncinata* has characteristic thin growth rings and the fact that most of the archaeological deposits were located at more than 1800 m above sea level, we assigned the mountain type pines to *Pinus* type *uncinata* (Euba 2008). Probably most of the analyzed charcoal corresponds to this species although some individuals could correspond to *Pinus sylvestris* (especially those recovered from sites located in the montane stage).

The results of the anthracological analysis showed a dominant presence of *Pinus* type *uncinata* and a very scarce presence of other taxa, so we decided to interpret the results based on the presence/absence of each taxon. This type of quantification is preferred by some authors (Smart and Hoffman 1988).

## RESULTS OF THE ANTHRACOLOGICAL ANALYSES

### Taxonomical analyses

We analyzed 3388 fragments from the Madriu valley (Figure 4). *Pinus* type *uncinata* is clearly dominant in the record, making up more than 50% of the charcoal in every structure, except for structures C2 and S13 at the altitudinal transect. In C2 60% corresponded with *Betula* sp. and 35% with *Pinus* type *uncinata*, and in S13 only *Abies alba* was identified. In structure C3 a large part of the charcoal corresponded to *Betula* sp. (38%).

In the rest of the analyzed structures *Pinus* type *uncinata* was the most represented taxon, and taxa other than *Pinus* comprised less than 8%. Apart from *Pinus* type *uncinata*, in structure S52 at Pla de l'Ingla *Abies alba*, *Juniperus* sp. and *Buxus sempervirens* were identified, in S104 at Bosc dels Estanyons we identified Ericaceae and in S110 at this site *Betula* sp. was identified. In S27 at Orris de Setut III the identified minority taxa were Ericaceae and *Juniperus* sp and in S152 of Pleta de les Bacives Monocotiledonea and *Juniperus* sp. In S153 of the same site two large-sized wood fragments were found, one corresponded to *Pinus* type *uncinata* and the other to *Juniperus* sp. Besides *Pinus* type *uncinata* in S121 of Basses de Setut *Salix* sp. was identified, in S137 of the same site Ericaceae and *Betula* sp., in C5 *Betula* sp. and *Salix* sp., and in C6 *Juniperus* sp. was identified.

In most of the analyzed structures the percentage of indeterminable charcoals did not exceed 15%. However, in S64 and S66 of Bosc dels Estanyons, S50 of

| Site | Structure | *Pinus* type *uncinata* | *Abies alba* | *Juniperus* sp. | *Betula* sp. | *Buxus sempervirens* | Ericaceae | Monocotiledonea | *Salix* sp. | *Pinus* sp. | Indeterminable | Total analysed |
|---|---|---|---|---|---|---|---|---|---|---|---|---|
| Sobre els Orris de Setut I | 10 | 93.33 | | | | | | | | | 6.66 | 90 |
| Sobre els Orris de Setut II | 15 | 76.66 | | | | | | | | 13.33 | 9.99 | 30 |
| Pla de l'Ingla I | 53 | 100 | | | | | | | | | | 9 |
| | 58 | 95 | | | | | | | | 1 | 4 | 100 |
| | 59 | 92.94 | | | | | | | | 2.9 | 4.13 | 241 |
| Pla de l'Ingla II | 60 | 82.73 | | | | | | | | 2.38 | 14.87 | 168 |
| Pla de l'Ingla III | 52 | 85.97 | 0.72 | 0.72 | | 0.24 | | | | 2.17 | 10.16 | 413 |
| | 102 | 95.92 | | | | | | | | | 4.06 | 221 |
| Bosc dels Estanyons | 64 | 72.5 | | | | | | | | 5 | 22.5 | 40 |
| | 65 | 88 | | | | | | | | 8 | 4 | 50 |
| | 66 | 70 | | | | | | | | 7.5 | 22.5 | 40 |
| | 67 | 82.5 | | | | | | | | 12.5 | 5 | 40 |
| | 104 | 84.99 | | | | | 0.83 | | | | 14.15 | 120 |
| | 110 | 85.87 | | | 3.52 | | | | | | 10.58 | 85 |
| Setut I | 1 | 94 | | | | | | | | 4 | 2 | 50 |
| | 2 | 77.46 | | | | | | | | 14.08 | 8.45 | 71 |
| Portella de Setut I | 47 | 98 | | | | | | | | | 2 | 50 |
| Carboneres de Setut | 48 | 88 | | | | | | | | 4 | 8 | 50 |
| | 50 | 61.11 | | | | | | | | 2.77 | 36.09 | 36 |
| | 129 | 74 | | | | | | | | | 26 | 50 |
| Orri del Riu dels Orris | 82 | 80.73 | | | | | | | | 0.74 | 18.51 | 135 |
| Orris de Setut III | 27 | 87.52 | | 0.21 | | | 0.63 | | | 3.8 | 7.81 | 473 |
| | 29 | 78.66 | | | | | | | | 0.01 | 13.32 | 75 |
| | 35 | 77.33 | | | | | | | | 9.33 | 13.33 | 75 |
| Pleta de les Bacives | 151 | 90 | | | | | | | | | 10 | 40 |
| | 152 | 62.22 | | 2.22 | | | | 7.77 | | | 27.77 | 90 |
| | 153 | 50 | | 50 | | | | | | | | 2 |

| Site | Structure | *Pinus* type *uncinata* | *Abies alba* | *Juniperus* sp. | *Betula* sp. | *Buxus sempervirens* | Ericaceae | Monocotiledonea | *Salix* sp. | *Pinus* sp. | Indeterminable | Total analysed |
|---|---|---|---|---|---|---|---|---|---|---|---|---|
| Basses de Setut II | 121 | 86.79 | | | | | | | 1.88 | | 11.31 | 53 |
| | 133 | 92 | | | | | | | | 0.8 | 7.2 | 125 |
| | 135 | 91.74 | | | | | | | | 1.17 | 7.05 | 85 |
| | 137 | 82.5 | | | 1.25 | | 3.52 | | | | 12.5 | 80 |
| Altitudinal transect | C1 | 57.14 | | | 38.09 | | | | | | 4.76 | 21 |
| | C2 | 35 | | | 60 | | | | | | 5 | 20 |
| | C3 | 95 | | | 5 | | | | | | | 20 |
| | C4 | 90 | | | | | | | | | 10 | 20 |
| | C5 | 75 | | | 5 | | | | 5 | | 15 | 20 |
| | C6 | 95 | | 5 | | | | | | | | 20 |
| | C7 | 90 | | | | | | | | | 10 | 20 |
| | C8 | 100 | | | | | | | | | | 20 |
| | C9 | 90 | | | | | | | | | 10 | 20 |
| | 13 | | 90 | | | | | | | | 10 | 20 |

Figure 4. Presence/absence of taxa in structures from the Madriu valley

| Site | Structure | *Pinus* type *uncinata* | *Abies alba* | *Juniperus* sp. | *Buxus sempervirens* | Ericaceae | Monocotiledonea | Pomoideae | *Salix* sp. | *Pinus* sp. | Indeterminable | Total analysed |
|---|---|---|---|---|---|---|---|---|---|---|---|---|
| Cerneres | 35 | 60.18 | 26.85 | 1.84 | | | | | | 0.92 | 10.17 | 100 |
| | 36 | 64.45 | 21.68 | 1.8 | | | | | | 4.21 | 6.6 | 166 |
| | 37 | 82.16 | | | | | | | | | 5.28 | 359 |
| | 39 | 93.26 | 0.19 | 0.19 | 0.39 | | | | | 0.39 | 5.12 | 506 |
| | 51 | 72.5 | | | | | | | | 2.5 | 25 | 40 |
| Goleró | 2 | 79.48 | | 2.56 | 0.85 | | | | | 11.11 | 5.97 | 117 |
| | 11 | 88.52 | | | | | | | | | 11.47 | 61 |
| | 12-1 | 87.8 | | | | | | | | 2.43 | 9.75 | 41 |
| | 16 | 90 | | | | | | | | 5 | 5 | 60 |
| | 42 | 54.18 | | 1.11 | | | | 0.55 | | 9.49 | 34.62 | 179 |
| | 43 | 91.41 | 0.09 | 0.27 | 0.09 | | | | | 1.89 | 6.22 | 1107 |
| | 46 | 85.96 | 0.87 | 1.75 | | | | | | 7.89 | 3.5 | 114 |
| | 48 | 92.91 | | 0.41 | | | | | | | 6.66 | 240 |
| | 75 | 88.97 | | 0.73 | 0.73 | | | | | | 9.55 | 136 |
| | 79 | 91.33 | | | | | | | 1.57 | 0.78 | 6.29 | 127 |
| | 80 | 97.5 | | | | | | | | | 2.5 | 80 |
| Pradell | 30 | 88.88 | | | | | 4.44 | | | | 6.66 | 45 |
| | 31 | 83.73 | | 0.81 | | 1.62 | | | | | 13.81 | 123 |
| | 81 | 91.25 | | 1.25 | | | | | | 1.25 | 6.25 | 80 |
| Carboneres | 82 | 85 | | | | | | | | | 15 | 20 |
| | 83 | 80 | | | | | | | | | 19.99 | 60 |
| | 84 | 95 | | | | | | | | | 4.99 | 120 |

Figure 5. Presence/absence of taxa in structures from the Vansa valley

Carboneres de Setut, S82 of Orri del Riu dels Orris and S152 of Pleta de les Bacives this percentage was between 18 and 36% (Figure 4).

A total of 3881 charcoal fragments from the Vansa valley were analyzed. Like in the Madriu valley, the most represented taxon was *Pinus* type *uncinata*. The lowest

| Fragments with morphology | Number of fragments | Site | Structure | Taxon |
|---|---|---|---|---|
| Fragments with ax-cut | 4 | Bosc dels Estanyons | 110 | *Pinus* type *uncinata* |
| | | Pla de l'Ingla | 52 | *Pinus* type *uncinata* |
| Utensile handle | 1 | Pla de l'Ingla | 52 | *Pinus* type *uncinata* |
| Fragment of enclosure | 2 | Pradell | 31 | *Pinus* type *uncinata* |
| Lid of containers | 2 | Pradell | 30 | *Pinus* type *uncinata* |
| | | | 31 | *Pinus* type *uncinata* |
| Fragments of roof | 2 | Pleta de les Bacives | 153 | *Pinus* type *uncinata* and *Juniperus* sp. |
| Shavings | More than 20 | Cerneres | 35 | *Abies alba* |

Figure 6. Charcoal fragments with identified morphologies

proportion was found in structure S42 of Goleró (54%). This structure was also the one with the highest fraction of indeterminable charcoal (35%). In the rest of the analyzed structures the proportion indeterminable charcoal was not higher than 20%.

In Cerneres, in S35 and S36, 60-64% corresponded to *Pinus* type *uncinata*, 22-27% to *Abies alba* and 2% to *Juniperus* sp. In S39 of this site apart from *Pinus* type *uncinata*, a scarce presence of *Buxus sempervirens*, *Abies alba* and *Juniperus* sp. was recorded. In S37 and S51 only *Pinus* type *uncinata* was identified. At the Goleró site, besides *Pinus* type *uncinata* the following taxa were identified: In S2 *Juniperus* sp. and *Buxus sempervirens*, in S42 *Juniperus* and Pomoideae, in S43 *Juniperus* sp., *Abies alba* and *Buxus sempervirens*, in S46 *Juniperus* sp. and *Abies alba*, in S48 *Juniperus* sp., in S75 *Juniperus* sp. and *Buxus sempervirens* and in S79 *Salix* sp. In the rest of the structures at this site (S11, S12-1, S16 and S80), only *Pinus* type *uncinata* was identified. At the Pradell site, apart from *Pinus* type *uncinata*, the following taxa were identified: In S30 Monocotiledonea was identified, in S31 Ericaceae and *Juniperus* sp., and in S81 *Juniperus* sp. In the three structures from Carboneres, only *Pinus* type *uncinata* was identified (Figure 5).

### Macroscopic observation

The studied material presents a series of macroscopical alterations as well, related to human activities. Woodworking comprises several processes such as the preparation of timber and the development of it to a utensil (Piqué 1999).

In our material, four fragments with axe cut marks, one possible utensil handle, two fragments of enclosures, two lids for containers, two elements of roofing material and more than 20 pieces of shavings have been identified. These have enabled us to recognize objects and activities and thus to understand aspects of the use of wood in the structures we studied. These forms could be observed without the need of a microscope because they were large enough to retain some morphology (Figure 6).

## DISCUSSION

### Vegetation reconstruction

For the subalpine stage the anthracological data indicated that Mountain pine forest dominated at both sunny and shady slopes, in structures from all the periods and functionalities. The undergrowth of these forests consisted of boxwood, bearberry and juniper in sunny areas. Pastures were abundant because in open patches grasses and weeds dominated over trees. In shady areas, Mountain Pine forest co-existed with blueberry and rhododendron. Also Fir Tree was present at the inferior part of this height zone, as well as stream bank vegetation species like birch, willow and rowan. At altitudes lower than 1900 m.a.s.l. Scots pine was dominant instead of Mountain pine (Blanco *et al.* 1998).

According to palynological data, high mountain meadows were occupied by birch and Mountain pine following the climate improvement 11,000 years ago during the Preboreal (Miras *et al.* 2007; Ejarque *et al.* 2010). These species have been identified in Bosc dels Estanyons and in the charcoal kilns that were located at the altitudinal transect, all near the river Madriu. The expansion of birch is favored by the impact of anthropogenic activities such as grazing and charcoal production, because birch is a colonizing species (Blanco *et al.* 1998).

Fir forest has been identified at the bottom of the Perafita Valley, at 1500 m.a.s.l, near Entremesaigües (S13 at the altitudinal transect). This species is very scarce in the Madriu valley and in the eastern Pyrenees in general. In the Madriu valley, between 9400 and 9100 cal. BP (i.e. during the transition from the Boreal to the Atlantic), more humid climatic conditions favor the introduction of Fir Tree into the subalpine stage (Miras *et al.* 2007). In the Central Catalan Pyrenees a decline of this species in favor of beech and pine has been documented in deposits resulting from charcoal kiln activity that took place in montane stages during the 5th century AD (Pèlachs *et al.* 2006).

**Use of wood**

The introduction of roots and sediment into the charcoal fragments dominates at the pastoral structures. These alterations have been especially observed in small charcoal fragments, probably because they occur in the oldest structures which generally suffered a more intense fragmentation and a longer sedimentary process. Archaeological evidences show that the first traces of human presence in high mountain environments go back to the Neolithic (Pleta de les Bacives, Madriu Valley) (Palet *et al.* 2006; 2007). It concerns structures related to herding. Palynological datings from Bosc dels Estanyons placed these first anthropic indicators in the 5th millennium BC (Miras *et al.* 2007; Ejarque *et al.* 2010). Shepherds ascended to these altitudes in search of summer meadows and constructed perennial structures with pine and juniper wood, to spend the night (cabins) and to keep the cattle (enclosures). In addition they burned parts of high altitude forests in order to obtain fields (Miras *et al.* 2007; Ejarque *et al.* 2010). Probably they used Gramineae (Monocotyledons) as a bed (S152 at Pleta de les Bacives).

Pastoral structures from the 14th and 15th century have been identified in the Madriu valley. In the Cadí Mountain Range, several structures from this period were documented in the deposits of the Pradell and Goleró sites. Until now we have not identified 16th and 17th century pastoral structures. This can be related to a diminution of pastoral activity at higher altitudes during this period, due to an increase in the demand of (stable) cattle supply. Nevertheless, in the 18th century agriculture entered a socio-economical crisis and herding in high mountain environments returned to be of commercial importance. This agricultural crisis was a general phenomenon in all Europe (Codina 2004). From the Early Middle Ages to Modern Age pine was the most abundant wood source, as in previous periods, but also other available bush resources were exploited, such as Ericaceae and juniper. The different wood species were used for (animal) bedding or food, as fuel for domestic fires and construction.

The major part of anthracological evidences of wood use is related mainly to roman and modern metalworking activities. In the Cadí Mountain Range, the furnaces of Goleró were used for the first steps of iron processing. It is possible that in these furnaces pine coal was used because charcoals fragments had acute angles, came from big calibre branches and had crystallized appearance (Euba 2008); it was a local but quite intensive artisan activity which required a large quantity of fuel. This type of furnace requires a minimum temperature of 600°C and pine wood quickly reaches high temperatures (Magnin 2004). Until now roman and medieval charcoal kilns have not been located in this area, but we do not discard that they exist and can be discovered in next excavation campaigns.

The furnace at Pla de l'Ingla (Madriu valley) is different from those of Goleró. It was used to obtain resin (Orengo 2010). According to our data the furnace of Pla de l'Ingla used altered wood (probably fallen wood) as combustible. We identified large-sized branches (over 10 cm in diameter) with a lot of micro-organism alterations, probably caused during the degradation of the wood after falling. Like in the Cadí Mountain Range, pine seems to be the main fuel source.

Also Fir- and boxwood were identified in the furnaces of Pla de l'Ingla and Goleró. Possibly it concerns objects that were carbonized either accidentally or deliberately. These two taxa are especially suitable for specific purposes like tool construction. They were not used as a combustible; because we only identified a few individuals of these species.

Activities related to charcoal kilns had a pronounced impact on forests of the Pyrenees (deforestation) according to palynological and archaeological data (Miras *et al.* 2007; Ejarque *et al.* 2009; 2010; Palet *et al.* 2006; 2007). The charcoal kilns documented in the Cadí Mountain Range were operational from the 14th to the 17th century and in the Madriu valley from the 16th to the 19th century. Birch, fir and especially pine are very suitable fuel to feed the fire of forges because they provide a good flame, they burn easily, and they produce a lot of heat. In addition these species are available in montane and subalpine areas. The presence of other taxa in the anthracological record of the charcoal kilns, such as willow or juniper, is not directly related to the fuel of the kilns, probably they had another function. They could have been used to cover the kiln, to ignite the flame, they could correspond to small branches that were accidentally dragged along with the firewood, or perhaps they were objects that burned accidentally in the kiln (Euba 2008). From a charcoal kiln in Cerneres remains of shavings were recovered which were probably a by-product of the shaping of Fir wood related to the drawing of combustion from the charcoal kiln (Figure 6).

In charcoal kilns a slow combustion without sharp rises in temperature is produced. Vitrification has been linked by several authors to a slow and anaerobic combustion (Fabre 1996; Marguerie and Hunot 2007). In charcoal kilns this type of alteration is not found very frequent, but more frequent than in other structure types. In addition the amount of cracks related to violent combustion and quick increase of temperatures (Théry-Parisot 2001) is lower in charcoal kilns compared to other types of structures.

We do not know the exact size of the wood used in charcoal kilns due to fragmentation of the charcoal, which mainly takes place during post-depositional processes. Nevertheless, most of the charcoal fragments that preserved the entire width (some with axe-cut evidence) had an average diameter of 10 cm. We know that the best wood for charcoal kilns are medium-sized branches (more or less 8 – 20 cm in diameter) and our anthracological results are concordant with this (Euba 2008).

The very low amount of alterations caused by micro-organisms in all the structures is an indication of the use of green wood and of logging at the time of use, instead of storage of the wood. This is due to the fact that most of the archaeological structures correspond to charcoal kilns, for which good quality timber is required. If we only take the other types of structures into account the relative amount of alterations caused by micro-organisms is higher, specially by xylophages (Euba *et al.* 2010).

The morphological characteristics of the charcoal kilns in our study area are very similar to those of charcoal kilns in the Ariège (eastern French Pyrenees). The kilns were mainly built on slopes, with retaining walls of dry stone and terraces which were elaborated to flatten the territory. We also identified charcoal kilns on plane surfaces in our study area (Pla de l'Ingla and Carboneres sites). The charcoal kilns located on slopes were smaller and more elliptical than those on the planes, whereas the kilns located in flat areas were larger, rounder and without retaining walls.

In the 19th century, socio-economic changes caused a sudden decrease of activities in the Cadí Mountain Range and the Madriu valley. Industrial activities gained much importance against agro-livestock activities that became increasingly expensive. In high mountain areas mining and pasturing, the main reasons of human presence, were no longer economically profitable. This caused these activities to come to an end and subsequently the forests could regenerate.

## CONCLUSION

Antracological studies played a fundamental role in this High Mountain Landscape Archaeology project – focused on two valleys in the eastern Pyrenees – since charcoal residues were present and particularly abundant in all the archaeological structures.

From the Neolithic through the Antique to the Modern Age, people ascended to high altitudes in search of various resources for different activities. High mountain environments, contrary to what is traditionally believed, have never been marginal areas. During times when agricultural activities were the principal means of existence, such as the roman or medieval period, the occupation of high mountain areas did not decrease and even stabilized, since this environment offered a range of much appreciated specific resources since prehistory (i.e.: forest, minerals and summer meadows). Charcoal kilns, furnaces and cattle herding-related constructions required a great amount of raw material which was obtained from the immediate surroundings (principally pine, but also juniper and Ericaceae).

In the Cadí Mountain Range primary transformation of iron ore took place during Roman times, for which pine coal have been used as fuel. In the Madriu valley resin production was a very important activity for which fallen and (therefore) altered pine wood was used as a combustible.

Nevertheless, the majority of the data related to the use of wood was obtained from the remnants of charcoal kilns dated between the 14th and 18th century. The wood types found were freshly cut pine, birch and fir. Probably medium-sized branches were used as fuel. This charcoal production caused an important deforestation in the Pyrenees and nowadays we can still see its impact on the High Mountain forests.

## References


ALLUÉ, E. 2002. *Dinámica de la vegetación y explotación del combustible leñoso durante el Pleistoceno Superior y el Holoceno del Noreste de la Península Ibérica a partir del análisis antracológico.* PhD thesis, Universitat Rovira i Virgili.

BADAL, E. and CARRIÓN, Y. 2004. La presencia de hongos e insectos xilófagos en el carbón arqueológico. Propuestas de interpretación. In *Avances en Arqueometría,* 98-106. Cádiz, Servicio de Publicaciones de la Universidad de Cádiz.

BLANCO, E., CASADO, M.A., COSTA, M., ESCRIBANO, R., GARCÍA, M., GÉNOVA, M., GÓMEZ, A., GÓMEZ, F., MORENO, J.C., MORLA, C., REGATO, P. and SAINZ, H. 1998. *Los bosques ibéricos. Una interpretación geobotánica.* Madrid, Planeta.

BONHÔTE, J. and VERNET, J.L. 1988. La mémoire des charbonnières. Essai de reconstitution des milieux forestiers dans une vallée marquée par la métallurgie (Aston, Haute-Ariège). *Revue Forestière Française* 40, 197-212.

BROCHIER, J.E., CLAUSTRE, F. and HEINZ, C. 1998. Environmental impact of Neolithic and Bronze Age farming in the eastern Pyrenees forelands, based on multidisciplinary investigations at La Caune de Bélesta (Bélesta Cave), near Perpignan, France. *Vegetation History and Archaeobotany* 7, 1-9.

CARRIÓN, Y. 2007. Dendrología y arqueología: las huellas del clima y de la explotación humana de la madera. In Molera, J.; Farjas, J.; Roura P.; Pradell, T. (eds) *IV Congreso Ibérico de Arqueometría,* 273-282. Girona, Fundació Privada Girona, Universitat i Futur.

CODINA, O. 2004. *De fer et de laine. Les vallées andorranes du XVIe au XIXe siècle.* Perpignan, Presses Universitaires de Perpignan, Fundació Julià Reig.

DAVASSE, B. 2000. *Forêts, charbonniers et paysans dans le Pyrénées de l'est, du moyen âge à nos jours. Une approche géographique de l'histoire de l'environnement.* Toulouse, GEODE.

DAVASSE, B., GALOP, D. and RENDU, C. 1997. Paysages du Néolithique à nos jours dans les Pyrénées de l'Est d'après l'écologie historique et l'archéologie pastorale. In *Dynamique des paysages protohistoriques, antiques, médiévaux et modernes*, 577-599. Sophia Antipolis.

Di LERNIA, S. and FIORENTINO, G. 1995. Decoding an open-air settlement of Terragne (Manduria Taranto, South eastern Italy); formation processes and spatial analyses. *Origini. Preistoria e Protoistoria delle civiltà antiche* XIX, 7-53.

DUBOIS, C., MÉTAILIÉ, J.P. and IZARD, V. 1997. Archéologie de la forêt charbonnée: questions et méthodes, illustrées par l'exemple du site de Lercoul (Ariège). *Dynamique des paysages protohistoriques, antiques, médiévaux et modernes*, 525-540. Sophia Antipolis.

DUFRAISSE, A. 2006. Charcoal anatomy, wood diameter and radial growth. In A. Dufraisse (ed) *Charcoal Analysis: New Analytical Tools and Methods for Archaeology, British Archaeological Reports* 1483, 47-60.

DURAND A. 1998. *Les paysages médiévaux du Languedoc (Xe- XIIe siècles)*. Toulouse, Presses Universitaires du Mirail.

EJARQUE, A., JULIA, R., RIERA, S., PALET, J.M., ORENGO, H.A., MIRAS, Y. and GASCON, C. 2009. Tracing the history of highland human management in the eastern Pre-Pyrenees: an interdisciplinary palaeoenvironmental study at the Pradell fen, Spain. *The Holocene* 19 (8), 1241-1255.

EJARQUE, A., MIRAS, Y., RIERA, S., PALET, J.M. and ORENGO, H.A. 2010. Testing micro-regional variability in the Holocene shaping of high mountain cultural landscapes: a palaeoenvironmental case-study in the eastern Pyrenees. *Journal of Archaeological Science* 37 (7), 1468-1479.

EUBA, I. 2008. *Análisis antracológico de estructuras altimontanas en el valle de la Vansa Sierra del Cadí (Alt Urgell) y en el valle del Madriu (Andorra): explotación de recursos forestales del Neolítico medio a época moderna*. PhD thesis, Universitat Rovira i Virgili.

EUBA, I., ALLUÉ, E. and PALET, J.M. 2010. Alteraciones biológicas de la madera: El caso de los materiales constructivos del yacimiento de Estall Serrer (Valle del Madriu, Andorra). In *Actas del VIII Congreso Ibérico de Arqueometría (Terol, 19 – 21 d'octubre de 2009)*, 305-314. Terol.

FABRE, L. 1996. *Le charbonnage historique de la chênaie à Quercus ilex L. (Languedoc, France): conséquences écologiques*. PhD thesis, Sciences et techniques su Languedoc, Université de Montpellier II.

HEINZ, C. 1996. Les transformations du paysage végétal pyrénéen au cours de l'épipaléolithique et du mésolithique. In Bintz, P. and Thérevin, A. (eds), *L'Europe des derniers chasseurs. Epipaléolithique et Mésolithique. Actes du 5e Colloque International UISPP. Commission XII. Grenoble, 18-23 Septembre 1995.* Paris, CTHS.

LLOVERA, X., BOSCH, J.M., RUF, Mª.A., SOLÉ, X. and VILA, A. (coord) 1997. *Roc d'Enclar. Transformacions d'un espai dominant. Segles IV-XIX.* Andorra, Govern d'Andorra-Ministeri de Cultura-Servei de Recerca Històrica.

MAGNIN, M.D. 2004. *Le fer.* Archéologiques. Paris. Errance.

MARGUERIE, D. and HUNOT, J.Y. 2007. Charcoal analysis and dendrology: data from archaeological sites in north-western France. *Journal of Archaeological Science* 34 (9), 1417-1433.

MIRAS, Y., EJARQUE, A., RIERA, S., PALET, J.M., ORENGO, H. and EUBA, I. 2007. Dynamique holocène de la végétation et occupation des Pyrénées andorranes depuis le Néolithique ancien d'après l'analyse pollinique de la tourbière de Bosc dels Estanyons (2180 m, Vall del Madriu, Andorre). *Paléontologie humaine et préhistoire* 6 (4), 291-300.

ORENGO, H.A. 2010. *Arqueología de un paisaje cultural pirenaico de alta montaña. Dinámicas de ocupación del valle del Madriu-Perafita-Claror (Andorra).* PhD thesis, Institut Català d'Arqueolofia Clàssica – Universitat Rovira i Virgili.

PALET, J.M., RIERA, S., MIRAS, Y., EJARQUE, A. and EUBA, I. 2006. Estudi i revalorització dels paisatges culturals d'alta muntanya: els projectes vall de Madriu (Andorra) i la Vansa-Serra del Cadí (Alt Urgell). *IBIX Annals 2004-2005 (Actes IV Colloqui d'Estudis Transpirinencs)* 4, 89-107.

PALET, J.M., EJARQUE, A., MIRAS, Y., RIERA, S., EUBA, I. and ORENGO, H. 2007. Formes d'ocupació d'alta muntanya a la vall de la Vansa (Serra del Cadí-Alt Urgell) i a la vall del Madriu-Perafita-Claror (Andorra): estudi diacrònic de paisatges culturals pirinencs. *Tribuna de Arqueología*, 229-253.

PÈLACHS, A. 2004. *Deu Mil Anys de Geohistòria Ambiental al Pirineu Central Catalá. Aplicació de técniques paleogeogràfiques per a l'estudi del territori i el paisatge a la Coma de Burg i a la Vallferrera.* PhD Thesis, Departament de Geografia, Universitat Autònoma de Barcelona.

PÈLACHS, A., SORIANO, J.M., ESTEBAN, A., PIQUÉ, R. and MATAMALA, N. 2006. Antracología y palinología como herramientas complementarias para la caracterización de una perturbación en el Pirineo Central Catalán en época tardo-romana. *III Congreso Español de Biogeografía/Espainiako III. Biogeografia Biltzarra*, 134-138.

PIQUÉ, R. 1999. La gestión de los recursos leñosos en la Cova des Càrritx. In Llull, V., Micó, R., Rígete, C. and Risch, R. (eds), *Ideología y Sociedad en la Prehistoria de Menorca. La Cova des Càrritx y La Cova des Mussol,* 489-520. Barcelona, Consell Insular de Menorca, Ajuntament de Ciutadella, Fundació Rubío Tudurí-Andrómaco.

ROS i MORA, M.T. 1995. Datos antracologicos sobre la diversidad paisajística de Catalunya en el Neolítico. *Rubicatum* 1, 43-56.

SCHWEINGRUBER, F.H. 1990. *Anatomie europäischer Hölzer. Ein Atlas zur Bestimmung europäischer Baum-, Strauch- und Zwergstrauchhölzer./ Anatomy of European woods. An Atlas for the identification of European trees, shrubs and dwarf shrubs.* Sttutgart, Paul Haupt.

SCHWEINGRUBER, F.H. 1996. *Tree Rings and Environment Dendroecology*. Bern, Stuttgart, Vienna, Swiss Federal Institute for Forest, Snow and Landscape Research.

SMART, T.L. and HOFFMAN, E.S. 1988. Environmental interpretation of archaeological Charcoal. In Hastorf, C.A. and Popper, V. (eds), *Current Paleoethnobotany. Analytical methods and cultural interpretations of archaeological plant remains,* 167-205. Chicago, University of Chicago Press.

THERY-PARISOT, I. 2001. *Économie des combustibles au Paléolithique. Experimentation, taphonomie, anthracologie.* Paris, Centre National de la Recherche Scientifique. Centre d'études Préhistoire, Antiquité, Moyen Âge.

THIEBAULT, S. 2005. L'apport du fourrage d'arbre dans l'élevage depuis le Néolithique. *Anthropozoologica* 40(1), 95-108.

YAÑEZ, C., BURJACHS, F., JUAN-TESSERRAS, J. and MESTRES, J.S. 2001-2002. La fosa de Prats (Andorra), un jaciment del bronze mitjà al Pirineu. *Revista d'Arqueologia de Ponent* 11-12, 123-150.

ZAPATA, L. 2002. El origen de la agricultura en el País Vasco y transformaciones en el paisaje: análisis de restos vegetales arqueológicos. *Kobie* Anejo 4. Bilbo, Bizkaiko Foru Aldundia.

# THE ANSWER IS DEEP DOWN IN THE WELL: CHARCOAL AND WATERLOGGED WOOD FROM ARCHAEOLOGICAL WELLS IN SOUTHERN FRANCE

Isabel FIGUEIRAL and Laurent FABRE

Inrap, DIR Méditerranée, 561, rue Etienne Lenoir, Km Delta, 30900 Nîmes, France;
Centre de Bio-Archéologie et d'Ecologie, (USTL/CNRS/EPHE), Institut de Botanique,
163 rue Auguste Broussonet, 34090 Montpellier, France

Lucie CHABAL

Centre de Bio-Archéologie et d'Ecologie, (USTL/CNRS/EPHE), Institut de Botanique,
163 rue Auguste Broussonet, 34090 Montpellier, France

**Abstract**: *Waterlogged wood and charcoal preserved in 21 archaeological wells (Neolithic – Roman) from the Languedoc region (southern France) are studied, in an attempt to understand their provenance and the information they can provide in terms of the environment and function at the site. The amount and variety of debris (lost accidentally, fallen naturally, but mostly voluntarily disposed of) suggest that wells were used as dumps and were ultimately unusable as fresh water sources. Simple and / or sophisticated objects from day-to-day life, imported or made locally, provide information on the development of carpentry procedures / skills. A large variety of wood species was used, some eventually chosen based on their characteristics and properties. The identification of charcoal provides a diversified taxonomic spectrum and a similar pattern of vegetation composition and plant use to the one suggested by domestic dispersed charcoal. The unexpected abundance of reeds in rural sites suggests specific uses in the farming system. The remarkable taxonomic diversity found in some wells (including species rarely identified in habitat layers) may reflect a direct relationship between taxonomic diversity and mechanical constraints. The absence of trampling and the presence of water may favour the survival of smaller fragments / rarer species.*

**Key-words**: *Archaeology, Neolithic – Roman period, Well, Anaerobic conditions, Waterlogged wood, Charcoal, Southern France*

## INTRODUCTION

In southern France, increasing attention has been given recently to the contents of archaeological wells, especially those dating back to the Roman period, but also from Proto and Prehistory. The interest of digging this type of structure has been pointed out by different authors, in France and elsewhere (Greig 1988; Matterne 2000; Giraud *et al.* 2005; Auxiette *et al.* 2003; Zwierzinski 1999, Knörzer 1984, among others). However, the success of these previous studies was sometimes hampered by technical difficulties while digging the wells, relying on artificial stratigraphy, and unable to reach the lowermost levels of the structures. Furthermore, attention was focussed mainly on seed / fruit remains, discussed in terms of economy and phytosociology.

Wells usually provide a remarkable wealth of information, due to the preservation of remains of day-to-day material life, which quickly degrade in less favourable conditions. The anaerobic environment inhibits micro-biological activity and ensures perfect conditions for the protection of organic material, both waterlogged and charred, provided that the water-table level remains stable. However, and as a result of their singularity, these contexts have not been systematically studied and little is known of the meaning of remains preserved.

We are particularly interested in the waterlogged wood and wood charcoal and in trying to understand (1) how the organic remains were transported, (2) where they came from, (3) their function or use and (4) their archaeological meaning. In short, our study aims to understand the provenance of discarded woody elements, and to recognize the potential information they may provide in terms of the environment and function of the site.

Our discussion will be based on the evidence assembled from 21 wells of the Languedoc region, in southern France (Figure 1).

## ARCHAEOLOGICAL BACKGROUND

The Languedoc region is characterized by a long history of human settlement going back to the Palaeolithic period. Year-round access to water was essential for human activities. Under environmental constraints or when lacking direct access to this vital resource, man was able to dig for underground / runoff water.

The earliest known wells in this region date back to the Middle Neolithic period ("Chasséen", *c.* 4000 cal BC). Clear examples are rare (Mas de Vignolles 9, Nîmes) as the exact function of structures identified is not always clear. Some, such as the "wells with offerings" discovered at Villeneuve-Tolosane, in the Haute-Garonne

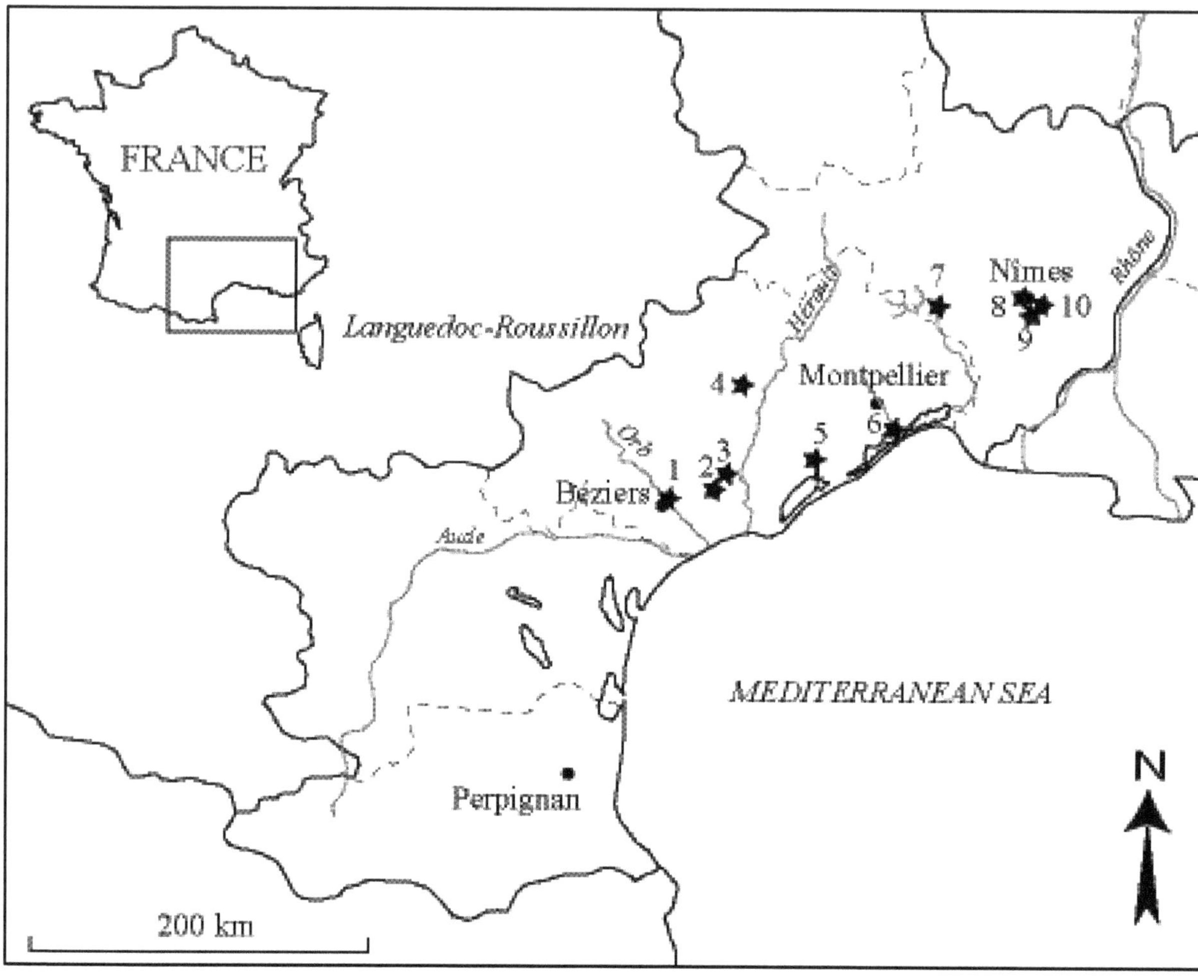


Figure 1. Map showing the location of archaeological sites referred to in the text. 1. Le Gasquinoy (Béziers), 2. Rec de Ligno (Valros), 3. Mont Ferrier (Tourbes), 4. Peyre Plantade (Clermont l'Hérault), 5. Roumèges (Poussan), 6. Lattara (Lattes), 7. Ambrussum (Viletelle) 8. Av. Jean Jaurès (Nîmes), 9. Mas de Vignolles 9 et 13 (Nîmes), 10. Moulin Villard (Caissargues)

region (Clottes *et al.* 1980), may have had a funerary connotation.

As far as we know, well PT1731 (Middle Bronze Age), from Mas de Vignolles 9, constitutes the only example uncovered from the Bronze Age. Up to present, this type of structure has not been identified in any of the protohistoric sites excavated. Wells appear more frequently from the 5th century BC onwards (Mas de Vignolles 9 at Nîmes, Castelnau-le- Lez and *Lattara* at Lattes). During Antiquity (after The Republic), they become a common feature in both rural and urban sites.

## MATERIAL AND METHODS

### The wells

The wells under study have been excavated as part of both long-term programmed excavations and rescue work carried out prior to new redevelopments (motorways, railway lines, housing and commercial estates). This is a difficult and dangerous work, made possible by recent safety developments, and mostly carried out by specialists (Archéopuits), able to spot areas of special interest and distinguish stratigraphic units. It is also an expensive exercise but the variety and interest of data obtained seems to justify such investment.

The wells were uncovered either in non-sedentary sites (Mas de Vignolles 9 – Middle Neolithic and Bronze Age), domestic sedentary sites (Mas de Vignolles 9 – Iron Age), rural sites (Le Gasquinoy, Mont Ferrier, Rec de Ligno, Mas de Vignolles 9 – Roman, Mas de Vignolles 13, *Ambrussum*), small towns (*Lattara*, av. Jean Jaurès Nîmes) spanning the period Middle Neolithic – Roman period (Figure 1). The roman wells were located in household areas, farmyards, streets / squares, agricultural areas.

The majority of the structures was relatively deep and lined with stone. At *Lattara*, from the middle of the 1st century BC, most of the wells were lined with stones, others used a *dolium* (without its bottom) as a border (Piqués and Buxo 2005) and one was 'lined' with wood (a big barrel in its bottom). Some of the Roman wells present a border (parapet) around the well's entrance. Depth varies between 3 and 13 m. Five of the twenty-two wells excavated at *Lattara* (especially the most ancient), and the five wells from Mas de Vignolles 9 (Neolithic-Roman), were unlined. Those from Mas de Vignolles 13 were also unlined, but presented a walled up upper part, probably used to hold a parapet, since dismantled.

Different categories of wells are distinguished, according to function:

1. "true" wells, which collect water from a low or high water-table. For example, the wells from *Lattara* (28 structures identified, covering the period 1st century BC / 2nd century AD), were fed by the river water table (Lez) situated at around +0,5 to +1 m ASL. The wells, however, were dug deeper down (most of them at 4 m) to reach the receding water table in the summer months. At present, the water table is stable at around 0,5 m ASL, a little higher than sea level (0,11 m ASL). All the wells studied here were true wells.
2. citerns / water tanks, which stock channelled water (ex: rain water). Clear examples are difficult to identify and proper citerns have not been identified in our sites. It is however curious that at least some of our wells are located in patios / yards or in the immediate vicinity of walls, perpendicular to roofs. Rain water may thus have trickled down to the wells. Furthermore, the wells from Rec de Ligno and Mont Ferrier were not dug deep enough to reach the ground water table. They were fed, apparently, by runoff water, thus fulfilling the role of a citern.

Some wells have certainly been conceived to both collect the water from the water table and stock it during the periods of low water level (during the summer) (Bakalowicz and Blanchemanche 2005). It is however difficult to prove this double function. Well known examples have been identified in the Near East (Bakalowicz, oral communication).

It might be interesting to mention that the channelling of rain water towards amphorae has been clearly identified at *Lattara* (Piqués and Buxo 2005).

### Recovery of material

Sediments retrieved from the structures were processed by water sieving (2, 1 mm mesh) and flotation (2, 1 and 0,5 mm mesh). The uncharred organic material recovered (waterlogged wood) was preserved wet, while carbonized fragments were left to dry. Both categories of material are considered here. The good preservation of the wood structure allowed their identification, following standard procedures described previously (Figueiral 1999, Figueiral and Willcox 1999).

## RESULTS AND DISCUSSION

### The organic residues

Waterlogged wooden objects are amongst the most remarkable remains recovered: pieces of furniture (*Lattara, Ambrussum,* Mont Ferrier) domestic or agricultural tools (*Lattara*, *Ambrussum*, Mont Ferrier, Le Gasquinoy), building residues such as planks, beams, posts and rafters (*Ambrussum*, *Lattara*, Montferrier, Mas de Vignolles 9, Roumèges), cutlery (*Ambrussum*, Le Gasquinoy), basketry (Roumèges, Mont Ferrier, *Lattara*), toiletries (*Lattara*, Le Gasquinoy, *Ambrussum*), writing tablets (*Ambrussum*) as well as the buckets used to draw the water (Roumèges, *Ambrussum*).

Objects recovered are very sophisticated / high quality or of more modest quality. Up to present, the most remarkable find concerns that of the *mensae tripedes*, uncovered at *Lattara* (Chabal and Feugère 2005). This is the name given to small three legged tables, made out of heavy wood (*Buxus*, *Acer monspessulanum* and *Malus* or *Pyrus*), turned, moulded and carved following roman models, as represented in funerary *stellae*. These high quality objects could have been imported or made locally by skilled artisans, using plant species available in the area.

Less sophisticated objects were made locally using a large variety of available plant resources (Chabal and Feugère 2005). Plant species identified include *Buxus*, *Juglans*, *Fagus*, *Malus* / *Pyrus*, *Acer monspessulanum*, *Quercus* (deciduous and evergreen) *Fraxinus*, *Ulmus*, *Cornus*, *Juniperus*, *Salix*, *Abies*, *Picea* / *Larix*. Any type of wood coming to hand could be used. However, objects for particular purposes might require wood with specific characteristics and properties. For instance, material from a single well testifies to the constant use of *Cornus*, *Quercus* (evergreen) and *Fraxinus* in the manufacture of ladders and chairs. *Salix* is particularly suitable for basketry but *Cornus* and *Clematis*, were also used.

The study of these objects / implements goes beyond their immediate significance as multipurpose artefacts. They also provide information on the development of carpentry procedures / skills (traces of wood saw or of other cutting tools, turning).

In addition to objects, wood remains also comprise waste from wood working, wood fuel cuttings and fruit tree pruning. Fragments of twigs and small branches are particularly conspicuous at av. Jean Jaurès (Nîmes), *Lattara, Ambrussum,* Mont Ferrier, and Le Gasquinoy.

Other waterlogged material often found comprises fruit and seeds (wild or cultivated), insects, fauna remains (bones of large and small animals, whole or single pieces), rope, leather, objects made of bone, metal and textiles. Human remains are found more sporadically at *Lattara* (Duchesne and Treil 2005), *Ambrussum* (Fiches, unpublished report) and Le Gasquinoy (Forest, inédit).

Large quantities of charred wood are found in many wells, including small fragments of charcoal, remains of twigs and branches, fragments of burnt objects. Some remains of seeds / fruits have also been preserved by charring.

**The wells used as rubbish pits**

Depending on the sites, these remains appear to be linked with diverse activities (domestic, farming tools, handcraft, food, fuel ...) and testify to the use of wells as rubbish dumps. It seems evident that very little of the fill of these structures results from accidental falling, such as windblown seeds, material transported by running water, drop of small animals, accidental drop of objects by man.

The wide range of material assembled (not only organic material but also sediment, stones, pottery, fragmented votive plaques and altars), clearly shows that these are not natural deposits and that all this material was voluntary dumped in the wells, soon after their primary use, as shown by the tightened chronology generally obtained.

As a result, the use of wells as a rubbish dump can be easily recognizable. It is however, difficult to understand why this happened while the structure was still filled with water and usable. Was it by pure negligence or did the process start only when it was decided to stop using the well as source of water? It is difficult to answer this question, even considering the example of some wells at *Lattara* which contained, for instance, 30 whole dogs or 102 rabbits (Gardeisen 2005). These contents do not result from a single event, but a series of sporadic events (Gardeisen 2005). Was the resulting insalubrity of the well suddenly responsible for its use as a rubbish dump? Or, on the contrary, was its use as a rubbish dump responsible for the insalubrity of the well and its subsequent filling? Were people even conscious of the link between the dropping of organic matter (including excrements) and poisoning of the water? It might be interesting to mention that Pline (Hist. Nat. XXXI, 40) recommended boiling the water from rivers and wells, until reduced to half. However, throughout history, the water from wells was never considered as particularly suitable for drinking.

There is clear evidence that some wells were periodically cleaned (they were empty of residues), but apparently they were never used during a long time. Taking into account the effort needed to build stone-lined wells, and the danger involved, it is particularly surprising that they were not looked after and used for a long time. For example, according to the evidence provided by the pottery, the wells from Montferrier and Rec de Ligno were not used for more than 40 years. Those from *Lattara* were used for less than 50 years, certainly lesser for the unlined wells (Buxo and Piqués 2005, 68). Lined wells (wood + stone) used during a longer time span have been identified in northern France (75 – 100 years) (Giraud *et al.* 2005).

**Significance of charcoal spectra and frequencies**

Important concentrations of charcoal are found in some of the structures studied (*Lattara*, Mas de Vignolles 9 (Iron Age), Mas de Vignolles 13, Mont Ferrier). On the whole, their identification provides a diversified taxonomic spectrum and a similar pattern of vegetation composition and plant use to the one suggested by domestic dispersed charcoal. The richness of the floral list identified / well varies according to the number of fragments available: abundant charcoal fragments, such as at *Lattara,* provided 42 taxa in PT471 and 24 taxa in PT234 and PT348. Even, when few fragments are available, as at Rec de Ligno (Valros), the diversity of taxa identified can be surprisingly high: 14 species for 45 fragments (PT2005, Us 2593). Does this make this charcoal suitable for environmental reconstructions? What type of overall picture do we obtain?

We will focus first on the roman wells (urban and rural contexts) as they constitute the majority of the structures studied. The two main urban examples available so far are *Lattara* and avenue Jean Jaurès (Nîmes).

At *Lattara*, the charcoal contents from three roman wells (PT234, PT471 and PT348) are compared with those from the interior of a contemporaneous domestic unit (Zone 4, Us 4310) (Chabal 2005). The taxonomic lists obtained are very similar, and include plants representing the four major biotopes: the Mediterranean oak woodland, the river/ river plain vegetation, the shores of the coastal lagoon, and the orchard trees (Figure 2).

No significant differences of taxa frequency are noticed, with the exception of those of Monocotyledonae, possibly *Phragmites australis* (according to the size of the stems). These remains are not frequent in the site, but they are a little more abundant in the well PT471 (1%) and in the houses destroyed by fire (13%) (Buxo *et al.* 1996). They are interpreted as roofing material destroyed by fire, some of it discarded in the well.

The overall similarity of the diagrams obtained in the wells and in the domestic deposits appears to indicate that charcoal found in these wells comes from the clearing of domestic units and should therefore provide reliable data concerning local plant resources and vegetal environment.

At av. Jean Jaurès (Nîmes), the charcoal assemblage from well PT5053 (located in the courtyard of a *domus*) is dominated by fragments of small branches and twigs of evergreen *Quercus*: 93,5% in Us 5813 (formed while the well was used as such), 64,5% in Us 5800, 92,2% in Us 5777 and 82,3% in Us 5729 (fill and disuse). This suggests the existence of woodland management in the neighbouring limestone slopes, with the exploitation of coppiced *Quercus ilex*, thus ensuring the stability of the fuel supply to the urban area (Figueiral, unpublished report).

Further relevant and interesting data are provided by the rural contexts. The sites of Mont Ferrier (Tourbes) and Peyre Plantade (Clermont l'Hérault) are located c. 28 km apart, in the Hérault valley. Data obtained in the well PT2052 (Mont Ferrier, end of 1st century AD) and in the habitat levels (Peyre-Plantade, 1st/2nd centuries) are

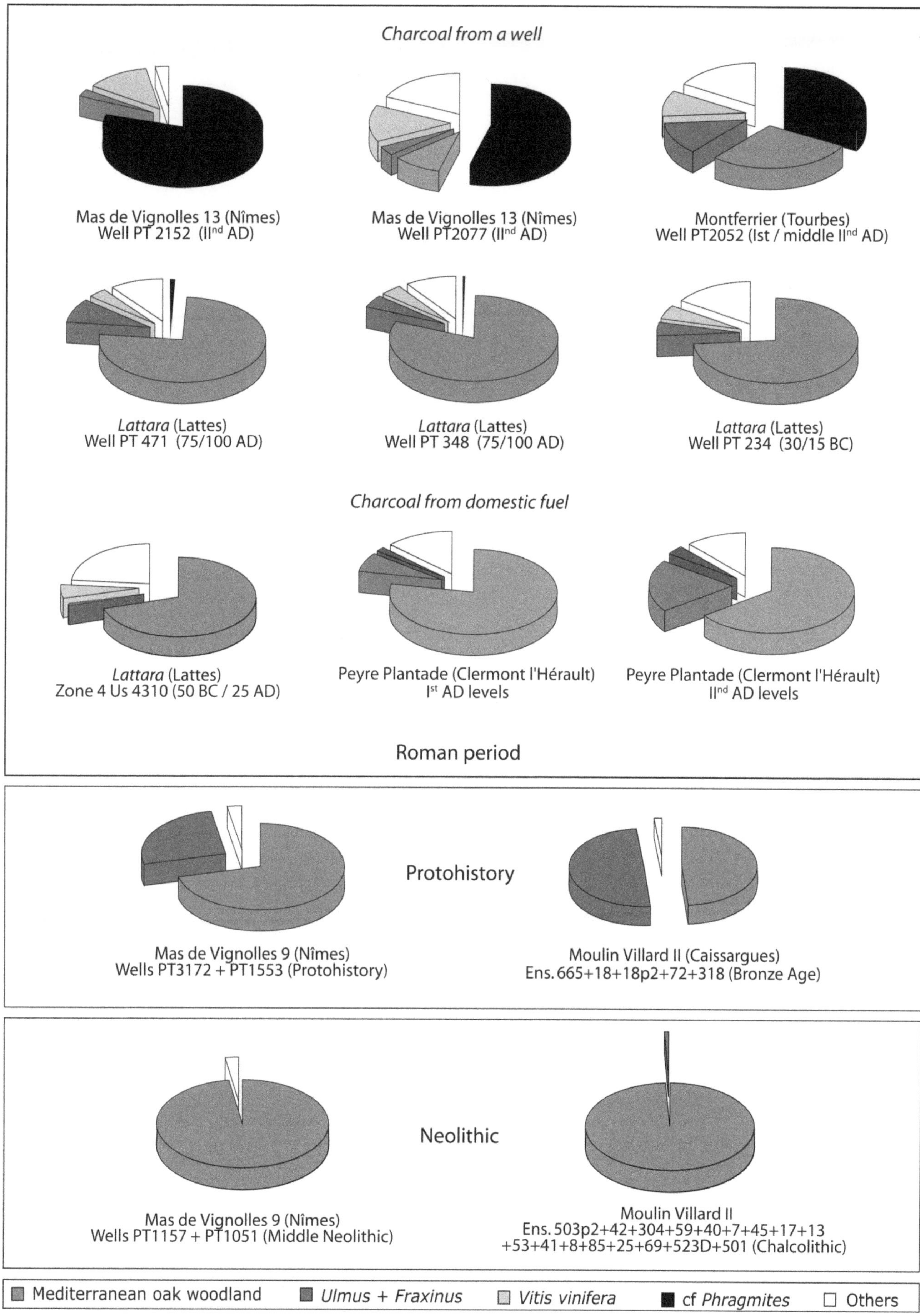


Figure 2. Ratios of taxa identified based on the study of (1) eight wells and (2) scattered charcoal from five domestic contexts

summarised in Figure 2 (Figueiral, unpublished report; Fabre, unpublished report). On the whole, the major biotopes represented are the same, mainly composed of the mixed oak woodland, the alluvial forest, and the vineyards. For this reason, we can assume that these deposits both provide a reliable representation of the vegetation of the valley and its borders. Different local topographic / edaphic conditions may explain the minor differences be detected concerning both the plant spectrum and relative frequencies of taxa identified. The presence of present-day exogenous species, such as *Fagus*, *Abies*, *Larix* / *Picea*, may result either from trade or from their sporadic survival (especially for *Fagus*) in the more favourable locations of the lowlands since the last Glacial Maximum as shown by previous studies (Triat-Laval 1978, 1982; Durand and Vernet 1987; Chabal 1997, 2001; Durand 1998; Puertas 1998; Fabre 2004).

However, the (very) high frequencies (33%) of monocotyledonous stems (*Arundo* / *Phragmites*) recorded in the well are more difficult to explain, based on local ecology alone, and arise further questions. Is it just a matter of chance? Or is the same phenomenon recorded elsewhere? We recall that, with the exception of the roofing material from *Lattara*, important frequencies of monocotyledonous stems are never recorded in habitat deposits, especially in those resulting from domestic fuel.

A survey of data obtained in all the structures studied shows that high frequencies of *Arundo* / *Phragmites* are also noticed in other roman wells, such as those at Mas de Vignolles 13 (PT2077: 52%, PT2152: 77,3%, PT2176: 14 fragments in 52) (Figueiral unpublished report) and Mas de Vignolles 9 (49 fragments in 63) (Figueiral unpublished report) and Le Gasquinoy (PT3407: 13 fragments in 58; PT5027: 47,6%) (Figueiral *et al.* 2010).

Does this abundance simply reflect the better preservation conditions provided by these structures? Or could it reflect specific activities carried out in the near vicinity? If so, what do these sites have in common to justify the special use of these plants?

We recall that the hollow monocotyledonous stems of both *Arundo* and *Phragmites* are very fragile due to the absence of secondary growth. Xylem and phloem are only present in the individual vascular bundles. The lignin content, in the stem, is therefore very low in comparison with that of cellulose. As a result, when in contact with fire, stems become brittle and powdery. Reeds burn like straw, usually down to ashes. Their preservation requires exceptional taphonomic conditions, such as those provided by wells, where trampling is absent.

The abundance of reeds in the wells could thus result from differential conservation. But wouldn't this also mean that fragments of these stems would be easily preserved in fireplaces? And yet, this is not the case.

Another very plausible explanation concerns the potential use of these plants in the farming system. Aligned plantation pits have been uncovered in the vicinity of all these wells. By their general size, shape and spacing they seem to provide convincing evidence of vine growing. It is therefore interesting to recall that according to Columella (Brun 2003), reeds were used to support vine plants. Although this particular practice is not clearly referred to in our region, it could explain the abundance of these plants in our agrarian sites. This ancient tradition still survives in Morocco (El Faïz 2005).

Furthermore, fragments of waterlogged reeds were also recovered from wells PT2077, PT2152, PT2176 (Mas de Vignolles 13) and PT2052 (Mont Ferrier); we can thus envisage that, in farming contexts, bundles of reeds (to be used as vine support) could be assembled in the vicinity of the wells. Debris and used / damaged bundles ready for replacement could be burned and discarded in the wells. This would explain both the abundance of reeds in the wells and their absence from the protected environment of fireplaces.

The examples presented so far, illustrate a single chronological period (Roman). However earlier periods are also represented by the structures studied.

The Iron Age wells from Mas de Vignolles 9 (PT3172, PT1553) also provided abundant charcoal fragments. The taxonomic list comprises elements affiliated to the two major local habitats: the mixed oak woodland and the alluvial forest, where *Fraxinus* and *Ulmus* grow abundantly. In this case, as in the roman wells, charcoal reflects a coherent ecological representation of local woodlands.

Unfortunately no contemporaneous habitat level has been excavated nearby to establish a direct comparison. However, it might be interesting to take into account the data from Moulin Villard II, a nearby site, occupied during the Bronze Age (habitat layers 665 + 18 + 18p2 + 72 + 318). The plant assemblage identified in the habitat layers is similar to that from Vignolles 9 and in conformity with the surrounding vegetation (Figure 2). As previously, minor frequency variations can be easily explained taking into account the different chronology, very local particularities, or may even be interpreted as resulting from temporary land clearance (the local prime agricultural land of the alluvial plain could be more intensely used).

Two Middle Neolithic "Chasséen" wells were uncovered at Mas de Vignolles 9 (PT1157 and PT1051). The relative scarcity of plant material preserved in both wells (52 charcoal fragments, no waterlogged wood, 115 seeds) may result from the very nature of the human settlement, i.e. a non-permanent habitat. The plant assemblage identified includes elements of the deciduous oak woodland, *Erica* and *Arbutus unedo,* characteristic of the Middle – Late Neolithic Mediterranean vegetation. Although the number of charcoal available is low, results obtained are coherent with those from Moulin Villard II (Late Neolithic/Chalcolithic) (Figure 2). During these early periods, the reduced sedimentation of the alluvial

| WELL | PT234 | PT471 | | | | PT348 | |
|---|---|---|---|---|---|---|---|
| AREA | 3 | 9 | | | | 8 | |
| STRATIGRAPHIC UNIT | 3456 | 9168 | 9167 | 9165 | 9164 | 8178 | 8176 |
| CHRONOLOGY | -30/-15 | 50/75 | 75/100 | | | 75/100 | |
| TAXA | | | | | | | |
| Angiospermae | | 9 | | | | | |
| *Abies* | 1 | | 1 | | | 1 | 1 |
| *Acer monspessulanum* | | | 1 | | | | |
| *Acer* | | | | | | 2 | |
| *Acer cf. E. campestre* | 1 | | 1 | 2 | | | |
| *Alnus* | | 1 | | | | 1 | 2 |
| *Arbutus unedo* | 13 | 3 | 5 | 3 | 3 | 3 | |
| *Buxus sempervirens* | | | | 3 | 5 | | 1 |
| *Chenopodiaceae* | 2 | | | | 1 | | |
| *Cistaceae-Ericaceae* | 5 | 11 | | | | | |
| *Clematis* | 2 | | | | | 1 | |
| *Cornus* | 1 | | | | | | 1 |
| *Corylus avellana* | | | 1 | | | | |
| *Cupressus* | | | 2 | | | | |
| *Daphne gnidium* | | 1 | | | | | |
| *Erica cf. E. arborea* | | | | 1 | | | |
| *Erica arborea/multiflora/scoparia* | 12 | 15 | 18 | 18 | 18 | 11 | 8 |
| *Erica (rays 3-s)* | | | 3 | | | | |
| *Evonymus europaeus* | 1 | | | | | | |
| *Fagus sylvatica* | | | | | | | 1 |
| *Ficus carica* | 4 | 1 | | 2 | | | |
| *Fraxinus* | 10 | 8 | 13 | 6 | 1 | 5 | 2 |
| *Gymnospermae* | | | | | | 1 | |
| *Juglans regia* | | | | 1 | 3 | | |
| *Juniperus* | | 1 | | 1 | 1 | | 1 |
| *Fabaceae* | | | | | | | |
| *Monocotyledonae* | | 1 | 3 | | | 1 | |
| *Olea europaea* | | | | 1 | | | |
| *Phillyrea-Rhamnus alaternus* | 3 | 7 | 11 | 3 | 2 | 2 | 8 |
| *Phragmites australis* | | 1 | | | | | |
| *Pinus* | | | | | | | |
| *Pinus halepensis* | | | 9 | 8 | 2 | | |
| *Pinus cf. P. halepensis* | 1 | 9 | | | | 1 | |
| *Pinus mugo/nigra/sylvestris/uncinata* | | 3 | 4 | | | | 1 |
| *Pistacia* | 1 | | | | 3 | | |
| *Pistacia lentiscus/cf lentiscus* | 5 | 2 | 3 | 1 | | | 7 |
| *Pistacia terebinthus* | 1 | | | | | | |
| *Populus* | 3 | | 4 | | 1 | | |
| *Prunus avium + cf. Prunus avium* | 1 | 1 | | 1 | | | |
| *Quercus* deciduous | 10 | 15 | 19 | 13 | 1 | 4 | 16 |
| *Quercus ilex/coccifera* | 54 | 30 | 64 | 114 | 33 | 52 | 121 |
| *Quercus* | | 1 | | | | | |
| *Rosaceae Maloideae* | | 2 | | | 2 | 1 | 3 |
| *Rosaceae Prunoideae* | | 1 | | 1 | 1 | 2 | |
| *Salix* | 2 | | | | | | |
| *Tamarix* | | 1 | 15 | 1 | | 11 | 4 |
| *Ulmus* | 6 | 6 | 13 | 1 | 6 | 1 | 3 |
| *Viburnum tinus* | 4 | | | | | | |
| *Vitis vinifera* | 7 | 4 | 11 | 1 | 4 | 6 | 5 |
| *Number of fragments* | 150 | 134 | 213 | 182 | 87 | 106 | 185 |
| Minimal number of species | 22 | 21 | 17 | 19 | 17 | 17 | 17 |
| Number of taxa represented by 1 - 3 fragments | 13 | 14 | 8 | 15 | 12 | 11 | 10 |

Figure 3. Charcoal assemblages from three wells located at Lattara (Lattes, Roman Period): taxa represented by very few specimens (1-3 fragments) occur very frequently, more than in aerobic deposits

plain explains the negligible representation of the affiliated vegetation.

**Is archaeological charcoal better preserved in wells?**

As seen in Figure 2, the data obtained in wells PT471, PT234 and PT348 from *Lattara*, are very similar to those provided by charcoal sampled in the dwellings (Chabal 2005). As already remarked above, these results suggest that the charcoal found in the wells is derived from domestic fuel, and should be interpreted as resulting more from household activities (clearing of fireplaces) than from other combustions. However, these wells also appear to contain a greater range of taxa in comparison with the habitat layers (Figure 3). Species represented by very few specimens (1-3 fragments) occur very frequently, sometimes adding up to more than half of the taxonomic spectrum.

We believe this may reflect a direct relationship between taxonomic diversity and mechanical constraints. Well deposits are not subject to trampling and the presence of

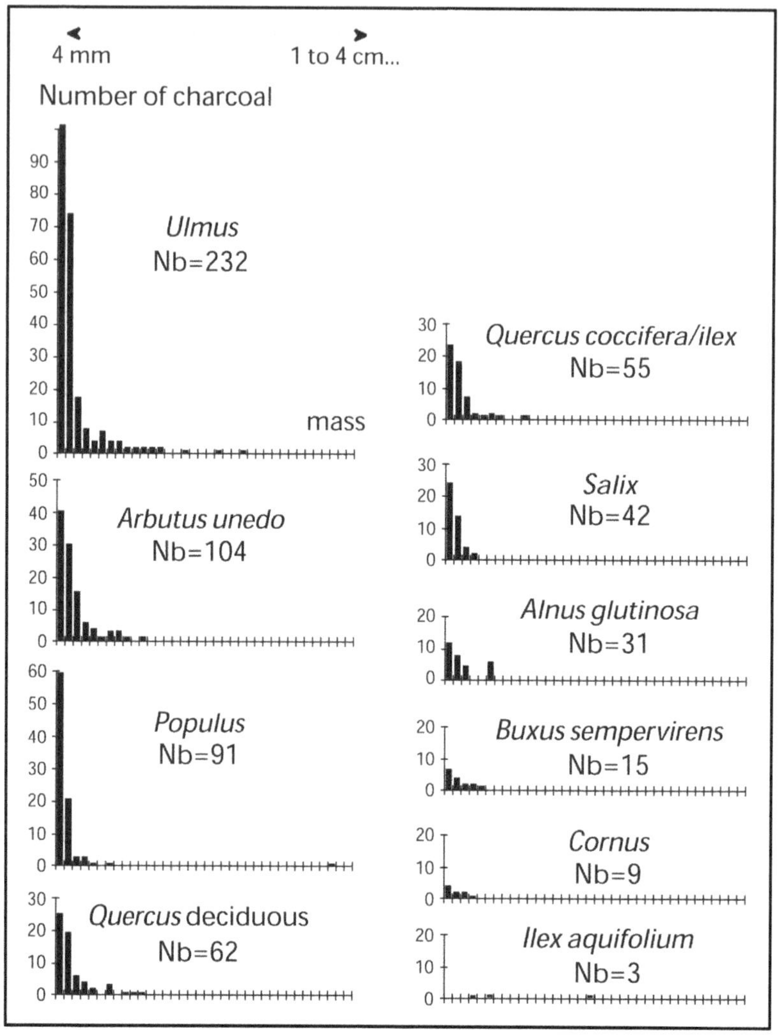


Figure 4. Fragmentation of charcoal leads to a statistical process, similar to a 'Poisson distribution', identical for all the species within a single archaeological sample, providing that material is recovered by sieving (Z.11 dec.21A, Le Marduel, Iron Age 450/425 BC, mesh 4 mm)

water, which is uncompressible, limits the effects of the sediment weight and abrasion on charcoal. We could argue that less mechanical constraints equal more charcoal fragments rather than more species. However, it has been frequently noticed that sporadic taxa tend to be identified while analysing the smaller fragments of a given sample (for instance, sieved with a 4 mm mesh) for the only reason that small fragments are more numerous (Chabal 1990). Furthermore, the fragmentation of the biggest fragments increases the number of fragments, but of course not the number of taxa (Chabal 1989). This leads us to ask if outside anaerobic contexts, mechanical constraints prompt the destruction of the smaller fragments, and thus of the rarer taxa.

We recall that charcoal fragmentation in a sieved sample always follows the same pattern: frequent taxa are identified among both small and large fragments (from 4 mm to about 4 cm) while rarer plants are only present in the smaller fragments (about 4 mm) (Chabal 1990, 1997). This is not a result of species properties as might be expected. Data obtained demonstrate that fragmentation is identical for all the species of a given archaeological sample, as shown by the example illustrated in Figure 4, (Z.11 dec.21A, Le Marduel, 450/425 BC). For instance, *Buxus sempervirens*, a species well known for the density of its wood, is only sporadically recorded in this sample. As a result it has survived only among the small fragments. But in other samples, where *Buxus* is abundant, it is identified among both small and large fragments. Furthermore, and as demonstrated by Badal Garcia (1990) the identification of the smaller fraction of samples (2-5 mm) yields similar results to those obtained with the larger fraction (>5 mm). The number of taxa identified is comparable. It is for this reason that, when scattered charcoal is abundant, specialists tend to focus on the fraction > 4 or 5 mm, making sure however that the smaller fragments included are analysed.

The rarer taxa are recorded only among the smaller fragments because of the fragmentation process. When an accumulation of charcoal fragments is subjected to post-depositional mechanical constraints, resulting in further fragmentation, it is the smaller fragments (rarer species) which disappear in priority. The smaller charcoal fragments are less crushed under the low mechanical constraints exerted in the wells, the taxa represented by few specimens are better protected and consequently the number of taxa identified is higher.

The same phenomenon is observed when sieving is carried out. Flotation sieving provides more species than manual sieving, because small fragments are better protected from complete destruction, and because big fragments are less fragmented (Chabal 1989).

## CONCLUSION

The diversity of waterlogged wood residues in the wells studied is remarkable. Their function and provenance vary between wells and in the same well. Remains of domestic, agricultural and craft industry implements abound in some structures. Most of the objects would be made locally, with available woods, but the exquisite quality of some pieces also suggests importation.

The exploitation of local woods for fuel and raw materials is illustrated by the profusion of wood waste and charcoal. The study of these charcoal fragments provides information, concerning the vegetal environment of the site, its use and management, similar to the one suggested by domestic dispersed charcoal. Within a single site, similar results are obtained when studying charcoal from different contemporaneous wells.

More specific information such as eventual fire events and the particular use of plants, such as reeds, in the farming system is also obtained.

The remarkable taxonomic diversity found in some wells (including species rarely identified in habitat layers) may reflect a direct relationship between taxonomic diversity and mechanical constraints. The absence of trampling and the presence of water may favour the survival of smaller fragments and correlatively, of rarer species.

Although some of the debris might have fallen naturally or been lost accidentally, it is clear that most of the remains preserved result mainly from the use of wells as a rubbish dump and have little or no relationship with the drawing of water. Each well containing organic material, is a case apart, providing specific information, which needs to be analysed, deciphered and explained.

### Acknowledgements

The authors would like to thank the excavation directors, who allowed us to study the material from the wells: R. Buxo, J.-L. Fiches, G. Piques (UMR 5140 CNRS), L. Bouffat, J.Y. Breuil, J.-B. Chevance, M. Compan, M. Piskorz, P. Séjalon (INRAP), R. Bourgaut (Communauté de Communes Nord du Bassin de Thau). The authors gratefully acknowledge J.-M. Femenias (ARCHEOPUITS).

### References

AUXIETTE, G., BOULEN, M., DESENNE, S., MATTERNE, V., ROCQ, C., YVINEC, J.H., PONEL, P. and PERNAUD, J.-M. 2003. Un site du Hallstatt à Villeneuve-Saint-Germain « Les Etomelles » (Aisne). *Revue Archéologique de Picardie* 3 / 4, 21-65.

BADAL GARCIA, E. 1990. Méthode de prélèvement et paléoécologie du Néolithique d'après les charbons de bois de "La Cova de les Cendres" (Alicante, Espagne), In T. Hackens, A.V. Munaut and C. Till (eds), Wood and Archaeology, Bois et archéologie, First European Conference, Louvain-la-Neuve, October 2nd-3rd 1987, *PACT* 22. Louvain-la-Neuve, 231-243.

BAKALOWICZ, M. and BLANCHEMANCHE, P. 2005. Le contexte hydrogéologique des puits et l'économie de l'eau douce à Lattara. *Lattara* 18, A.D.A.L. Ed., Lattes, 7-12.

BUXO, R., CHABAL, L. and ROUX, J.-C. 1996. Toiture et restes carbonisés d'une maison incendiée dans l'habitat de Lattes au IVe s. av. n. ère. In M. Py (ed), Architecture et urbanisme de Lattes antique, *Lattara* 9, A.R.A.L.O. Ed., Lattes, 373-400.

BRUN, J.P. 2003. *Le vin et l'huile dans la Méditerranée antique. Viticulture, oléiculture et procédés de fabrication.* Coll. des Hespérides, Ed. Errance, Paris.

CHABAL, L. 1989. Perspectives anthracologiques sur le site de Lattes (Hérault). *Lattara* 2, A.R.A.L.O. ed., Lattes, 53-72.

CHABAL, L. 1990. L'étude paléoécologique de sites protohistoriques à partir des charbons de bois: la question de l'unité de mesure. Dénombrements de fragments ou pesées? In T. Hackens, A.V. Munaut and C. Till (eds), Wood and Archaeology, Bois et archéologie: First European Conference, Louvain-la-Neuve, October 2nd-3rd 1987, *PACT* 22. Louvain-la-Neuve, 189-205.

CHABAL, L. 1997. Forêts et sociétés en Languedoc (Néolithique final, Antiquité tardive). L'anthracologie, méthode et paléoécologie. *Documents Archéologie Française* 63, Paris, MSH.

CHABAL, L. 2001. Les Potiers, le bois et la forêt à Sallèles d'Aude (I-IIIe s. ap. J.-C.). In 20 ans de recherches à Sallèles d'Aude: le Monde des potiers gallo-romains, Colloque 27-28 sept. 1996, Sallèles d'Aude, Annales Littéraires de l'Université de Besançon, Série Amphores 5, 93-110.

CHABAL, L. 2005. Charbons de bois et bois gorgé d'eau des puits antiques: des jardins de *Lattara* aux forêts du delta du Lez. *Lattara* 18, A.D.A.L. Ed., Lattes, 221-234.

CHABAL, L. and Feugère, M. 2005. Le mobilier organique des puits antiques et autres contextes humides de Lattara. *Lattara* 18, A.D.A.L. Ed., Lattes, 137-188.

COTTES, J., COULAROU, J., GIRAUD, J.-P., ROUZAUD, F., DUDAY, H. and VAQUER, J. 1980. Nouvelles découvertes à Villeneune-Tolosane (Haute-Garonne), *Archeologia* 142, 61-62.

DUCHESNE, S. and TREIL, J. 2005. Analyse de trois squelettes humains et de restes de nouveaux-nés, *Lattara* 18, A.D.A.L. Ed., Lattes, 335-344.

DURAND, A. 1998. *Les paysages médiévaux du Languedoc (Xe-XIIe siècles)*. Presses Universitaires du Mirail (ed.), Toulouse.

DURAND, A. and VERNET, J.L. 1987. Anthracologie et paysages médiévaux: à propos de quatre sites languedociens, *Annales du Midi* 4, nov. – déc., 397-405.

El FAÏZ, M. 2005. Les vignobles de l'Oudaya de Marrakech (Maroc). In Les paysages culturels viticoles, Etude thématique dans le cadre de la Convention du Patrimoine mondial de l'UNESCO, Juillet 2005, ICOMOS.

FABRE, L. 2004. Les restes de bois. In R. Thernot., V. Bel and S. Mauné (Dir.), L'établissement rural antique de Soumaltre à Aspiran (Hérault), Ferme, auberge, nécropole en bordure de la voie *Cessero-Condatomagus* (Ier-IIe s. ap. J.-C.), *Archéologie et Histoire Romaine* 13, Mergoil (ed.), Montagnac, pp. 262-265.

FIGUEIRAL, I. 1999. Fossil wood and charcoal identification techniques. In T.P. Jones and N.P. Rowe (eds), Fossil Plants and Spores. Modern Techniques. Geological Society of London, 92-96.

FIGUEIRAL, I. and WILLCOX, G. 1999. Archaeobotany and sub-fossils: collecting and analytical techniques. In T.P. Jones and N.P. Rowe, (eds), Fossil Plants and Spores. Modern Techniques. Geological Society of London, 290–294.

FIGUEIRAL, I., BOUBY, L., BUFFAT, L., PETITOT, H., TERRAL, J.-F. 2010. Archaeobotany, vine growing and wine producing in Roman Southern France: The site of Gasquinoy (Béziers, Hérault). *Journal of Archaeological Science*, 37, p. 139-149.

GIRAUD, P., LOUIT, S., GIRARDCLOS, O., ROCQ, C., MATTERNE, V., MILLE, P., DUPERON, M. and DUPERON, J. 2005. Un puits cuvelé de la Tène finale à Saint-Denis, In O. Buchsenschutz, A. Bulard, and T. Lejars (eds), L'âge du Fer en Ile-de-France, *Revue archéologique du Centre de la France* 26e Suppl., 55-71.

GREIG, J. 1988. The interpretation of some Roman well fills from the midlands of England, in H. Küster (Dir.), *Der prähistorische Mensch und Seine Umwelt.* Festschrift für Prof. U. Körber-Grohne. *Forschungen und Berichte zur Vor- und Frühgeschichte in Baden-Württemberg* 31, 367-378.

KNÖRZER, K.-H. 1984. Veränderungen der Unkrautvegetation auf Rheinischen Bauernhöfen seit der Römerzeit. Bonner Jahrbücher 184, 479-503.

MATTERNE, V. 2000. Etude de restes végétaux provenant des puits antiques du site du Palais de Justice à Melun (Seine-et-Marne). Archéopages 1, INRAP, Paris, 10-19.

PIQUES, G. and BUXO, R. (eds) 2005. Onze puits gallo-romains de *Lattara* (I^er^ s. av. n. è. – II^e^ s. de n. è.). *Lattara* 18, A.D.A.L. Ed., Lattes.

PUERTAS, O. 1998. Palynologie dans le delta du Lez. Contribution à l'histoire du paysage de Lattes. *Lattara* 11, A.D.A.L. Ed., Lattes.

TRIAT-LAVAL, H. 1978. *Contribution pollenanalytique à l'histoire tardi- et postglaciaire de la végétation de la basse vallée du Rhône.* Thèse Doctorat ès Sciences, Aix-Marseille III.

TRIAT-LAVAL, H. 1982. Pollenanalyse de sediments quaternaires récents du pourtour de l'Etang de Berre. *Ecologia Mediterranea* VIII, 4, 98-115.

ZWIERZINSKI, E. 1999. *Apport de la carpologie à la caractérisation des espaces: l'exemple des remplissages de deux puits de l'agglomération gallo-romaine de Jouars-Ponchartrin (78, Yvelines).* D.E.A, Université Paris X, Nanterre.

# USE OF WOOD AND ENVIRONMENT IN BRONZE AGE EBLA (NW SYRIA): RESULTS OF THE ANTHRACOLOGICAL ANALYSES

Girolamo FIORENTINO and Valentina CARACUTA
Laboratory of Archaeobotany and Palaeoecology. University of Salento, Italy

**Abstract**: *Starting from 2003, several archaeobotanical campaigns have been carried out on the Ebla site, NW Syria, in order to shed light on the use of plants between the Early Bronze Age IIIa and Middle-Late Bronze Age. The results of the anthracological analysis, here presented for the first time, following a continuous chronological pattern, reveal changes in plants exploitation according to cultural aspects, functions of use and climate. Several contexts were investigated and the anthracological analysis shows changes in the use of wood between the Early and Middle-Late Bronze Age. Wood species found in the EBA Royal palace G were, for the major part, beams and poles of* Cedrus *and* Abies *employed as carpentry. In the same layers luxury artefacts of Pomoideae,* Fraxinus *sp. and* Olea *wood were found. More complex is the picture that results from the study of material coming from M/LBA phases due to the variety of contexts investigated. The anthracological analysis, carried out on private houses, cooking ovens (type* tannurs*), dumps and ritual wells, has revealed, indeed, a complex pattern of wood exploitation which varied according to cultural habits and natural resources.*

**Key words**: *Anthracology, Ebla, Bronze Age, Syria, Woodland*

## INTRODUCTION

The study of raw materials used in ancient times has long been the exclusive preserve of historians and philologists, especially in those areas characterized by availability of administrative texts and figurative reliefs. The effect of this tradition has been especially significant in Near Eastern archaeology, where until recently all aspects of the supply system, especially those related to scarcely-visible archaeological raw materials such as plant remains, were investigated by means of textual or artistic evidence (Archi 1991, 1999; Linder 1986; Milano 1981, 1987; Rowton 1967; Wiseman 1952).

In the last few decades, advances in archaeobotanical analysis (of both seeds/fruits and charcoals) have opened up new perspectives in the study of agricultural systems and human eating habits (Costantini and Dyson 1990; De Moulins 1997; Fiorentino and Caracuta 2010; Hald and Charles 2008; Riehl 2008; Van Zeist and Bakker-Heeres 1985) and triggered investigations of fuel selection and its related effects on forest cover (Deckers 2005; Deckers and Riehl 2007; McCorriston 2007; Miller 1990; Neef 2001; Pessin 2007; Willcox 1974, 1991, 1992, 1999).

In this study we intend to shed light on the use and provenance of wood resources in the protohistoric site of Ebla, and propose a reconstruction of the ancient environment based on the results of the archaeobotanical campaigns carried out by the Laboratory of Archaeobotany and Palaeoecology, University of Salento, between 2003 and 2007.

## ARCHAEOLOGICAL CONTEXT AND ENVIRONMENTAL SETTING

The city of Ebla (Tell Mardick – Idlib region) is well known in Near Eastern archaeology, since it played an important role in the history of the region in the 3rd and 2nd millennia BC. The first nucleus developed around 2600-2400 BC (Early Bronze Age II-III) when the city became the economic and administrative center of an area in northern Syria bounded by the Euphrates and the Orontes river valleys and the southern slopes of the Taurus (Matthiae 1989; Pettinato 1986).

The rise of Ebla is confirmed by the huge amount of administrative texts discovered in Royal Palace G dated to the Early Bronze Age IVa, they give direct information on the system of taxation and shed light on the social organization of the kingdom. The archaeological evidence for that period is limited to public contexts, both palatine and cultic, which were preserved by a fire which probably broke out during the destruction of the town by its enemies.

The centuries following the breakdown of Ebla are often characterized as a "dark age" (Early Bronze Age IVb), at least as far as the textual evidence is concerned (Kengler 1992).

The re-organization of the town occurred between 2000 and 1800 BC (Middle Bronze Age), when the city's layout underwent considerable modification.

A new palace complex (Palace FF) was built over the ancient ruins to a new design, while places of worship were built in the lower town and residential quarters in the eastern part. The powerful position achieved in this period was thrown into crisis in the following centuries by social upheavals and climatic changes which led to the definitive abandonment of the town (1700-1600 BC ca) (Fiorentino *et al.* 2008).

The influence of environmental factors in Ebla's history is closely related to its geographical position on the boundary between two different ecological zones, the arid

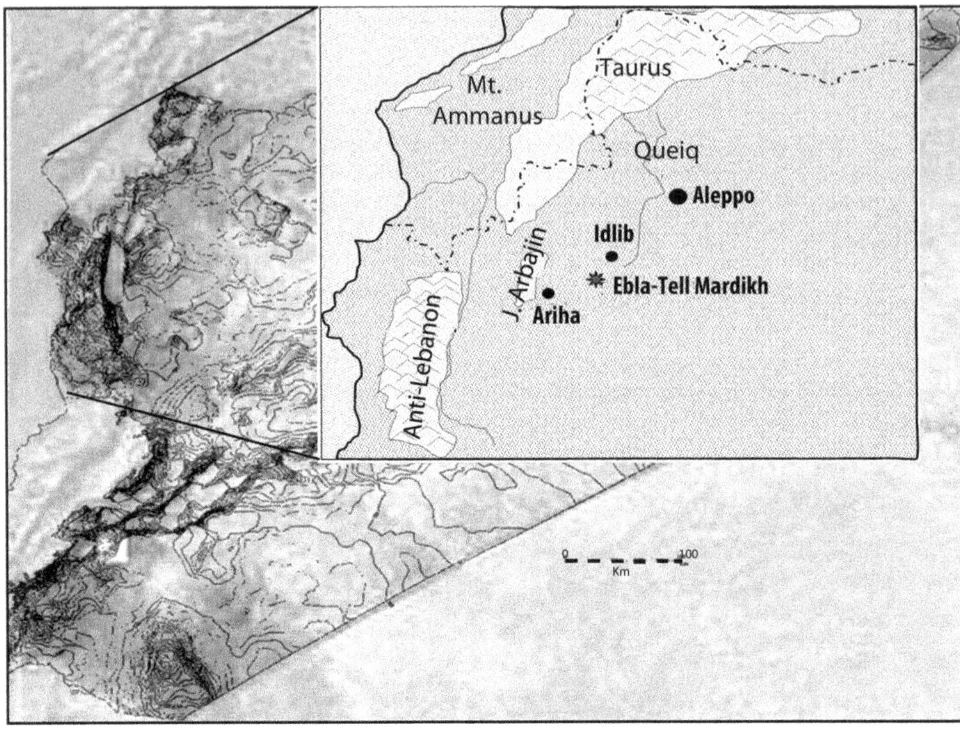


Figure 1. The context of study

eastern steppes and the more humid western belt, influenced by the Mediterranean (Fiorentino *et al.* 2008).

This dichotomy is partially mitigated by the presence on the flat steppe of the Jebel Arbajin, a hill covered by Mediterranean woodland, and of the river Qoueiq, a seasonal stream now almost dried up (Figure 1).

In good climatic conditions, water, wood and land for cultivation were available nearby. Nevertheless, drought and anthropogenic factors (i.e. deforestation) drastically reduced these resources and limited the settlement's development.

Considering that the Ebla kingdom developed over a time period of significant climate change (for further details see Anderson *et al.* 2007; Bar Mathwes *et al.* 1997; Courty 1994; Cremaschi 2007; Cullen *et al.* 2000; De Menocal 2001; De Menocal *et al.* 2000; Fiorentino *et al.* 2008; Frumkin *et al.* 1999; Issar 2003; Horowitz 1992; Kuzucuoglu and Marro 2007; Riehl *et al.* 2007; Valsecchi 2007; Weiss *et al.* 1993; Wilkinson 2004), analysis of the anthracological evidence and changes in its composition might tell us much about the city's relationship with local resources. Nevertheless, a good deal of caution is required before drawing paleoclimatical conclusions from anthracological analyses. The complex supply system makes it hard to discern local environmental patterns, since particular species could have been chosen for utilitarian purposes, even when not locally available.

Based on these assumptions, the archaeobotanical analysis was combined with the archaeological contextualization of the charred remains in order to distinguish allochthonous species from autochthonous and cultivated ones on the basis of their ecological character and the use they were intended for.

## METHODS AND MATERIALS

The anthracological remains analysed here were recovered during the 2003-2007 field campaigns, as charcoal dispersed in strata, concentrations of charred material, and burnt carpentry.

A small quantity of anthracological remains were recovered from both modern and old excavations of the EBA Royal Palace G. The majority were collected from MBA contexts such as the sacred area HH and the residential quarter B-est.

Special care was taken during the excavation of the midden EE, situated on the eastern rampart, which contained the rubbish of the entire MBA settlement. In this case we adopted a microstratigraphical excavation strategy which enabled us to distinguish between episodes of charred material being brought from ovens and discarded there, and occasions when fresh material was burnt *in situ*.

All the plant material, both from modern and old excavation, was separated from the mineral matrix of soil by using dry and wet sieving and flotation by means of meshes of 2,0 and 0,5 mm.

The flots were dried, labelled and identified, observing the main anatomical features on fresh fractures by light microscopy with dark and light fields. Identification was carried out by consulting different atlases of wood anatomy (Fahn *et al.* 1990; Greguss 1955, 1959; Schweingruber 1990) and making comparisons with reference material collected in the field by the authors from a number of locations in Syria and stored in the Laboratory of Archaeobotany and Palaeoecology, University of Salento. The use of such reference material was essential because of the large number of endemic species in the Near East (Willcox 1999).

## RESULTS

A total of 2727 anthraco-remains were analysed: 458 for the EBA and 2269 for the MBA contexts. Twenty-two taxa were identified: *Olea europaea* was the most attested, but significant differences were found between the EBA and MBA phases and between different contexts within the same phase (Figure 2; Figure 3).

Most of the material recovered from EBA Palace G belongs to *Abies* sp., *Cedrus* sp. and *Arundo donax* sp., with slightly smaller amounts of *Olea europaea.* In addition, some fragments of *Platanus* sp., *Populus/Salix*, *Fraxinus* sp. and Pomoideae were found.

The MBA anthraco-remains were more varied and depended on the type of context investigated.

The votive well, full of clay figurines and miniaturized pots, of analysed in sacred area HH was characterized by a large number of wood types, with *Olea europaea* and deciduous *Quercus* species as the main taxa. A good deal of fruit trees such as *Pistacia vera*, *Prunus* cf. *avium* and *Prunus* cf. *dulcis* were also well attested, together with the semi-deciduous *Quercus ithaburensis*.

Other species of *Pistacia* (*P. khinjuk/atlantica)* and oak (*Q. calliprinos*) were found but in smaller quantities.

Concerning residential contexts, the analysis carried out in the kitchens and ovens (*tannur*) discovered in quarter B-east revealed a predominance of *Olea europaea* and *Prunus* cf. *avium*, followed by *Fraxinus* sp. and *Pinus halepensis* sp.

Some of those taxa were also found in the midden EE, where *Olea europaea* and *Prunus* cf. *avium*, were once again the most frequent. Within the dump we also found three different oaks: *Q. calliprinos*, *Q. ithaburensis* and a deciduous species, together with the pistachio *Pistacia vera.*

Several other wood species were recovered from the midden, such as *Pistacia palaestina* and *P. khinjuk/atlantica* and indeterminate Ulmaceae and Leguminosae.

Ecologically, the twenty-two identified taxa vary in their provenance. Thus, before drawing any conclusions regarding the local environment, we needed to demonstrate the origin of the wood, distinguishing between timber imported for special uses, timber from locally growing wild species, and wood from trees cultivated locally for their fruit.

## DISCUSSION

The information from the anthracological analysis was studied with reference to their archaeological contexts and assumed uses of the wood, as revealed during the excavations, in order to correlate specialized uses of the wood to environmental assumptions.

The high frequency of *Abies* sp. and *Cedrus* sp. in the EBA Palace was explained by the need for luxury raw material for carpentry in royal palaces.

Large beams and columns of cedar wood were found in Palace G together with smaller beams of fir, and several fragments of reed interpreted as part of the roof cover.

There is also evidence from other archaeological contexts that poplar was used for roof timbers (Willcox 1992), but the few fragments found in Ebla do not allow certain attribution.

The furniture was made especially from *Fraxinus* sp., and was chosen together with a kind of Pomoideae to make decorative elements for a mixed wood-mother pearl relief to create chromatic contrast.

Olive wood was used in luxury manufacturing too, to make gold-laminated statuettes (Figure 4).

The anthracological remains from the EBA IVa Palace G are completely different to those of the MBA contexts. It is difficult to make direct comparisons, since the analysis of the remains from the MBA Palace FF is still in progress, but the charcoals collected in domestic and ritual contexts are totally different.

In domestic contexts (Quarter B-est), the structural timbers were most frequently made from *Olea* and *Prunus* cf. *avium*, as was almost all the furniture found. In addition, analysis of ovens showed that branches of both species were used as fuel, together with *Pinus halepensis* and *Fraxinus* sp.

It cannot be excluded that deciduous *Quercus* species were also used for fuel: a few fragments were found at the bottom of the ovens suggesting that oak fuel had been discarded before reusing the structures. The discovery of several branches of deciduous oak in layers of midden EE seems to confirm this hypothesis.

*Quercus ithaburensis*, *Q. calliprinos* and *Pistacia khinjuk/atlantica* may also have been used as fuel as they were found in ritual hearths and in midden layers, the latter as a result of being discarded from ovens.

Remains of fruit trees such as *Prunus* cf. *dulcis* and *Pistacia vera*, found together in the ritual well of the sacred area HH, can be interpreted as votive offerings.

Far less clear was the meaning of the other minor species such as *Capparis* sp., Leguminosae, *Rhamnus/Phillyrea*,

| Phase | | Archaeological Context | Function | Use of wood | *Abies* sp. | *Cedrus* sp. | *Pinus halepensis* | *Pinus sylvestris* type | *Cupressus* sp. | *Arundo* cf. *donax* | *Capparis* sp. | *Fraxinus* sp. | Leguminosae | *Olea europaea* | *Pistacia* cf. *vera* | *Pistacia* cf. *palaestina* | *Pistacia khinjuk/atlantica* | *Platanus* sp. | Pomoideae | *Populus/Salix* | *Prunus* cf. *avium* | *Prunus* cf. *dulcis* | *Quercus* deciduous | *Quercus* cf. *calliprinos* | *Quercus* cf. *ithaburensis* | *Rhamnus/Phillyrea* | Ulmaceae | |
|---|---|---|---|---|---|---|---|---|---|---|---|---|---|---|---|---|---|---|---|---|---|---|---|---|---|---|---|---|
| EBA | Royal palace G | L.2982 | treasure room | carpentry/furniture | 4 | 26 | 6 | - | - | - | - | 1 | - | - | - | - | - | - | 9 | 1 | - | - | - | - | - | - | - | |
| | | L.2980 | storage room | " | 115 | - | - | - | - | 61 | - | - | - | 17 | - | - | - | - | 2 | - | - | - | - | - | - | - | - | |
| | | Liv.4D | stairs | " | 1 | - | - | - | - | - | - | - | - | - | - | - | - | - | 1 | - | - | - | - | - | - | - | - | |
| | | 1030 | unknown | " | 4 | - | - | - | - | - | - | - | - | - | - | - | - | - | 1 | - | - | - | - | - | - | - | - | |
| | | L.8496 | administrative room | " | 12 | 31 | - | - | - | - | - | - | - | 1 | - | - | - | - | - | - | - | - | - | - | - | - | - | |
| | | L.8495 | " | " | 39 | 29 | - | - | - | - | - | - | - | 10 | - | - | - | - | - | - | - | - | - | - | - | - | - | |
| | | L.2752 | audience hall | " | 48 | - | - | - | - | - | - | - | - | - | - | - | - | - | - | - | - | - | - | - | - | - | - | |
| | | L.8606 | staircase | " | - | - | - | - | - | - | - | - | - | - | - | - | - | 30 | - | - | - | - | - | - | - | - | - | |
| | | L.8612 | room | " | 1 | - | - | - | - | - | - | 1 | - | - | - | - | - | - | - | - | - | - | - | - | - | - | - | |
| | | L.2601 | boardroom | " | - | - | - | - | - | - | - | 2 | - | - | - | - | - | 5 | - | - | - | - | - | - | - | - | - | |
| **Total** | | | | | **224** | **86** | **6** | **-** | **-** | **61** | **-** | **4** | **-** | **28** | **-** | **-** | **-** | **35** | **13** | **1** | **-** | **-** | **-** | **-** | **-** | **-** | **-** | **458** |
| MBA | Domestic context B-est | L.8454 | courtyard | carpentry/furniture | - | - | 16 | - | - | - | 3 | - | - | 135 | - | - | - | - | 7 | - | 29 | - | 1 | 1 | - | - | - | |
| | | T.8662 | cooking oven | fuel | - | - | 21 | - | - | - | - | 31 | - | 317 | - | - | - | - | - | - | 69 | - | 4 | 1 | - | - | - | |
| | | L.8455 | kitchen | carpentry/furniture | - | - | 1 | - | - | - | - | - | - | 61 | - | - | - | - | - | - | - | - | 1 | - | - | - | - | |
| | | T.8673 | cooking oven | fuel | - | - | 50 | - | - | - | 25 | - | - | 68 | - | - | - | - | - | - | 8 | - | 2 | - | - | - | - | |
| | | L.8650 | room | carpentry | - | - | - | - | - | - | - | - | - | 15 | - | - | - | - | - | - | - | - | - | - | - | - | - | |
| | | L.9512 | storage room | carpentry/furniture | - | - | 1 | - | - | - | - | - | - | 9 | - | - | - | - | - | - | - | - | - | - | - | - | - | |
| | Midden EE | Liv.2-5 | pits | oven fuel | - | - | - | 1 | - | - | - | - | 5 | 136 | 18 | 12 | 15 | - | 2 | - | 33 | - | 24 | 24 | 19 | 1 | 1 | |
| | | Liv.1-3-4-6-7 | hearths | carpentry/furniture | - | - | - | - | - | - | - | - | - | 78 | 25 | 7 | 19 | - | 3 | - | 48 | - | 47 | 39 | 38 | - | 6 | |
| | Temple HH | F.9308/F.9301 | bi-partite votive well | ritual fuel | - | 2 | - | 2 | 1 | - | 5 | 2 | - | 283 | 39 | - | 7 | - | - | - | 63 | 31 | 204 | 26 | 115 | 4 | 8 | |
| **Total** | | | | | **-** | **2** | **89** | **3** | **1** | **-** | **33** | **33** | **5** | **1102** | **82** | **19** | **41** | **-** | **12** | **-** | **250** | **31** | **283** | **91** | **172** | **5** | **15** | **2269** |

Figure 2. An overview of the wood species, on the basis of the context of finding, and their possible use

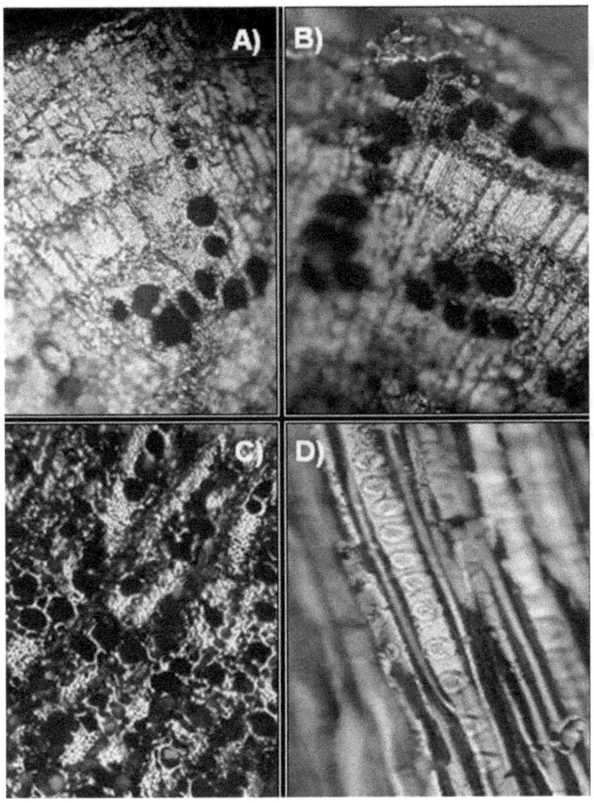


Figure 3. A) *Quercus deciduous*, transversal section (100x); B) *Pistacia khijuk/atlantica*, transversal section; C) *Olea europaea*, transversal section; D) *Cedrus* sp., radial section (500x)

Ulmaceae, *Cupressus* sp. and *Pinus silvestris*, found as discarded material in the votive well and the midden.

The use of timber is strictly related to its provenience. The above-mentioned 'functional' species such as cedar and fir were not necessarily local since, having particular qualities, they may have been traded across long distances (Willcox 1999).

Cedar and fir may have been imported from the Anti-Lebanon mountain range or the Amanus mountains (Zone C, Figure 5) via river systems, as documented by the inscription of Gudea of Lagash dated to 2200 BC (Linder 1986; Rostow 1967).

*Cupressus* sp., which was sought-after for luxury manufacturing and cosmetics (Kupper 1992), could also have originated from the western end of the Syrian-Turkish border. An even closer origin can be posited for *Q. ithaburensis* and *Q. calliprinos*, which could have been taken from the nearby Jebel Arbajin, where oaks benefit from favourable growing conditions even today (Zone B, Figure 5).

*Olea europaea*, *Prunus* cf. *avium*, *Prunus* cf. *dulcis*[1] and *Pistacia vera* almost certainly derived from local orchards, as documented by the administrative reports available for the EBA period in Ebla (Archi 1991; Milano 2006), although they may have been preferentially selected from local wild stands, deliberately preserved because of their function as a source of food.

The Aleppo pine, yielding poor quality timber, may have been cultivated for ornamental purposes (Willcox 1992) and collected for fuel.

Timber for construction may thus have been the object of trade over relatively long distances, while fruit-trees may have been cultivated in the surroundings of the site or exploited as part of the available natural resources.

Concerning wood used for fuel, it is commonly accepted that this reflects local vegetation, since the nearest available ligneous species to a site are the most frequently exploited for domestic use (Chabal 1997).

Excluding the cultivated fruit trees, the trees most widely used for fuel were deciduous *Quercus* species *Q. ithaburensis*, *Q. calliprinos* and *Pistacia khinjuk/atlantica,* which may have been the most readily available timber resources (Zones B-D, Figures 5 & 6).

While *Quercus ithaburensis* and *Q. calliprinos* could have originated from the Mediterranean belt, the eastern edge of which is marked by the Jebel Arbajin, the deciduous *Quercus* species and *Pistacia khinjuk/atlantica* originated further East.

Phytogeographical analyses have shown that the current range of *Pistacia khinjuk/atlantica*, now limited to a chain of hills stretching across the steppes of the Syrian desert, is the residue of a steppe-forest originating in the Zagros mountains which has been deprived of its oaks. The presence of a steppe-forest dominated by *Quercus brantii* associated with *Pistacia atlantica* and *P. khinjuk.* about 150 km east of the current distribution of this woodland is evidence for this assumption (Zohary 1973).

An extensive survey of texts dating back to 1250-550 BC gives a picture of the woodlands available in the region as a whole and suggests that the *Quercus-Pistacia* association extended over a vast area in ancient times, covering all the high ground of the Syrian desert with a mean annual rainfall of 450 mm (Zone D, Figure 5).

Today the region's woodland coverage is considerably smaller than in the past (compare Figure 5 to 6). Climatic factors and deforestation have left the Syrian territories almost without forest cover.

As well as in recent times, deforestation also had a great impact in the period from 2600 to 1250 BC. In this period Syria was affected by rapid urbanization and intensive forest clearance for cultivation, as well as climate instability, all of which led to a reduction of forest cover (Fall 2002; Miller 1990).

[1] *Prunus* cf. *avium* was distinguished from *P.* cf. *dulcis* on the base of distribution and dimension of pores, width of rays, and by the presences of uniseriate rays and libriform fibres.

Figure 4. A) Gold laminated harm with wooden core from EBA Palace G; B) mixed wood-mother pearl relief from EBA Palace G; C) Tannur from MBA residential quarter B-est. D) Piles and beams from MBA Palace FF

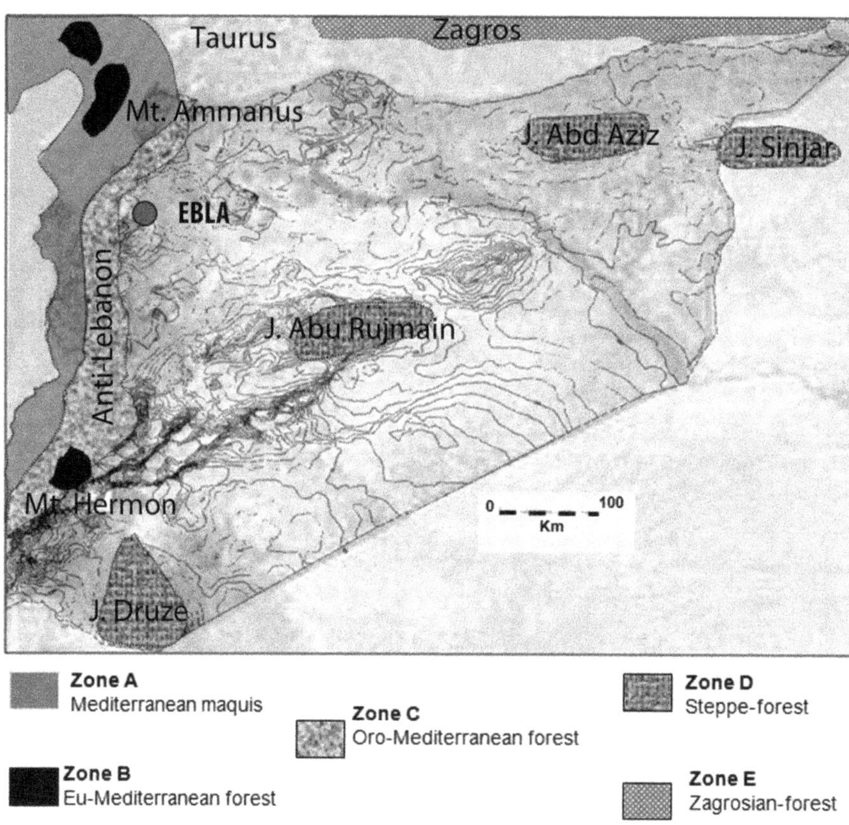


Figure 5. Modern woodland distribution in Syria (redrawn from Zohary 1973)

Anthracological analyses of Bronze Age remains carried out in the Khabur region (Deckers 2005; Deckers and Riehl 2004), the Euphrates Valley, the north-western end of the Syrian-Turkish border (Pessin 2007) and southern Syria (Willcox 1999) confirm that the *Quercus-Pistacia* association had a wider distribution than in the Iron Age, reaching the southern limit of the steppes and the Taurus mountain range in the west.

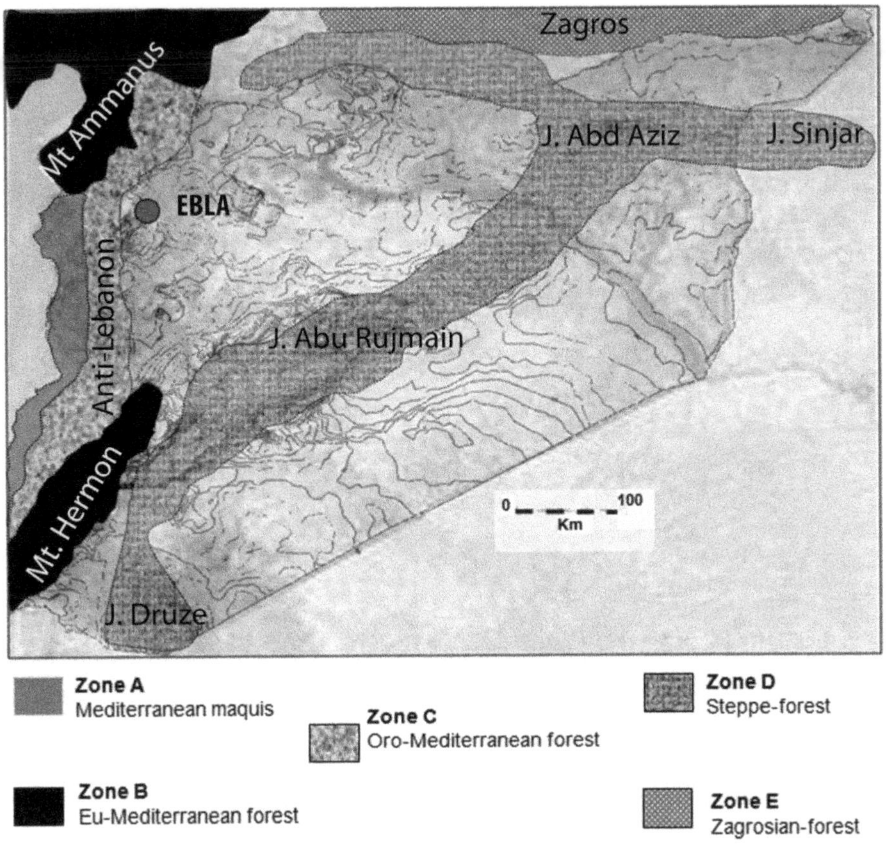


Figure 6. Distribution of woodland according to textual evidences dating back to 550-1250 BC (redrawn from Rowton 1967)

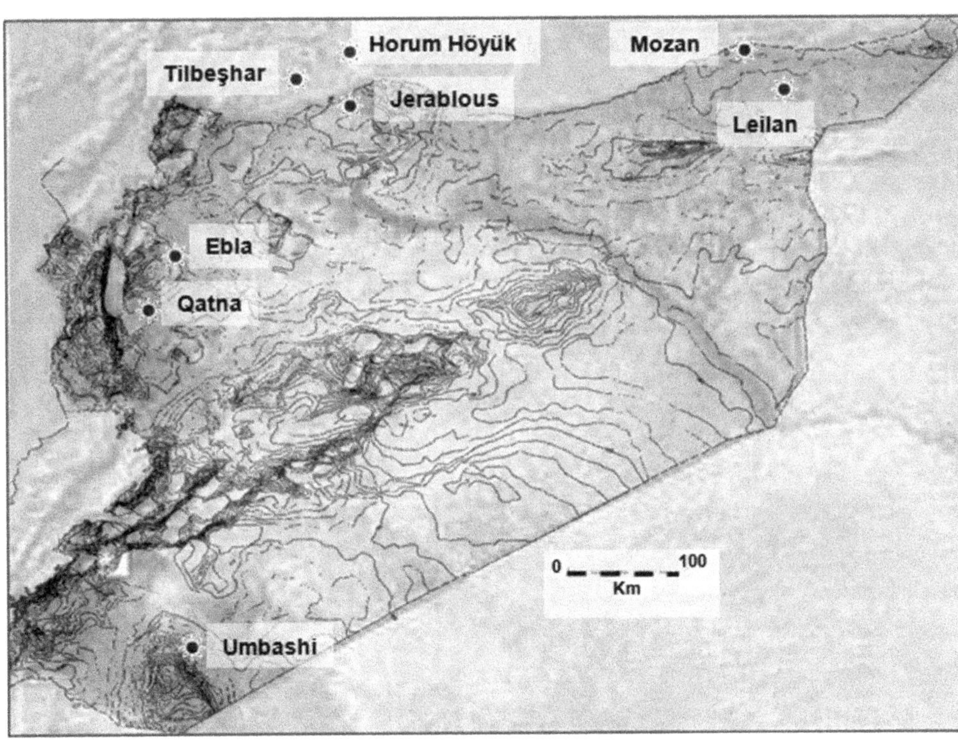


Figure 7. Map of the Bronze Age attestation of *Quercus* f.c.-*Pistacia* woodland in Syria

The discovery of deciduous *Quercus* species and *Pistacia khinjuk/atlantica* in Ebla add new information concerning the original distribution of the woodland, which extended much further to the West, probably reaching Qatna (for further detail see Deckers, http://www.geo.uni-tuebingen.de/?id=2609) (Figure 7).

The lack of remains of that particular species in the EBA contexts of Ebla tells us less about the environment than about the type of context investigated: palatine carpentry employed raw materials chosen for their status, regardless of the transport costs.

The only environmental information that can be inferred from the EBA contexts concerns the presence of a riverine forest, probably on the banks of the nearby Qoueiq, which included *Populus/Salix*, *Platanus* sp. and *Fraxinus* sp.

Residues of this woodland appear to have survived until the MBA, since *Fraxinus* sp. was used as fuel in domestic ovens and Ulmaceae were discarded in the ritual well.

Compared to the present situation, which is almost entirely depleted of woodland cover, the ancient Ebla district would have been quite heavily forested. The progressive deforestation which has taken place in Syria since the Late-Holocene has wiped out all the ancient woodland and very little woodland is found in the Ebla district today. Nevertheless, thanks to the anthracological analyses, which are able to identify charcoals of species that have long since disappeared from this area, it is now possible to describe the environment of a crucial area in north-western Syria.

## CONCLUSION

The study of the exploitation of ligneous species by analysing charcoal remains from the Ebla archaeological site has thrown light on the manner in which wood was used, the origin of the raw material and consequently the distribution of natural woodland.

As result, we have shown that timbers for luxury carpentry, such as *Abies* sp. and *Cedrus* sp., were imported from nearby mountains, while fruit trees, such as *Olea europaea* and *Prunus* cf. *avium*, were preferred for domestic carpentry and furniture. Together with *Prunus.* cf. *dulcis* and *Pistacia vera*, which had cultic significance, the fruit trees were cultivated in orchards or preferentially selected from among the available wild resources.

Finally, the natural woodland was characterized to the east by *Quercus-Pistacia* steppe-forest, while Mediterranean trees originated in the western hills. In the middle of these two ecological zones there was riparian vegetation, probably on the banks of the nearby river Qoueiq.

The anthracological analyses have confirmed the key role of Ebla in the local and regional context, indicating a complex pattern of supply which entailed both trading over long distances and exploiting local resources.

### Acknowledgment

The authors want to thanks Prof. Paolo Matthiae, head of the Ebla archaeological team, for allowing the study of the anthracoremains. Special thanks also to Luca Peyronel, Francesca Baffi, Marco Ramazzotti and Alessandra Enea, for the care showed in sampling during the archaeological excavation. Without their helpful work this research would have never been possible.

### References

http://www.geo.uni-tuebingen.de/?id=2609

ANDERSON, D.G., MAASCH, K.A., and SANDWEISS, D.H. 2007. *Climate change and cultural dynamics: a global perspective on Mid-Holocene transitions*. Amsterdam, Academic Press.

ARCHI, A. 1991. Culture de l'olivier et production de l'huile à Ebla. *Mélanges P. Garrelli*, Paris.

ARCHI, A. 1999. Cereal at Ebla. *Archív Orientální* 67, 503-518.

BAR-MATTHEWS, M., AYALON, A., KAUFMAN, A. 1997. Late Quaternary palaeoclimate in the eastern Mediterranean region from stable isotope analysis of speleothems at Soreq Cave, Israel. *Quaternary Research* 46, 155-168.

CHABAL, L. 1997. Forêts et sociétés en Languedoc (Néolithique final, Antiquité tardive). L'anthracologie, méthode et paléoécologie. DAF, 63, Paris, MSH.

COSTANTINI, L. and DYSON, R.H. 1990. The ancient agricolture of the Damghan Plain: the archaeobotanical evidence from Tepe Hissar. In N. Miller (ed.), *Economy and Settlement in the Near East: analysis of the ancient sites and* materials, 47-69. Masca Research Papers in Science and Archaeology, supplement to Volume 7. Philadelphia, University of Pennsylvania.

COURTY, M.A. 1994. Le cadre paléogéographique des occupations humaines dans le basin du Haut-Khabur (Syrie du nord-est): Premiers résultants. *Paléorient* 20, 21-59.

CREMASCHI, M. 2007. The environment of ancient Qatna. Contribution from natural science and landscape archaeology. In D. Morandi Bonacossi (ed.), *Urban and Natural Landscapes of an Ancient Syrian Capital. Settlement and Environment at Tell Mishrifeh/Qatna and in Central-Western Syria,* 331-336. Studi archeologici su Qatna 1. Udine, Forum.

CULLEN, H.M., DEMENOCAL, P.B., HEMMING, S., HEMMING, G., BROWN, F.H., GUILDERSON, T., SIROCKO, F. 2000. Climate change and the collapse of the Akkadian empire: Evidence from the deep sea. *Geology* 28(4), 379-382.

DECKERS, K. and RIEHL S. 2004. The development of economy and environment from the Bronze Age to the Early Iron Age in Northern Syria and the Levant. A case-study from the Upper Khabur region. *Antiquity* 78 (on the web site).

DECKERS, K. 2005. Anthracological research at the archaeological site of Emar on the Middle Euphrates, Syria. *Paléorient* 31, 152-166.

DECKERS, K. and RIEHL S. 2007. An evaluation of botanical assemblage from the Third to Second millennium BC in Northern Syria. In C. Kuzucuoğlu and C. Marro (eds.), *Sociétés humaines et changement climatique à la fin du troisiéme millénaire: une crise a-t-elle eu lieu en Haute Mésopotamie?* 481-502. Paris, De Boccard.

De MENOCAL, P., ORTIZ, J., GUILDERSON, T., SARNTHEIN, M. 2000. Coherent High- and Low-Latitude Climate Variability During the Holocene Warm Period. *Science* 288, 2198-202.

De MENOCAL, P.B. 2001. Cultural response to climate change during the Late Holocene. *Science* 292(27), 667-673.

De MOULINS, D. 1997. *Agricultural changes at Euphrates and steppe sites in the mid-8th to the 6th millennium BC.* British Archaeological Reports 683. Oxford, BAR Publishing.

FAHN, A., WERKER, E., BASS, P. (1990) *Wood anatomy and identification of trees and shrubs from Israel and adjacent regions.*

FALL, P.L., FALCONER S.E., LINES, L. 2002. Agricultural intensification and the secondary products revolution along the Jordan Rift. *Human Ecology* 30(4), 455-481.

FIORENTINO, G. and CARACUTA V. 2010. The use of plants in a ritual well at Ebla (Tell Mardikh)- North-western Syria. In Matthiae P., Pinnok F., Nigro L., Marchetti N., (eds.) *Proceedings of the 6th International Conference of Archaeology of Ancient Near East (ICAANE)* Volume 1, 307-320. Wisbaden: Harrassowitz Verlag..

FIORENTINO, G., CARACUTA, V., CALCAGNILE, L., D'ELIA, M., MATTHIAE, P., MAVELLI, F., QUARTA, G. 2008. Third millennium BC climate change in Syria highlighted by carbon stable isotope analysis of $^{14}$C-AMS dated plant remains from Ebla-Tell Mardikh. *Palaeogeography, Palaeoclimatology, Palaeoecology* 266(1-2), 51-58.

FRUMKIN, A., CARMI, I., GOPHER, A., FORD, D.C., SCHWARCZ, H.P., TSUK, T. (1999) A Holocene millennial-scale climatic cycle from a speleotheme in Nahal Qanah Cave, Israel. *The Holocene* 9(6), 677-682.

GREGUSS, P. 1955. Identification of living gymnosperms on the basis of their xylotomy. Budapest.

GREGUSS, P. 1959. Holzanatomie der Europäischen Laubhölzen und Sträucher. Budapest.

HALD, M.M. and CHARLES, M. 2008. Storage of crops during the fourth and third millennia BC at the settlement mound of Tell Brak, northeast Syria. Vegetation History and Archaeobotany 17(1), 35-41.

HOROWITZ, A. 1992. *Pollen in Arid Land.* Amsterdam, Elsevier Inc.

ISSAR, A. 2003. *Climate Changes During the Holocene and Their Impact on Hydrological Systems.* International hydrology series. New York/ Cambridge, University Press.

KLENGEL, H. 1992. *Syria. 3000 to 300 BC.* Berlin, Akademie Verlag.

KUPPER, J.R. 1992. Le bois à Mari. *Bulletin of Sumerian Agriculture* 6, 163-170.

KUZUCUOĞLU, C. and MARRO, C. (eds.) 2007. *Sociétés humaines et changement climatique à la fin du troisiéme millénaire: une crise a-t-elle eu lieu en Haute Mésopotamie?* Paris, De Boccard.

LINDER, E. 1986. The Khorsabad wall relief: a Mediterranean seascape of river transport of timbers? *Journal of the American Oriental society.* 106(2), 273-281.

MATTHIAE, P. 1989. Ebla. Un impero ritrovato. Torino.

McCORRISTON, J. 2002. Spatial and temporal variation in Mesopotamian agricultural practices in the Khabur Basin, Syrian Jazira. *Journal of Archaeological Science* 29, 485-498.

McCORRISTON, J. 2007. Cultural and environmental history in archaeological charred wood from the Khabur Drainage, Upper Mesopotamia. In C. Kuzucuoğlu and C. Marro (eds.), *Sociétés humaines et changement climatique à la fin du troisiéme millénaire: une crise a-t-elle eu lieu en Haute Mésopotamie?* 503-522. Paris, De Boccard.

MILANO, L. 1981. Alimentazione e regimi alimentari nella Siria preclassica. *Dialoghi di Archeologia* 3, 85-121.

MILANO, L. 1987. Barley for ratios and barley for sowing (ARET II 51 and related matters). *Acta Sumerologica* 9, 177-201.

MILANO, L. 1990. *Testi amministrativi: assegnazioni di prodotti alimentari (archivio L. 2712 – Parte I).* Roma, Missione archeologica italiana in Siria.

MILANO, L. 2006. La Siria. In M. Montanari and F. Sabban (eds.), *Storia e geografia dell'alimentazione.* Torino.

MILLER, N. 1990. Clearing land for farmland and fuel: Archaeobotanical studies of the ancient near east. In N. Miller (ed.), *Economy and Settlement in the Near East: analysis of the ancient sites and* materials, 70-78. Masca Research Papers in Science and Archaeology, supplement to Volume 7. Philadelphia, University of Pennsylvania.

NEEF, R. 2001. The plant remains. In Z.A. Kafafi (ed.), *Jebel Abu Thawwab (Er-Rumman), Central Jordan. The Late Neolithic and Early Bronze Age I Occupation*, 203-209. Berlin, Ex oriente.

PABOT, H. 1957. *Rapport au gouvernement de Syrie sur l'écologie végétale et ses applications.* Rapport n. 663, FAO, Rome.

PESSIN, H. 2007. Analyses anthracologique de deux sites du Moyen Euphrate: Tilbeşar et Horum Höyük. Contribution à la problématique paléoclimatique de l'Holocène moyen. In C. Kuzucuoğlu and C. Marro (eds.), *Sociétés humaines et changement climatique à la fin du troisiéme millénaire: une crise a-t-elle eu lieu en Haute Mésopotamie?* 557-572. Paris, De Boccard.

PETTINATO, G. 1986. Ebla. Nuovi orizzonti della storia. Milano.

RIEHL, S. 2008. Climate and agricolture in the ancient Near East: a synthesis of the Archaeobotanical and stable carbon isotope evidence. *Vegetation History and Archaeobotany* 17(1), 43-51.

RIEHL, S., BRYSON, R., PUSTOVOYTOV, K. 2007. Changing growing conditions for crops during the Near Eastern Bronze Age (3000-1200 BC): the stable carbon isotope evidence. *Journal of Archaeological Science* 20, 1-12.

ROWTON, M.B. 1967. The woodlands of ancient Western Asia. *Journal of Near Eastern studies* 26, 261-277.

SHWEINGRUBER, F.H., (1990) Mikroskopische Holzanatomie. Bern, Springer Verlag.

VALSECCHI, V. 2007. Vegetation and environmental changes during the Middle-Late Holocene at Tell Mishrifeh/Qatna: climate versus land-use. In D. Morandi Bonacossi (ed.), *Urban and Natural Landscapes of an Ancient Syrian Capital. Settlement and Environment at Tell Mishrifeh/Qatna and in Central-Western Syria,* 93-104. Studi archeologici su Qatna 1. Udine, Forum.

Van ZEIST, W. and BAKKER-HEERES, J.A.H. 1985. Archaeobotanical studies in the Levant. 4. Bronze Age sites on the Northern Syrian Euphrates. *Paleohistoria* 27, 247-316.

WEISS, H., COURTY, M.A., WETTERSTROM, W., GUICHARD, F., SENIOR, L., MEADOW, R., CURNOW, A. 1993. The Genesis and collapse of third millennium North Mesopotamian civilization. *Science* 261(5124), 995-1004.

WILKINSON T.J. 2004. *On the margin of the Euphrates. Settlement and land use at Tell es-Sweyhat and in the Upper Lake Assad Area, Syria.* The University of Chicago Oriental Institute publication 124.

WILLCOX, G. 1974. A history of deforestation as indicated by charcoal analysis of four sites in Eastern Anatolia. *Anatolian Studies* 24, 117-133.

WILLCOX, G. 1991. Exploitation des espèces ligneuses au Proche-Orient: données anthracologiques. *Paléorient* 17(2), 117-126.

WILLCOX, G. 1992. Timber and trees: ancient exploitation in the Middle East: evidence from plant remains. *Bulletin on Sumerian Agriculture* 6, 1-31.

WILLCOX, G. 1999. Charcoal analysis and Holocene vegetation history in southern Syria. *Quaternary Science Reviews* 18, 711-716.

WISEMAN, D.J. 1952. A new stela of Aššur-nasir-pal II. *Iraq* 14 (1), 24-44.

ZOHARY, M. 1973. *Geobotanical foundations of Middle East.* 2 Volumes. Amsterdam: Gustav Fischer Verlag, Stuttgart / Swets and Zeitlinger.

# WOODY PLANTS IN THE SEMI-ARID SOUTH-EASTERN IBERIAN PENINSULA DURING THE BRONZE AGE: CHARCOAL ANALYSIS FROM PUNTA DE LOS GAVILANES (MAZARRÓN, MURCIA, SPAIN)

María Soledad GARCÍA MARTÍNEZ
Dpto. de Biología Vegetal (Botánica), Facultad de Biología, Universidad de Murcia.
Campus de Espinardo 30100- Espinardo, Murcia, Spain
msgarmar@um.es

Elena GRAU ALMERO
Dpto. Prehistòria i Arqueologia, Facultat de Geografia i Història,
Universitat de València.Avgda. Blasco Ibàñez, 28. 46010 València, Spain

Maria Milagrosa ROS SALA
Dpto. Prehistoria, Arqueología, Historia Antigua, Historia Medieval y CC.TT. Historiográficas,
Facultad de Letras, Universidad de Murcia. C/ Santo Cristo, 1. 30001 Murcia, Spain

**Abstract**: *The findings of an anthracological study of the Bronze Age levels from Punta de los Gavilanes, a site located in a coastal environment of the southeastern Iberian peninsula, are reported. The results record the increase of sclerophyllous Mediterranean matorral vegetation in this environment, along with species of North African optimum, and xerophytes, indicators of great aridity. The forested strata would have been strongly reduced to isolated pines. The coastal mountain chains and the spaces influenced by the salinity of the coast would have been characterized by significant halophile vegetation, composed of several species of Chenopodiaceae. These conditions of progressive environmental degradation occurred during a large-scale episode of aridification that affected the entire Southeastern Iberian peninsula, beginning around c. 4500 BP. However, effects of anthropogenic action on the vegetation are not clearly noticeable around the site during the Bronze Age.*

**Keywords**: *Charcoal, Bronze Age, Palaeoecology, Fuelwood, Southeastern Iberia*

## PUNTA DE LOS GAVILANES. LOCATION, CLIMATE AND PRESENT-DAY VEGETATION

Punta de los Gavilanes is a rocky promontory that belongs to the coast of Mazarrón, in Murcia province, located on the Southeastern Iberian Peninsula (Figure 1). The climate of this area is typically Mediterranean, with a severe summer drought. Mean annual temperature is 16-19°C, reaching maximum values in summer, which can be higher than 40°C. The minimum values are hardly ever lower than 0°C. The precipitation regime is arid to semi-arid, with an annual rainfall between 200 mm and 350 mm.

The site is located in an ecologically singular area, marked by the great aridity recorded from Cabo de Gata (Almería province) to Cartagena (Murcia province). This ecology supports the development of a large number of endemic plants of North African optimum, such as *Tetraclinis articulata*, *Maytenus senegalensis*, *Periploca angustifolia*, *Withania frutescens,* etc. (Sánchez Gómez and Guerra Montes 2003).

However, the present-day vegetation in the Mazarrón area is strongly modified, mainly by anthropogenic action. Since frosts rarely occur during the year, agricultural exploitation has been favoured, particularly the cultivation of fruit trees. Cattle and grazing also contribute strongly to this damage. In addition, mining activity in Mazarrón has caused negative repercussions in the environment, especially during the last century. Finally, nitrophylle vegetation communities proliferate today due to the activities of coastal tourism.

In the areas closest to Punta de Los Gavilanes, the vegetation is dominated by plants adapted to saline and gypsum soils and ruderals. Above all, the abundance of Chenopodiaceae is highlighted, such as *Anabasis articulata, Arthrocnemum macrostachyum, Sarcocornia fruticosa, Suaeda vera, Atriplex halimus, A. prostrata, A. glauca,* etc. The only arboreal element found in such a saline soils is the genus *Tamarix*, which is also the main colonizer of nearby streams. The zone of former saltworks has been widely urbanized, and, even though some natural vegetation remains, mainly composed of Chenopodiaceae and phreatophytes such as *Phragmites australis,* there are also elements as *Asphodelus fistulosus* or *Teucrium*, very common in areas of dry pasture.

Further inland, in the local mountains, matorral vegetation is found. Xerophytic grasses, such as *Stipa tenacissima* or *Lygeum spartum,* are abundant, as well as leguminous plants (e.g., *Anthyllis cytisoides, Coronilla juncea, Genista umbellata* or *Spartium junceum*). Shallower soils are marked by species of Labiatae (*Lavandula dentata, Lavandula multifida, Thymus*

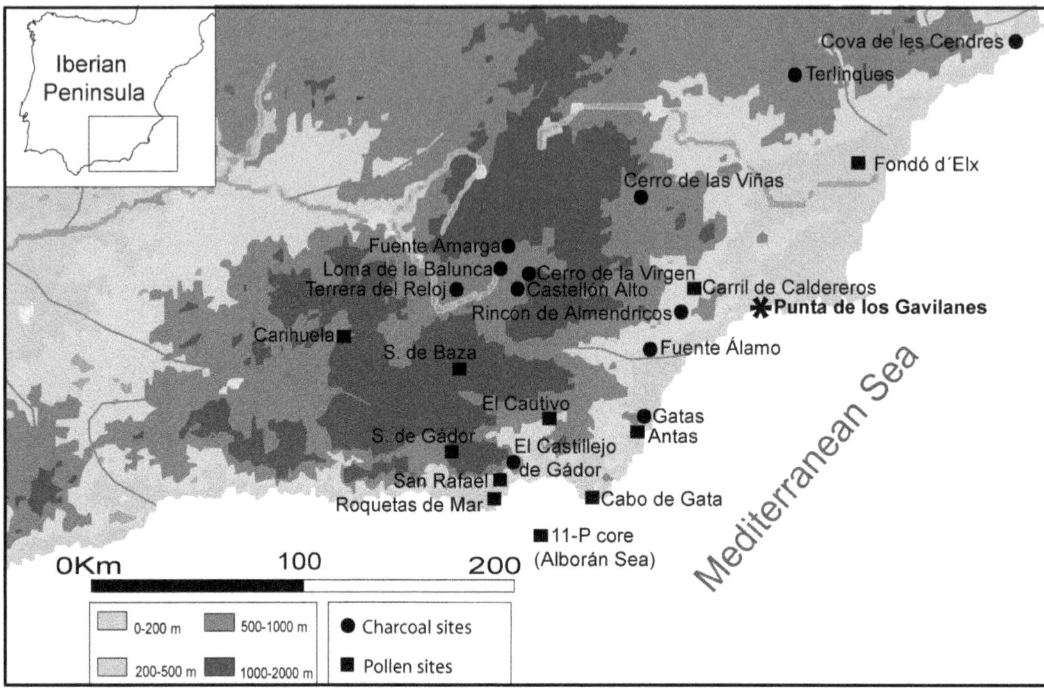


Figure 1. Location of Punta de los Gavilanes and the main sites mentioned in this paper

| Phase | Stratigraphic Unit | Laboratory Code | Material | $^{14}$C age years BP | Calibrated age Cal years BC (range 1σ) | Calibrated age Cal years BC (range 2σ) |
|---|---|---|---|---|---|---|
| IV | GV-2033 | KIA-32355 | charcoal | 3730 ± 30 BP | 2090-2040 (30,1%) | 2210-2030 (95,4%) |
| IV | GV-1743 | KIA-32366 | seeds | 3385 ± 35 BP | 700-1630 (48,6%) | 1770-1600 (92,3%) |
| IV | GV-1597 | KIA-32357 | charcoal | 3370 ± 40 BP | 1700-1610 (55,5%) | 1750-1530 (95,4%) |

Reference: Atmospheric data from Reimer *et al.* (2004); OxCal v3.10 Bronk Ramsey (2005);cub r:5 sd:12 prob usp [chron].

Figure 2. Radiocarbon dates from the phase IV of Punta de los Gavilanes (GV-IV)

*hyemalis, Rosmarinus officinalis*, etc*)*, Cistaceae (*Helianthemum almeriense* and *Fumana ericoides*) or Asteraceae (*Launaea arborescens or Helichrysum stoechas*). Other species are widely represented, such as *Thymelaea hirsuta, Globularia alypum, Rhamnus lycioides* or *Sedum sediforme*, together with species of the genus *Artemisia* and plants which are also present farther inland, but are here affected by the local salinity, such as *Lycium intricatum*. An edafoxerophitic vegetation, dominated by *Juniperus phoenicea,* is present in remnants throughout the Sierra de las Moreras.

## BRONZE AGE SETTLEMENT OF THE SITE

The archaeological interventions in Punta de los Gavilanes began in 1998. These excavations are part of an interdisciplinary research project whose aim is to study the process of the settlement in and around the mouth of the Rambla de Las Moreras River since the II millennium BC.

The first settlement of Punta de los Gavilanes (Phase GV-IV) is chronologically associated with the Argaric Bronze period, by the following reference radiocarbon dating: 3730 ± 30 BP, 3385 ± 35 BP and 3370 ± 40 BP (Figure 2). In this phase, the promontory was settled by a group who likely originated far away from the coast. They would have been a quite small population, due to the spatial limitations of this site, although apparently with the intent to establish a permanent settlement. This is demonstrated by a highly-structured urban settlement pattern incorportated all of the available space, and in a construction style based on stone that was designed to last over the long-term. (Ros Sala 2005b).

The main motivations for this installation may have been resulted from the improved and expanded exploitation of natural resources in the coastal environment, where the promontory is located (Ros Sala 2005a, 2005b). Evidence for fishing activities are, in fact, the best recorded for the site during this period, since the artefacts recovered indicate that the entire *chaine opératoire* of activities was developed at that moment. Fishing nets, made of esparto grass, have been recorded with a great deal of icthyofauna associated. In addition some structures related to the preservation of fish both systems for drying as well as smoking, have been recorded. However, there is also evidence of others subsistence activities, such as cattle breeding and cereal cultivation at the site.

## MATERIALS AND METHODS

Our study is based on the analysis of 2528 charcoal fragments recovered from 56 stratigraphic units associated with the Bronze Age. We have established a systematic sampling based on the recuperation of approximately 20-50 litres of sediment from each stratigraphic unit, increasing this quantity in stratigraphic units with high levels of organic material.

The sediment samples were processed by a machine flotation system (Buxó 1990), using a 1 mm sieve for the heavy fraction collection, and a 0.25 mm sieve for the light fraction.

The taxonomic identification of each charcoal fragment was carried out through the study of the three anatomical diagnostic planes: transversal, tangential and radial. For this purpose we used a metallographic Leica DM 2500 M microscope, using 100x to 500x magnification. Our identifications were supported by comparison between the archaeological charcoal and the reference collections of modern carbonized wood managed by the Laboratori de Prèhistoria i Arqueologia Milagro Gil Mascarell of the University of Valencia, and by the Archaeology Laboratory of the University of Murcia (LABAUMU). We also used wood anatomy atlases (Schweingrüber 1978, 1990) and a charcoal identification guide (Vernet *et al.* 2001).

## ANTHRACOLOGICAL RESULTS AND CHARACTERIZATION OF THE PAST ENVIRONMENT OF PUNTA DE LOS GAVILANES

The study of the anthracological samples obtained for the Bronze Age (2528 charcoal fragments) allowed us the identification of 27 taxa: *Atriplex halimus*, Chenopodiaceae, Cistaceae, *Daphne gnidium/Thymelaea hirsuta*, *Ephedra* sp., *Erica* sp., cf. *Fumana* sp., *Juniperus* sp., Labiatae, Leguminosae, *Lycium intricatum,* Maloideae, *Maytenus senegalensis*, Monocotyledoneae, *Olea europaea*, cf. *Periploca angustifolia*, *Pinus halepensis*, *Pinus pinea/pinaster*, *Pinus* sp., *Pistacia lentiscus*, *Prunus* sp., evergreen *Quercus*, *Rhamnus/Phillyrea* sp., *Rosmarinus officinalis*, *Tamarix* sp., cf. *Tetraclinis articulata* and cf. *Withania frutescens*. Moreover, nine fragments of unidentified Angiosperms and nineteen of conifers were recorded for which further determination was not possible. Finally, due to poor conditions of conservation, 190 indeterminate fragments were registered, which means 7.52% of the total (Figure 3).

| Punta de Los Gavilanes – Phase IV (GV-IV) | | |
|---|---|---|
| ***Taxa*** | **N°** | **%** |
| *Juniperus* sp. | 3 | 0,11 |
| cf. *Tetraclinis articulata* | 7 | 0,28 |
| *Pinus halepensis* | 170 | 6,72 |
| *Pinus* type *pinea/pinaster* | 32 | 1,27 |
| *Pinus* sp. | 21 | 0,83 |
| *Ephedra* sp. | 31 | 1,23 |
| Monocotyledoneae | 4 | 0,16 |
| *Atriplex halimus* | 33 | 1,3 |
| Chenopodiaceae | 112 | 4,43 |
| Cistaceae | 47 | 1,86 |
| *Daphne gnidium/Thymelaea hirsuta* | 9 | 0,36 |
| *Erica* sp. | 32 | 1,27 |
| cf. *Fumana* sp. | 4 | 0,16 |
| Labiatae | 169 | 6,68 |
| Leguminosae | 34 | 1,34 |
| *Lycium intricatum* | 2 | 0,08 |
| Maloideae | 6 | 0,24 |
| *Maytenus senegalensis* | 2 | 0,08 |
| *Olea europaea* | 378 | 14,95 |
| cf. *Periploca angustifolia* | 20 | 0,79 |
| *Pistacia lentiscus* | 1010 | 39,95 |
| *Prunus* sp. | 19 | 0,75 |
| *Quercus ilex/coccifera* | 18 | 0,71 |
| *Rhamnus/Phillyrea* sp. | 43 | 1,7 |
| *Rosmarinus officinalis* | 53 | 2,1 |
| *Tamarix* sp. | 34 | 1,34 |
| cf. *Withania frutescens* | 17 | 0,68 |
| Unidentified Angiosperm | 9 | 0,36 |
| Unidentified Gymnosperm | 19 | 0,75 |
| Indeterminable | 190 | 7,52 |
| **Total** | **2528** | **100** |

Figure 3. Anthracological results of the phase IV of Punta de los Gavilanes (GV-IV)

In general, the taxa show a very uneven percentage distribution (Figure 4). The taxon used most frequently as fuel at the site during the Bronze Age, with extremely high levels compared to the rest of the taxa, was *Pistacia lentiscus* (almost the 40% of the charcoal analyzed).

Although in much lower levels, *Olea europaea* makes up almost 15% of the assemblage. This taxon merits some additional remarks. Discrimination between wild and cultivated variety of *Olea europaea*, which was already discussed in a preliminary work by Vernet *et al.* (1983), has most recently been tackled by J.F. Terral (1996, 1997), who has managed to distinguish between the two using morphometric differences. However, this type of analysis was not applied in the case of Punta de los Gavilanes. Our identifications were made to species level, without further discrimination of the variety. Nevertheless, Olea europaea occurs naturally in the south of the Iberian Peninsula (in the thermomediterranean stage) since the Upper Palaeolithic (Aura Tortosa *et al.* 2002) and especially during the Neolithic. Its cultivation cannot be confirmed until the Roman era when the plant is widespread in macroremains deposits from mesomedi-

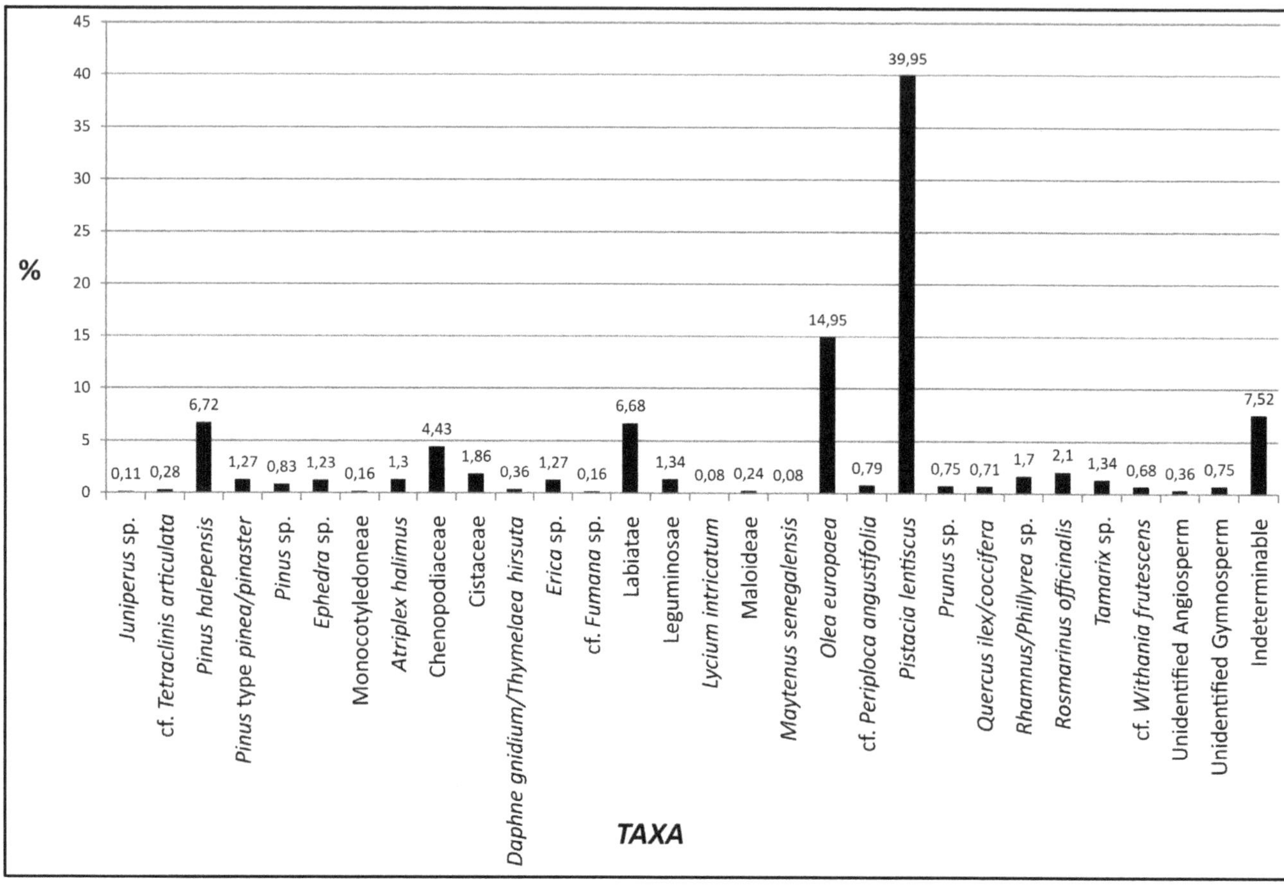


Figure 4. Histogram in relative values of the phase IV of Punta de los Gavilanes (GV-IV)

terranean stage (Rodríguez-Ariza and Montes Moya 2005, Carrión *et al.* 2010).

Each of the remaining individually recorded *taxa* contributes less than 10% of the total analyzed. Among the *taxa* with intermediate percentages the most abundant ones are the labiates (Labiatae + *Rosmarinus officinalis*), whose values jointly make up almost the 9% of the total and the Chenopodiaceae (Chenopodiaceae + *Atriplex halimus)*, which account for more than 5% of the analyzed remains. The main arboreal taxon recorded is *Pinus halepensis*, with values a bit higher than 6%. The punctual representation of the remained *taxa*, with percentages that never reach 2% of the total, indicates the variety of wood resources of the environment from a qualitative point of view. However, it provides scant information about the importance of these taxa in the Bronze age environment.

In general, this record implies a high taxonomic variability consisting of families, genus and species, at arboreal scale as well as shrubs. Due to the different ecological requirements of each taxon, their development would have been favored or limited by the different ecosystems within the environment, and in particular, by the degree of salinity in the soils.

The main contributors of fuel recorded in the site come from the species of Mediterranean character, associated with the Thermomediterranean bioclimatic stage. Their major development would have occurred in slightly further inland, such as the nearby mountains, or the inland plains, avoiding the saline soils of the coast. According to our results, these areas would have been dominated by vegetation mainly consisting of Mediterranean sclerophile matorral, composed by species with a great resistance to hydric stress such as *Pistacia lentiscus* and *Olea europaea.* These species would have generated dense formations that would have included shrubby elements, such as *Rhamnus/Phillyrea* sp. (especially *Rhamnus lycioides)*, some heathers (*Erica* sp.), and above all, large amount of species from the Cistaceae, Labiatae and Leguminosae families. The presence of xerophytes such as *Ephedra*, which are common in this type of matorral in arid environments, stony lands and coast dunes, verifies that certain conditions of aridity had been already established in this area and continued progressively to the modern day. Open spaces would have been common in all these areas, and would have been mainly occupied by grasses such as the esparto grass (*Stipa tenacissima*). Under these conditions, the arboreal stratum would have appeared quite reduced, and would have been composed basically by scarce and isolated pines that constituted the beginning or the remnants of Mediterranean pre forest. Among them we have been able to verify the major presence of *Pinus halepensis*, and in lesser proportions the *Pinus* type *pinea/pinaster*. *Pinus halepensis* would have occupied the calcareous, stony and dry points, whereas the edaphic requirements of *Pinus* type *pinea/pinaster* would have placed it in siliceous soils or even in areas of coast dunes. Some shrubs, such as *Juniperus phoenicea* which are interspersed in the surroundings of the site in the present-day, would have also grown, above all under the most rigorous conditions.

The development of the Mediterranean shrubby communities would have occurred together with several elements of the North African optimum in this area. Among them we have been able to identify the presence of *Lycium intricatum*, *Maytenus senegalensis*, cf. *Periploca angustifolia* and cf. *Withania frutescens*, which general occurs in very low percentages (less than 1% of the charcoal identified). However, these findings are interesting because environmental sensitivity to the low temperatures of the majority of these species is high, so that they provide a quite exact sign of the thermicity of this environment during the Bronze Age. In general, these taxa usually co-occur with *Tetraclinis articulata*, whose development also was recorded in the vicinity of Punta de los Gavilanes. The reduced Iberian distribution of this species in the present-day, constrained in some areas of the Sierra de la Unión (Cartagena), seems to indicate that *Tetraclinis articulata* was characterized by a very strict ecological range. However, there are several factors that contradict this assertion: On the one hand, the abundance in which it can occur within northern Africa at altitudes higher than 1500 m and with annual precipitations up to 700 mm; secondly, the fact that specimens planted in inland areas of the Iberian Peninsula can tolerate temperatures as low as -10ºC (Costa *et al.* 2001); and finally, palaeovegetal studies that attest to its presence since the Chalcolithic in an extensive area of the southeastern peninsular, in some cases under more rigorous bioclimatic parameters (Carrión Marco 2004; Grau 1990; Rodríguez-Ariza 1992a; Schoch and Schweingrüber 1982). Plausibly this difficulty in colonizing the southeastern peninsula was due to problems in its regeneration caused by edaphic factors which are not present in more humid areas of the North of Africa (Costa *et al.* 2001). In the case of Punta de los Gavilanes, the retraction of this species relatd to competition with the predominant arboreal element, *Pinus halepensis* must be highlighted. With respect to this point, several studies conducted on modern populations of *Tetraclinis articulata* and *Pinus halepensis* in the Sierra de la Unión (Nicolás *et al.* 2004) indicate that the presence of the latter species has an important influence on the distribution of *Tetraclinis articulata.* According to these studies, when the cover of *Pinus halepensis* increases, both the number of trees as well as the total area covered by *Tetraclinis articulate* decrease. Conversely, when *Pinus halepensis* is absent, the highest densities of *Tetraclinis articulata* are found in direct sunlight, although its major cover occurs in shady places.

Despite the arid conditions in the environment of Punta de los Gavilanes, some nearby points, especially at high elevation, or in shady zones, may have shown some isolated specimens of evergreen *Quercus*, although these are completely extinct nowadays in this area. A few relict evergreen oaks (*Quercus ilex*) are preserved in the Sierra de Cartagena, as a limit of aridity in the Iberian Peninsula (Costa *et al.* 2001). Some studies about the potential of the species *Quercus rotundifolia* in this area emphasize its double physiological tolerance, so that it would be able to develop in places where extremely cold winters occur, while it can also be found in semi-arid areas, with marked hydric stress. Its maximum tolerance for aridity would have been at approximately 350 mm of annual rainfall (Chaparro Fuster 1996). In the Bronze Age, the presence of this species would have been anecdotal and reduced in the presence of anthropogenic felling of trees or constant burning of determinate spaces. In this respect, the residual presence of elements such as *Quercus coccifera*, *Erica* sp. or some species of the Cistaceae family (especially the genus *Cistus*), that usually function as elements of substitution of the evergreen-oak grove formations, would have been indicative of this condition of degradation.

The supply areas of wood resources closest to the promontory would have been located on the coastline itself and in its surroundings, where great resources of several halophile taxa, such as Chenopodiaceae or *Tamarix* sp., would have existed. However, these halophile trees or shrubs occur in very low percentages in this occupation phase of the site, which indicate they were not plentiful in the environment. Likely, these species were not selected because they are poor resources for fuel, in contrast to the wide formations of Mediterranean matorral that would have been the primary choice for fuel in the domestic activities of the group. In addition, we know that the use of halophytes as fuel increases remarkably at this site around the 4th-3rd centuries BC., when a strong reduction of resources from Mediterranean species is recorded, likely induced by production activities in this area and its surroundings (García Martínez *et al.* 2007).

Finally, the riparian ecosystems around Punta de los Gavilanes should have been almost nonexistent. In this environment, permanent watercourses do not appear, although the drainage system consists of a great number of streams, highly affected by the salinity of the soil. Due to their high level of salinity, the vegetation of these nonpermanent watercourses would have presented a similar structure to that of the coastline. It would have fundamentally been composed of species of the genus *Tamarix*, which would have appeared both on banks as well in the river bed, and these would have been accompanied by a great variety of Chenopodiaceae, and also probably by several haloresistant grasses. However, the procurement of wood resources from the nearby Rambla de las Moreras should have been guided byo the same selective process discussed before for the coastal mountainous vegetation.

In conclusion, the interpretation of the results provides an image of the environment of Punta de los Gavilanes marked by a strong process of "matorralization", where the Mediterranean vegetation would have suffered severe degradation and the pre forest elements would have been very scarce. It would have been a process with a trend toward greaterenvironmental aridity, resulting in the increase of the protagonist of the taxa typical of steppic formations such as *Ephedra,* and of the growing extension of Chenopodiaceae due to the increase of environmental salinity.

## DISCUSSION OF THE RESULTS AT A REGIONAL SCALE

The anthracological results from Punta de los Gavilanes indicate strong environmental degradation, where the pre-forest and forest component is clearly retracted. Most of the anthracological studies for other Bronze Age sites in the dry-semi-arid environment of the southeastern Iberian Peninsula indicate a similar grade of deterioration in its immediate surroundings.

The Argaric sites from Fuente Álamo and Gatas (Almería) would have been situated in an environment similar to that of Punta de los Gavilanes because of their proximity to the sea. The data obtained from Fuente Álamo give evidence of a great aridity environment, and also salinity, where the humidity would not have been higher than that of the present-day. Like the environment of Punta de los Gavilanes, this context would have also been dominated by an open matorral with thermophilous and xerophilous species, together with a sparse arboreal element, composed of pines (Carrión Marco 2004). In Gatas, Gale (1999) also records a great majority of thermophilous taxa, widely dominated by *Olea europaea.*

Even in the higher altitude areas, located in the modern Mesomediterranean bioclimatic stage, a marked environment degradation has been demonstrated for the Bronze Age. The Argaric site of Cerro de las Viñas (Lorca, Murcia), indicates the dominance of *Pinus halepensis* in the environment, together with a great abundance of legumous species, and lesser quantities *Quercus ilex* (Grau 1990). In the Mesomediterranean Andalusian zone, the analysis of several sites near Granada, such as Cerro de la Virgen, Fuente Amarga, Castellón Alto, Loma de la Balunca or Terrera del Reloj show the same phenomenon. These data allow us to confirm that during the Chalcolithic period the environment would have been dominated by mesophilous species, principally evergreen *Quercus*. However, in the early Bronze Age (Cerro de la Virgen), the first retraction of *Quercus ilex/coccifera* and the progressive increase of *Pinus halepensis* are observed. The Argaric period eventually records a quite degraded landscape, mainly composed of heliophilous matorral, even a pre forest where the arboreal stratum would have been represented by *Pinus halepensis*, and to a lesser extent, by evergreen *Quercus*, which points to increasing permanency in these newer regional climatic conditions (Rodríguez-Ariza 1992a, 1992b; Rodríguez-Ariza and Esquivel 2007).

Farther North, in the semi-arid region of Alicante, anthracological study of the Bronze Age site of Terlinques (Villena) (2100 cal yr BC – 1830 cal yr BC) records the importance of a substitution matorral with *Pinus halepensis*, *Rosmarinus officinalis* and *Viburnum*, suggesting a degradation of the Mediterranean forest, which led to collecting processes for wood resources in very diversified areas (Machado Yanes *et al.* 2004). The Bronze Age levels from Cova de les Cendres (Teulada) show an expansion of the matorral, probably associated with agricultural activities. As a result of this expansion, there is a reduction of elements such as *Olea* or *Pinus halepensis*, the latter being the predominant arboreal component. In this chronology, *Quercus ilex-coccifera* and *Q. faginea* are hardly represented at Cendres, and the shoreline vegetation is not important either (Badal *et al.* 1994).

However, some studies of Argaric sites indicate a permanence of evergreen-oak groves at a local scale, probably as a result of the location of the site, or of a lower intensity in the anthropogenic activities of the groups. In Rincón de Almendricos (Lorca, Murcia), the environment would have been an altered forest of *Quercus ilex/coccifera*, with *Pinus halepensis, Olea europaea* and *Erica multiflora* (Grau 1990). In the Argaric levels of Castillejo de Gádor (Almería) the scarce presence of *Pistacia lentiscus* and *Olea europaea* var. *sylvestris*, is notable and *Pinus halepensis* is not recorded. However, there is a predominance of evergreen *Quercus* and *Rosmarinus officinalis*, interpreted as indicating that the Mesomediterranean stage at this time occurred at a level lower than in the present day (Rodríguez-Ariza 2001).

The local significance of these data, together with those obtained for Punta de los Gavilanes, are inserted inside of an aridification dynamic at a regional scale which could have been elucidated by some Holocene sequences in the southeast of Iberian Peninsula. These sequences would place the beginning of the arid conditions in this environment at about 4500 BP (Pantaleón-Cano *et al.* 2003). In this sense, the Bronze Age would have represented an abrupt start to increasing deforestation in the environment, such that this Peninsular region should be considered the last stage of clear development of the forestal system in the III millennium BC, as it has been observed in the sequence of Carril de Caldereros (Lorca, Murcia) (Fuentes *et al.* 2005).

The causes of this process have been based on two complementary points of view: a natural trend toward the aridification or anthropogenic inducing of this phenomenon. In this sense, the wide diachronic vision offered by some regional pollen sequences, in addition to the important anthropogenic signals of several palynomorphs and concentrations of microcharcoal provide a more detailed perspective of the development of these mechanisms.

The pollen sequences obtained from the coastal area of Almería through studies in Antas, San Rafael and Roquetas de Mar (Pantaleón-Cano *et al.* 2003) are coherent and indicate an abrupt change of the vegetation around 4500 BP, at which time begins a trend towards more steppic characteristics, indicated mainly by the presence of *Artemisia* and Chenopodiaceae. These changes would have originated, according to the authors, from modifications of the hydrologic balance, generated by an increase of environmental aridity, and probably not by human actions, because significant modifications in the palinomorph, directly associated with these kinds of activities, are not recorded. This trend toward a more open landscape is also obvious at Cabo de Gata (Burjachs *et al.* 1996), where halophytes become more important, particularly during the last 3000 years. In nearby areas such as El Cautivo (Tabernas, Almería) these changes are interpreted as the result of climate oscillations, and not as a result of activities such as agriculture or the management of fuel, which would have played a limited role in this process (Nogueras *et al.* 2000). Similarly, the pollen analysis of Fondó d'Elx (Burjachs *et al.* 1997) provides signs of the depletion of the ecosystem since the Neolithic, when a structural change of the vegetation was produced, with a progressive diminishing of the arboreal stratum, dominated by *Juniperus* and *Pinus*, meaning increasing environmental aridity, without associated signs of anthropisation, such an increase of Cerealia or ruderals.

However, it has been also reported that this process of aridification could have been anthropogenically induced at some point by the strong economic activity which characterized the productive system of the Bronze Age in the southeast Peninsula (Argar Culture). According to some authors, this system depleted the existent natural resources, by massive occupation of the land, both for the cereal cultivation, as well as for other economic activities (Castro *et al.* 1999). This could have produced an ecological collapse (Lull 1983) because of the inadequacy of a semi-arid environment to satisfy with this tremendous demand. In fact, large-scale irrigation systems, whose construction and control would have been the beginning of strong social hierarchy, have been reported (Chapman 1991; Gilman and Thornes 1985). However, the analysis of carbon isotopes in seeds from several Argaric sites does not support the use of irrigation practices at these sites, with the exception of *Vicia faba*, (Araus *et al.* 1997). On the other hand, the paleocarpologic studies in the area point to the utilization of the streambeds for farming, using their periodic flooding or the surface or underground water resources, and probably ephemeral irrigation structures that are not preserved in the archaeological record (Buxó 1997; Rovira 2007).

In some pollen sequences, however, a direct relation between the frequencies of several palynomorphs and/or the concentrations of the microcharcoal and this great intensity of the exploitation of natural resources has been established for the Bronze Age in the southeastern Iberian Peninsula. With respect to this, the recent study of the sequence of Sierra de Baza (Granada) (Carrión *et al.* 2007) reveals that during the Argaric period an important environmental change was produced in this area, which is shown in a strong reduction of the arboreal cover, where mesophytic trees, such as *Pinus* (high mountainous types) or deciduous *Quercus*, disappear, while evergreen *Quercus* expands, along with with sclerophyllous matorral species. This change correlates with a great increase of microcharcoal in the sequence, which would have indicated the presence of burning, both anthropogenic as well as wildfires. According to the authors, activities such as mining or grazing might have generated successive burning of the vegetation, producing a progressively diminishing forest, and accelerating the increase of the sclerophilous matorral and its invasive species such as *Quercus ilex/coccifera*, *Erica*, *Pistacia lentiscus*, *Phillyrea angustifolia* and especially species of the genus *Cistus*. This phenomenon is also shown in the Sierra de Gádor, with a decrease of deciduous *Quercus*, together with a strong increase of *Pinus* and evergreen *Quercus*, at the same time that a significant increase of microcharcoals occurs, similar to that recorded for the Sierra de Baza. This could have been related to the intensity of anthropogenic action (Carrión *et al.* 2003) beginning by 3940 cal yr BP. The Holocene sequence of Cueva de la Carihuela (Píñar, Granada), also records a process of matorralization with a strong reduction of the arboreal cover in its pollen zone 22 (*c.* 5470-1250 BP), which, even if there are difficulties in the interpretation of its archaeological materials, the effect of agriculture and grazing through a progressive increase of some genera such as *Pinus* and *Juniperus*, along with a decrease of mesophytic trees (deciduous *Quercus*), a minor representation of evergreen *Quercus*, and an increase of xerophiles *taxa* (Fernández *et al.* 2007). The marine core 11-P of the Alborán Sea (Targarona *et al.* 1996), records a deforestation of the surroundings of Almería related to anthropogenic activities. This deforestation is verified here between the 8000 and 3000 BP, although a denudation of the soil is demonstrated to have occurred later (*c.* 2000 BP).

The pressure on the wood resources produced by the inhabitants of Punta de los Gavilanes during the Bronze Age would have been moderately intense. The economic activities were fundamentally of a domestic nature, related to fishing and fish processing through drying or smoking. In addition, the unfavorable environment for agriculture would have greatly limited the establishment of a large-scale productive system. The provisioning of fuel for these activities, together with the degradation of the soils caused by a significant agro-pastoralism, could have generated some deforestation in its local field, aggravating the great sensitivity of the ecosystem and its reduced capacity for regeneration. Despite a stage of progressive xerophitisation of the vegetation, which is typical of the natural dynamic throughout the southeastern Iberian Peninsula, the evidence of the significant impact caused by anthropogenic pressure over the structure of the vegetation is not detected at Punta de los Gavilanes. This strong change is perceptible, according to the anthracological indicators, beginning in the 4th to 3rd centuries BC, related to a reduction of resources, as well as the depletion of the system due to strong demands for fuel generated by the important development of metallurgic activities (García Martínez *et al.* 2007).

## CONCLUSIONS

The anthracological study of the first phase of the settlement in Punta de los Gavilantes (GV-IV) gives information related to the palaeoenviromental features of this area and to the patterns of the fuel management by the group during the Bronze Age.

- During this period, the characteristic vegetation was dominated by a sclerophilous Mediterranean matorral, mainly composed with *Pistacia lentiscus* and *Olea europaea*, along with arboreal-shrubby elements of North African optimum and with xerophyte indicators of environment aridity conditions.
- In this context, the arboreal stratum was almost nonexistent, with a few sporadic examples of of *Pinus halepensis* Mediterranean pre forest, although in protected areas some isolated specimens of evergreen *Quercus* may have remained, perhaps evidence of an early forest stage during a past wet period or more probably from a period of less anthropic pressure where the vegetation dynamics were active.
- The vegetation of the saline areas and the streams was marked by massive development of communities represented by a great number of species belonging to the Chenopodiaceae family and some trees of the genus *Tamarix*. However strong selection for halophilous plants is not evident, in light of a sufficient presence of Mediterranean matorral species.
- The ecological context suggested by the anthracological results of the Punta de los Gavilanes supports the regional dynamic of the progressive aridification reported for the southeastern Iberian Peninsula since *c.* 4500 BP.
- Domestic activities during the Bronze Age, principally fishing and fish treatment, would not have generated signficant deforestation. This process is more visible, according to the anthracological data, in the IV-III centuries BC, as a result of the metallurgic activities of an intense character recorded on the promontory.

### Acknowledgements

We are grateful to Michelle Elliott and Azucena Avilés for their help in the translation of this paper.

### References

ARAUS, J.L., FEBRERO, A., BUXÓ, R., RODRÍGUEZ-ARIZA, M.O., MOLINA, F., CAMALICH, M.D., MARTÍN, D. and VOLTAS, J. 1997. Identification of ancient irrigation practices based on the Carbon isotope discrimination or plant seeds: a case study from the South-East Iberian Peninsula. *Journal of Archaeological Science* 24, 729-740.

AURA TORTOSA, J.E., JORDÁ PARDO, J.F., PÉREZ RIPOLL, M., RODRIGO GARCÍA, M.J., BADAL, E. and GUILLEM CALATAYUD, P. 2002. The far south: The Pleistocene-Holocene transition in Nerja Cave (Andalucía, Spain). *Quaternary International* 93-94, 19-30.

BADAL, E., BERNABÉU, J. and VERNET, J.L. 1994. Vegetation changes and human action from the Neolithic to the Bronze age (7000-4000 BP) in Alicante, Spain, based on charcoal analysis. *Vegetation History and Archaeobotany* 3, 155-166.

BRONK RAMSEY, C. 2005. *OxCal Program v3.10.* Oxford, Oxford Radiocarbon Accelerator Unit.

BURJACHS, F., GIRALT, S., RIERA, S., ROCA, J.R. and JULIÀ, R. 1996. Evolución paleoclimática durante el último ciclo glaciar en la vertiente mediterránea de la Península Ibérica. *Notes de Geografia Física* 25, 21-39.

BURJACHS, F., GIRALT, S., ROCA, J.R., SERET, G. and JULIÀ, R. 1997. Palinología holocénica y desertización en el Mediterráneo Occidental. In J. J. Ibáñez, B.L. Valero and C. Machado (eds.), *El paisaje mediterráneo a través del espacio y del tiempo. Implicaciones en la desertificación,* 379-394. Logroño, Geoforma.

BUXÓ, R. 1990. *Metodología y técnicas para la recuperación de restos vegetales (en especial referencia a semillas y frutos) en yacimientos arqueológicos.* Girona, Ajuntament de Girona.

BUXÓ, R. 1997. *Arqueología de las plantas. La explotación económica de las semillas y los frutos en el marco mediterráneo de la Península Ibérica.* Barcelona, Crítica.

CARRIÓN, J.S., SÁNCHEZ-GÓMEZ, P., MOTA, J.F., YLL, R. and CHAÍN, C. 2003. Holocene vegetation dynamics, fire and grazing in the Sierra de Gádor, southern Spain. *The Holocene* 13 (6), 839-849.

CARRIÓN, J.S., FUENTES, N., GONZÁLEZ-SAMPÉRIZ, P., SÁNCHEZ QUITANTE, L., FINLAYSON, J.C., FERNÁNDEZ, S. and ANDRADE, A. 2007. Holocene environmental change in a montane region of southern Europe with a long history of human settlement. *Quaternary Science Reviews* 26, 1455-1475.

CARRIÓN, Y., NTINOU, M. and BADAL, E. 2010. *Olea europaea* L. in the North Mediterranean Basin during the Pleniglacial and the Early-Middel Holocene. *Quaternary Science Reviews* 29, 952-968.

CARRIÓN MARCO, Y., 2004. Análisis antracológico del yacimiento de Fuente Álamo (Cuevas de Almanzora, Almería). Usos de la madera y paleovegetación. In L. Hernández Alcaraz and M. S. Hernández Pérez (eds.), *La Edad del Bronce en tierras valencianas y zonas limítrofes,* 477-486. Alicante, Ayuntamiento de Villena.

CASTRO, P.V., CHAPMAN, R.W., SURIÑACH, S., LULL, V., MICÓ, R., RIHUETE, C., RISCH, R. and SANAHUJA, M.E. 1999. *Proyecto Gatas 2. La dinámica arqueoecológica de la ocupación prehistórica*. Sevilla, Junta de Andalucía.

CHAPARRO FUSTER, J. 1996. *Distribución potencial del bosque y de sus especies arbóreas en zonas mediterráneas semiáridas: modelos y aplicaciones*. Unpublished PhD thesis, University of Murcia.

CHAPMAN, R.W. 1991. *La formación de las sociedades complejas. El sureste de la Península Ibérica en el marco del Mediterráneo occidental*. Barcelona, Crítica.

COSTA, M., MORLA, C. and SAINZ, H. (eds) 2001. *Los bosques ibéricos. Una interpretación geobotánica*. Barcelona, Planeta.

FERNÁNDEZ, S., FUENTES, N., CARRIÓN, J.S., GONZÁLEZ-SAMPÉRIZ, P., MONTOYA, E., GIL, G., VEGA-TOSCANO, G. and RIQUELME, J.A. 2007. The Holocene and Upper Pleistocene pollen sequence of Carihuela Cave, southern Spain. *Geobios* 40, 75-90.

FUENTES, N., GARCÍA MARTÍNEZ, M.S., GONZÁLEZ-SAMPÉRIZ, P., FERNÁNDEZ, S., CARRIÓN, J.S., LÓPEZ-CAMPUZANO, M. and MEDINA, J. 2005. Degradación ecológica y cambio cultural durante los últimos cuatro mil años en el sureste ibérico semiárido. *Anales de Biología* 27, 69-84.

GALE, R. 1999. Gatas: Análisis antracológico. In P.V. Castro, R.W. Chapman, S. Suriñach, V. Lull, R. Micó, C. Rihuete, R. Risch and M.E. Sanahuja (eds), *Proyecto Gatas 2. La dinámica arqueoecológica de la ocupación prehistórica*, 296-311. Sevilla, Junta de Andalucía.

GARCÍA MARTÍNEZ, M.S., GRAU ALMERO, E. and ROS SALA, M.M. 2007. Paisaje y gestión de los recursos vegetales en la costa de Mazarrón (Murcia), según el antracoanálisis de Punta de los Gavilanes. In J. Lario and P. Silva (eds.), *Contribuciones al estudio del período Cuaternario*, 107-108. Ávila, Aequa.

GILMAN, A. and THORNES, J.B. 1985. *Land use and Prehistory in Southeast Spain*. London, The London Research Series in Geography 8.

GRAU, E. 1990. *El uso de la Madera en Yacimientos Valencianos de la Edad del Bronce a la Época Visigoda. Datos Etnobotánicos y Reconstrucción Ecológica según la Antracología*. Unpublished PhD thesis, University of Valencia.

LULL, V. 1983. *La "cultura" de El Argar. Un modelo para el estudio de las formaciones económico-sociales prehistóricas*. Madrid, Akal.

MACHADO YANES, M.C., JOVER MAESTRE, F.J. and LÓPEZ PADILLA, J.A. 2004. Primeras aportaciones antracológicas del yacimiento de Terlinques (Villena, Alicante). In L. Hernández Alcaraz and M. S. Hernández Pérez (eds.), *La Edad del Bronce en tierras valencianas y zonas limítrofes*, 363-368. Alicante, Ayuntamiento de Villena.

NICOLÁS, M.J., ESTEVE, M.A., PALAZÓN, J.A. and LÓPEZ, J.J. 2004. Modelo sobre las preferencias de hábitat a escala local de *Tetraclinis articulata* (Vahl) Masters en una población del límite septentrional de su área de distribución. *Anales de Biología* 26, 157-167.

NOGUERAS, P., BURJACHS, F., GALLART, F. and PUIGDEFÀBREGAS, J. 2000. Recent gully erosion in the Cautivo badlands (Tabernas, SE Spain). *Catena* 40, 203-215.

PANTALEÓN-CANO, J., YLL, E.I., PÉREZ-OBIOL, R. and ROURE, J.M. 2003. Palynological evidence for vegetational history in semi-arid areas of the western Mediterranean (Almería, Spain). *The Holocene* 13(1), 109-119.

REIMER, P.J, BAILLIE, M.G.L., BARD, E., BAYLISS, A., BECK, J.W., BERTRAND, C.J.H., BLACKWELL, P.G., BUCK, C.E., BURR, G.S., CUTLER, K.B., DAMON, P.E., EDWARDS, R.L., FAIRBANKS, R.G., FRIEDRICH, M., GUILDERSON, T.P., HOGG, A.G., HUGHEN, K.A., KROMER, B., MCCORMAC, G., MANNING, S., RAMSEY, C.B., REIMER, R.W., REMMELE, S., SOUTHON, J.R., STUIVER, M., TALAMO, S., TAYLOR, F.W., Van der PLICHT, J. and WEYHENMEYER C.E. 2004. IntCal04 Terrestrial Radiocarbon Age Calibration, 0-26 cal kyr BP. Radiocarbon 46 (3), 1029-1059.

RODRÍGUEZ-ARIZA, M.O. 1992a: *Las relaciones hombre-vegetación en el Sureste de la Península Ibérica durante las Edades del Cobre y Bronce a partir del análisis antracológico de siete yacimientos arqueológicos*. Unpublished PhD thesis, University of Granada.

RODRÍGUEZ-ARIZA, M.O. 1992b. Human-plant relationships during the Copper and Bronze Ages in the Baza and Guadix basins (Granada, Spain). *Bulletin de la société botanique de France. Actualités Botaniques* 139, 451-464.

RODRÍGUEZ-ARIZA, M.O. 2001. Análisis antracológico de El Castillejo de Gádor (Almería). In B. Gómez Tubío, M.A. Respaldiza and M.L. Pardo (eds.), *III Congreso Nacional de Arqueometría*, 173-182. Sevilla, Universidad de Sevilla.

RODRÍGUEZ ARIZA, M.O. and ESQUIVEL, J.A. 2007. Una valoración de la paleovegetación del sureste de la Península Ibérica durante la Prehistoria Reciente a partir de aplicaciones estadísticas en Antracología. In J. Molera, J. Farjas, P. Roura and T. Pradell (eds.) *Avances en Arqueometría 2005. Actas del VI Congreso Ibérico de Arqueometría*, 263-272. Girona, Universitat de Girona.

RODRÍGUEZ-ARIZA, M.O. and MONTES MOYA, E. 2005. On the origin and domestication of Olea europaea L. (olive) in Andalucía, Spain, based on biogeographical distribution of its finds. *Vegetation History and Archaeobotany* 14 (4), 551-561.

ROS SALA, M.M. 2005a. La Punta de los Gavilanes en el contexto histórico de Mazarrón. In J. M. López Ballesta (coord.), *Carlantum, II Jornadas de Estudio sobre Mazarrón*, 43-70. Mazarrón, Universidad Popular de Mazarrón.

ROS SALA, M.M. 2005b. Metalurgia y sociedad en el Sureste prerromano. In M. Parra Lledó (dir.), *Bocamina. Patrimonio minero de la Región de Murcia,* 39-58. Murcia, Museo de la Ciencia y el Agua, Ayuntamiento de Murcia.

ROVIRA, N. 2007. *Agricultura y gestión de los recursos vegetales en el sureste de la Península Ibérica durante la Prehistoria Reciente.* Unpublished PhD thesis, University of Pompeu Fabra of Barcelona.

SÁNCHEZ GÓMEZ, P. and GUERRA MONTES, J. 2003. *Nueva Flora de Murcia –Plantas Vasculares-.* Murcia, Diego Marín.

SCHOCH, W. and SCHWEINGRÜBER, F.H., 1982. Holzkohlenalytische ergebnisse aus der bronzezeitlichen siedlung Fuente Alamo. Prov. Almeria. Spanien. Archäologisches *Korrespondenzblatt* 12, 451-455.

SCHWEINGRÜBER, F.H. 1978. *Mikroskopische Holzanatomie. Anatomie microscopique du bois. Microscopic wood anatomy.* Birmensdorf, Swiss Federal Institute for Forest, Snow and Landscape Research.

SCHWEINGRÜBER, F.H. 1990. *Anatomie europäischer Hölzer. Ein Atlas zur Bestimmung europäischer Baum-, Strauch- und Zwergstrauchhölzer. Anatomy of European woods. An atlas for the identification of European trees shrubs and dwarf shrubs.* Stuttgart, Verlag Paul Haupt.

TARGARONA, J., ALONSO, B., CACHO, I., CANALS, M. and LÓPEZ SÁEZ, J.A. 1996. Climatic changes during the Late Quaternary in South East Spain: Implications for the stablishment of the Murciano-Almeriense bioprovince. *Estudios palinológicos,* 119-123.

TERRAL, J.F. 1996. Wild and cultivate olive (*Olea europaea L.)*: a new approach to an old problem using inorganic analyses of modern wood and archaeological charcoal. *Review of Palaeobotany and Palynology* 91, 383-397.

TERRAL, J.F. 1997. *La domestication de l'olivier (Olea europaea L.) en Méditerranée nord-occidentale: Approche morphométrique et implications paléoclimatiques.* Thèse de Doctorat. Université de Montpellier II.

VERNET, J.-L., BADAL, E. and GRAU, E. 1983. La végétation néolithique du sud-est de l´Espagne (Valencia, Alicante) d'après l'analyse anthracologique. *C.R. Académie des Sciences* 296, 669-672.

VERNET, J.L., OGEREAU, P., FIGUEIRAL, I., MACHADO, C. and UZQUIANO, P. 2001. *Guide d'identification des charbons de bois préhistoriques et récents. Sud-Ouest de l'Europe: France, Péninsule Ibérique et îles Canaries.* Paris, CNRS.

# THE ABUNDANCE OF CHARCOAL FRAGMENTS EMPHASIZES THE ASSUMPTION OF HUGE PALAEOFIRES IN THE MIXED MOIST SEMI-EVERGREEN RAINFOREST OF THE NORTHERN REPUBLIC OF CONGO

Jean-François GILLET
University of Liège, Gembloux Agro-Bio Tech, Unit of Forest and Nature Management, Laboratory of Tropical and Subtropical Forestry, Nature plus asbl, B-5030 Gembloux, Belgium
jf.gillet@natureplus.be

Jean-Louis DOUCET
University of Liège, Gembloux Agro-Bio Tech, Unit of Forest and Nature Management, Laboratory of Tropical and Subtropical Forestry, B-5030 Gembloux, Belgium

**Abstract**: *In this paper, we study the origins of the northern Congo Republic rainforests. Macroscopic charcoal fragments were systematically recorded through auger investigations and pits observations in four forest types: the open canopy Marantaceae forest, the dense forest with Marantaceae, the* Gilbertiodendron dewevrei *forest and the* Triplochiton scleroxylon *forest. In addition, the charred* Elaies guineensis *seeds were distinguished from the other charcoals in the pits. Ten selected charcoals, including charred* E. guineensis *seeds, were dated by AMS. Abundance of charcoal fragments in the soils at various depths indicated several episodes of fires in the region. A dryer climatic phase, between 2320 and 1330 BP, associated with a large scale human occupation related to the important harvesting of oil palm nuts, could explain the widespread Marantaceae forests. Recent slash-and-burn shifting cultivation, c. 200 BP, would allow* T. scleroxylon *installation. Implications for forest management are discussed.*

**Keywords:** *$^{14}C$ dating, charred* Elaeis guineensis *seed, Marantaceae,* Gilbertiodendron dewevrei, Triplochiton scleroxylon, *slash-and-burn shifting cultivation, human impact*

## INTRODUCTION

Forests of Northern Congo Republic cover an area of 16 million ha and are of major interest for conservation, wood production, and survival of local population (de Wasseige *et al.* 2009). Considered by White (1983) as a complex mosaic of temporally flooded and terra firma forests, they include the largest area of open canopy forests with a dense understorey of giant herbs belonging to Marantaceae family (de Namur 1990). Origins of these Marantaceae forests remain uncertain (Brncic 2002; Harris 2002). They could result from particular soil conditions as well as past or recent disturbances (Brncic 2002; Rogers and Williamson 1987; Swaine 1992). Due to the high herbaceous cover, tree regeneration is known to be very difficult (Brncic 2002; Lejoly 1996).

Northern Congo closed canopy forests include large patches of gregarious species. Two of them, *Gilbertiodendron dewevrei* (De Wild.) J.Léonard and *Triplochiton scleroxylon* K. Schum. have high local densities. While *G. dewevrei* is an evergreen shade-bearer species (Hart *et al.* 1989), *T. scleroxylon* is a deciduous light-demanding species (Aubréville 1959). Coexistence of such different guilds let suppose local past disturbances whose main drivers could be linked to climate fluctuations combined with the human activities (Hart *et al.* 1996; Brncic *et al.* 2007).

Indeed, Northern Congo forests could have experienced frequent and sometimes dramatic disturbances over the last 4000 years, associated with climate and human demographic changes (Willis *et al.* 2004; Brncic *et al.* 2007). About 2500 years ago, a climate-driven deforestation is thought to have split up the forest block (Maley 2002). A dryer and more seasonal climate would have been favourable to the expansion of pioneer species. Among them, the spread of the edible oil palm *Elaeis guineensis* Jacq., could even have facilitated the Bantu migration (Schwartz 1992; Maley and Chepstow-Lusty 2001). Over last centuries, Bantu shifting cultivators could have facilitated the regeneration of light-demanding species, including commercial species nowadays exploited for their high value timber (van Gemerden 2003).

Understanding past forest dynamics is of major interest in predicting effects of both present climate change and human activities. Consequently, the objective of this study is to evaluate if past fires occurred in Northern Congo and to which extent they could explain the present forest composition. It was conducted in a 800,000ha forest concession. There, we investigated presence and abundance of charcoal and charred *E. guineensis* seeds through auger sampling and pits observations in the main vegetation types.

## STUDY AREA

The study area (0.5-2.5°N 16.0-17.5°E) is located in the northern Republic of Congo (Figure 1), at the north-

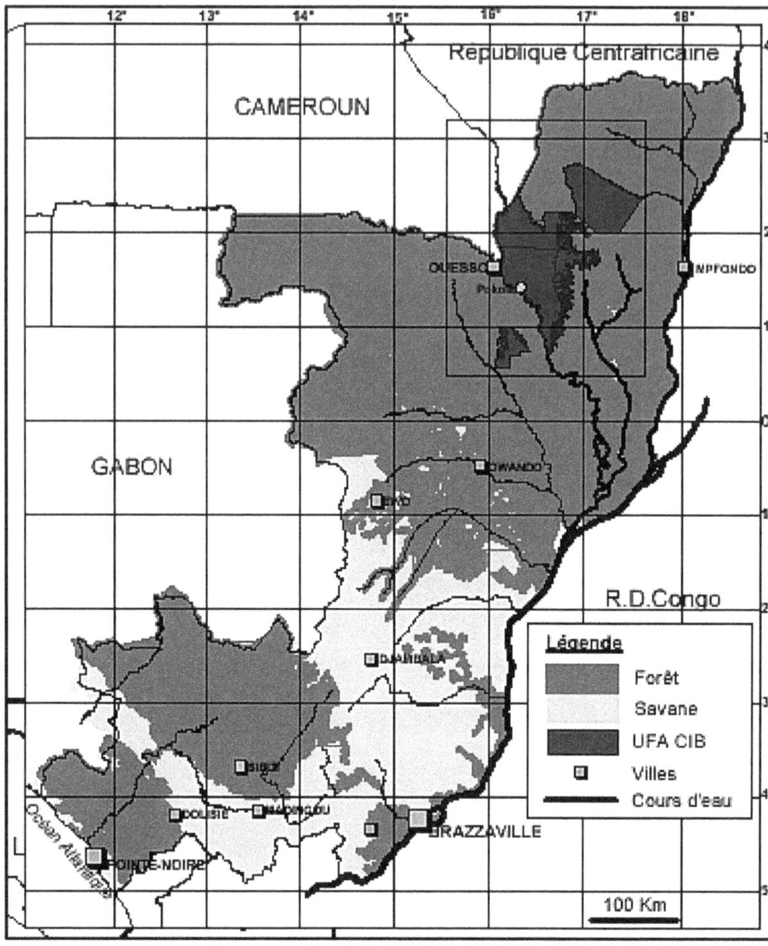


Figure 1. Location of the study area and distribution of the forest/savannah in the Republic of Congo (from CIB-management cell)

western limit of the Congo basin, a wide and flat continental depression (Léonard 1947). The site, covering 800,000 ha, includes three Forest Management Units (Pokola, Kabo and Toukoulaka) granted to the CIB company. It is bordered by the Sangha River to the west and by the Likwala-aux-Herbes swamps and the Tele Lake to the east (Figure 2). Several forest types occur (Laporte and Lin 2004): swampy forests and periodically flooded forests (24%), dense semi-deciduous forests (including *T. scleroxylon* patches) and dense forests with Marantaceace (26%), evergreen *G. dewevrei* forests (10%), as well as open canopy Marantaceae forests (36%, 4% are not defined). The current Bantu population is concentrated at the boundaries of the site (Figure 2), principally along the Sangha River and in the 'Terre of Kabounga' (Lewis 2002). In addition, many temporary settlements of Pygmies hunter-gatherers are present inside the forest (Lewis 2002). The northern Congo climate is equatorial, without real dry season (Vennetier 1963). The average monthly temperatures fluctuate slightly around 25°C. The mean annual hygrometry is 85% (Paget and Desmet 2003). The average annual precipitations in Ouesso (1.616°N 16.318°E) is comprised between 1547 mm (1945-65; Letouzey 1968) and 1685 mm (1961-90; data from ASECNA in Paget and Desmet 2003). Altitudes are between 330 and 460 meters.

The site is mainly covered by Quaternary alluvial deposits, except the extreme North, which is composed of Mesozoic sandstones (Schwartz and Lanfranchi 1993) and Precambrian schisto-quartzitic series (Vennetier 1963).

## MATERIAL AND METHOD

Tree line-transects totalling 10 km in length were installed (T1 to T3) in the largest patch of open canopy Marantaceae forest in the south-eastern part of the study area (Figure 2). Along these transects, thirty-seven soil samplings of 2m depth were carried out with a pedological auger using a toposequency method, i.e. considering slopes and vegetations changes (Barnerias *et al.* 2004; Gillet *et al.* 2008). Three main vegetations types were crossed: the open canopy Marantaceae forest (71%), the dense forest with Marantaceae (22%) and the evergreen *G. dewevrei* forest (4%). The other 3% of swampy forest were not considered. The dense forest with Marantaceae had a dense understorey of giant herbs such as the open canopy Marantaceae forest but the canopy is most often closed.

Thirteen complementary soil samplings (1 to 13) were made until 80 cm depth in old *Triplochiton scleroxylon*

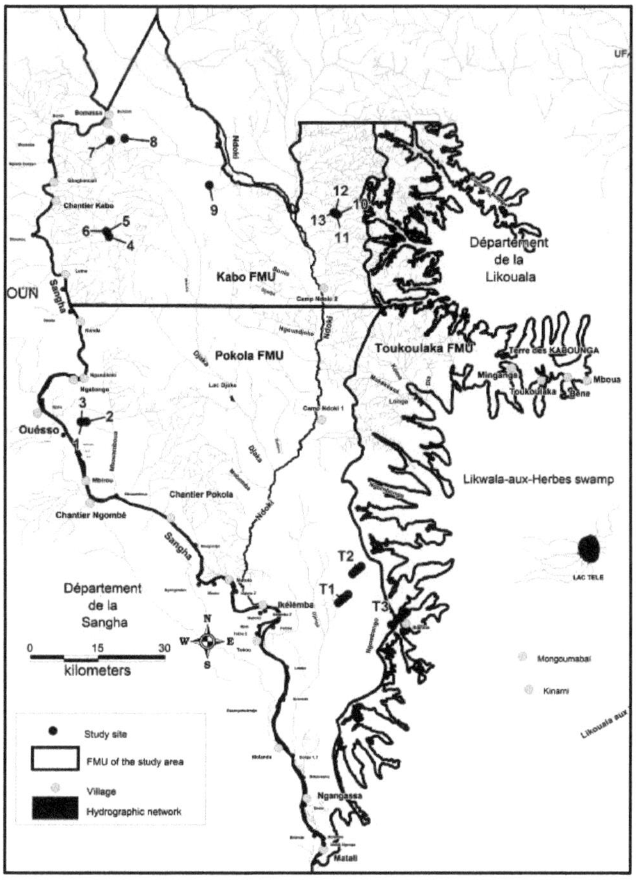


Figure 2. Close-up of the study area, location of the study sites, zoning, villages and hydromorphic network (from CIB-management cell)

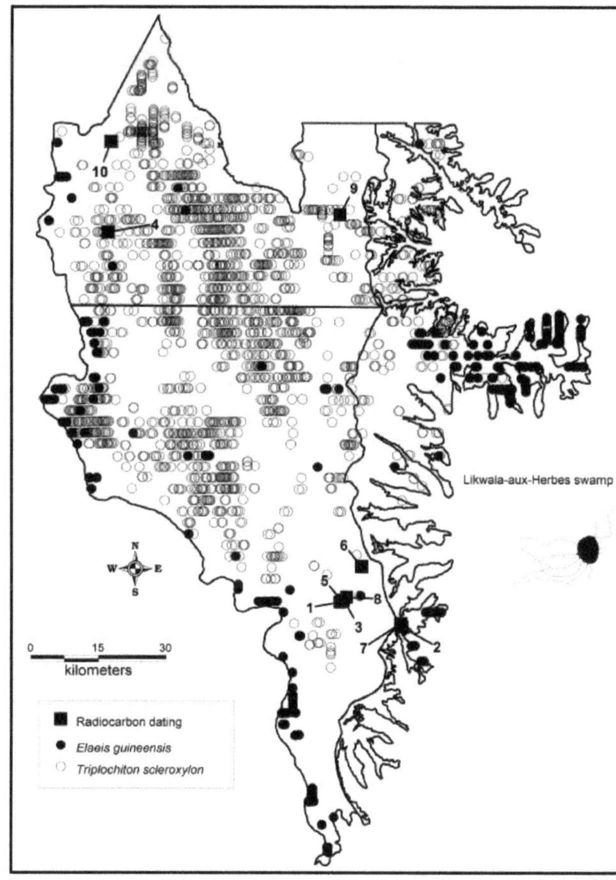


Figure 3. Location of the ten radiocarbon dating, current distribution of *Elaeis guineensis* and *Triplochiton scleroxylon* (from CIB-management cell)

stands in the northern part of the study area (Figures 2 and 3). Furthermore, to have a better spatial view of macroscopic charcoal main levels linked to the major disturbances, and to differentiate the associated charred *E. guineensis* seeds from the other charcoals (Figure 4), ten pedological pits were dug in the following sampled vegetation types: four in the open canopy Marantaceae forest; one in the dense forest with Marantaceae, two in *Gilbertiodendron dewevrei* forests, and three in the *T. scleroxylon* stands.

The macroscopic charcoal fragments of a minimum size of 1 mm were systematically recorded with the auger. Indeed from that size, they are visually identifiable and remain near the source of fire and therefore represent a local record of fire events (Clark and Patterson 1997). Each sample was dried and cleaned before being placed in a sealed packet. A rating system was used to quantify the abundance of charcoal and charred oil palm nuts: 0 (no charcoal); 1 (scarce and small charcoal fragments) and 2 (profuse charcoal). Estimations were made by 20 cm depth intervals (auger) or by pedological layer (pits).

Ten charcoals were chosen for dating considering the most recent disturbed horizon of each pit. Moreover, nuts were preferred since the age of the tree is included in the tree charcoal dating. Each sample was processed and analysed individually. The acid-base-acid pre-treatment allowed respectively to eliminate carbonates, humic acids and to neutralize the sample. Between each step of one hour, the charcoal was rinsed with water Mq. Finally, it was oven dried (KIK-IRPA Brussels). The dating was done by AMS Accelerator Mass Spectrometry (Leibniz-Laboratory for Radiometric Dating and Stable Isotope Research, University of Kiel, Germany). Calibrated $^{14}C$ dates were obtained with "References – Northern Hemisphere Atmospheric data from Reimer *et al.* (2004); OxCal v3.10 Bronk Ramsey (2005)".

Management inventories of the logging company were used to map the current distribution of targeted species. They were carried out along parallel transects divided in geo-referenced contiguous plots. The sampling rate of mixed terra firma forests was 0.97% for *T. scleroxylon* from the timber tree inventories and 0.19% for *E. guineensis* from the non-timber forest products inventories (Paget and Desmet 2003). Thematic maps (Projection UTM north zone 33, ellipsoid WGS 84) were developed with MapInfo 8.0 professional software (Figures 1 to 3).

The average cumulative abundances of charcoal were compared until 2 m depth for the 37 first auger samplings, and until 80 cm depth to compare the auger surveys in Marantaceae and *T scleroxylon* forests (n = 43) with analyses of variance (ANOVA), after verification of

Figure 4. Fragments of charred oil palm seeds *Elaeis guineensis* of c. 2000 BP found again at 50 cm depth in a Marantaceae forest

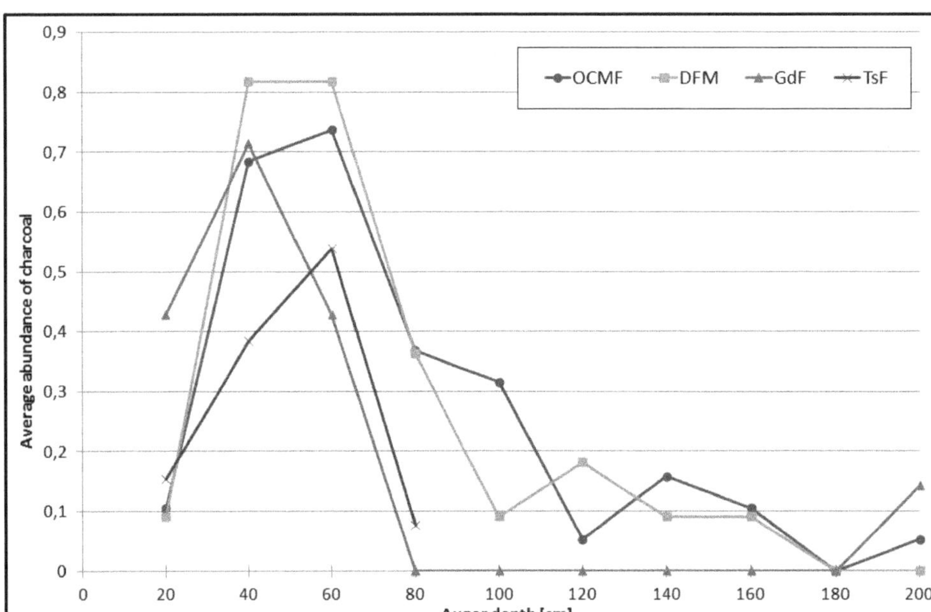


Figure 5. Distribution of the average abundance of charcoal fragments for the four vegetation types and according to the depth intervals, for 50 auger's sampling. OCMF = Open canopy Marantaceae forest, DFM = Dense forest with Marantaceae, GdF = *Gilbertiodendron dewevrei* forest and TsF = *Triplochiton scleroxylon* forest

assumptions of normality and homoscedasticity. Post-hoc tests (Newmann-Keuls) allowed detecting significant differences.

## RESULTS

Charcoal fragments were omnipresent (80%) in the soil of the four forest types sampled with the auger. A high concentration, more extensive in the Marantaceae forests, is observed below 20 cm and up to 60 cm (Figure 5). In the south-eastern part, there is no difference of cumulative abundance of charcoal between the three sampled vegetation types (ANOVA, $F = 0.74$, $p = 0.484$). Nevertheless, cumulative abundance of charcoal varied by location (ANOVA, $F = 8.04$, $p = 0.007$); Marantaceae forests of the SE contain more charcoals that northern stands of *T. scleroxylon* (Newmann-Keuls post-hoc test, $p<0.01$).

In the pits, the soil depth corresponding to the greatest concentration of charcoals was always the second layer,

| N° | KIA # | Calibrated date (CI: 95.4%) | Depth [cm] | Charcoal type dated | Vegetation type | Oil palm nut abund. | Real charcoal abund. | Location |
|---|---|---|---|---|---|---|---|---|
| 1 | 34143 | 2320-2100 BP (92.6%) | 40 | Wood | OCMF | 0 | 2 | SE |
| 2 | 34144 | 2150-1990 BP | 50 | Palm nut | DFM | 1 | 2 | SE |
| 3 | 34147 | 1870-1690 BP | 30 | Wood | GdF | 0 | 1 | SE |
| 4 | 37687 | 1830-1690 BP (82.7%) | 30 | ND seed | TsF | 1 | 2 | N |
| 5 | 37684 | 1540-1410 BP | 50 | Palm nut | GdF | 2 | 1 | SE |
| 6 | 34141 | 1530-1380 BP | 55 | Palm nut | OCMF | 1 | 1 | SE |
| 7 | 37686 | 1520-1330 BP | 40 | Palm nut | OCMF | 2 | 1 | SE |
| 8 | 34142 | 660-550 BP | 35 | Palm nut | OCMF | 2 | 2 | SE |
| 9 | 37685 | 220-140 BP (45.9%) | 30 | ND seed | TsF | 0 | 1 | N |
| 10 | 37683 | 220-140 BP (51.9%) | 20 | Wood | TsF | 0 | 1 | N |

Figure 6. Main anthracological characteristics of the second soil horizon of the ten pedological pits. The numbering corresponds to dating (see Figure 3 for location). The vegetation types are OCMF = Open canopy Marantaceae forest, DFM = Dense forest with Marantaceae, GdF = *Gilbertiodendron dewevrei* forest and TsF = *Triplochiton scleroxylon* forest

i.e. between 20 and 55 cm (Figure 6), roughly corresponding to the auger results (Figure 5). Charcoal fragments and charred fruits were found in inclusions of variable size, never in continuous layers. Nuts of *E. guineensis* are present in 60% of the pits (Figure 6). Present in the four vegetation types, they appear more abundant in the Marantaceae forests (Figure 6). Nowadays the species is rare in the study area and is principally located not far from the current villages along the Sangha River and in the 'Terres des Kabounga' (Figures 2 and 3).

The ten AMS radiocarbon dating are listed in Figure 6. The seven oldest dating fluctuated between 2320 and 1330 BP and were found in the Marantaceae forests and in associated *G. dewevrei* forests. An intermediary dating (660-550 BP) was observed in an open canopy Marantaceae forest. The two youngest charcoals (220-140 BP) were found in old stands of *T. scleroxylon* (Figure 6).

## DISCUSSION

The abundance of charcoal fragments in the soils at various depths indicates past episodes of fires in the region. While the highest concentrations of charcoal were found between 20 and 55 cm, charcoal fragments were still observed at a depth of 200 cm. Thus fires have probably influenced several times the structure and composition of the vegetation.

The major perturbations were dated between 2320 and 1330 BP. They were often related to charred *E. guineensis* seeds. In southern Nigeria, charred *E. guineensis* nuts were the most frequent charcoal found (White and Oates 1999). Next to our study area, Fay (1997) observed numerous *E.guineensis* seeds dated between 2300 and 900 BP. In fact, until the Late Stone Age, the edible endocarp of *E. guineensis* was one of the most important products for the survival of local populations (Sowunmi 1999).

The late Holocene, characterized by a phase of forest decline around 2500 BP would have facilitated the Bantu migration through the central African forests (Schwartz 1992). This period was probably favourable to oil palm expansion (Maley and Chepstow-Lusty 2001). The species regenerates in post-cropping fallow (Lebrun and Gilbert 1954) and the seed dormancy is broken by fire (Sowunmi 1999). These characteristics associated with preferences for river banks and periodically flooded land (Ergo 1997; Ataga and van der Vossen 2007) could explain today's distribution in the study area.

The greatest abundance of charcoals associated with charred *E. guineensis* seeds in Marantaceae forests could be explained by an old, intensive and widespread human occupation as supposed by Brncic (2002). Large opened areas associated with particular soil proprieties could have facilitated Marantaceae propagation. Indeed, like *E. guineensis*, Marantaceae could have a preference for humid soils (Letouzey 1983).

While present in *G. dewevrei* patches, charcoals were aged between 1870-1410 BP. In the NE part of the Congo Basin, charcoals under *G. dewevrei* dated 2180-1750 BP were slightly older (Hart *et al.* 1996).

Old stands of *T. scleroxylon* probably result from younger fires since charcoals were dated to *c.* 200 BP as reported by Hart *et al.* (1996) for the youngest disturbance in the Ituri forests. The local abundance of *T. scleroxylon*, like other light-demanding and anemochorous species of emergent trees, would be the result of human fires linked to the slash-and-burn shifting cultivation as suggested by Brncic *et al.* (2007), Hallé (1961), Letouzey (1968), van Gemerden (2003) and White and Oates (1999). This hypothesis is consistent with the high species growth rate (around 1 cm in diameter by year) and the diameter of the species up to 2 m (Durieu de Madron *et al.* 2000; Palla and Louppe 2002). The present scarcity of seedlings and saplings (Paget and Desmet 2003) is probably linked to less mobile human populations than before (Bahuchet and Guillaume 1979; Vennetier 1963). The bell-shaped population structure (stems distribution by diametric classes) in its whole distribution range (Forni 1997; Gillet 2004; Hall and Bada 1979; Kouadio 2009) is characteristic of old stands of a gregarious and light-demanding species (Rollet 1974). It is really concerning due to its high logging rates (ATIBT 2008; Bosu and Krampah 2005; Palla and Louppe 2002). Most of the commercial species being light-demanding (Doucet 2003), further investigations on the origin and evolution of their populations are definitely needed to achieve sustainable management of the African moist forests.

### Aknowledgments

We would like to thank the donor 'Fonds Français pour l'Environnement Mondial', the logging company 'Congolaise Industrielle des Bois' and the NGO Nature +. Indeed, the data presented were collected during the three-year project 'Suivi du Programme Dynamique Forestière – Agroforesterie – Inventaires Faune' of the FFEM grant under the financing agreement N° CCG 1071.01 A. We also thank Dr Bock Laurent, Dr Langhor Roger, Dr Paget Dominique, Dr Monticelli David, Vau-travers Emilie and Federspiel Michèle, and the field team Ati Ngouabe Maurice Ndambilo, Luc Mouyamah, Fidèle Zoung, Bokoye Mbongo and the deceased Justin Beloua.

### References

ATAGA, C.D. and van der VOSSEN, H.A.M. 2007. *Elaeis guineensis* Jacq. [Internet] Fiche de Protabase, in H.A.M. van der Vossen and G.S. Mkamilo (eds). PROTA (Plant Resources of Tropical Africa), Wageningen, Pays-Bas. <http://database.prota.org/recherche.htm>. Visited August 14, 2008.

ATIBT, 2008. Statistics. La Lettre de l'ATIBT, 35-55.

AUBRÉVILLE, A. 1959. *La Flore de la Côte d'Ivoire, 2.* Nogent-sur-Marne, Centre Technique Forestier Tropical.

BAHUCHET, S. and GUILLAUME, H. 1979. Relation entre chasseurs-collecteurs pygmées et Agriculteurs de la Forêt du Nord-Ouest du Bassin Congolais. In S. Bahuchet (ed.), *Pygmées en Centrafrique: études ethnologiques, histoires et linguistiques sur les pygmées Ba-Mbenga, Aka/Baka du Nord-ouest du Bassin congolais,* 109-139. Paris, Selaf.

BARNERIAS, C., BÖHM, F., DUBOIS, S., JUNOD, G., le TREIS, M. and POUSSE, J.S. 2004. Effets de l'engorgement hydrique sur la couleur des sols et la distribution des espèces du BAFOG (Guyane Française), Ecole Nationale du Génie Rural des Eaux et Forêts.

BOSU, P.P. and KRAMPAH, E., 2005. *Triplochiton scleroxylon* K. Schum. In. Louppe, D. (Eds). Prota 7: Timbers/Bois d'œuvre. [CD-Rom]. PROTA, Wageningen, Pays-Bas.

BRNCIC, T.M., 2002. *Ecology and patch dynamics of* Megaphrynium macrostachyum *(Benth.) Milne–Redh. (Marantaceae) in the south-west Central African Republic.* Unpublished PhD thesis, University of Oxford.

BRNCIC, T.M., WILLIS, K.J., HARRIS, D.J. and WASHINGTON, R. 2007. Culture or climate? The relative influences of past processes on the composition of the lowland Congo rainforest. *Philosophical Transactions of the Royal Society B* 362, 229-242.

BRONK RAMSEY, C. 2005. Improving the resolution of radiocarbon dating by statistical analysis, in T.E. Levy and T.F.G. Higham (eds.), *The Bible and Radiocarbon Dating: Archaeology, Text and Science,* 57-64. London, Equinox.

CLARK, J.S. and PATTERSON, W.A.I. 1997. Background and local charcoal in sediments: scales of fire evidence in the paleorecord, in J.S. Clark, H. Cachier, J.G. Goldammer and B. Stocks (eds), *Sediment Records of Biomass Burning and Global Change,* 23-48. Berlin, Springer.

de NAMUR, C. 1990. Aperçu sur la végétation de l'Afrique centrale atlantique, in Lanfranchi R. & Schwartz D. (eds). *Paysages quaternaires de l'Afrique centrale atlantique,* 60-67. Paris, ORSTOM.

de WASSEIGE, C., DEVERS, D., MARCKEN, P., EBA'A ATYI, R., NASI, R. and MAYAUX, P. (eds) 2009. *Les Forêts du Bassin du Congo – Etat des Forêts 2008.* Office des publications de l'Union européenne, Luxembourg

DOUCET, J.L. 2003. *L'alliance délicate de la gestion forestière et de la biodiversité dans les forêts du centre du Gabon.* Unpublished PhD thesis, Faculté Universitaire des Sciences Agronomiques de Gembloux.

DURRIEU de MADRON, L., NASI., R. and DETIENNE, P. 2000. Accroissements diamétriques de quelques essences en forêt dense africaine. *Bois et Forêts des Tropiques* 263, 63-74.

ERGO, A.B. 1997. Nouvelle évidence de l'origine africaine de l'*Elaeis guineensis* Jacq. par la découverte de graines fossiles en Uganda. *Annales de Gembloux* 102, 191-201.

FAY, J.M. 1997. *The ecology, social organization, populations, habitat and history of the western lowland gorilla* (Gorilla gorilla gorilla *Savage and Wyman 1847).* Unpublished PhD thesis, University of Washington.

FORNI, E. 1997. *Type de forêts dans l'Est du Cameroun et étude de la structure diamétrique de quelques essences.* Unpublished DEA, Faculté Universitaire des Sciences Agronomiques de Gembloux.

GILLET, J.F. 2004. *Dynamique forestière et régénération naturelle en essences commerciales et espèces concurrentes au sein des formations végétales à Marantaceae intactes et exploitées – Programme d'intervention sylvicole de l'ayous Triplochiton scleroxylon K. Schum. (Sterculiaceae).* Internal report. Congolaise Industrielle des Bois, Office Allemand de la Coopération Technique (GTZ) and Nature +.

GILLET, J.F., BEECKMAN, H. and DOUCET, J.L. 2008. The origin of the forests in the North of the Congo Republic: an anthraco-pedological contribution, in F. Damblon and M. Court-Picon (eds.), *4th International Meeting of Anthracology,* 67. Brussels, Royal Belgian Institute of Natural Sciences.

HALLÉ, N. 1961. *Sterculiacées. Flore du Gabon, 2.* Paris, Muséum National d'Histoire Naturelle.

HALL, J.B. and BADA, S.O. 1979. The distribution and ecology of Obeche (*Triplochiton scleroxylon*). *Journal of Ecology* 67, 543-564.

HARRIS, D.J. 2002. The vascular plants of the Dzanga-Sangha Reserve, Central African Republic. *Scripta Botanica Belgica* 23, 1-274.

HART, T.B., HART, J.A. and MURPHY, P.G. 1989. Monodominant and species-rich forests of the humid tropics: Causes for their co-occurrence. *The American Naturalist* 133 (5), 613-633.

HART, T.B., DECHAMPS, R., FOURNIER, M. and ATAHOLO, M. 1996. Changes in forest composition over the last 4000 years in the Ituri Basin, Zaire. In L..J.G. van der Maesen, XM. van der Burgt et JM. van Medenbach de Rooy (Eds.). The biodiversity of African plants. Proceedings XIVth AETFAT Congress. Kluwer Academic, Dordrecht, The Netherlands, 545-563.

KOUADIO, Y.L. 2009. *Mesures sylvicoles en vue d'améliorer la gestion des populations d'essences forestières commerciales de l'Est du Cameroun.* Unpublished PhD thesis, Faculté Universitaire des Sciences Agronomiques de Gembloux.

LAPORTE, N. and LIN, T. 2004. *Utilisation de la télédétection pour le développement des plans d'aménagement et le suivi de l'exploitation forestière au Nord Congo.* Woods Hole Research Center, Wildlife Conservation Society and Congolaise Industrielle des Bois.

LEBRUN, J. and GILBERT, G. 1954. *Une classification écologique des forêts du Congo.* Institut National pour l'étude Agronomique du Congo Belge.

LEJOLY, J. 1996. *Biodiversité végétale dans le Parc National d'Odzala (Congo).* Projet ECOFAC – Composante Congo AGRECO – CTFT, Université Libre de Bruxelles.

LÉONARD, J. 1947. Contribution à l'étude des formations ripicoles arbustives et arborescentes de la région d'Eala. *INEAC Publication (Special issue)*, C. R. Semaine Agric., Yangambi.

LETOUZEY, R. 1968. *Etude phytogéographique du Cameroun.* Paris, Paul Lechevalier.

LETOUZEY, R. 1983. *Manuel de botanique forestière – Afrique Tropicale* 2B, CTFT, Nogent-sur-Marne.

LEWIS, J. 2002. *Forest hunter-gatherers and their word: A study of the Mbendjele Yaka Pygmies of Congo-Brazzaville and their secular and religious activities and representations.* Unpublished PhD thesis, University of London.

MALEY, J. 2002. A catastrophic destruction of African forests about 2,500 years ago still exerts a major influence on present vegetation formations. *Bulletin Inst. Development Studies* 33, 13-30.

MALEY, J. and CHEPSTOW-LUSTY, A. 2001. *Elaeis guineensis* Jacq. (oil palm) fluctuations in central Africa during the late Holocene; climate or human driving forces for this pioneering species? *Vegetation History and Archaeobotany* 10, 117-120.

PAGET, D. and DESMET, O. 2003. *Inventaire des ressources forestières de l'Unité Forestière d'Aménagement de Pokola (Congo).* Unpublished Report. Pokola, Congolaise Industrielle des Bois.

PALLA, F. and LOUPPE, D. 2002. *Obeché.* FORAFRI, CIFOR and CIRAD, Montpellier, France.

REIMER, P.J., BAILLIE, M.G.L., BARD, E., BAYLISS, A., BECK, J.W., BERTRAND, C., BLACKWELL, P.G., BUCK, C.E., BURR, G., CUTLER, K.B., DAMON, P.E., EDWARDS, R.L., FAIRBANKS, R.G., FRIEDRICH, M., GUILDERSON, T.P., HUGHEN, K.A., KROMER, B., McCORMAC, F.G., MANNING, S., BRONK RAMSEY, C., REIMER, R.W., REMMELE, S., SOUTHON, J.R., STUIVER, M., TALAMO, S., TAYLOR, F.W., van der PLICHT, J. and WEYHENMEYER, C.E. 2004. *Radiocarbon* 46, 1029-1058.

ROGERS, M.E. and WILLIAMSON, E.A. 1987. Density of herbaceous plants eaten by gorillas in Gabon: some preliminary data. *Biotropica* 19, 278-281

ROLLET, B. 1974. L'architecture des forets denses humides sempervirentes de plaine. CTFT., Nogent-sur-Marne.

SCHWARTZ, D. 1992. Assèchement climatique vers 3000 BP et expansion Bantu en Afrique centrale atlantique: quelques réflexion. *Bulletin de la Société Géologique de France* 3, 353-361.

SCHWARTZ, D. and LANFRANCHI, R. 1993. Les cadres paléoenvironnementaux de l'évolution humaine en Afrique Centrale Atlantique. *L'Anthropologie* 97, 17-50.

SOWUNMI, M.A. 1999. The significance of the oil palm (*Elaeis guineensis* Jacq.) in the late Holocene environments of West and west central Africa; a further consideration. *Vegetation History and Archaeobotany* 8, 199-210.

SWAINE, M.D. 1992. Caracteristics of dry forest in West Africa and the influence of fire. *Journal of Vegetation Science* 3, 365-374.

van GEMERDEN, B.S., OLFF, H., PARREN, M.P.E. and BONGERS, F. 2003. The pristine rain forest?

Remnants of historical human impacts on current tree species composition and diversity. *Journal of Biogeography* 30, 1381-1390.

VENNETIER, P. 1963. *Les Hommes et leurs Activités dans le Nord du Congo-Brazzaville.* Unpublished PhD thesis, University of Bordeaux.

WHITE, F. 1983. *The vegetation of Africa*, Paris, Unesco.

WHITE, L.J.T. and OATES, J.F. 1999. New data on the history of the plateau forest of Okomu, southern Nigeria: an insight into how human disturbance has shaped the African rain forest. *Global Ecology and biogeography* 8, 355-361.

WILLIS, K.J., GILLSON, L. and BRNCIC, T.M. 2004. How 'Virgin' is Virgin Rainforest? *Science* 304, 402-403.

# AN EXPERIMENTAL AND ECOLOGICAL APPROACH TO MODELING ANCIENT FUEL USE

David John GOLDSTEIN
United States National Park Service, Christiansted National Historic Site, Christiansted, Virgin Islands, 00820, USA
David_Goldstein@nps.gov

Izumi SHIMADA
Southern Illinois University at Carbondale, Department of Anthropology, Faner Hall, Carbondale, IL, 62901-4208, U.S.A.
ishimada@siu.edu

**Abstract**: *This paper presents a case study for the development of paleoethnobotanical data together with a focus on ancient charcoal use to interpret the organization of craft production and domestic economy. In 1999 and 2001 the Sicán Archaeological Project, working in the La Leche Valley, excavated mains from a ceramic and metalworking facility active during the tenth century AD in Northern Peru. Additional experimental work, including recreating the pyrotechnologies used in production, a survey of the local forests of the area, and a detailed examination and determination of fuel remains from the site, all aided in the reconstruction of primary material allocation across contexts. This work highlights the necessary coordination among a variety of anthropological and other scientific methods of inquiry to interpret adequately the carbonized production remains. Although wood charcoal can address some of the variables critical to production, we are only able to reconstruct site-wide activities when these data are interpreted together with the feature-to-feature remains of seeds, flowers and fruits. Our analyses permit the description and assessment of recycling and fuel supplements, solutions seen in ethnographic examples of fuel use that typifies some pre-industrial models of resource allocation. These integrated analyses generate baseline data for ecological studies to understand human-environment interactions over the long term and assess sustainable resource use in the past and present.*

**Key Words**: *Ethnoarchaeology, Forest Ecology, Wood Anatomy, Craft Production, South America*

## INTRODUCTION

Several years of research at Huaca Sialupe on the north coast of Peru provide new insight into the technology and organization of craft production that spanned an estimated 100 years during Middle Sicán period (1000-1100 AD). For this study, the excavations and experimental work by the Sicán Archaeological Project (SAP) focused on the production debris from a multi-craft workshop. This paper demonstrates how the interpretation of archaeological charcoal provides an innovative way to examine issues of resource management and labor organization. Specifically, we employ a paleoethnobotanical approach to understand fuel use by Middle Sicán metalsmiths and potters who shared the workshop space 1000 years ago. Using experimental firing of archaeological features, in replica, provided a clearer understanding of how fuel was used at the site. Our experience demonstrates the extent that fuel use between these two industries, and other activities at the site, was complementary.

## RESEARCH ISSUES AND CONSIDERATIONS

The workshop at Huaca Sialupe (Figure 1) manufactured a variety of products in at least two media, metal and ceramics, for a brief 100-year period. In terms of ceramics, artisans relied on molds and small, manageable kilns to produce both oxidized and reduced fine wares. Metalsmiths in adjacent workspaces fabricated, among other activities, thin sheets of arsenical copper and gold alloys using simple, but efficient furnaces. They cut, shaped and/or joined together the resultant sheet metal to produce predominantly small utilitarian objects and ornaments. We have no evidence of on-site smelting. Our reconstruction of the activities at Sialupe is based on excavation of the site in 1999 and 2001 and the experimental reconstruction of the kilns and furnaces, described below, during July 2000.

Our research focuses on the importance of fuel use in ancient craft production. Matson (1989, 23-24) and Wertime (1964) have each argued that the pyrotechnologies used in ceramic and metal production developed through a series of synergistic processes. Matson argues that metal furnaces and ceramic kilns developed parallel to one another technologically. Both industries, fuel intensive and fuel dependent, are tied to the overriding concern regarding fuel consumption and gathering. Our study of fuel remains and their contexts at the craft production site of Huaca Sialupe permitted us to evaluate the relationship between these and other production activities at the site.

Previous work in Sicán period contexts clearly demonstrates that ceramic and metal products played different roles and manifested different values in the Middle Sicán culture. Black ceramic bottles bearing the image of the Sicán Deity, the principal deity of the Sicán

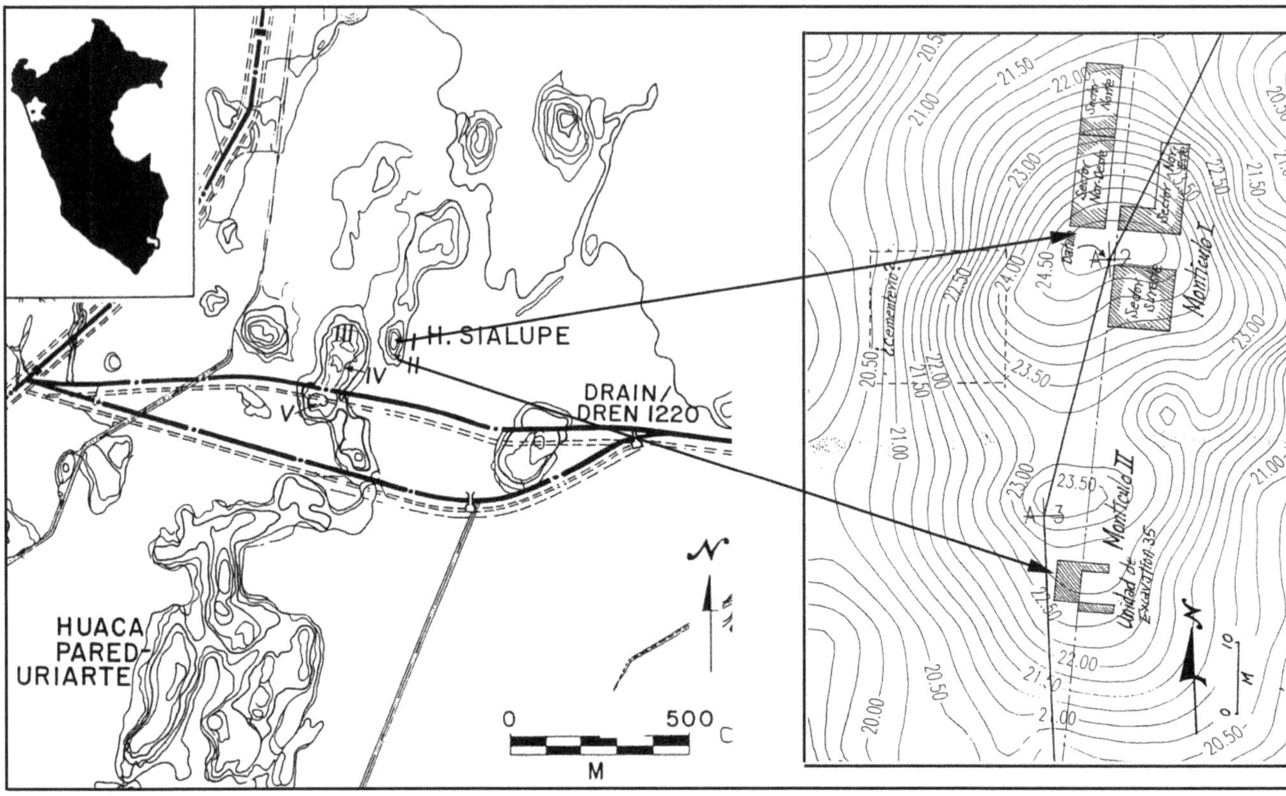


Figure 1. Map of Northern Perú Indicating Location of Huaca Sialupe and Surrounding Study Area (Goldstein 2007)

religion were a signature craft product for middle and high status burials of the culture. These items were clearly valued, exchanged, and even imitated in form and finish well outside of the Sicán territory (Shimada 1990; Shimada and Wagner 2000). In fact, the large-scale production of true black pottery is one of the major Sicán legacies. Yet, the Middle Sicán artistic and technical standard bearers were metal objects, particularly those of precious metals that were available only to the social elite (Cleland and Shimada 1998; Shimada 1994a, 1994b). The black pottery, in essence, was a simplified version of precious metal vessels in form, color and iconographic contents.

Given the higher social value metalwork held relative to pottery and the diverse role metals played in the Middle Sicán, we have elsewhere expected that metal production received preferential material and labor support from the social elite (Cleland and Shimada 1998; Shimada 1994a, 1994b). More specifically, we argued that Sicán metalworkers had greater access than potters to supplies of high-quality hardwood fuels, a potentially limited resource in the region's ecologically specialized dry tropical forests (Proyecto Algarrobo 1997; Shimada and Wagner 2000). These conclusions, however, were established without a clear understanding of the archaeological production contexts. This study examines the charcoal and non-charcoal remains from these contexts and interprets them in conjunction with experimental reconstructions of firing events. In contrast to the previous model, this research suggests that ceramic and metal craft production at Huaca Sialupe were integrated technologies that relied on multiple strategies of artisans for local fuel procurement.

## METHODOLOGY

Few studies of pre-Hispanic craft production have examined fuel-use in any depth, with most projecting ethnographic or ethnoarchaeological observations into the past (e.g., Arnold 1985; Sillar 2000; Winterhalder *et al.* 1974). These studies do not benefit from the examination of excavated production contexts. Thus, the deliberate sampling and examination of fire-use contexts excavated from Huaca Sialupe's production features is unique to our study. We have identified three production features at the site: ceramic kilns, metallurgical furnaces, and hearths. We also consider that charcoal production on site was a potential activity associated with ceramic and metal making. We contextually interpret the hearth features as the remains of cooking fires, not directly related to ceramic or metal production fire-use, but important complements to the economic processes of artifact production (Goldstein 2007). To aid in our archaeological interpretations, we conducted firing experiments in the field using both a full-scale replica ceramic kiln and metalworking furnace (Goldstein and Shimada 2007; Shimada *et al.* 2003a, 2003b; Shimada and Wagner 2000; Wagner *et al.* 2001). Additionally, we made our own ethnographic study of traditional charcoal making in the study area to understand the potential impact of these activities on the archaeological record (Goldstein 2007).

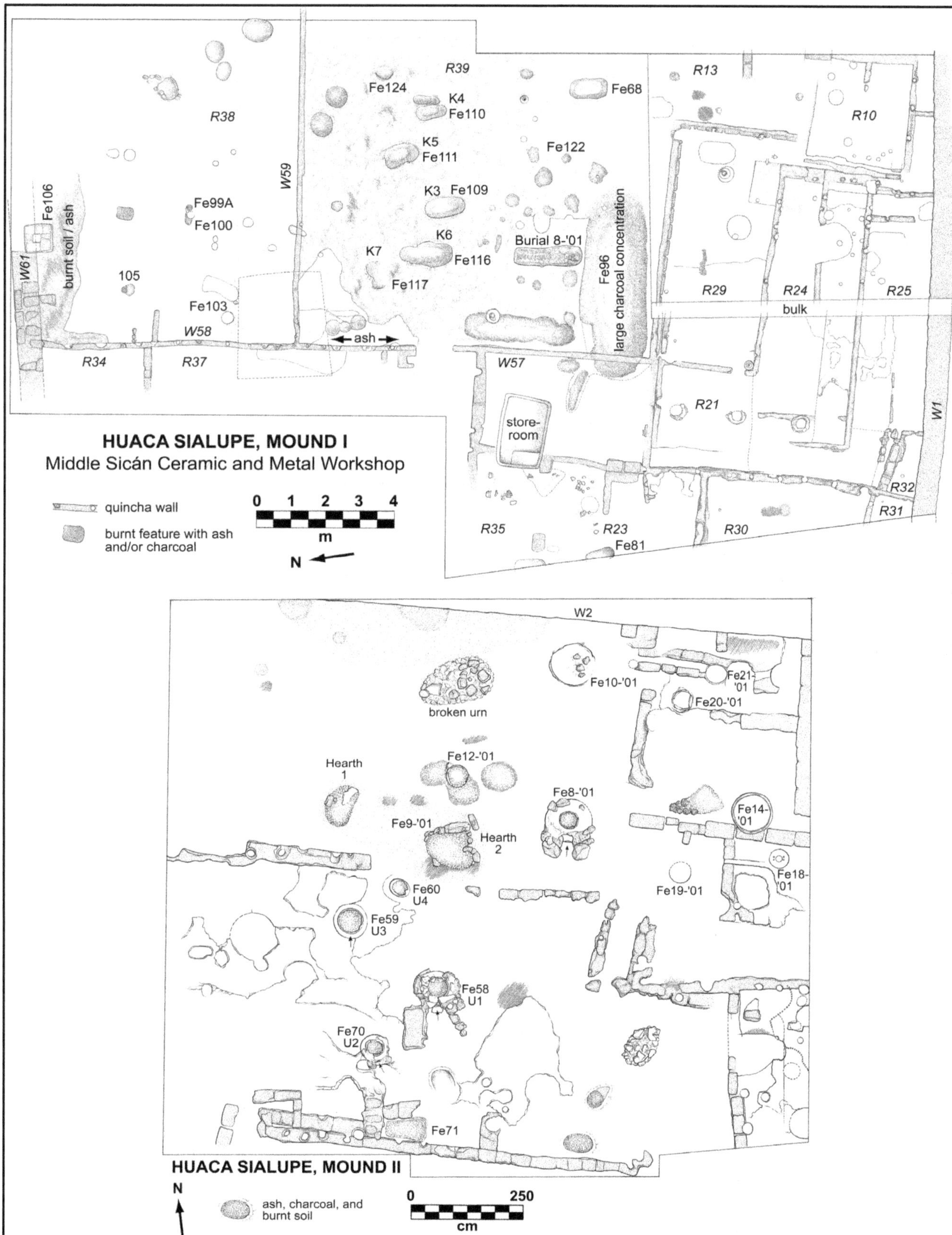


Figure 2. Plan of Middle Sicán Workshop at Huaca Sialupe (Shimada and Montenegro 2001)

Fifty one of the total excavated fire-use features (n = 153) from the 1999 and 2001 field seasons provide the materials analyzed in our discussion. Excavation at the site included the recovery of nearly two-thirds of the entire Sicán period workshop (Figure 2). The workshop areas discussed here consist of two conjoined mounds within the five mounds that comprise the site. The preserved remains of the workshop cover an estimated

| Common Names | Scientific Name |
|---|---|
| Alcapparillo | *Cassia bicapsularis* L. |
| Algarrobo (macho) | *Prosopis pallida* (H. and B. ex Willd.) H.B.K. |
| Angolo | *Pithecellobium multiflorum* (H.B.K.) Benth. |
| Cabelludo, Palo Verde | *Parkinsonia acuelata* L. |
| Cuncun | *Valesia glabra* (Cav.) Link. |
| Faique | *Acacia macarantha* H. and B. ex Willd. |
| Hualtaco | *Loxopterygium huarango* Spruce ex. Engler |
| Overo | *Cordia lutea* Lam. |
| Pacay, Guaba | *Inga feuillei* DC. |
| Pajaro Bobo | *Tessaria integrifolia* R. and P. |
| Palo Santo | *Bursera graveloens* (H.B.K.)Triana and Planch |
| Palo Verde | *Cerciduim praecox* subsp. *praecox* (R. and P.) Harms. |
| Sapote | *Capparis angulata* R. and P. ex DC. |

Figure 3. List of Dominant Woody Species of La Leche River Valley Dry Tropical Forests (Goldstein 2007)

1750 $m^2$ (ca. 70 m N-S x 28 m E-W). While production materials were scattered throughout the five-mound complex, the areas excavated contained the highest concentrations of fuel and manufacturing debris on the surface. In other words, it is likely that artifact production took place throughout the entire five-mound site, although perhaps at a smaller scale in the areas outside Mounds I and II.

We completely sampled features associated with production for examination. However, where it was difficult to define the feature completely (n = 32), we removed a measured quantity relative to the total feature volume as recovered by excavation, 25% of the total volume. All feature samples were dry sieved through a standard series of fine grade geological sieves (4.0-0.425 mm). We sorted the materials under magnification using incident light and used standard anatomical techniques for wood identification (Wheeler and Baas 1998) and also those used techniques for seed identification (Martin and Barkley 1961). This includes comparisons with modern thin sections from systematically identified wood voucher specimens collected in the field and samples of known charred material. Our identifications of the ancient charcoal specimens are based on magnified views (<180x) of the standard three dimensions (tangential, radial, and cross sections) of charcoal greater than 2 mm in size. We examined the remains from the experimental firing of full-scale replica kilns and metalworking furnaces, produced in 2000, in tandem and in the same manner as the archaeological materials. This process is ongoing and the results presented here are the first in a series of evaluations of the fuel remains.

## NORTH COASTAL ENVIRONMENTAL DYNAMICS PAST AND PRESENT

The site of Sialupe is located just southeast of the modern pottery-making town of Morropé, in the Department of Lambayeque, at the southern boundary of the Sechura desert. Dry tropical forest dominates the desert's southern edge (Tosi 1960; Proyecto Algarrobo 1997). The regional woody flora consists of xerophytic trees, low shrubs, and cacti. The tree species are adapted to an extremely dry environment (precipitation <10 mm/year), and typically the majority of the area's woody plants come from two main families Capparidaceae (*Capparis* spp., four species), and Fabaceae (*Prosopis* spp., three species) and *Acacia* spp., two species) as seen in Figure 3 (Proyecto Algarrobo 1997). These three tree genera comprise the majority of dominant mature stature trees in the modern forest in the region. Presently, the cessation of charcoal production and agricultural clearing are at the heart of a regional campaign to stop deforestation in the area to preserve the remaining areas of dry coastal forest (Risco 1998; Cuba Salerno *et al.* 1998). Dry coastal forests are extremely vulnerable to deforestation by ecosystem disruption due to slow annual growth of dominant tree species, uncertain periodic water and nutrient recharge, and low biodiversity (Bullock *et al.* 1995; Cuba Salerno *et al.* 1998). Modern forest resources in the study area are quickly disappearing as they are cut to create agricultural fields and supply fuel or building materials (Risco 1998).

Earlier phases of the SAP focused on the development of a profile of the regional paleoecology (Shimada 1981). This work determined that the modern climate and environmental processes do not differ significantly from that of the Sicán period, save for the impact of 500 years of landscape modifications by humans, e.g. overall decrease in plants/animal populations and community biodiversity (Shimada 1995). Work by Craig and Shimada (1986) indicates that periodic inundations, due to El Niño/Southern Oscillation (ENSO) rains, have been consistently episodic during the more recent geological epoch.

Previously published archaeological and experimental research in the area identifies only two species of primary

| Tree Species | *Acacia* spp., Corresponds to local *A. macracantha* | *Capparis* spp., Corresponds to local *C. angulata* or *C. ovalifolia* | *Cercidium* spp., Corresponds to local *C. praecox* | *Prosopis* spp., Corresponds to three species of *Prosopis* spp. available locally | *Quercus alba* | *Eucalyptus globulous* |
|---|---|---|---|---|---|---|
| Caloric Value (Dry Wood) KJ/kg (Harker *et al.* 1992) | 19054 (n = 18) | 20101 (n = 2) | 18782 | 20737 | 17340 | 20719 |

Figure 4. Comparison of KJ/Kg Values of Select Wood Species (Goldstein and Shimada 2007). N.B. White Oak is not native to the area and is presented here as a temperate hardwood for comparison. Likewise, Eucalyptus is not native to the area but is presently in common use in the area for reforestation and wood lot cultivation

wood fuel from select contexts, *Prosopis pallida* and *Capparis angulata* (Shimada *et al.* 1998; Shimada and Merkel 1991; Wagner *et al.* 1998). These studies recognize *Prosopis* spp.,[1] locally *algarrobo*, as the primary high quality fuel resource available in the area; *Capparis angulata,* locally *sapote*, while releasing comparable energy as fuel wood (Figure 4), does not make good charcoal and is rapidly consumed in the fire (Shimada 1981; Shimada 1985; Shimada and Wagner 2000). These two genera represent only seven of the woody plant species dominant in the region today (Cuba Salerno *et al.* 1998; Proyecto Algarrobo 1997). Because *Prosopis* spp. meets the physical demands of both sustained ceramic firing and metal smelting and smithing, the SAP proposed that it was the cornerstone of Middle Sicán production (Shimada 1998b; Shimada and Merkel 1991; Shimada and Wagner 2000). From a purely ecological standpoint this genus also served as the backbone of the ancient dry tropical forests as the dominant and most long-lived species. If this genus served as the principal source for craft production fuel, the implication for forest destruction is profound.

The three *Prosopis* spp. species (*P. pallida, P. juliflora, P. affinis*) provide more than just high-grade wood and charcoal fuel, they are the backbone of the mature dry forest ecosystem (Delvaud 1984; Proyecto Algarrobo 1997). Dry tropical forests are highly specialized ecosystems that rely on a limited number of tree species to maintain soil structure and fertility, and provide shelter to other plants and animals critical to forest regeneration and regrowth (Bullock *et al.* 1995; Cuba Salerno *et al.* 1998). In particular, *Prosopis* spp. trees require nearly 100 years to attain mature stature (Ocampo 1991). Modern charcoal makers, however, frequently cut them before they reach even 30 years (Risco 1998).

The propensity of charcoal makers and agriculturalists to cut these trees has affected significantly forest regeneration throughout the colonial and modern era (Delvaud 1984; Risco 1998). As a result, *Prosopis* spp. charcoal production has become a critical deforestation problem over the past 200 years and has forced the regulation of forest products in the coastal areas of Peru (Ferreyera 1987). At present, rural forest residents produce *Prosopis* spp. charcoal illegally for additional household income to be sold in local and regional markets, and/or for domestic use for fuel (Risco 1998). At the beginning of the twenty-first century the protection of this species has become a source of local political and social friction between illegal farmsteaders and governmental protection agencies.

## PRODUCTION FEATURE RECOVERY AND DEFINITION

Key to our interpretations is the reconstruction of ancient technology based on experimenting with ancient kiln and metal furnace technologies. Experiments provide opportunities to understand how these features functioned given different fuel loads and production capabilities. In the end, they provide a clearer idea of how fuel can be used either singly for each particular type of production, or in tandem. Eventually, we can obtain a more holistic idea of how fuel use and quality influences the different forms of production encountered at the site.

One category of features critical to our study is ceramic kilns (Figure 5). Although remains of both oxidation and reduction kilns with distinct morphologies were found at the downwind, north end of the workshop, (Shimada and Wagner 2000, 2007), the current discussion refers exclusively to reduction kilns. Their downwind location is logical for removing heat, smoke, and debris generated by kilns and furnaces.

Field and laboratory firing experiments, as well as the comprehensive archaeometric analysis of samples of excavated ceramics, have shown the widely valued glossy Middle Sicán black pottery owed its distinctive appearance to a number of factors. The vessel to be fired was first thoroughly burnished in the leather hard state, necessary for creating a charged surface onto which a layer of amorphous carbon with occasional crystals of graphite formed during firing under reducing conditions at temperatures of 800-900°C (Shimada *et al.* 2003; Shimada and Wagner 2000). Our excavations indicated that the sort of reduction firing described above was conducted in the four evenly spaced kilns set into the floor of the large walled but unroofed area designated as

[1] There are at least three known genera of *Prosopis* spp. known in the research area, *P. pallida, P. juliflora, P. affinis*. Present research worldwide continues to have difficulty classifying these species (Pasiecznik *et al.* 2001). Here we use *Prosopis pallida* species considered the most common, and *Prosopis* spp. as a generic term that encompasses inclusive of the three species as they are indistinguishable anatomically (Pasiecznik *et al.* 2001).

Figure 5. Middle Sicán Blackware Reduction Kiln (Goldstein and Shimada 2007)

Room 39 (Figure 2). The kilns were elliptical or teardrop-shaped and had their large firing chamber oriented downwind. Their slightly concave, sloping clay-lined bottoms were recessed ca. 15 cm below floor level and measured around 100 cm x 45 cm wide, with some degree of variation. Fragments of collapsed linings indicate that the kilns originally had incurving sidewalls to create a dome-like firing chamber with a height of ca. 30 cm above the floor (Figure 6). Most of these kilns were relatively clear of charcoal larger than 4mm, indicating a cleaning episode prior to abandonment. Additionally, we recovered one feature, number 68 (described elsewhere), which still had a full charcoal charge within, yet with its ceramics and refractory sidewalls removed.

We also recovered the remains of metal working furnaces at the north end of the workshop (Mound I). An additional set was found embedded well into the earthen floors of Room 39 towards the south end of the workshop (Figure 2; Mound II). The furnaces consist of large inverted, ceramic urns typically around 40 cm in diameter and 45 cm high. The urn bottoms had been removed to create opening ca. 20-25 cm in diameter at the top (Figure 6). The exterior was covered with an insulating layer of clay and pieces of adobe block up to 12 cm thick. As excavated the furnaces contained a mix of carbonized organic fuel materials (ca. 8L). Each of these furnaces had a round 8-9 cm draft-hole cut into the body near floor level facing the direction of the prevailing wind. Similar to the ceramic kilns, these furnaces were positioned to maximize the natural draft. Additionally, the furnaces included two short flanking walls to funnel wind towards the furnace's fuel charge. The overall charcoal quality was clean, with little ash, potentially indicating that the furnaces were charged for use just prior to abandonment.

## FIRING EXPERIMENTS

Based on the excavated fuel-use contexts, we conducted a series of firing experiments at the southern edge of Huaca Sialupe in July 2000 to gain comparative data for understanding the material and behavioral correlates of Middle Sicán ceramic firing and metalworking and the physical performance of the associated kilns and furnaces. The experiments afforded a unique opportunity to observe the decision-making processes and behaviors involved in successful firing, which is commonly regarded as the most challenging stage of pottery making (Shimada 2005). The 2000 experiments built on experience and knowledge gained from similarly conceived research conducted in 1992 using a 3000-year old kiln and its modern replica (e.g., Shimada *et al.* 1998, 2003; Wagner *et al.* 1998).

The master craftsman José Sosa prepared the full-scale replica kiln and pottery used in the 2000 experiments. He is a professional ceramist from Chulucanas, a pottery-making town in the Department of Piura, with a sound understanding of pre-Hispanic ceramic technology. The

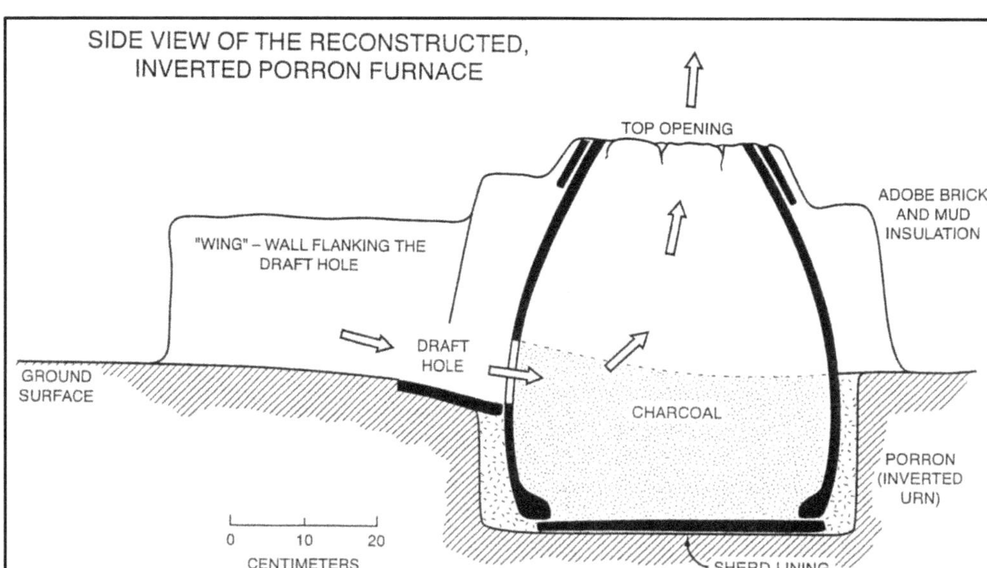


Figure 6. Plan of Reconstructed Middle Sicán Metalworking Furnace (Goldstein and Shimada 2007)

replicas approximated measurements, materials, and techniques as determined by our prior archaeometric analysis of excavated materials (Shimada *et al.* 2003; Shimada and Wagner 2000).

Fuels for the experimental firings consisted of dry, fallen hardwood branches (1-4 cm in diameter) collected from the nearby settlement and ground surface. *Prosopis* spp. was used at a ratio of 5:1 by weight to *Capparis angulata.* Both species of wood exhibited bark beetle burrowing on the outermost wood surfaces of the collected materials. Jose Sosa, our ceramist, was responsible for fuel wood collection, and selected what he felt was a good mix between hotter *Prosopis* spp. fuel and *Capparis angulata*, which while burns hot, is less long lasting in firing. Sosa used leaves and small twigs as kindling and occasionally to boost combustion in later stages of firing.

The reducing conditions produce the carbon/graphite surface on the ceramics and reduce the hardwood fuels to charcoal that remains in the kiln after firing. Our experiments produced glossy black pottery that closely approximates that produced 1000 years ago at the Huaca Sialupe workshop. They also yielded vessels with incomplete reduction and mottled appearance just as we have documented among wasters recovered from Huaca Sialupe. We collected the fuel residues in the form of charcoal from the experimental kilns for analysis. These fuel residues were clean, intact, large pieces, similar to those removed from the archaeological ceramic kilns and the experimental charcoal kiln described below.

We made our replica metal working furnace with locally made adobe bricks, clays and a *porrón* (large ceramic storage urn) that closely matched in size, shape, materials and techniques of the Middle Sicán originals previously documented at Huaca Sialupe. We charged the furnace with *Prosopis* spp. charcoal to the height just below the bottom of the circular draft hole that faced the prevailing wind. Our experiment showed that, with only natural wind fanning the charcoal bed, the furnace is capable of achieving an internal temperature between 1000°C to 1100°C within 20 minutes from the time of ignition. Egressing air at the top of the furnace also reached 800°C. We could control easily the furnace temperature to suit a variety of metal working tasks such as melting of ingots, alloying, and annealing of metals found in association with the original furnaces, by adjusting the draft entering the furnace with a small potsherd. The experiment effectively showed that the Middle Sicán furnaces made with inverted urns were highly efficient and flexible devices, well-suited for a variety of metalworking tasks. Additional details of the experimental set-ups and results of archaeometric analysis of fired vessels and kiln lining are described elsewhere (Shimada and Wagner 2007; Shimada *et al.* 2003a, 2003b, Wagner *et al.* 2001). Unspent charcoal fuels from these kilns indicated a covering of ash, and scanning electron microscopy observed the presence of ash in the pores of unspent charcoal fuels from these experiments.

## CHARCOAL FURNACES

Work completed by Goldstein in 2002, observations by Shimada (1985), and recent work by Cayo (2003) demonstrates that charcoal can be produced in quantity at the household or small scale. Goldstein (2007) observed in the field that one person could collect easily 500 kg of *Prosopis pallida* and construct a small charcoal kiln to produce around 250 kg of hardwood charcoal. The labor investment to set the kiln and collect the wood requires a single day for one person, and the firing takes four days. The labor investment and productivity of these small-scale furnaces depends on the quantity of wood used, but this work demonstrates the ease with which charcoal can be prepared.

The wood is arranged in a set of parallel lines on an area of ground cleared of loose soil and debris. Once the wood is arranged, a layer of *poña*, or kindling, collected from the surrounding area is laid on top of the logs. *Poña* consists of leaves, refuse, small twigs and animal dung gathered from the below the trees in the area surrounding the kiln. The entire pile is covered with compact peds of the soil dug from the perimeter of the piled logs and kindling. Hardpan clods are preferred, and loose dirt is saved for the last, uppermost covering of the kiln above the clods. Overall, the processing of 500 kg of wood into charcoal results in a pile 1.5 m high with a surface area that covers 4 x 6 m.

A small area is left open at the downwind side where a small initial fire is started, called the *caja*. Once the *caja* catches fire, it too is covered, and the fire is left for several days to burn through the entire pile. During the second day, small vents are made at the end of the kiln opposite to the *caja* by inserting long straight sticks about 1 cm in diameter to let the smoke and gases to escape. When the firing is complete and successful, the charcoal makers carefully excavate the pile and the carbonized branches are removed intact, clean of ash, and often in single pieces.

Processes such as charcoal making may have generated the features excavated at the very northern end of Mound I. In 1999, in a series of excavated test pits, we encountered at least two pits reaching to a depth of 70 cm below surface that contained an estimated 30-40 l or more of seemingly fresh charcoal lumps (1-2 cm diameter). In light of the ethnoarchaeological correlate for small-scale charcoal making, these pits could be the remains of the perimeter trenches similarly the result of dirt clod sections from the soil from around the base of the charcoal kiln. Their contents may be explained by leveling of the area after the kiln was dismantled, during which the dregs from the charcoal kiln ended up as perimeter pit infill. There is no doubt that ash and burnt soil mixed with some small charcoal lumps were discarded in this sector. In particular, we recovered numerous, superimposed lenses containing such mixtures from the NE corner of Mound I. The depth and quality of the charcoal recovered from these features may be the remains of local, small-scale, charcoal furnaces associated with metal production at Huaca Sialupe.

| Feature Type | Blackware Kilns (n = 13) | Preheating Pits (n = 5) | Metalworking Furnaces (n = 8) | Domestic Hearths (n = 25) |
|---|---|---|---|---|
| Principal Fuels (Percent g/L of Like Features) | *Prosopis* spp. 21.5%<br>*Capparis angulata* 15.3%<br>*Acacia macracantha* 12.3%<br>*Capparis ovalifolia* 3.8%<br>*Tessaria integrifolia* 3.8%<br>Other Wood 18.5%<br>Other Non-Wood 24.8% | *Prosopis* spp. 46.5%<br>*Acacia macracantha* 6.3%<br>*Capparis angulata* 0.3%<br>Other Wood 40.3%<br>Other Non-Wood 5.8% | *Acacia macracantha* 19.7%<br>*Prosopis* spp. 10.6%<br>*Capparis angulata* 1.2%<br>*Valesia glabra* 0.6%<br>Other Wood 52.3%<br>Other Non-Wood 15.6% | *Prosopis* spp. 38.5%<br>*Acacia macracantha* 24.5%<br>Other Wood 5%<br>Other Non-Wood 32% |
| Est. Average Diameter of Carbonized Twig | 0.5 cm< x < 5.0 cm | 0.5 cm< x < 1.5 cm | 0.5 cm< x< 2.0 cm | 0.5 cm < x < 8.0 cm |
| Est. Life Stage of Stemwood | Branch 32%<br>Juvenile 11%<br>Mature 5%<br>Unclassified 51% | Juvenile 78%<br>Branch 12%<br>Unclassified 10% | Branch 79%<br>Juvenile 11%<br>Unclassified 10% | Branch 43%<br>Juvenile 2%<br>Mature 1%<br>Unclassified 46% |
| Degree of Combustion | Incomplete/Cleaned with Small Carbonized Twigs Remaining (Except for #68, see discussion) | Incomplete/Small Twig Charcoal Remains/Ash | Complete/Some Charcoal Leftover | Near Complete/Ash |

Figure 7. Summary Table of Contexts Sampled at Huaca Sialupe and Their Wood Fuel Contents. Percentages Reflect Density (Total Taxon Weight (g)/Total Feature Volume(L))

## PALEOETHNOBOTANICAL ANALYSIS AND INTERPRETATION

A summary table of the wood charcoal remains from the different fire use features at Huaca Sialupe is presented in Figure 7. This composite picture of all the fuel remains collected from the various features includes those items that we could classify based on recurring feature morphology and our experimental work. The percentages correspond to an overall assessment of the representative quantities (charcoal weight to volume) of the dominant wood species recovered and determined. The validity of this exercise for interpreting the variability of taxa present in charcoal assemblages is argued elsewhere, both experimentally and archaeologically, by Chabal *et al.* (1999); a more detailed report of our findings appears elsewhere as well (Goldstein 2007). Critical to our interpretations, however, are Chabal's observations about the *law of fragmentation*, where archaeological charcoal assemblages tend to represent the relative abundance of the materials collected and utilized in the case of fuel use features. From this standpoint we use relative abundance of charcoal (Figure 7) to describe fuel use for ceramic production at the site.

### Fuelwood Choice at Huaca Sialupe

To begin we see across all of the contexts, save for the metalworking furnaces, a preference for *Prosopis* spp. fuels. The charcoal recovered from the blackware reduction ceramic kilns demonstrates the preference for *Prosopis* spp. hardwood that we assumed for the ethnoarchaeological reconstructions of both ceramic and metal furnace firings. What is interesting, from the point of view of Figure 8, however, is the presence in varying quantities of different taxa in distinct relative quantities per fire use context. This diversity highlights how ethnoarchaeological experimentation ignores a diverse range of available fuel resources (including non-wood) in the past, which severely limits our ability to interpret ancient fuel use. Also evident is the comparative importance of non-wood combustible materials across all of the different feature types, a topic that will be discussed below in more detail.

Overall, we see that all of the materials present represent the local flora as reconstructed from both the modern ethnobotanical survey and the paleoecological reconstruction, indicating that artisans mediated their local resources to meet production needs. In other words, no fuel was imported to the site to meet production needs. Across the various contexts we additionally see that different technological needs were likely met through variations in how these resources were allocated. We see that *Prosopis* played a dominant role in blackware ceramic production, both in the kilns and in the preheating fires. This same genus, however, played a considerably lesser role as a fuel in metal production contexts. Also notable is that for ceramic production, we see the importance of mixing leguminous hardwoods, *Acacia* and *Prosopis*, with faster burning and cooler heat of combustion woods such as *Capparis angulata* and *Capparis ovalifolia* (Figure 4). This may have been to keep combustion going in these low oxygen kilns, while trying to temper firing intensity over the long 17 hour firing process. In contrast, where intense heat was needed, in the metal furnaces or cooking fires, the leguminous hardwoods were favored. This last observation is in complete accord with Matson's (1989) hypothesis regarding the importance of ceramics as a hub for high quality fuels in pyrotechnological contexts.

### Fuel Source Maturity and Life Form

Almost unequivocally the indications from all of the main features is that the none of the fire use features relied on

| Family | Determination | Kiln | Preheating Fire | Furnace | Hearth | Grand Total |
|---|---|---|---|---|---|---|
| Verbenaceae | *Lippia* sp. | 2339 | 8042 | 5 | 502 | 10888 |
| Fabaceae | *Prosopis* sp. | 266 | 26 | 26 | 65 | 383 |
| Poaceae | *Zea mays* | 87 | 2 | 24 | 86 | 199 |
| Fabaceae | cf. Fabaceae | 135 | 10 | 7 | 8 | 160 |
| Boraginaceae | *Heliotropum* sp. | 117 | 13 | – | 10 | 140 |
| Fabaceae | *Acacia* sp. | 52 | 43 | – | 24 | 119 |
| Chenopodiaceae | *Chenopodium* sp. | 64 | 10 | 1 | 4 | 79 |
| Malvaceae | *Gossypium* sp. | 12 | – | 11 | 23 | 46 |
| Capparidaceae | *Capparis crotonides* | – | 28 | – | – | 28 |
| Capparidaceae | *Capparis ovalifolia* | – | 25 | – | – | 25 |
| Solanaceae | cf. Solanaceae | 14 | 2 | 3 | 2 | 21 |
| Capparidaceae | *Capparis* sp. | 14 | – | – | – | 14 |
| Solanaceae | *Nicotiana* sp. | 10 | – | 3 | – | 13 |
| Solanaceae | *Capsicum* sp. | 6 | 1 | 1 | 3 | 11 |
| Euphorbiaceae | *Euphorbia* sp. | 7 | 1 | 1 | – | 9 |
| Cucurbitaceae | *Lagenaria* sp. | 1 | – | – | 7 | 8 |
| **Grand Total** | | **3124** | **8203** | **82** | **734** | **12143** |

Figure 8. Some Select Non-Wood (Seeds/Fruits/Flowers) Determinations from Pyrotechnology Features. Numbers Reflect Density (Count/L) For All Sampled Features

wood taken from mature stature tree trunks. There is a marked, but extremely limited presence across the assemblage of any large tree ring diameter segments. The presence of few large bole indicators suggests that cutting trees for fuel use was not a common practice at Sialupe. Additionally, almost all of the materials (Figure 7) clearly demonstrate the use of juvenile or small branches in all instances. Of particular note is the preference for the use of juvenile *Prosopis* branches, less than 1 cm in diameter, in ceramic preheating fire pits. Elsewhere, larger diameter branch wood served as the fuel for the other fire use features. Additionally, larger branches used as fuel were not present in the residues of the metalworking furnaces. When used, these branches were mostly 3-5 cm in diameter, based on comparing modern charred materials with ancient ring diameter estimates.

Almost all of the branch wood shows evidence of heavy insect infestation: channels riddle the wood structure of even the smallest twigs, with evident exit holes and below bark disturbance. Based on channel morphology, 1-3 mm across and full of pulverized sawdust-like frass, this damage can be linked to beetles from the Burpestidae or Scylotidae families (Borror and White 1970; Rogers 1999). These species feed on sick, felled, or dead trees. The thoroughness of consumption, coupled with the absence of tool marks on the examined pieces, strongly argues for post-mortem infestation. The patterning of the infestation is similar to modern samples of dead and fallen branch wood gathered in the field and used in the firing experiments discussed above. The nature of the insect activity and the apparently low representation of carbonized mature tree parts indicates that gathered fuel wood from the forest floor was likely the prime fuel source in all three types of fire-use features. The exploitation of gathered materials is a common strategy for traditional metal workshops in the modern Western Asia as documented in the ethnographic record (Matson 1966).

Another observation regarding the fuels used by the artisans at Huaca Sialupe is the role that low shrubs served in instances where fuel sources could be more generalized and less intensively focused on high caloric/more stable fuels. In the cases of metalworking furnaces and kilns, low shrubs like *Valesia glabra* and *Capparis ovalifolia*, as well as a variety of other smaller taxa (Table 4; Other Non-Wood), provided up to half of the fuel load in some instances. The presence of these small shrubs indicates that this vegetation, found at heights up to 1m above the forest floor in healthy dry forests, were easily accessible to the artisans. The use of smaller taxa together with fallen branch wood suggests that a healthy dry forest ecosystem was likely within proximity to the Huaca Sialupe settlement. This forest provided enough mature stature cover to generate sufficient deadwood and shrubs to fuel their production needs.

## Non-Wood Components of Combustion

An additional drawback to focusing exclusively on charcoal in archaeological contexts and wood fuel in ethnoarchaeological experiments, is the failure to recognize the contribution that non-wood fuels make to production. In the instance of both the blackware kilns and, as to be expected, in the cooking hearths, we see that 24-32% of the assemblage represents non-wood remains.

While the hearths are mostly ephemeral features without solid enclosing structures, their location and morphology leads us to believe that they were ancillary to industrial production and related to the domestic economy of the site (Figure 2). Their insubstantial nature permits the following observations regarding domestic fire-use relative to craft production. While burning tended to be complete, leaving a shallow ashy lens, we recovered a wide variety of materials, a sample of which are enumerated in Figure 8.

The non-wood materials recovered indicate the presence of three complexes: herbaceous plants, crop plants, and potential fodder species. The herbaceous plants include seeds and leaves from the families Verbenaceae, Boraginaceae, Chenopodiaceae, and. Euphorbiaceae; crop plant macroremains include *Zea mays, maiz* or corn, *Gossypium* sp., cotton, *Capsicum* sp., chili peppers, and *Lagenaria* sp, bottle gourds. Remains of animal fodders include leaves, pods, and seeds from the same trees used as fuels: *Prosopis* spp., and *Acacia macracantha.*

The presence of these the herbaceous and food species in the hearth contexts is to some degree expected; some of the fodder and crop macroremains also occur in the production contexts, however. These seed and fodder materials are probably entering the industrial contexts through the secondary deposition of domestic garbage as fuel, or the use of camelid dung as fuel (Hastorf and Wright 1998). For instance, in the experimental reconstructions of blackware firing performed by Jose Sosa (Shimada and Wagner 2000), and replication experiments of Formative Period kilns (Shimada *et al.* 1998), the potters used dry herbaceous plants and tree leaves as supplementary tinder to aid hardwood to catch fire.

## CONCLUSIONS AND IMPLICATIONS FOR FUTURE INVESTIGATIONS

Combining paleoethnobotanical and ethnoarchaeological approaches to the analysis of fuel remains enhances our ability to examine the decision-making processes involved in craft production. Evidence from the workshop at Huaca Sialupe clearly supports Matson's assertion that ceramic production fuel is central to a variety of pyrotechnic activities. Moreover, our charcoal analysis has enabled us to locate a variety of other ecological and economic patterns within the data. It is only through the examination and determination of the charcoal taxa that we can plausibly reconstruct a clear connection between local resources and economic activity in the archaeological record. The experimental approach provided a complementary perspective: it allowed us to understand the different fuel requirements of each technology, as well as to recognize the potential for resource sharing, a pattern that we can see clearly in the non-charcoal fuel remains of the archaeological dataset.

The sustained clearance of *Prosopis* spp. in antiquity, or the concentrated removal of any dominant tree species, was likely deleterious to the forest ecosystem. The SAP's archaeological and experimental work suggests that metalworking and smelting alone may have adversely affected regional *Prosopis* spp. populations and regional forest structure. For instance, smelting experiments indicate that 306,600 kg of hardwood was required to produce 219 kg of raw copper alloy ingot, an estimate that does not include fuel required for working and forming the ingots (Shimada and Merkel 1991). Given that a single elite Middle Sicán shaft tomb contained nearly two metric tons of metalwork, one of dozens, if not hundreds, in the area, and that Sicán period tombs were mined systematically for precious metalwork from the Inka Empire through modern day, the draw on fuel resources for Sicán production was considerable (Shimada 1995). From an ecological understanding of *Prosopis* spp., dry tropical forests, and experimental production estimates, artisans likely used specific fuel-use strategies to maintain production to mitigate impact on fuel resources during the Middle Sicán.

The SAP's previous work modeled the potential of Middle Sicán period metalwork to overtax dry tropical forest resources, assuming fuel resources depended exclusively on the one or two taxa that were significant contributors to ecological stability. These types of models, however, failed to recognize the flexibility of fuel use from the artisan's point of view and can misidentify resource extraction in the past as unsustainable. The dataset from Huaca Sialupe demonstrates that variables such as life form/stage and available non-wood supplements play substantial roles in the mediation of fuel resources between different fuel use contexts. Future work should focus on locating minimum inputs of quality fuels balanced against non-wood or fast growing fuel supplements, e.g., shrubs.

Although charcoal evidence is a key to understanding fuel resource use, the contribution of other archaeobotanical and archaeological data to a more holistic view of ancient production cannot be underestimated. This study illustrates the importance of integrating multiple analytical avenues, including the excavation of production sites, charcoal analysis, experimental archaeology, ethnobotanical survey, etc, to reconstruct the production processes and technology of multi-craft interaction.

### Acknowledgements

The 1999 and 2001 fieldwork at Huaca Sialupe was supported by grants to Izumi Shimada from the H. John Heinz III Fund for Latin American Studies, Committee for Research and Exploration of the National Geographic Society, and the Wenner-Gren Foundation for Anthropological Research. The 2000 field experiments were made possible by a Southern Illinois University Faculty Research grant to Izumi Shimada. NSF Dissertation Improvement Grant #0211364 provided the support for David Goldstein's 2002 field research. His ongoing analysis is supported by a Doctoral Fellowship from the Graduate School at Southern Illinois University at Carbondale. Archaeometric analyses by Ursel Wagner supported by a research grant from the German Research Council. Additional archaeometric analyses and consultation was offered by Aniko Bezúr, Department of Conservation, Buffalo State College. The authors are indebted to members of the 1999, 2000, and 2001 seasons of the Sicán Archaeological Project for their assistance in data and sample collection, as well as the staff at the Museo Nacíonal Sicán in Ferreñafe, Peru. Participation of Caesar Samillán and the potter, José Sosa, in our 2000 field furnace and kiln firing experiments was critical to the success of our experiments.

**References**

ARNOLD, D.E. 1985. *Ceramic Theory and Cultural Process.* Cambridge, Cambridge University Press.

BORROR, D.J. and WHITE, R.E. 1970. *Peterson Field Guide to Insects.* Boston, Houghton Mifflin Company.

BULLOCK, S.H., MOONEY, H.A., and MEDINA, E. 1995. *Seasonally Dry Tropical Forests.* Cambridge, Cambridge University Press.

CAYO, L. 2003. *Informe: Proceso de Elaboracíon de Carbón en el Departmento de Potosí.* Proyecto Arqeologico Porco-Potosi. Report on file with Dr Mary Van Buren, Colorado State University, Ft. Collins.

CHABAL, L., FABRE L., TERRAL, J.F., and THÉRY-PARISOT, I. 1999. L'Anthracologie., in A. Ferdiére (eds.), *La Botanique: Collection "Archéologiques",* 43-104. Paris, Editions Errance.

CLELAND, K.M. and SHIMADA, I. 1998. *Paleteada* Pottery: Technology, Chronology and Sub-Culture, in I. Shimada (ed), *Andean Ceramics: Technology, Organization and Approaches*, 111-150. Philadelphia, Museum Applied Science Center for Archaeology, The University Museum of the University of Pennsylvania.

CRAIG, A.K., and SHIMADA, I. 1986. El Niño Flood Deposits at Batán Grande, Northern Peru. *Geoarchaeology* 1(1):29-38.

CUBA SALERNO, A.B., SILVA PERALTA, A., and CORNEJO FLORES, C. 1998. *Bosques Secos y Desertificacion.* Lima, Insitutio Nacional del Recursos Naturales.

DELVAUD, C.C. 1984. *Las Regiones Costeñas del Peru Septentrional.* CIPCA/Pontifica Universidad Catolica del Peru, Lima.

F.A.O. 1983. *Simple Technologies for Charcoal Making.* Rome, Food and Agricultural Organization of the United Nations.

FERREYRA, R. 1987. *Estudio Sistemático de los Algarrobos de la Costa Norte del Perú.* Lima, Ministerio de Agricultura.

GOLDSTEIN, D.J. 2007. *Forests and Fires: A Paleoethnobotnical Assessment of the Impact of Middle Sicán Pyrotechnology on the Dry Tropical Forests of the La Leche River Valley, Lambayeque, Peru (950-1050 C.E.).* Ph.D. Thesis, Department of Anthropology, Southern Illinois University at Carbondale. Ann Arbor, University Microfilms International.

GOLDSTEIN, D.J. and SHIMADA, I. 2007. Middle Sicán Multi-Craft Production: Resource Management and Labor Organization. In I. Shimada (ed.), *Craft Production in Complex Societies: Multi-Crafting, Sequential Production, and Producers*, 44-67, Salt Lake City, University of Utah Press.

HARKER, A.P., SANDELS, A., and BURLEY, J. 1982. *Calorific Values for Wood and Bark and a Bibliography for Fuel Wood.* Tropical Products Institute Technical Report G 162. London, Overseas Development Agency.

HASTORF, C.A. and WRIGHT, M.F. 1998. Interpreting Wild Seeds from Archaeological Sites: A Dung Charring Experiment from the Andes. *Journal of Ethnobiology* 18(2):211-227.

LUNA-SUÁREZ, S., LUNA-GUIDO, M.L., FRIAS-HERNÁNDEZ, J.T., OLALDE-PORTUGAL, V., and DENDOOVEN, L. 1998. Soil Processes as Affected by Replacement of Natural Mesquite Ecosystem with Maize Crop. *Biology of Fertilized Soils* 27:274-8.

MATSON, F.R. 1966. Power and Fuel Resources in the Ancient Near East. *The Advancement of Science,* 23(10):146-53.

MATSON, F.R. 1989. Ceramics: The Hub of Ancient Craft Interplay, in P.E. McGovern (ed.), *Cross-Craft and Cross-Cultural Interactions in Ceramics*, 13-28. Ceramics and Civilization Vol. IV. Westerville, Ohio, The American Ceramic Society.

MARTIN, A.C. and BARKLEY W.D. 1961. *Seed Identification Manual.* Berkeley, University of California Press.

OCAMPO, M. 1991. *Metodología para la Determinación de la Edad y las Tasas de Crecimiento para las Especias Forestales de los Bosques Secos de la Costa Norte del Perú.* Chiclayo, Proyecto Algarrobo with the Fundación para EL DESARROLLO NACIONAL.

PASIECZNIK, N.M., FELKER, P., HARRIS, P.J.C., HARSH, L.N., CRUZ, G., TEWARI, J.C., CADORET, K., and MALDONADO, L.J. 2001. *The* Prosopis juliflora- Prosopis pallida *Complex: A Monograph.* Coventry, Henry Doubleday Research Association.

Proyecto Algarrobo. 1997. *Inventorio Forestal de los Bosques Secos de Batán Grande.* Lima, Universidad Nacional Agraria la Molina.

RAMÍREZ, S.E. 1996. *The World Turned Upside Down: Cross Cultural Contact and Conflict in Sixteenth-Century Peru.* Stanford, Stanford University Press.

RISCO, J.J. 1998. Tecnologías Locales Mejoradas en el Uso de los Rescuros del Bosque Seco en un Contexto Familiar: Cocinas Mejoradas y Algarroberas, in A.B. Cuba Salerno, A. Silva Peralta, and C. Cornejo Flores (eds.), *Bosques Secos y Desertificacion*, 129-138. Lima, Instituto Nacionl de Recursos Naturales.

ROGERS, K.E. 1998. *The Magnificent Mesquite.* Austin, University of Texas Press.

ROSTWOROWSKI de DIEZ CANSECO, M. 1981. *Recursos Naturales Renovables y Pesca Siglos XVI y XVII.* Lima, Instituto de Estudios Peruanos.

SHIMADA, I. 1990. Cultural Continuities and Discontinuities on the Northern North Coast, Middle-Late Horizons, in M.E. Moseley and A. Cordy-Collins (eds.), *The Northern Dynasties: Kingship and Statecraft in Chimor,* 297-392. Washington DC, Dumbarton Oaks.

SHIMADA, I. 1994a. The Role of Metals in Middle Sicán Society, in G Burenhult (ed.) *The Illustrated Encyclopedia of Humankind, Vol. 4: New World and Pacific Civilizations*, 94-95, Sydney, Weldon Owen Pty Limited.

SHIMADA, I. 1994b. Prehispanic Metallurgy and Mining in the Andes: Recent Advances and Future Tasks, in A.K. Craig and R. West (eds.), *In Quest of Mineral Wealth: Aboriginal and Colonial Mining and Metallurgy in Spanish America,* 37-73. *Geoscience and Man* Vol. 53. Baton Rouge, Louisiana State University.

SHIMADA, I. 1995. *Cultura Sican: Dios, Riqueza y Poder en la Costa Norte del Peru.* Fundacion del Banco Continental para el Fomento del la Educacion y la Cultura, Edubanco, Lima.

SHIMADA, I., CHANG, V., WAGNER, U., GEBHART, R., NEFF, H., GLASCOCK, M., and KILLICK, D. 1998. Formative Ceramic Kilns and Production in Batán Grande, North Coast of Peru, in I. Shimada (ed), *Andean Ceramics: Technology, Organization and Approaches*, 23-62. Philadelphia, Museum Applied Science Center for Archaeology, The University Museum of the University of Pennsylvania.

SHIMADA, I., and MERKEL, J. 1991. Copper Alloy Metallurgy in Ancient Peru. *Scientific American* 265(1):80-6.

SHIMADA, I., MONTENEGRO, J.A., HÄUSLER, W., JAKOB, M., RIEDERER, J. and WAGNER, U. 2003. Early Pottery Making in Northern Coastal Peru: Part IV: Mössbauer Study of Pottery from Huaca Sialupe, in U. Wagner (ed.) *Mössbauer Spectroscopy in Archaeology,* 91-105, *Hyperfine Interactions* 150(1-4).

SHIMADA, I., SCHAAF, C.B., THOMPSON, L.G., and MOSLEY-THOMPSON, E. 1991. Cultural Impacts of Severe Droughts in the Prehistoric Andes: Application of a 1,500-year Ice Core Precipitation Record. *World Archaeology* 22(3):247-70.

SHIMADA, I. and WAGNER, U. 1998. Peruvian Black Pottery Production and Metal Working: A Middle Sicán Craft Workshop at Huaca Sialupe. *Materials Research Society Bulletin* 26(1); 25-30.

SHIMADA, I. and WAGNER, U. 2007. A Holistic Approach to Pre-Hispanic Craft Production, in J.M. Skibo, M.W. Graves, and M.T. Stark, *Archaeological Anthropology: Perspectives on Method and Theory,* 163-197, Tucson, University of Arizona Press.

SHIMADA, I., WAGNER, U., BEZÚR, A., and GOLDSTEIN, D.J. 2003. A Sicán Metalworking Furnace: A Holistic Understanding. Paper presented at the 51st International Congress of Americanists, Santiago, Chile, 14-18 of July, 2003.

SILLAR, B. 2001. Dung by Preference: The Choice of Fuel as an Example of How Andean Pottery Production is Embedded within Wider Technical, Social, and Economic Practices. *Archaeometry* 42(1):43-60.

TAYLOR, S. 2002. *Artisan Autonomy in the Middle Sicán State: Variability in Mold-Made Ceramic Production.* Unpublished M.A. Thesis, Southern Illinois University, Department of Anthropology.

TOSI, Jr., J.A. 1960. *Zonas de Vida Natural en el Peru: Memoria Explicativa sobre el Mapa Ecologico del Peru.* Instituto Interamericano de Ciencias Agricolas de Organización de los Estados Americanos.

WAGNER, U., SHIMADA, I., GOLDSTEIN, D.J., and HÄUSLER, W. 2001. Sicán Kilns and Furnaces: Field Firings and Archaeometric Studies. *Hyperfine Interactions* C: 1-4. New York, Kluwer Publishing.

WAGNER, U., GEBHART, R., MURAD, E., RIEDERER, J., SHIMADA, I., ULBERT, C., and WAGNER, F.E. 1998. Production of Formative Ceramics: Assessment by Physical Methods, in I. Shimada (ed), *Andean Ceramics: Technology, Organization and Approaches*, 173-198. Philadelphia, Museum Applied Science Center for Archaeology, The University Museum of the University of Pennsylvania.

WERTIME, T.A. 1964. Man's First Encounters with Metallurgy. *Science* 146(3649): 1257-1267.

WHEELER, E.A. and BAAS, P. 1998. Wood Identification-A Review. *IAWA Journal* 19(3):241-264.

WINTERHALDER, B., LARESEN, R. and THOMAS, R.B. 1974. Dung as an Essential Resource in a Highland Community. *Human Ecology* 2(2):89-104.

# VEGETATION AND WOOD EXPLOITATION AT HARAPPA, PUNJAB (PAKISTAN): PRELIMINARY RESULTS OF THE CHARCOAL ANALYSIS

Carla LANCELOTTI
George Pitt-Rivers Laboratory, Department of Archaeology, Downing Street, Cambridge CB2 3DZ, England

Margareta TENGBERG and Stéphanie THIÉBAULT
UMR 7209 – Archéozoologie, archéobotanique: Sociétés, pratiques et environnements, Muséum national d'Histoire naturelle, 55 rue Buffon, 75005 Paris, France

**Abstract**: *Preliminary results of an anthracological analysis carried out at Harappa, Punjab on samples dated to the 3B and 3C periods of the Harappan phase (c. 2450-1900 BC) show that wood resources were mainly exploited from two local vegetation communities: dry tropical thorn woodlands with* Acacia *sp.,* Prosopis cineraria, Salvadora *sp. and* Ziziphus *sp. and riverine forests with* Acacia nilotica, Dalbergia sissoo *and* Tamarix *sp. Two allochthonous taxa were also found in the charcoal assemblages – walnut* (Juglans regia) *and pine* (Pinus *sp.*) *– indicating that timber was imported from the temperate mountain areas to the north, perhaps by floating down rivers.*

**Key words**: *Anthracological analysis, Harappa, riverine forest, dry thorn woodland, Juglans regia, Pinus, wood importation*

## INTRODUCTION

The archaeological site of Harappa is situated in proximity of the homonymous modern town, in the Sahiwal (former Montgomery) district of the Pakistani province of Punjab (Figure 1). Since its discovery in 1872, Harappa has been one of the most intensively excavated settlements of the Indus Valley Civilisation. This culture developed in the territories of modern Northern India and Pakistan during the Bronze Age (3500-1900 BC) and it was one of the most complex ancient societies. Its development is still under investigation, as are the causes for its emergence and collapse; indeed, one of the most interesting aspects, in this respect, is the understanding of human-environment interactions. Palaeoenvironmental evaluations have become a focal point in the study of the Indus Civilisation, especially in the light of new studies that confuted the climatic explanation for its rise and fall (Madella and Fuller 2006).

This paper explores what was the vegetation around Harappa in the 3rd millennium BC and how the inhabitants of the site exploited it. The present investigation is based on the analysis of charcoal fragments retrieved from various contexts during excavations conducted between 1988 and 1995 in the framework of the *Harappa Archaeological Research Project* (HARP), under the joint direction of R.H. Meadow (Harvard University), J.M. Kenoyer (University of Wisconsin) and R.P. Wright (New York University). The anthracological analysis is part of a larger set of palaeoenvironmental studies aiming at understanding the interactions between the ancient Harappans and the biotic and mineral components of their milieu (Amundson and Pendall 1991; Fuller 2001; Meadow 1991; Meadow and Kenoyer 1995; Miller 1991; Schuldenrein *et al.* 2004).

The charcoal assemblage presented in this paper covers periods 3B (*c.* 2450-2200 BC) and 3C (*c.* 2200-1900 BC) (Kenoyer 1998, 24). Together with period 3A (*c.* 2600-2450 BC) these correspond to the so-called Mature or Urban Phase when the city knew its largest extension and covered an area of over 150 hectares. This was a period of increasing urbanisation and social complexity, including craft specialisation and long-distance trade (Kenoyer 1991, 1997; Meadow and Kenoyer 1995; Possehl 1191).

Several charcoal studies have previously been conducted at pre- and proto-historic sites in Pakistan at Kot Diji in the province of Sindh (Castelletti *et al.* 1994), at Mehrgarh, Lal Shah and Nausharo in central Baluchistan (Tengberg and Thiébault 2003; Thiébault 1988a, 1988b, 1989, 1992) and at Miri Qalat and Shahi Tump in the district of Makran in south-western Baluchistan (Tengberg and Thiébault 2003). To these we can add the analysis of wood samples found during the early excavations at Harappa (Chowdury and Ghosh 1951; Vats 1940, 162) and Mohenjo-Daro (Marshall 1931).

The above-mentioned anthracological studies have concentrated mainly on botanical identification of charcoal assemblages and on the reconstruction of past vegetation. The results have shown the preferential exploitation of two different ecosystems, both present on the alluvial plains. The identified taxa are, on the one hand, characteristic of hygrophilous riverine formations for example tamarisk (*Tamarix* sp.), poplar (*Populus euphratica*) and sissoo (*Dalbergia sissoo*). On the other hand, the inhabitants of these sites exploited taxa from dry open woodlands or so-called pseudo-savannas. Among these are members of the Mimosoidae (*Prosopis cineraria*, *Acacia* spp.), the toothbrush tree (*Salvadora* spp.), jujube (*Ziziphus* spp.) and caper (*Capparis*

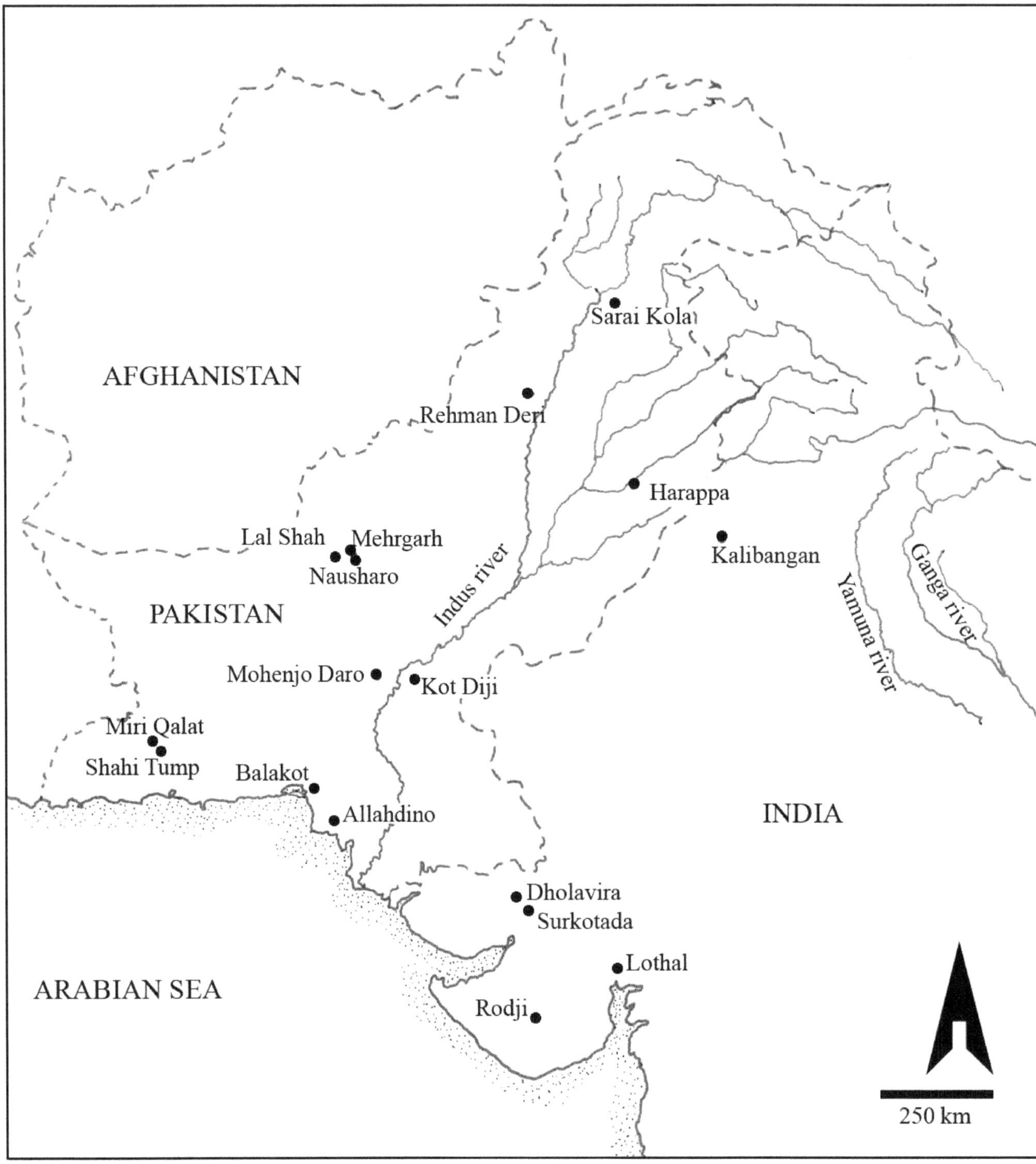


Figure 1. Map of South Asia, showing the location of Harappa (30°35'30.75"N, 72°54'59.54"E) and some of the sites mentioned in text as well as some of the most important sites pertaining to the Harappan Civilisation

*decidua*). The occasional presence of species that belong to other ecosystems, such as pine (*Pinus* sp.), Himalayan cedar (*Cedrus deodara*) and juniper (*Juniperus* sp.) indicates the selective use of some 'exotic' resources. Their presence is particularly interesting and will be considered more in detail in the discussion below.

From a purely floristic point of view, the taxa noted in the archaeological contexts are more or less the same as those which can be found in these areas today. Nonetheless, the vegetation cover seems to have undergone important changes with a decline in both the biodiversity and biomass through time, mainly due to intense human activities (Tengberg and Thiébault 2003 47). The decline of certain vegetation types, in particular the riverine forests, seems to be significant and is notable already in the past through the anthracological records from proto-historical Makran (Tengberg and Thiébault 2003).

## PRESENT-DAY VEGETATION AROUND HARAPPA

The floodplains of Punjab – 'the land of the five waters' – are today intensively cultivated and constitute the granary of the whole country. As elsewhere in the province, orchards and fields of cotton, wheat and sugarcane dominate the surroundings of the archaeological site of Harappa. Even though agriculture has constituted the main subsistence activity in this area since prehistory, as shown by archaeobotanical analysis (Weber 1999, 2003), the entirely anthropogenic aspect of the landscape that we observe today is a quite recent phenomenon and it is mainly the result of the introduction of intensive irrigation schemes since the end of the 19th century. Agriculture, herding and urbanisation have profoundly transformed the environment in Punjab during the last hundred years and not much unaltered vegetation remains. From a phytogeographical point of view the alluvial plains of the Indus and its tributaries constitute the easternmost extension of the subtropical Nubo-Sindian floral division that stretches from the southern Arabian Peninsula to the hot deserts of western India. The main vegetation types of this unit consist of xeric and open to very open thorn wood- and scrublands but a denser vegetation develops wherever water is available, for example along permanent or ephemeral watercourses (Aitchison 1869; Bandhari 1995; Puri 1960).

As said in the introduction, the environment surrounding the site of Harappa is highly influenced by human activities and it is characterised mainly by cultivated plants of various origins. However, the site itself is part of a botanical reserve, instigated by the Botany Department, Lahore University. Here, species belonging to the indigenous flora are protected and multiplied within the area of the archaeological site and their presence can give us an idea of the appearance of the potential vegetation in this part of the Punjab would look like.

During a short survey in the immediate surroundings of the site, undertaken in 2000, a limited number of species were observed and sampled: *Calotropis procera*, *Capparis decidua*, *Lycium* sp., *Salvadora oleoides* and *Suaeda* sp. grow on the alluvial and sandy soils of the site. *Acacia nilotica*, *Dalbergia sissoo*, *Prosopis cineraria* and *Tamarix aphylla* thrive along irrigation canals around the fields. The banyan tree (*Ficus benghalensis*) and the pipal (*Ficus religiosa*) are cultivated near houses and sanctuaries. It is noteworthy that even though *Ficus religiosa* has not yet been identified by the anthracological study, its characteristic heart-shaped leaves are frequently represented on Harappan pottery and must thus have been familiar to the early inhabitants of the Indus Valley.

## MATERIAL AND METHODS

The charred wood fragments analysed in this preliminary study result either from manual sampling carried out during the excavation or from the flotation of sediment samples taken from different ashy contexts. It should be noted that the archaeological significance and thus the interpretation of these two categories of samples are not exactly the same.

Some studies (for a summary see Chabal 1997), conducted on the representative value of archaeological charcoal, have demonstrated that the hand picking of larger pieces of wood, easily observed in the archaeological layers, tends to favour certain taxa. Their identification will provide punctual information on the wood used at the site and thus contribute in a limited manner to the reconstruction of the plant communities that were exploited in the past. However, large fragments (>5-10 cm) and especially pieces of logs are likely to correspond to burnt construction elements and can thus inform us on the selection of wood for this specific purpose. Two thirds of the samples analysed so far at Harappa result from handpicking collections.

Charcoal fragments extracted from sediment samples by flotation or sieving are more likely to have accumulated during a longer period (for example, through the successive emptying of hearths and ovens) and thus to be more representative of the vegetation cover from which wood was collected (Chabal 1997, 26-38). However, in order to be interpreted from a palaeoecological point of view, these fragments should preferentially result from a use as domestic fuel as this type of wood is generally less subjected to a selection than for example construction wood or fuel used for specific craft activities.

This dichotomy in the sampling strategy is well reflected in the data analysed for the present study. Indeed, the represented percentages of the various species are very different in the handpicked samples compared to the flotation samples. A discussion of how this can be analysed in the broader context of the archaeological interpretation can be found in the following sections.

The charcoals from Harappa were collected from different contexts such as hearths, occupational floors, ashy deposits and fills. However, for many of the samples the exact context is not yet known by the present authors, rendering their precise interpretation somewhat uncertain. Therefore, we have chosen in this paper to present the general results of the botanical identification (for a more detailed study taking into account the archaeological interpretation see Lancelotti 2010).

The charred samples from Harappa are generally in a rather good state of preservation even though all the fragments resulting from flotation are very small. This can be a result of the high salinity of soil and water used for flotation which increase charcoal fragmentation. The botanical identification of charcoal was done under a reflecting light microscope at magnifications between 50x and 800x using a dark as well as a light field view. The fragments to be studied were fractured manually along the transversal, tangential and radial planes. The comparative tools used for the identification consisted of a reference collection of modern wood samples from the Middle East as well as wood anatomical atlases (Fahn *et al.* 1986; Jagiella and Kürschner 1987; Neumann *et al.* 2000; Schweingruber 1990).

## RESULTS

### Descriptive

The analysis of 552 fragments from 37 different samples has allowed the identification of 10 ligneous taxa, most of which still grow in the province of Punjab, some even in the immediate surroundings of the site: *Acacia* sp., *Acacia* cf. *nilotica, Dalbergia sissoo, Juglans regia, Phoenix dactylifera/sylvestris, Pinus* sp., *Prosopis cineraria, Salvadora oleoides/persica, Tamarix* sp. and *Ziziphus nummularia/spina-christi.*

As can be seen from the list above, the degree of precision attained by the botanical identifications is variable. Indeed, some of the wood types identified among the Harappan charcoal are represented by several species that are close anatomically and that we have not been able to separate with absolute certainty. However, the final interpretation is not significantly altered since closely related species in our record have comparable ecological requirements and are members of the same plant communities.

The different taxa were identified and described on the basis of personal observations compared with published material. Most of the identified taxa are expected in the region and their identification is beyond doubt, therefore, their description is not given. In contrast, pine and walnut are not native to the area and, as a matter of clarity, the description of the features observed in the samples are listed below.

#### *Pinus sp.*

One coniferous taxa, the pine tree, has been recognised in our samples. It represents little more than 11% of the handpicked fragments but is less common in the flotation samples (2.9%). This homoxylous wood is characterised in the transversal section by a rather abrupt early/latewood transition and the presence of resin ducts (Figure 2). The uniseriate rays are most often around 5-7 cells high but can occasionally attain up to 15 cells. The radiovascular pits are small and of pinoid type. Ray tracheids with dentate walls were observed in our samples.

Three pine species are present in Pakistan. The blue pine (*Pinus wallichiana*) and the longleaved pine (*P. roxburghii*), both called *chir* grow in the mountainous areas of northern Pakistan, the first at slightly higher altitudes than the second. Both these species are tall trees, producing a soft timber. Finally, *P. gerardiana* or the *chilgoza*, is more common in the hilly parts of inner Baluchistan, notably in the Quetta and Kalat divisions and is a medium sized tree, valued for its wood as well as for its seeds that contain oil and are edible. A few trees of this species have even been reported growing on the upper Ravi in the beginning of the 20th century (Brandis 1906, 691; Burkill 1909).

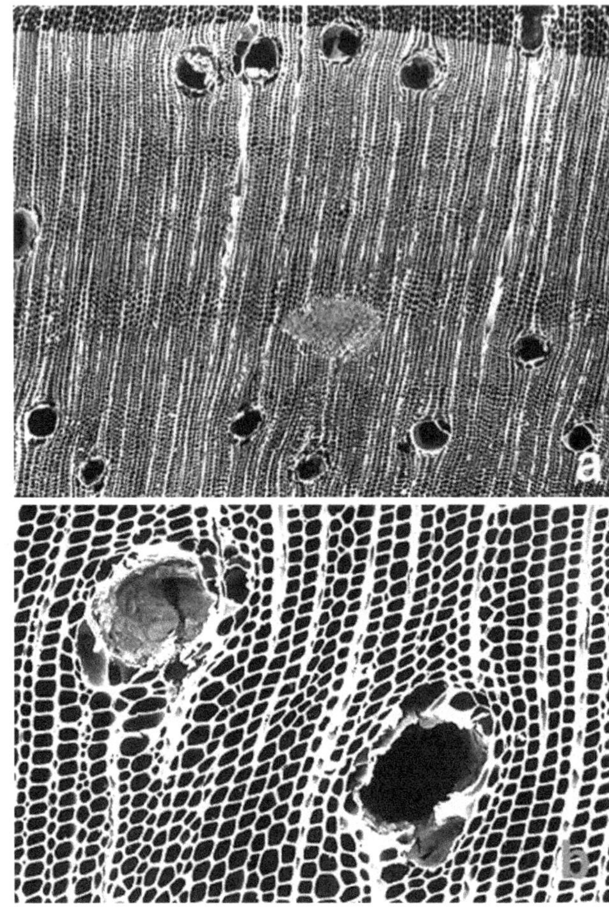


Figure 2. SEM photographs of pine wood (*Pinus* sp.) – (a) transversal section (30x) (b) detail of the resin ducts (140x)

None of these three pine species would have grown near Harappa in the past and must thus have been brought to the site from the Himalayan or sub-Himalayan tracts (for a discussion, see *infra*).

#### *Juglans regia L.*

A few but large fragments from feature 38 have been identified as belonging to *Juglans regia* or walnut (*akhrot* in Urdu and Punjabi) from the Juglandaceae family. This diffuse to semi-ring porous wood is characterised, in the transverse section, by the presence of large vessels mostly solitary but sometimes also disposed in radial rows of 2-3(4) cells. Axial parenchyma is scarce and mostly diffuse. In the tangential section, the rays appear to be 1-5 cells wide and the intervessel pits are small and alternate. The characteristic rays are composed of procumbent cells with one or a few marginal rows of square cells (Figure 3). This character is peculiar also to a local species *Azadirachta indica* (Neem). Nevertheless, the vessels arrangement and grouping, the distribution of parenchyma and the absence of two distinct vessel sizes

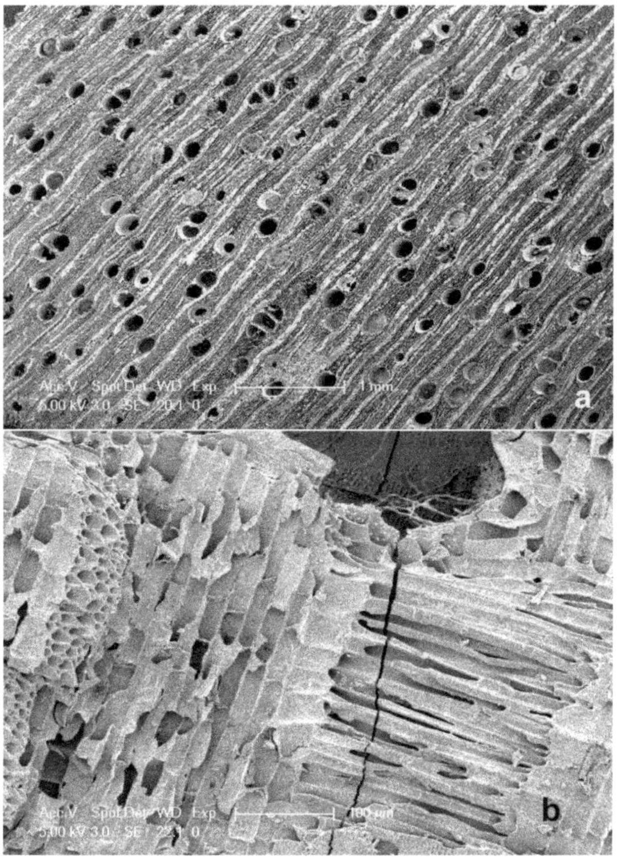


Figure 3. SEM photographs of the sample identified as *Juglans regia*. Although the transverse section (a) does not show the typical distinct ring boundaries, the detail of the radial section (b) shows the characteristic structure of *Juglans's* rays, composed of procumbent cell bordered by one row of upright/square cells (Schweingruber 1990, 431)

and spiral thickenings, exclude *Azadirachta* sp. from identification.

Walnut is a large deciduous and aromatic tree, indigenous to the trans-Indus in the Kuram valley, the North-West Himalaya, where it is also cultivated from 1000 to nearly 3350m (Brandis 1906, 620). Cultivation of this tree is reported also from Hazara and the Murree Hills in the North West Frontier Province, Pakistan (Baquar 1995, 337).

Walnut has several uses: on the one hand, the wood is excellent for furniture and carving (Baquar 1995, 337); on the other hand, it produces edible fruits (nuts) which are used both as food and as an oil extract or to produce liqueurs.

Archaeobotanical evidence of walnut comes from Kashmir where nuts have been identified in the Neolithic I layers at Burzahom (2375-1700 BC) as well as in later periods at the same site (600 BC – AD 200) and at Semthan (1500-600 BC; AD 1500) (Lone *et al.* 1993, 161).

**Quantification**

Although the total number of charcoal fragments, as well as the amount and nature of contexts analysed is limited, some preliminary observations on the changes in variability and quantity of species can be drawn. As shown in Figures 4 and 5, some care needs to be taken when interpreting the results from a quantitative point of view. In the first place, the difference in species representation, between the handpicked samples and the samples collected through flotation is noticeable (Figure 4). The graphic clearly shows the misrepresentation that can arise when only one of the techniques is used. Thanks to previous studies we know that the picture given by the samples collected through flotation is more reliable and represents the composition of the natural vegetation more accurately. However, in this case, as still in many excavations, the handpicking collection of samples is a common used practice. This is even truer on sites located in very arid and dry areas where, sometimes, water is not abundant and readily available for flotation. In this case, considering the high number of charcoal fragments collected by handpicking, the results of the two analyses have been merged and all the samples have been considered together.

The second major problem in the quantitative interpretation of the data, especially in regard to the diachronic variation of species, derives from the relative number of fragments analysed for each period as well as the fact that a great amount of them belongs to contexts that have not been definitively dated but appear to belong to a transitional phase (Figure 5). Indeed, on the total number of fragments analysed, the proportion between the periods is biased towards period 3C. In fact, *c.* 60% of the fragments analysed belongs to this period, whereas, only 33% of fragments belonging to period 3B were analysed. Therefore only differences in the range of 30% can be considered as significant.

With these remarks in mind, some interesting considerations on the changes of species over time can indeed be formulated. Figure 6 shows the trend of the different species over the three periods expressed as a percentage of the total. It is clear that all the species related to the exploitation of the riverine system, such as *Acacia nilotica*, *Dalbergia sissoo* and *Tamarix* sp. decrease steadily. In contrast, all tropical thorn woodland species remain almost stable over time and do not increase. This is in accordance with the conclusion reached by Madella and Fuller (2006) that the last period of the urban Harappan phase is set in progressively drying condition, but not in a period of complete and severe drought.

Particularly interesting is the appearance of the so-called 'exotic' taxa in the transition between period 3B and 3C. As explained above, neither *Pinus* sp. nor *Juglans* sp. could have grown near the site of Harappa. These essences had to be imported from upstream and were probably considered precious and used as luxury items. The presence of coniferous wood is not totally surprising:

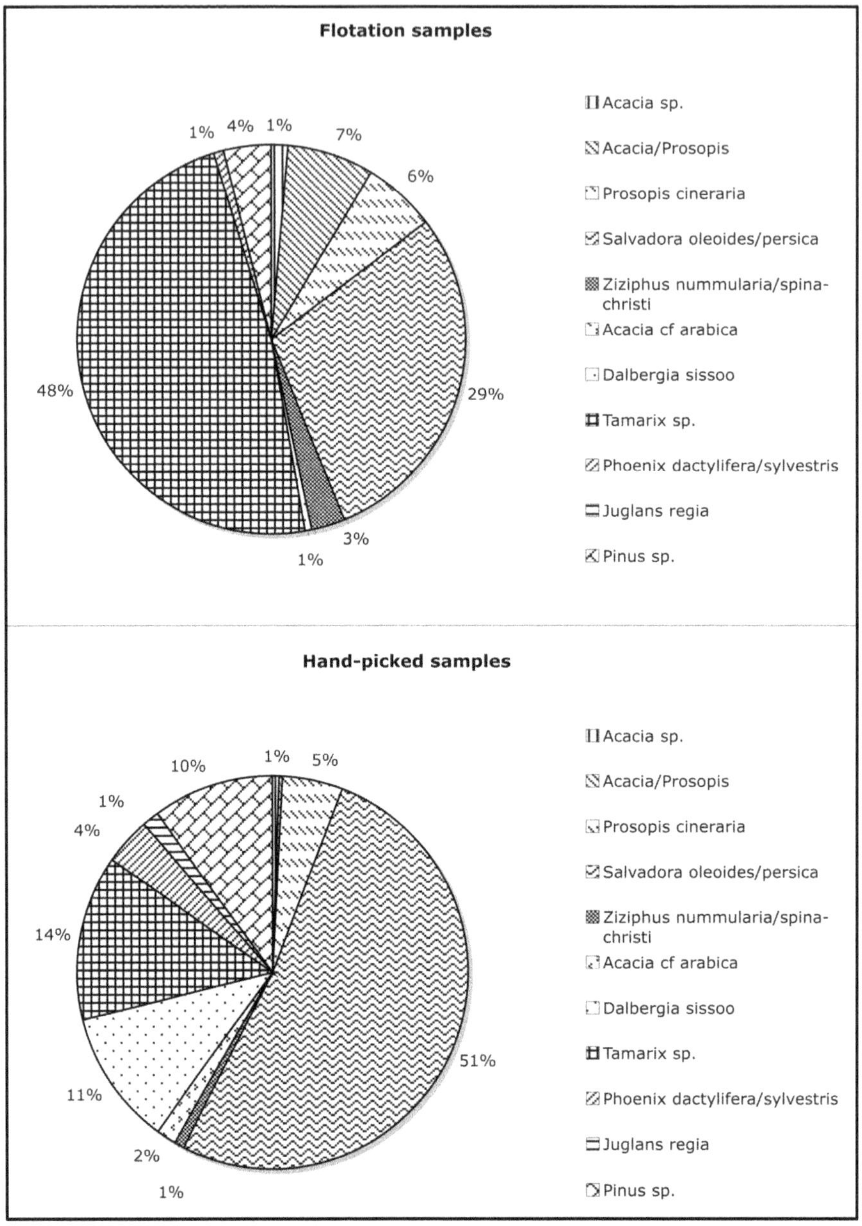


Figure 4. Differences between handpicked and flotation samples

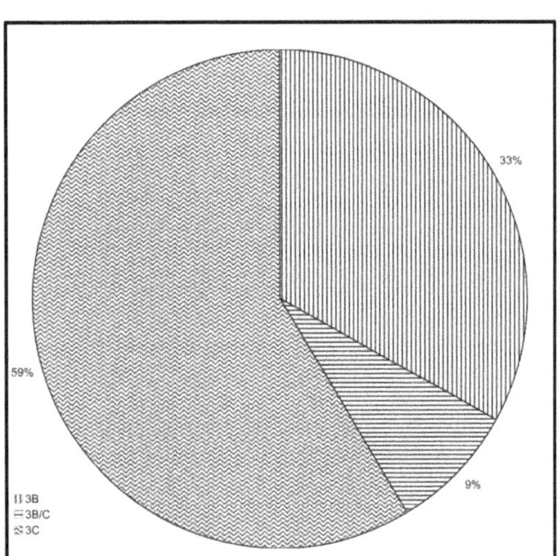


Figure 5. Number of fragments analysed from each chronological phase

a coffin has been found in Harappa and identified as being made of cedar wood (Chowdhury and Gosh 1951, 8-12).

## DISCUSSION

The first results of the charcoal analysis carried out at Harappa suggest, not surprisingly, that most of the wood used at the site was of local origin. This is in accordance with the results obtained at all the other Pakistani sites studied from this point of view. In every case, the ecosystems exploited are also similar even although the floristic composition varies somewhat from one site to another depending on local ecological conditions and probably also on different sampling techniques. Two main vegetation types are reconstructed from the archaeobotanical analysis of samples from Harappa: tropical thorn woodlands and riverine formations.

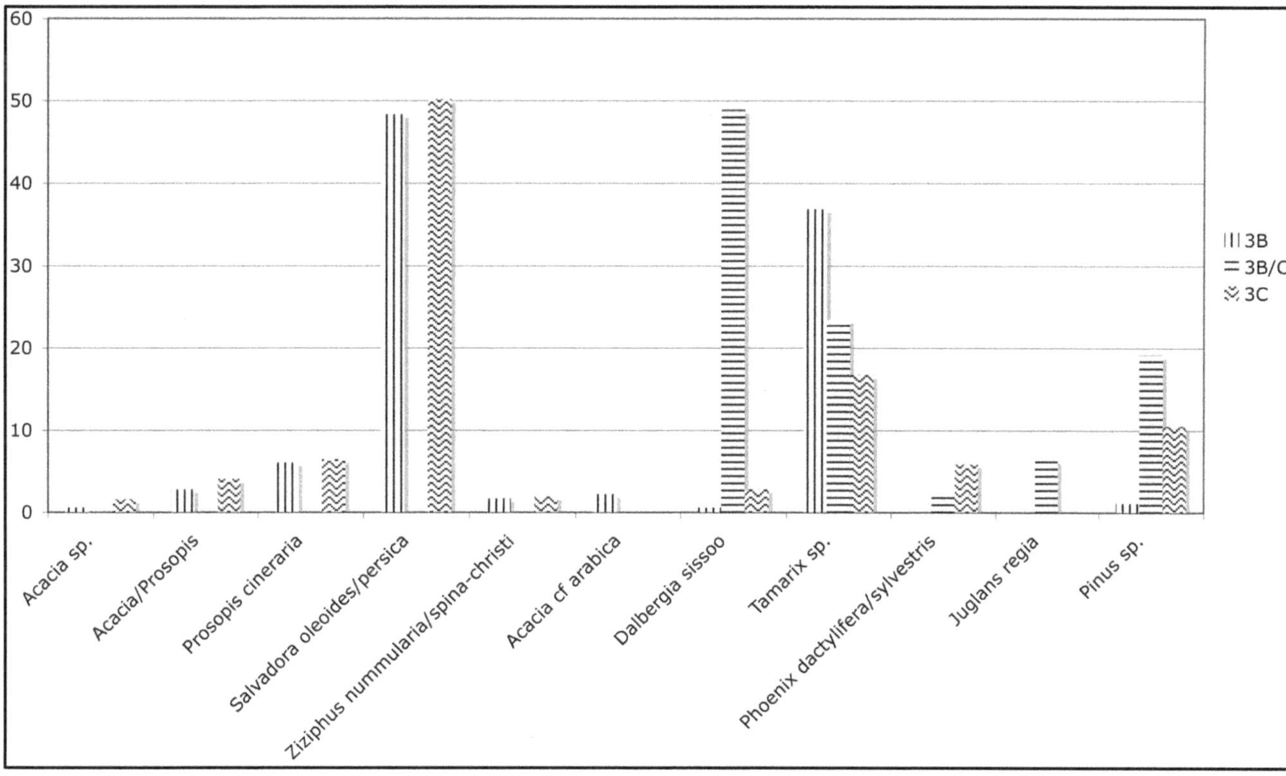


Figure 6. Diachronic variation of species (values expressed in percentage)

Tropical thorn woodlands are characterised by the presence of xerophilous taxa such as *Acacia* sp., *Prosopis cineraria*, *Salvadora* sp. and *Ziziphus* sp. These so-called *rakh* forests, generally rather low and scattered, nowadays occupy wide areas in the dry plains of Punjab, Sindh and Baluchistan. At present, they usually include a few other species that were not attested in the samples under study, notably *Capparis decidua* and *Dodonaea viscosa*.

Riverine or *bela* forests are represented at Harappa by *Acacia nilotica*, *Dalbergia sissoo* and *Tamarix* sp. They would have bordered the Ravi and probably formed a lush environment and an appropriate habitat for numerous animal species, of which some were exploited by the Harappans. In general, the results of the palaeoenvironmental analyses carried out in the Indo-Iranian borderlands show the importance of the alluvial ecosystems in the local subsistence economies. Indeed, these provided a number of resources (arable soils, water, clay, plants, mammals, birds, fish, shellfish, etc.) that were essential for the development of complex societies in an otherwise harsh environment.

One riverine species is surprisingly absent at Harappa so far: the Euphrates poplar (*Populus euphratica*). This species is attested at Kot Diji as well as at other sites in Baluchistan although it is not a dominant species in Baluchistan (Castelletti *et al.* 1994; Tengberg and Thiébault 2003). Poplars are tall trees, with characteristic white bark. They thrive along river courses and in wet-lands and have many uses: their wood is burned as fuel in Sindh whereas the leaves are fed to goats as fodder.

Among the genera identified by the charcoal study at Harappa a few do not belong to the subtropical vegetation communities described above. This is notably the case for pine (*Pinus* sp.) and walnut (*Juglans regia*), which both grow under temperate conditions in the mountain regions of northern Pakistan. Both these taxa must thus have been imported to the site, either by road or by floating down the river. The fragments of pine and walnut were in general bigger and better preserved than other charcoal pieces found at the site and seem to have resulted from the burning of construction material rather than their use as fuel.

These new data seem to confirm the initial studies that have been carried out on wood remains and are included in the first reports of excavations in Harappa and Mohenjo-Daro. These studies have shown evidence of the use of other temperate species. Chowdhury and Gosh (1951, 8-12) report the presence at Harappa of Himalayan cedar (*Cedrus deodara*, or *dayar*) and elm (*Ulmus* sp.), both growing above 1000 m as well as pine (Vats 1940, 162). Cedar has also been reported from Mohenjo-Daro (Marshall 1931, 19).

## CONCLUSION

The present study is based on a limited number of charcoal fragments. Still its results show the potential for this type of analysis at a site such as Harappa. Among the issues that could be addressed in a more detailed study, some may play an important role in our understanding of the Harappan Civilisation.

The question of wood importation from distant areas would enlighten the commercial relationships among the different communities and would give us potential information to understand the centralisation of the exploitation of natural resources. In fact, retrieving distant raw materials and organising their transportation to the site, implies a great degree of centralisation or, at least, a different labour organisation in respect to the use of local resources. The archaeological interpretation of the anthracological identifications would probably allow us to relate the exploitation of 'exotic' woods to the exploitation of other raw materials.

Another question that can be addressed through the study of charcoal is the utilisation of fuel resources. The site of Harappa has shown great potential in these terms: during excavations many different productive contexts have been identified that required the use of fire. Some very specialised craft productions, as steatite firing, etched bead production or black and red pottery craft, would have required selection of specific fuels for the crafts involved. The study of charcoal remains, supported by a detailed context can give insights on the selection of different species of wood in relation to different craft productions.

Related to the last issue, is the problem of the human impact on ancient vegetation. As discussed at the beginning of this report, the environmental conditions of modern Pakistan do not seem to have changed greatly from the Bronze Age. Nonetheless the archaeobotanical studies have detected a decrease in both the number of woody taxa present as well as a decline of the biomass. The variations in the species used by people towards taxa more easily adapted to depleted soils or taxa able to regenerate quicker after having been heavily exploited, as it is the case for *Tamarix* sp., will give us an idea on the human impact on the surrounding ecosystem.

All these issues will be taken into account in future studies; meanwhile the results presented in this paper can be regarded as the basis on which future investigations will be constructed.

### Acknowledgments

The authors are grateful to Dr Richard Meadow, Dr Mark Kenoyer and Dr Rita Wright for providing samples for this study and for their precious comments on the first draft of this article. We would also like to thank you all the colleagues who double-checked our identification of Walnut fragments. CL's research has been funded by AHRC, Isaac Newton European Trust and St John's College.

### References

AITCHISON, J.E.T. 1869. *A catalogue of the plants of the Punjab and Sindh (to which are added some others that, from their present geographical proximity, may be found hereafter to occur in the Punjab)*. London, Taylor and Francis.

AMUNDSON, R. and PENDALL, E. 1991. Pedology and late quaternary environments surrounding Harappa: a review and synthesis. In Meadow, R.H. (ed.), *Harappa Excavations 1986-1990*, 13-27. Madison, Prehistory Press.

BAQUAR, S.R. 1995. *Trees of Pakistan. Their natural history, characteristics and utilization.* Karachi, Royal Book Company.

BHANDARI, M.M. 1995. *Flora of the Indian desert.* Jodhpur, MPS Repros.

BRANDIS, D. 1906. *Indian trees. An account of trees, shrubs, woody climbers, bamboos and palms indigenous or commonly cultivated in the British Indian Empire.* London, Archibald Constable and Co. Ltd.

BURKILL, I.H. 1909. *A working list of flowering plants of Baluchistan.* Calcutta, Superintendent Government Printing.

CASTELLETTI, L., MADELLA, M. and QAYOM MAHAR, A. 1994. Charcoal analysis at Kot Diji: a preliminary report. *Ancient Sindh* 1, 37-41.

CHABAL, L. 1997. *Forêts et sociétés en Languedoc (Néolithique final, Antiquité tardive. L'anthracologie, méthode et paléoécologie.* Paris, Editions de la Maison des Sciences de l'Homme.

CHOWDURY, K.A. and GHOSH, S.S. 1951. Plant remains from Harappa. *Ancient India*, 3-19.

FAHN, A., WERKER, E. and BAAS, P. 1986. *Wood Anatomy and Identification of Trees and Shrubs from Israel and Adjacent Regions.* Jerusalem, Israel Academy of Sciences and Humanities.

FULLER, D.Q., 2001. Harappan seeds and agriculture: some considerations. *Antiquity* 75(228), 410-414.

JAGIELLA, C., KÜRSCHNER, H. 1987. *Atlas der Hölzer Saudi Arabiens. Die Holzanatomie der wichtigstens Bäume und Sträucher Arabiens mit einem holzanatomischen Bestimmungsschlüssel.* Wiesbaden, Dr Ludwig Reichert Verlag.

KENOYER, J.M. 1991. Urban process in the Indus tradition: a preliminary model from Harappa. In R.H. Meadow (ed.), *Harappa Excavations 1986-1990*, 29-60. Madison, Prehistory Press.

KENOYER, J.M. 1997. Trade and technology of the Indus Valley: new insights from Harappa, Pakistan. *World Archaeology* 29(2), 262-280.

KENOYER, J.M. 1998. *Ancient cities of the Indus valley civilization.* Karachi, American Institute of Pakistan Studies & Oxford University Press.

LANCELOTTI, C. 2010. *Fuelling Harappan Hearths: Human-Environment Relations as Revealed by Fuel Exploitation and Use.* Unpublished thesis (PhD), University of Cambridge (UK).

LONE, F.A., KHAN, M., BUTH, G.M. 1993. *Palaeoethnobotany. Plants and ancient Man in Kashmir.* Rotterdam, A.A. Balkema.

MADELLA, M. and FULLER, D.Q. 2006. Palaeoecology and the Harappan Civilization of South Asia: a reconsideration. *Quaternary Science Reviews,* 25, 1283-1301.

MARSHALL, J. 1931. *Mohenjo-daro and the Indus civilization (Being an official account of archaeological excavations at Mohenjo-daro carried out by the Government of India between the years 1922 and 1927).* London, Arthur Probsthain.

MEADOW, R.H. (ed.) 1991. *Harappa Excavations 1986-1990,* Madison, Prehistory Press.

MEADOW, R.H. and KENOYER, J.M. 1995. Excavations at Harappa 1994-1995: new perspectives on the Indus script, craft activities and city organization. In Allchin, R. and Allchin, B. (eds.), *South Asian Archaeology.* Cambridge, Science Publishers INC, 139-72.

MILLER, H.M.-L. (1991). Urban palaeoethnobotany at Harappa. In Meadow, R.H. (ed.), *Harappa Excavations 1986-1990,* 121-6. Madison, Prehistory Press.

NEUMANN, K., SCHOCH, W., DÉTIENNE, P. and SCHWEINGRUBER F.H. 2000. *Woods of the Sahara and the Sahel.* Bern, Stuttgart and Wien, Birmensdorf, Eidg. Forschungsanstalt WSL. Haupt.

POSSEHL, G.L. 1991. A short history of archaeological discovery at Harappa. In R.H. Meadow (ed.), *Harappa Excavations 1986-1990.* Madison, Prehistory Press, 5-11.

PURI, G.S. 1960. *Indian forest ecology.* New Delhi, Oxford Book and Stationery Co.

SCHULDENREIN, J.R., WRIGHT, P., RAFIQUE MUGHAL, M. and AFZAL KHAN, M. 2004. Landscapes, soils and mound histories of the Upper Indus Valley, Pakistan: new insights on the Holocene environments near ancient Harappa. Journal of Archaeological Sciences 31, 777-97.

SCHWEINGRUBER, F.H. 1990. *Anatomy of European Woods.* Bern, Stuttgart, Paul Haupt.

TENGBERG, M. and THIÉBAULT, S. 2003. Vegetation history and wood exploitation in Pakistani Baluchistan from the Neolithic to the Harappan period: the evidence from charcoal analysis. In S.A. Weber and W.R. Belcher (eds.), *Indus Ethnobiology. New perspectives from the field.* Oxford, Lexington Books, 21-63.

THIÉBAULT, S. 1988a. Palaeoenvironment and ancient vegetation of Baluchistan based on charcoal analysis of archaeological sites. *Proceedings of the Indian National Science Academy* 54, 501-509.

THIÉBAULT, S. 1988b. L'évolution de la végétation à Mehrgarh. *Les cités oubliées de l'Indus.* Paris, Musée National des Arts Asiatiques Guimet.

THIÉBAULT, S. 1989. A note on the ancient vegetation of Baluchistan based on charcoal analysis of the latest periods from Mehrgarh, Pakistan. In K. Frifelt, P. Sørensen (eds.) *South Asian Archaeology 1985.* Scandinavian Institute of Asian Studies, 186-188.

THIÉBAULT, S. 1992. Complementary results in anthracological analysis from sites in Baluchistan. In C. Jarrige (ed.), *South Asian Archaeology 1989.* Madison, Prehistory Press, 271-276.

VATS, M.S. 1940. *Excavations at Harappa (Being an account of archaeological excavations at Harappa carried out between the years 1920-21 and 1933-34).* Delhi, Government of India Press.

WEBER, S.A. 1999. Seeds of urbanism: palaeoethno-botany and the Indus civilization. *Antiquity* 73, 813-826.

WEBER, S.A. 2003. Archaeobotany at Harappa: indications for change. In S.A. Weber and Belcher W.R. (eds.) *Indus Ethnobiology. New perspectives from the field.* Oxford, Lexington Books, 175-198.

# NEW DATA ABOUT WOOD USE IN THE NORTHWEST OF THE IBERIAN PENINSULA

María MARTÍN SEIJO
Grupo de Estudos para a Prehistoria do Noroeste-Dep Historia I, Universidade de Santiago de Compostela, 15782 Santiago de Compostela, A Coruña, Spain,
maria.martin.seijo@gmail.com

Raquel PIQUÉ i HUERTA
Departament de Prehistòria, Universitat Autònoma de Barcelona, Edifici B, Campus de la UAB, 08193 Bellaterra (Cerdanyola del Vallès), Despatx: B9/119, Spain
raquel.pique@uab.cat

**Abstract**: *This paper presents new data about wood use in archaeological contexts in the Northwest of the Iberian Peninsula. The sites analysed and presented in this paper are: Montenegro (Neolithic), Os Remedios (Chalcolithic), Castro de Montealegre, Castro de Punta do Muiño, Castro de Navás, Castrolandín, Castro de O Neixón Grande, Castro de Zoñán, Corona del Cerco de Borrenes, El Castrelín de San Juan de Paluezas, Pedreiras del Lago, Chaos de Mourán, Castro Orellán, Cerro Pendón (Iron Age-roman period), Mourela (medieval-postmedieval period).*

*Charcoal and wood samples were analysed from these different sites, most of which were settlements dating from prehistoric to historical times. The results were recorded in a specific archaeobotany database, which was used in combination with a GIS program to analyse the different spatial contexts of the samples on both a microscale and intrasite level, and within a broader geographical context. The results show the different uses of wood within the settlements and the differences and continuities between geographical areas and chronological periods.*

**Key-words**: *Wood use. Charcoal Analysis. Neolithic. Iron Age. Roman Period. Medieval-Postmedieval Period. Northwest of the Iberian Peninsula*

## INTRODUCTION

The application of charcoal analyses (or anthracology) in the Northwest of the Iberian Peninsula has not been uniform. While in North Portugal anthracology have been systematically applied since 1980 (Figueiral 1993, 1994, 1996, Figueiral, Sanches 1998-9), in Galicia and the Northwest of Leon its application has been sporadic (Carballo 1989, Aira, Uzquiano 1996), with results remaining unpublished in many cases (Piqué unpublished, Uzquiano unpublished). However, in recent years archaeological interventions have begun to systematically incorporate archaeobotanical research, with reference to data on forest management and the uses of wood in the past (firewood, construction, and wood-working), as well as completing the palaeoenvironmental data by carrying out pollen and pedological analyses (Carrión 2003, Martin, 2005, 2006, 2007, 2008a).

This paper summarises the data obtained from the analysis of charcoal and wood samples from various sites in the Northwest Iberian Peninsula. The sites date from recent prehistory to the postmedieval period, with the majority belonging to the 1st millennium BC (Iron Age and roman period). The palaeoecological and palaeoeconomical significance of this data is also discussed. In most cases the remains of carbonised wood are the residues generated during the consumption of firewood, and represent the material remains of activities linked to various spheres of production. However, some of these remains are related to episodes of collapse and the destruction by fire of structural elements and furniture. Finally, although less common, some of the remains were found in a mineralized state inside metallic objects. The study of these charcoal and wood samples allows for an analysis of woodland management and provides insight into the species consumed, the areas of exploitation, the anatomical parts used and the characteristics of utilised wood. It also provides information on the immediate vegetational landscape, as species used for firewood were often collected in the immediate vicinity of a settlement.

## SITES

The samples examined represent 15 archaeological sites (Figure 1) from the Northwest of the Iberian Peninsula (Galicia and Northwestern León), and from two biogeographical regions (Atlantic and Mediterranean). In the Atlantic area, with its oceanic climate, the sites of Montenegro 1 y 2 (Moaña, Pontevedra), Os Remedios (Moaña, Pontevedra), Montealegre (Moaña, Pontevedra), Punta do Muiño (Vigo, Pontevedra), Navás (Nigrán, Pontevedra), Neixón Grande (Boiro, A Coruña) and Castrolandín (Cuntis, Pontevedra) are situated on hills, while the sites of Mourela 1 y 2 (As Pontes, A Coruña) and Zoñán (Mondoñedo, Lugo) are located at higher altitudes on mountains. In the Mediterranean region various sites are grouped in mesomediterranean

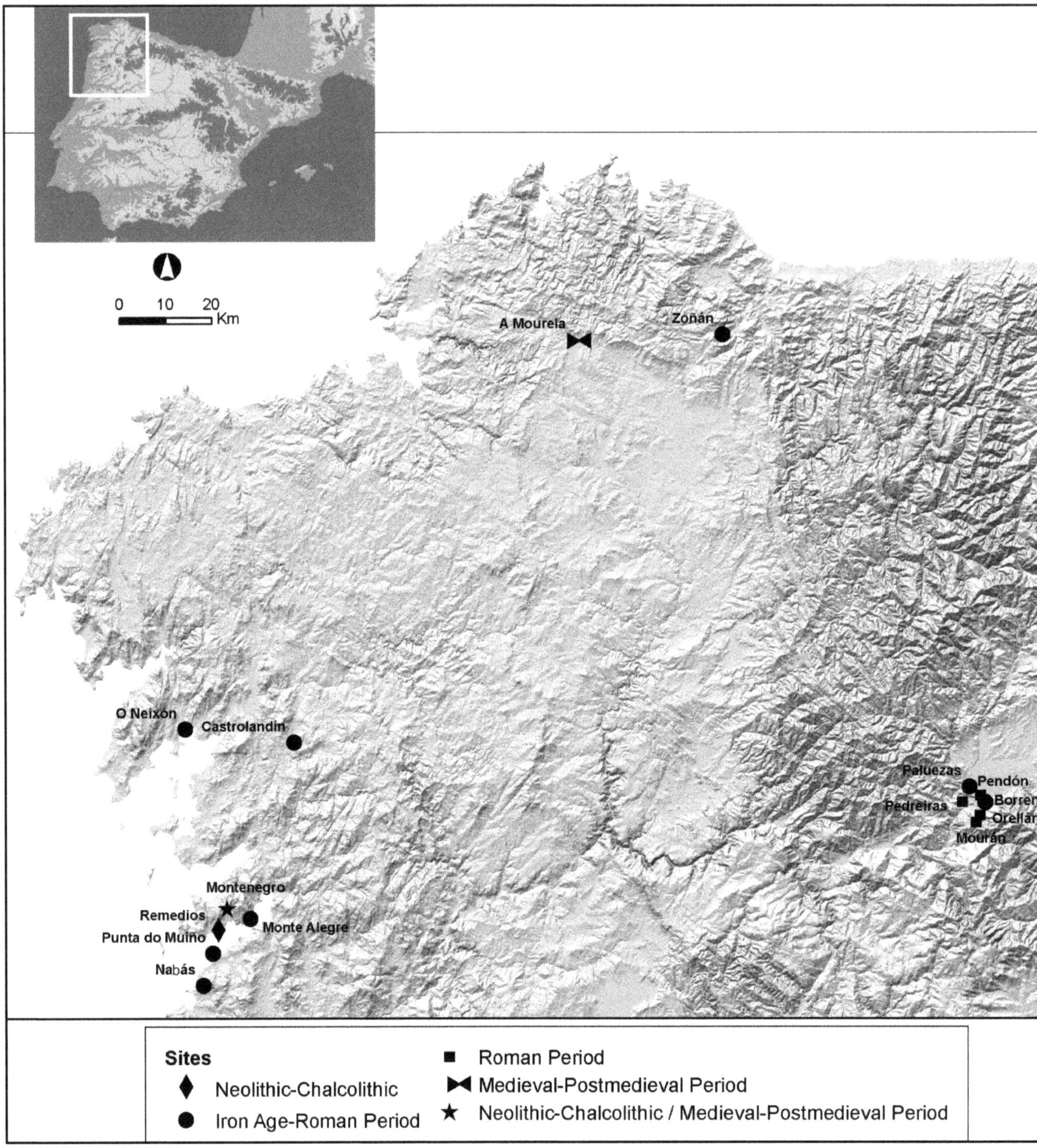


Figure 1. Location of the studied sites in the Northwest of the Iberian Peninsula

thermoclimatic zones: Cerro de Borrenes, Castrelín de San Juan de Paluezas, Cerro Pendón, Pedreiras del Lago, Castro Orellán and Chaos de Mourán (Archaeological Zone of Las Médulas, León).

The majority of the sites are open-air settlements of different characteristics and chronologies. The oldest sites (Remedios and Montenegro) are located in the *ría* (estuary) of Vigo. Remedios is an open-air settlement with a long occupation from the 5th until the 1st millennium BC. A systematic dating of different types of structures and deposits provided the following 2σ calibrated absolute dates: 4720-4530 BC, 4340-4150 BC, 4240-3980 BC, 3810-3705 BC, 3635-3375 BC, 3025-2880 BC, 3125-2195 BC, 2920-2740 BC y 845-767 BC. The settlement was enclosed by a palisade and the internal living area contained habitations made from perishable materials, hearths, pits, etc. (Bonilla y César 2005; Bonilla, César, Fábregas 2006 y 2007). Montenegro 1, an open-air settlement that was occupied in the 3rd millennium BC, contained pits, hearths and buildings made with perishable materials (as represented by postholes, occupation floors, and foundation trenches, *etc.*) and a circular enclosure delimited by two stone rings

(Gianotti, Cancela 2005; Gianotti, Mañana y Criado in press). The radiocarbon dates obtained for this site and calibrated at 2σ are: 2880-2570 BC, 2790-2570 BC y 2470-2130 BC.

All the analysed samples dating to the 1st millennium BC were recovered from *castros* (hillforts/fortified settlements). The chronological sequence of these sites range from the Early Iron Age to the Indigenous-Roman Period, with a variety of spatial locations (coastal, inland) and activities (commercial and mining activities, etc.) represented by these sites. The sites can be grouped according to geographic area and also in relation to the administrative divisions of the roman era. The first group consists of *castros* situated in the coastal and inland regions of *Gallaecia*, while the second group is comprised of settlements near the mining zone of Las Médulas in *Asturia*.

Within the *Gallaecia* group of sites we can differentiate between coastal and inland *castros*. The coastal sub-group includes the sites of Punta do Muiño, Neixón Grande and Montealegre, and also Navás, which is located near the coast. Castrolandín and Zoñán belong to the inland group. The *castro* of Punta do Muiño has a long sequence of occupation from the 7th to the 1st centuries BC, although due to the lack of stratigraphical contextualisation of the samples, it is impossible to know to which occupation period they belong, and as such they are presented as a group (Rey *et al.*, in press). The analysed archaeobotanical remains from the *castro* at Neixón Grande belong to the period between the 4th and 2nd centuries BC. The majority of the studied samples were recovered from different fills of the enclosure ditch and from one of the postholes situated beneath the ditch (Ayán 2008). Various construction phases were also identified at Castrolandín, which was occupied from the 2nd century BC to the 1st century AD (Criado y Ayán 2006). During excavation, dispersed charcoal samples were recovered from the interior and exterior areas. In one of the zones, and dating to the construction phase, remains of carpentry and incompletely carbonised wooden objects were found. The castro of Navás had a short occupation from the 1st century BC to the 1st century AD. In one structure a layer of burning was identified along with the remains of firewood that represented the last fire within the hearth (Martín 2008b). Although the *castro* of Montealegre represents a settlement with a long occupation, the analysed samples correspond to a period from the 2nd century BC to the 1st century AD (Aboal & Castro 2006; González-Ruibal, Rodríguez, Aboal, & Castro 2007). From the dating of samples the chronological range of occupation at the *castro* of Zoñán was found to extend from the 4th century BC to the 4th century AD: 388-204 BC, 357-57 BC, 180-1 BC, 20 BC-128 AD y 129-323 AD (2σ calibrated dates) (Vigo 2007).

All of the sites within *Asturia* are located near the Archaeological Zone of Las Médulas (Sánchez-Palencia 2000). Cerro de Borrenes has an occupation during the Late Iron Age, but was abandoned during the settlement's construction. As the studied samples of this site came from collapse levels few taxa were identified, most of which probably relate to construction material. During the excavation of San Juan de Paluezas, a *castro* that was occupied from the 3rd to the 1st centuries BC, a number of construction phases and a phase of complete abandonment of the site were identified. The *castro* at Orellán, which has a short occupation during the 1st century BC, was a metal-working site with evidence for smelting. Chaos de Mourán is a settlement with an occupation during the roman period and is located in the heart of the mining zone of Las Médulas. The site of Cerro Pendon was also occupied during the roman period. The domus of Pedreiras del Lago was built following the roman architectural model for a local aristocracy who were probably involved in the control and management of the mines.

The most recent dates recovered were from the sites of Montenegro 2 and A Mourela. Montenegro 2 was occupied during the medieval period, and contained various structures including a collapsed stone building, some pits and ditches. During the excavation of A Mourela, a *castro* situated on a mountain ridge, two occupation phases were identified (Fábregas y Bonilla 2008). The site had sporadic occupation between the 7th and 11th centuries AD (2σ calibrated dates: 604-778 AD y 772 -1048 AD), which was represented by a hearth and a circular stone-built enclosure that was probably connected with pastoral activities. Later, and in the vicinity of the enclosure, a shepherd's cabin was constructed, occupied and rebuilt over a number of centuries (14th -19th centuries AD) as is shown by the $^{14}C$ dates obtained (1391-1454 AD, 1393-1522 AD, 1397-1489 AD, 1465-1678 AD, 1483-1683 AD y1877-1917 AD).

## MATERIAL AND METHODS

### Samples

All the analysed charcoal and wood samples were collected during excavation work. In 80% of the sites the samples were collected by hand, while in the other ones (A Mourela, Navás, Castrolandín) a combination of manual collection and soil-sieving (water sieve and flotation) was employed. A combined collection strategy minimises the fragmentation of large charcoal samples by collecting by hand, and also allows for the identification of a greater variety of taxa through the use of soil-sieving and flotation. In the sites of Northwest Iberia, where the samples were collected by hand, there is an over representation of species such as *Quercus* sp deciduous, as opposed to shrub species (*Fabaceae*, *Corylus avellana*), or soft wood species (*Salix/Populus, Frangula alnus, Alnus* sp). This result bias could be derived from the collection method employed.

It must also be noted that the studied sample size varies between sites, although this does not appear clearly to be the cause of the greater or lesser diversity of the

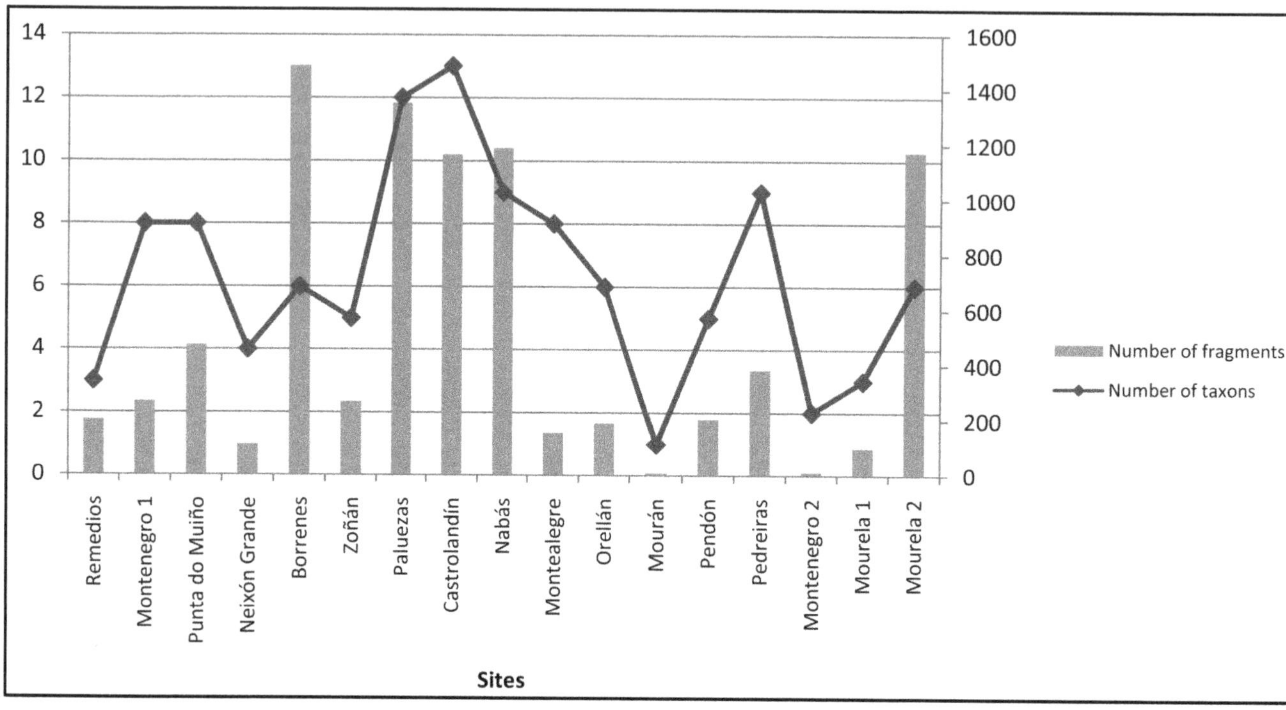


Figure 2. Number of fragments and number of taxons identified in the studied sites

documented taxa. As we can see in Figure 2 the diversity of taxa seems not have a correlation with the number of samples studied. But, if we except Mourela 2 Borrenes and Navas the correlation number of fragments/number of taxons remains possible.

**Charcoal/Wood analysis**

Each fragment of charcoal that was analysed was first prepared for identification by breaking by hand and observing the transversal, tangential and radial surfaces. In the case of mineralized wood, three anatomical cuts were made using a knife to enable identification. Samples were examined using a reflected/transmitted light microscope, with lenses of 20x to 400x. Atlases of wood anatomy by F.H. Schweingruber (1978, 1990), J.G. Hather (2000), and García *et al.* (2003) were used for comparative purposes, along with a charcoal reference collection.

As well as taxonomic identification, certain dendrological aspects (Marguerie, Hunot 2007) were examined, such as: (1) measuring complete branch diameters to obtain data about the size of firewood and timber used; (2) the presence of vitrification and radial cracks in the tissues, which can be related to the combustion of green wood or combustion in the absence of oxygen, (3) signs of entomofaunal action that can be related to consumption of dead firewood or a xylophagous attack on structural timbers, (4) the presence of scars, (5) compression and tension wood. This paper is concerned with the dendrologycal data from sites dating to the 1st millennium BC only.

**Archaeobotanical database and GIS**

The information for each fragment was inputted in an archaeobotanical database, in order to unify the variables and facilitate the quantification of the results. The results were then expressed spatially using a GIS model. Total fragment counts and a frequency analysis of different taxa were undertaken. A statistical analysis of different aspects of the samples, such as dendrological factors (vitrification, entomofaunal evidence, *etc.*) and palaeoenvironmental conditions (trees and shrubs selection and consumption, *etc.*), was also carried out. The results were analysed on both a microscale and intrasite level, and within a broader geographical context. Three levels of analysis were employed: (1) on a regional scale to observe the data within the area of the Northwest peninsula in general; (2) on an intersite level to compare various sites within a specific geographic area; (3) on a microspatial and intrasite level to observe the distribution and percentages of concentrations within a site in relation to its structures and deposits.

## FOREST MANAGEMENT

### 5th-3rd millennium BC

The data concerning woodland management during the Neolithic and Chalcolithic were recovered from the sites of Remedios and Montenegro (Figure 3). In both sites all of the analysed charcoal samples were recovered from structures, which could be the cause of the limited species variability noted, especially at Remedios. Pollen and pedological analysis indicates that the temperature began to drop at the end of the 4th millennium/beginning of the 3rd millennium BC and continued until 2500 BC. At the end of the 4th millennium BC erosive processes of an anthropogenic origin are detectable, with a decrease in forest coverage and the appearance of charcoal lines in the soil indicative of fires (Fábregas, Martínez-Cortizas,

| | Montenegro 1 | Remedios |
|---|---|---|
| *Quercus* sp. deciduous | 124 | 199 |
| Fabaceae/cf. Fabaceae | 99 | 1 |
| *Corylus avellana* | 11 | – |
| Rosaceae/Maloideae | 3 | – |
| *Quercus* sp. | 2 | – |
| *Betula* sp. | 2 | – |
| cf. *Salix/Populus* | 1 | – |
| *Laurus nobilis* | 1 | – |
| *Arbutus unedo* | 1 | – |
| Compositae | – | 1 |
| Undetermined | 28 | – |
| **Total fragments** | **269** | **201** |

Figure 3. Charcoal analysis from 4th-3rd millennium BC sites

Blanco, R., Chesworth, W. 2003). During the 3rd and 2nd millennia BC there was an increase in anthropogenic pressure on the woodland environment, with settlements becoming more permanent in character and fire being used to clear forests, obtain pastures and to favour the growth of certain species (Martínez-Cortizas, Franco, Fábregas 2000; Fábregas, Martínez-Cortizas, Blanco, R., Chesworth, W. 2003).

Notable in this period is the preferential exploitation of mixed deciduous forests, with the use of the wood of *Quercus* sp deciduous along with that of species indicative of the existence of forest clearings (*Maloideae*, *Corylus avellana*, *Arbutus unedo*) or pioneering species (*Betula* sp). Fuelwood was also collected from the scrubland that formed the forest limits, where the prominent species are the Fabaceae.

**1st millennium BC**

The significant number of studied sites dating to the first millennium BC enables the observation of how the dynamics of earlier periods continued with some transformations (Figure 4). These changes are related to more permanent occupations of settlements over longer periods, over various centuries, with a greater concentration of population and an intensive exploitation of the environs of the sites, all of which during this period produced a major impact on woodlands.

On a general level it can be considered that from 2700-2500 BP temperature and humidity are increasing (Fábregas, Martínez-Cortizas, Blanco, R., Chesworth, W. 2003) with respect to the previous period. The pollen data from the *Ría* of Vigo allows a reconstruction of t he evolution of the vegetation in this area, which coincides with many of the sites analysed in this study (Desprat, Sánchez-Goñi, Loutre 2003). The earliest dates from this analysis are 975 ± 36 cal BC, a moment in which a faint human impact on the vegetal environment is noted, reflecting small-scale cultivation. From 130 BC – 200 AD, during the galaico-roman period, a more perceptible human impact on the woodland resource is noticeable. This was followed by the appearance of *Juglans* and *Castanea c.* 1 BC and the beginning of the deforestation process, which the authors of the Vigo environmental analysis related to mining activity. The bogs of Penido Vello and Pena da Cadela (Lugo) demonstrate that during this period an atmospheric contamination is increasing caused by lead related to metallurgical activities, a process which had begun during the Late Bronze Age (Martínez-Cortizas *et al.* 1997a; Martínez-Cortizas *et al.* 1997b; Martínez-Cortizas *et al.* 2002). From *c.* 2500 BP until 1800 BP the presence of lead in the atmosphere increased in an almost continuous manner, with the exception of a relative minimum documented around 2100 BP. This process coincided with the beginning of the roman exploitation of the mineral resources of the northwest, which began at the end of the 2nd century BC, increased until the 1st century BC and reached its peak in the high imperial period (Martínez-Cortizas *et al.* 2002).

As well as mining and metallurgical activity, the important impact on woodlands caused by the establishment of *castro* settlements over long periods of time must also be evaluated (fuel requirements, cultivation fields, pastures, *etc.*). The charcoal samples reveal certain continuity in the exploitation of mixed deciduous forests in the sites of the Atlantic area, although from the 5th to 2nd millennia BC there was a significant increase in the exploitation of river-bank woodland species (*Salix/Populus, Alnus sp, Frangula alnus*). In the Atlantic zones with mediterranean influence there was a significant exploitation of *Quercus sp* evergreen, with a sporadic recovery from sites of carbonised fragments of the bark of *Quercus suber* and fragments of *Laurus nobilis*. The exploitation of the wood of wild fruit-bearing trees such as *Corylus avellana* and *Maloideae* is documented, as well as others such as *Castanea sativa* and *Juglans regia*, which were probably cultivated to avail both of the wood and fruit. Scrubland had a great importance, and is represented by the *Fabaceae* and in the zones of greatest degradation of arboreal coverage by *Erica* sp.

In the sites of the Mediterranean zone of *Las Médulas* (León) a preference for *Quercus* sp deciduous and *Quercus sp* evergreen was observed, the latter being most common in sites of the roman period. As in the Atlantic zone, there was a continued presence of river-bank woodland species (*Salix/Populus, Fraxinus sp, Alnus sp*). Species related to the existence of clearings in the deciduous forests such as *Prunus sp* and *Pyrus sp* appeared sporadically. In contrast to the Atlantic zone, there was a significant absence in the exploitation of the *Fabaceae*. In this zone scrubland species were rarely documented. *Fabaceae* was only identified at San Juan de Paluezas, a site with continued occupation of approximately two centuries, at the metallurgical site of Orellán, where the appearance of *Pistacia terbinthus* was probably related to the degradation of holm oak forests,

| | Punta do Muiño | Neixón Grande | Borrenes | Zoñán | Paluezas | Castrolandín | Nabás | Montealegre | Orellán | Mourán | Pendón | Pedreiras |
|---|---|---|---|---|---|---|---|---|---|---|---|---|
| *Quercus* sp. deciduous | 229 | 79 | 1429 | 251 | 634 | 807 | 713 | 44 | 85 | 2 | 192 | 100 |
| *Salix/Populus*/cf. *Salix/Populus* | 41 | – | 3 | 11 | 111 | 9 | 13 | – | 3 | – | 2 | – |
| *Fraxinus* sp. | – | – | 2 | 2 | 45 | 2 | – | – | 37 | – | 4 | 91 |
| *Alnus* sp. | 26 | – | – | – | 272 | 85 | 1 | – | – | – | 4 | 9 |
| *Quercus* sp. evergreen | 40 | – | – | – | 93 | 11 | 170 | 9 | 34 | – | – | 118 |
| Fabaceae/cf. Fabaceae | 114 | 6 | – | – | 42 | 114 | 191 | 7 | – | – | – | – |
| Rosaceae/Maloideae/cf. Rosaceae/Maloideae | 2 | 10 | – | – | – | 11 | 5 | 12 | – | – | – | 3 |
| *Corylus avellana* | 4 | – | – | – | – | 83 | 85 | 4 | – | – | – | – |
| *Prunus* sp. | – | – | – | – | 24 | – | – | – | 6 | – | – | 2 |
| *Taxus baccata* | – | – | 5 | 1 | 14 | – | – | – | – | – | – | – |
| *Quercus suber* | – | – | – | – | 41 | – | 8 | – | – | – | – | 11 |
| *Quercus* sp./cf. *Quercus* sp. | – | 2 | – | 2 | – | – | – | 14 | – | – | – | – |
| *Frangula alnus* | – | – | – | – | – | 12 | 4 | – | – | – | – | – |
| *Erica* sp./ cf. *Erica* sp. | – | – | 1 | – | – | – | – | 16 | – | – | – | – |
| *Laurus nobilis* | 8 | – | – | – | – | – | – | 9 | – | – | – | – |
| *Juglans regia* | – | – | – | – | 6 | 5 | – | – | – | – | – | – |
| *Arbutus unedo*/cf. *A. unedo* | – | – | – | 1 | 7 | – | – | – | – | – | – | – |
| *Hedera helix*/cf. *Hedera helix* | – | – | – | – | – | 1 | – | 1 | – | – | – | – |
| *Castanea sativa* | – | 15 | – | – | – | – | – | – | – | – | – | – |
| *Juniperus* sp. | – | – | 1 | – | – | – | – | – | – | – | – | – |
| *Pinus* tp. *sylvestris/nigra* | – | – | – | – | 10 | – | – | – | – | – | – | – |
| *Ilex aquifolium* | – | – | – | – | – | 4 | – | – | – | – | – | – |
| *Alnus/Corylus* | – | – | – | – | – | 1 | – | – | – | – | – | – |
| *Betula* sp. | – | – | – | – | – | 1 | – | – | – | – | – | – |
| *Pistacia terebinthus* | – | – | – | – | – | – | – | – | 4 | – | – | – |
| *Pyrus* sp. | – | – | – | – | – | – | – | – | – | – | 2 | – |
| *Viburnum* sp. | – | – | – | – | – | – | – | – | – | – | – | 1 |
| *Vitis vinifera* | – | – | – | – | – | – | – | – | – | – | – | 1 |
| Undetermined | 8 | 1 | 46 | – | 53 | 20 | 10 | 40 | 22 | 8 | 1 | 48 |
| **Total fragments** | **472** | **113** | **1487** | **268** | **1352** | **1166** | **1200** | **156** | **191** | **10** | **205** | **384** |

Figure 4. Charcoal analysis results from Iron Age and Roman Period sites

and at Borrenes, where other species indicative of soil deterioration, *Erica sp* or *Juniperus sp*, were identified.

Charcoal of arboreal and bushy species is well documented on the sites for the first millennium BC. Their extension is related to the adoption of silviculture practices as represented by *Juglans regia, Castanea sativa* and *Vitis vinifera*. *Vitis vinifera* was developed spontaneously in the Iberian peninsula on the banks of the *rías* (estuaries), Based on forestry practices a domestic variation was obtained (Precioso 2004). It is not very common during the first millennium BC, although it has been documented in other sites in the Northwest of the peninsula: Castro das Ermidas, Castro de Penices e Castro de Crastoeiro (Figueiral 1996). Pollen analysis shows that *Castanea sativa* is a species native to this zone, although in this period it extends in relation to silviculture practices (Conedera *et al.* 2004). Its presence has also been documented in the sites of Castelo de Matos, Castro de Penices, Castro de Terroso, San Juliao (Figueiral 1996, Figueiral, Bettencourt 2004). The case of *Juglans regia* is similar to that of *Castanea sativa*: charcoal of this species was identified in Castro das Ermidas y Castro de Penices (Figueiral 1996).

| | Montenegro 2 | Mourela 1 | Mourela 2 |
|---|---|---|---|
| *Quercus* sp. deciduous | – | 2 | 886 |
| Rosaceae/Maloideae/cf. Rosaceae/Maloideae | – | 2 | 20 |
| *Betula* sp. | – | – | 2 |
| *Ilex aquifolium* | – | – | 3 |
| Fabaceae/cf. *Fabaceae* | 12 | – | 21 |
| *Erica* sp./cf. *Erica* sp. | – | 23 | 216 |
| *Castanea sativa* | 2 | – | – |
| Undetermined | – | 74 | 27 |
| **Total fragments** | **14** | **101** | **1175** |

Figure 5. Charcoal analysis results from Medieval and Postmedieval sites

**Medieval and Postmedieval Period**

In the sites of the medieval to contemporary period there is a significant reduction in the number of taxa represented (Figure 5). In Montenegro 2 the determination of various fragments of *Castanea sativa* in the interior of one of the medieval ditches stands out. At Mourela 1, charcoal was recovered from a combustion structure. Although the degree of vitrification impeded the taxonomic identification of the majority of fragments, the predominant species was *Erica sp*., which is indicative of soil deterioration. At Mourela 2, mixed deciduous forest species (*Quercus sp* deciduous, *Maloideae, Betula sp* and *Ilex aquifolium*) and scrub species (*Erica sp* and *Fabaceae*) were represented.

## WOOD USES

Charcoal analysis and wood studies give us a better understanding of wood use in this area in the past. Fundamentally it provides information relating to fuels, wood for construction and for the elaboration of objects. As the majority of the analysed sites belong to the Iron Age and the roman period, this study puts the accent mostly on the analysis of data from sites of this chronology.

While limited, the data obtained from the analyis of charcoal samples from combustion structures indicate that *Quercus sp* deciduous was the principal fuel source. Occasionally it was used in combination with *Quercus sp* evergreen in certain zones with mediterranean influence, and with fuelwood obtained from bushes like *Fabaceae*. In the Chalcolithic site of Os Remedios the analysis of a combustion structure with a stone grill identified *Quercus* sp deciduous (125 fragments) and *Fabaceae* (1 fragment). At the site of Navás, which has a more recent chronolgy from the 1st century BC to the 1st century AD, similar results were observed for this type of structure, with a predominance of *Quercus* sp deciduous (29 fragments) and the marginal presence of other taxa like *Quercus* sp evergreen (2 fragments) and Fabaceae (8 fragments). The preferential use of oak, in this case for firewood, is not only documented in the study of charcoal related to combustion structures. It is also the most abundant taxa in general in habitation structures, as is the case in the site of Castrolandín. Within the study group of sites oak is the most recurrent species, regardless of chronology or location. In general the data points to a similar pattern throughout the study group in the provisioning of fuel, with a combined use of the wood of *Quercus* sp deciduous and *Fabaceae*, with occasional contributions of other arboreal species, especially bushes: *Alnus sp, Maloideae, Salix/Populus, Corylus avellana* and *Frangula alnus*.

The majority of the dates relating to construction timbers come of sites from the first millennium BC, with samples recovered from levels of burning or from posts carbonised *in situ*. During this period both arboreal and bushy species were used in the construction of habitation structures and for the delimitation of settlements. *Quercus* sp deciduous and *Quercus* sp evergreen were used to make supporting elements (posts or beams) in the enclosing structures of settlements, as was the case of the *castro* at Neixón Grande. Bushy or arboreal species with flexible and large branches were used to construct frameworks or coverings: *Fabacea* was used for coverings, *Corylus avellana* and *Salix/Populus* was used for frameworks. At the *castro* of Punta de Muiño branch impressions in clay were measured. Based on 265 measurements of the most habitual diameters, 43.01% measured between 0.30-0.79 cm, with 26.4% between 0.80 cm and 1.09 cm. These coincide with the measurements of the complete branch diameters of samples of *Corylus avellana* from Punta do Muiño (1 example) and Navás (12 examples), where the diameters of 91.6% of the examples oscillate between 0.60 cm and 1.5 cm. The results are suggestive of the practice of coppicing in species like *Corylus avellana* and probably in others like *Alnus sp* and *Salix sp*, in order to produce large, straight branches with a determined diameter.

A micro-spatial analysis of the data obtained from the *castro* of Navás permits the observation of the association and distribution of different taxa within the interior of a structure (Figure 6). Found beneath a structural stone collapse, charcoal samples from a layer of burning with an important concentration of archaeobotanical remains were manually collected and sieved in water. The fragments, which were concentrated beneath the collapse and close to the walls of the structure, were not abraded and were relatively large, with more than 85% measuring between 1 and 6 cm. Taxonomic identification showed that the charcoal samples that probably correspond to the covering structure consisted principally of *Quercus* sp deciduous and *Quercus* sp evergreen, along with *Fabaceae* and *Corylus avellana*, which are the species that appear associated in the majority of samples from

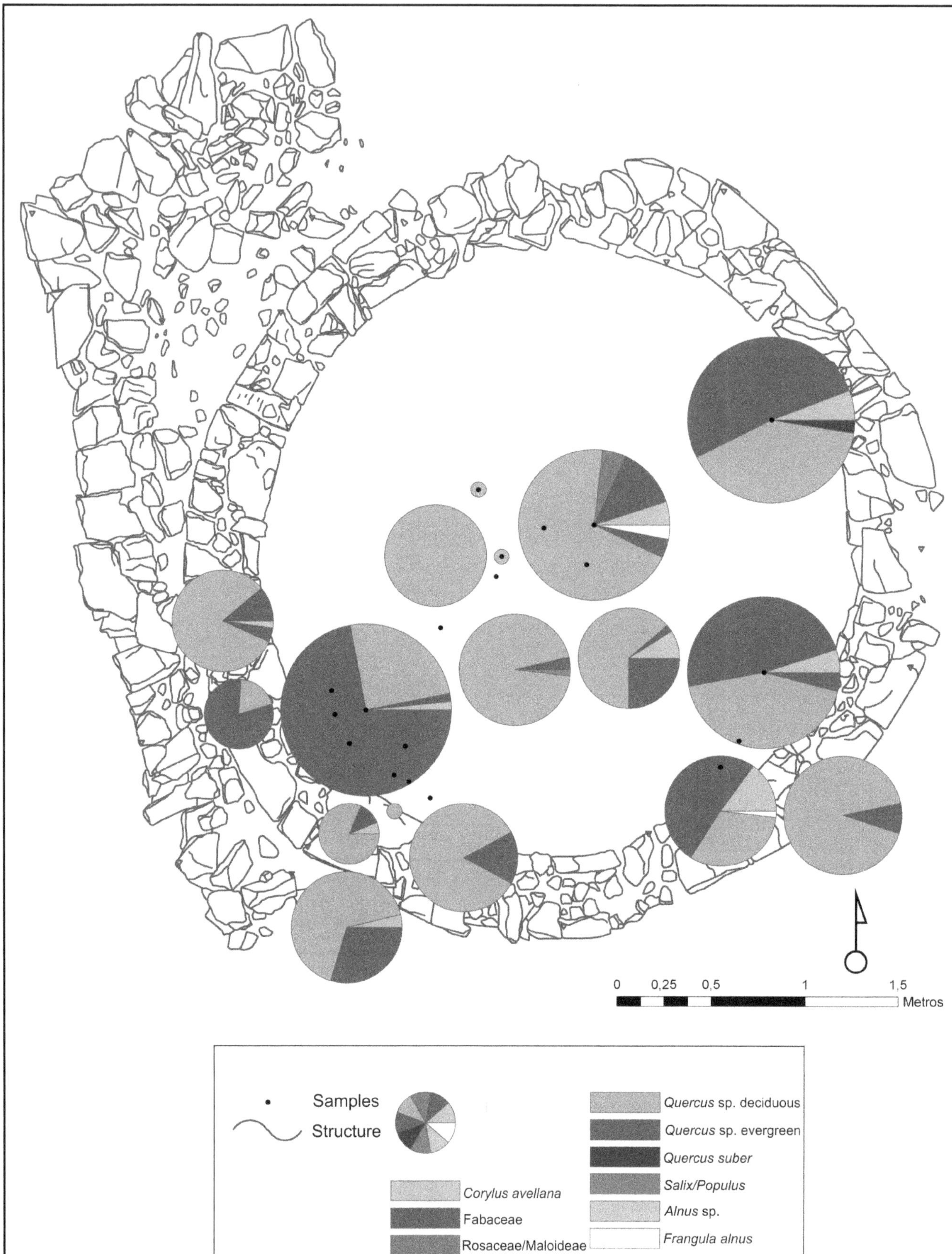


Figure 6. Microscale distribution of taxons inside the construction Castro of Navás

this deposit. *Alnus* sp, *Frangula alnus* and *Maloideae* appeared more sporadically and could correspond to other types of objects or carbonised structures.

The determination of entomofaunal action on wood was very occasional, with only four sites from the first millennium BC (Figure 7). In some cases the wood with

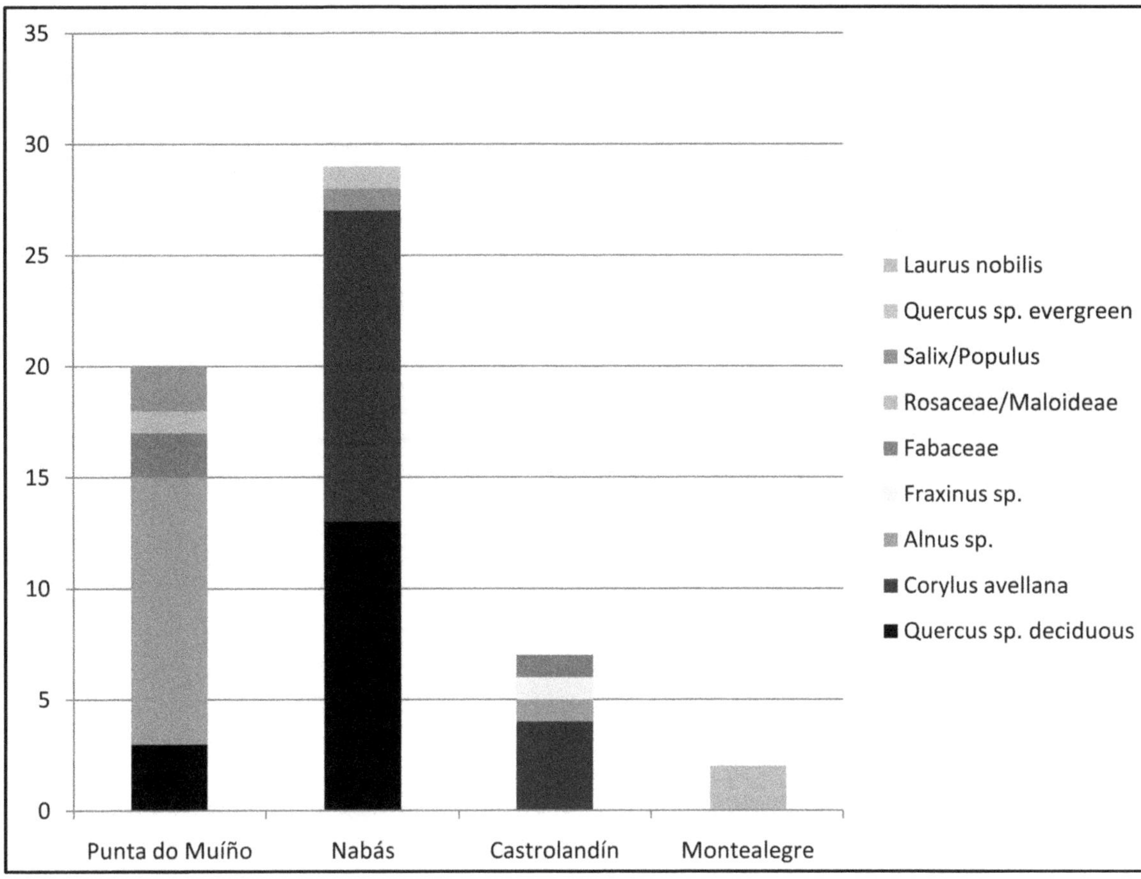


Figure 7. Number of fragments with entomofaunal evidence

evidence of entomofaunal action belongs to *Fraxinus* sp or *Alnus* sp and also records woodworking marks, as is the case in Castrolandín. In most of these samples, these could represent wood used for construction or for fuel wood.

Data about objects made of wood are still very scarce (Castromao, Cameixa, Castrovite, Zoñán, etc.). However, the appearance of fragments of mineralised or carbonised wood in the interior of metallic objects in the *castros* of Neixón Grande and Montealegre, and the identification of wood working marks and semi-elaborated objects of carbonised wood in Castrolandín, provides some insights into wood working and which species were used to make tools, receptacles, vessels *etc*.

In Montealegre a charcoal piece of *Corylus avellana* was recovered from the interior of a small bronze ferrule, while at Neixón a fragment of mineralised wood of *Quercus* sp deciduous was found in the interior of a similar object. However, the most complete data regarding wood working comes from the site of Castrolandín. At this site evidence for felling and cutting was identified (small branches with cutting surfaces/planes), and remains related to carpentry (splinters and small wedge-shaped fragments), objects with evidence of wood working (reductions or cuts to give shape to the pieces) and vessel fragments (rim and body fragments of a wooden vessel) were recovered. Most of these were found in a space between various structures that were probably dedicated to wood working activity.

The wood working evidence at Castrolandín shows that particular species were selected on the basis of the characteristics of the wood and its suitability for the final use of the object. Fragments of receptacles or vessels are smooth (*Alnus* sp), of a light wood, with a fine, soft grain, that is easily worked and polished (López 2002, Abella 2003). Small construction materials such as planks, wedges and joins for the union of pieces were elaborated in ash (*Fraxinus* sp) and oak (*Quercus* sp deciduous). On one fragment, a section of a branch of *Quercus* sp deciduous, two small branches of hazel (*Corylus avellana*) were attached. A small handle for an indeterminate object was also elaborated in *Quercus* sp deciduous. Oak wood is strong and very solid, resistant to rot and easy to polish (López 2002, Abella 2003). It can be used as a structural element and, because of its resistance to compression and flexion, it is used for tool handles. Ash wood, in which a small splint was elaborated, is suitable when elasticity and flexibility is necessary, however once dry it is exceptionally rigid, and as such is not suitable to remain exposed in the open (López 2002, Abella 2003). Various fragements of an indeterminate object in *Ilex aquilofolium* (holly) were also recovered. Holly is a very heavy wood, with a fine and uniform texture. It is hard to work, difficult to dry and in order to prevent torsion it is cut into small pieces and worked in the round (López 2002, Abella 2003).

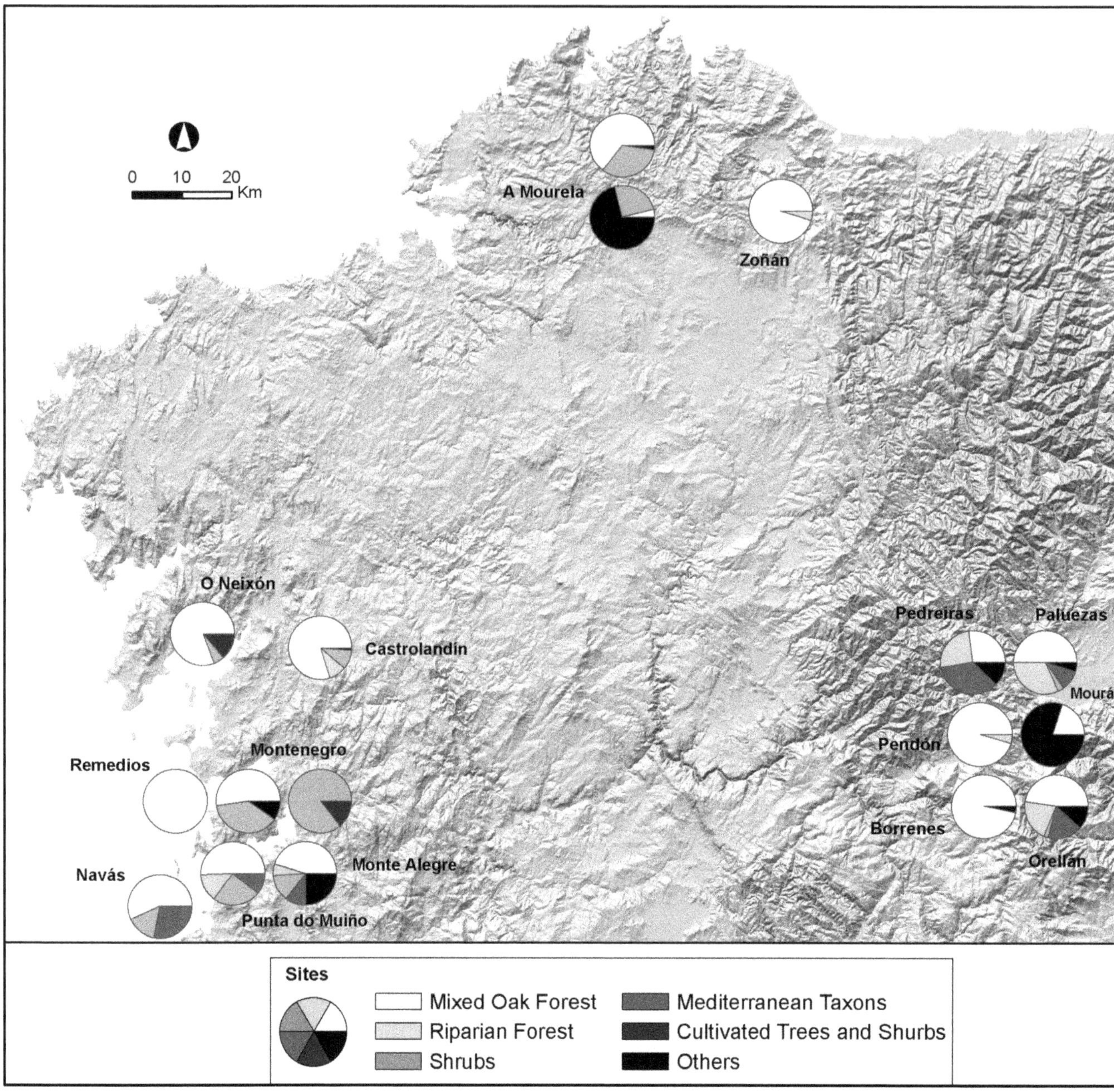


Figure 8. Charcoal analysis results of the studies sites in the Northwest of the Iberian Peninsula

## CONCLUSION

The results of this study have enabled an approximation of the palaeoecological and palaeoeconomical background to woodland management and wood use in the Northwest Iberian Peninsula. Since recent prehistory and until the post-medieval period, woodland management was based on the systematic use of deciduous species, especially of *Quercus* sp deciduous as well for fuelwood than for the elaboration of artefacts and structures (Figure 8).

Despite the quantitative limitations of the data set of analysed samples, the information retrieved coincides with the pollen and pedological data which also indicate a retreat of arboreal coverage from the 5th to 3rd millennia BC. Anthracology reveals the presence of mixed deciduous forests related to the existence of clearings (*Maloideae*, *Corylus avellana*, *Arbutus unedo*) and scrub species (*Fabaceae*).

In the first millennium BC the exploitation of mixed deciduous forest resources for timber and firewood continued, but there was also an important representation of riverbank species, especially *Salix/Populus* and *Fraxinus* sp. The exploitation of scrubland species, in particular *Fabaceae*, was also important in the Atlantic area. In sites with very long occupations and significant population concentrations, such as the *castro* of Montealegre, *Erica* sp was also represented. In the Mediterranean area however, scrubland species had a lesser importance and only appeared sporadically in sites with long occupations or in specialised sites with metallurgical activities. Coppicing of species like *Corylus avellana* and others trees or shrubs related to forestry practices (*Castanea sativa, Juglans regia, Vitis vinifera*) are documented.

The use of wood for construction during this period indicates that *Quercus* sp deciduous and *Quercus* sp evergreen were some of the species utilised for supporting elements (posts, beams, *etc*), while the vegetal framework was constructed using species like *Corylus avellana, Alnus sp* or *Salix/Populus*. With respect to manufacturing objects, selection of species was based on the wood intrinsic characteristics (easy to polish, tenacity, flexibility *etc.*) and its suitability for the different applications intended.

In the medieval and post-medieval periods the importance of scrublands is apparent, although there still exists a preference for mixed forest species for fuelwood, construction material *etc.* During this period it is documented that fire was systematically used to impede the development of woody vegetation, making scrubland widespread, especially from the 15th – 19th centuries AD (Guitián 2001).

**Acknowledgements**

We would like to thank the directors of the excavations, Roberto Aboal, Xurxo Ayán, Andrés Bonilla, Ramón Fábregas, Camila Gianotti, Carlos Otero, Fco.-Javier Sánchez-Palencia and Abel Vigo, for putting at our disposition both the study material and the archaeological information derived from them. Thanks also to Cliodhna ni Lionan for the translation of the texts and Emilio Abad for the elaboration of the maps for this article.

**References**

ABELLA, I. 2003. *El hombre y la madera*. Ed. Integral, Barcelona.

ABOAL, R., CASTRO, V. (coord.) 2006. *O Castro de Montealegre (Moaña, Pontevedra)*. Ed. Toxos Outos, Noia.

AIRA, M.J., UZQUIANO, P. 1996. Análisis polínicos e identificación de carbones en necrópolis gallegas de época romana. In A. Rodríguez (coord.) *Lucus Augusti I. El amanecer de una ciudad*, 47-53. Fundación Pedro Barrié de la Maza, A Coruña.

AYÁN, X.M. (coord.) 2008. *Os Castros de Neixón II*. Ed. ToxosOutos, Noia.

BONILLA, M., CÉSAR, M. 2005. Calcolítico: Monte dos Remedios. In Criado, F., Cabrejas, E. (Coord.) *Traballos de Arqueoloxía e Patrimonio*, 35, 55-58.

BONILLA, A., CÉSAR, M., FÁBREGAS, R. 2006. Nuevas perspectivas sobre el espacio doméstico en la Prehistoria reciente del NO: el poblado de Os Remedios (Moaña, Pontevedra). *Zephyrus*, 59, 257-273.

CARBALLO, X. 1989. *A cultura castrexa na bacía media do rio Ulla*. Unpublished PhD thesis, University of Santiago de Compostela.

CARRIÓN, Y. 2003. *Afinidades y diferencias de las secuencias antracológicas en las vertientes mediterránea y atlántica de la Península Ibérica.* Unpublished PhD thesis, University of Valencia.

CARRIÓN, Y., BADAL, E. 2004. La presencia de hongos e insectos xilófagos en el carbón arqueológico. Propuestas de interpretación. *Avances en Arqueometría* 2003, 98-106. Universidad de Cádiz, Cádiz.

CONEDERA, M., KREBS, P., TINNER, W., PRADELLA, M., TORRIANI, D. 2004. The cultivation of Castanea sativa (Mill.) in Europe, from its origin to its diffusion on a continental scale. *Vegetation History and Archaeobotany*, 13, 161-179.

CRIADO, F., AYÁN, J.M. 2006. Sondaxes arqueolóxicas no Castro de Castrolandín (Cuntis, Pontevedra): informe valorativo. *El Museo de Pontevedra*, nº60, 11-82.

DESPRAT, S., SÁNCHEZ-GOÑI, M.F., LOUTRE, M.F. 2003. Revealing climatic variability of the last three milennia in northwestern Iberia using pollen influx data. *Earth and Planetary Science Letters*, 213, 63-78.

FÁBREGAS, R., BONILLA, A. 2008. Escavación arqueolóxica en área do GA150700128, A Mourela 5, As Pontes de García Rodríguez (A Coruña). *Actuacións Arqueolóxicas: Ano 2006*, 187-188. Ed. Xunta de Galicia, Santiago de Compostela.

FÁBREGAS, R., BONILLA, A., CÉSAR, M. 2007. *Monte dos Remedios (Moaña, Pontevedra). Un asentamientos de la prehistoria reciente.* Ed. Tórculo, Santiago.

FÁBREGAS, R., MARTÍNEZ-CORTIZAS, A., BLANCO, R., CHESWORTH, W. 2003. Environmental change and social dynamics in the second-third millennium BC in NW Iberia. Journal of Archaeological Science, 30, 859-871.

FIGUEIRAL, I. 1993. Charcoal analysis and the vegetational evolution of North-West Portugal. *Oxford Journal of Archaeology*, 12 (2), 209-222.

FIGUEIRAL, I. 1996. Wood resources in north-west Portugal: their availability and use from the late Bronze Age to the Roman period. *Vegetation History and Archaeobotany*, 5, 121-129.

FIGUEIRAL, I., BETTENCOURT, A.M.S. 2004. Middle/Latte Bronze Age plant communities and their exploitation in the Cávado Basin (NW Portugal) as shown by charcoal analysis: the significance and co-occurrence of Quercus (deciduous)-Fabaceae. *Vegetation History and Archaeobotany*, 13, 219-232.

GARCÍA, L., GUINDEO, A., PERAZA, C., De PALACIOS, P. 2003. *La madera y su anatomía. Anomalías y defectos, estructura microscópica de coníferas y frondosas, identificación de maderas, descripción de especies y pared celular*, Ed. Mundi-Prensa, Madrid.

GIANOTTI, C., CANCELA, C. 2005. Testemuña da ocupación humana durante o Neolítico Final e o periodo Alto-Medieval na Península do Morrazo. In Criado, F., Cabrejas, E. (Coord.) *Traballos de Arqueoloxía e Patrimonio*, nº35, 50-54.

GIANOTTI, C., MAÑANA-BORRAZÁS, P., CRIADO, F. in press. Montenegro, a Neolithic enclosure in Galicia. Insights into Megalithic space, *BAR International Series*.

GONZÁLEZ-RUIBAL, A., RODRÍGUEZ, R., ABOAL, R., CASTRO, V. 2007. Comercio mediterráneo en el castro de Montealegre (Pontevedra, Galicia). Siglo II a.C. – inicios del s. I d.C. *Archivo español de arqueología*, 80, 43-74.

GUITIÁN, L. 2001. La destrucción histórica del bosque en Galicia. *Semata*, 13, 105-166.

HATHER, J.G. 2000. *The identification of the Northern European Woods. A guide for archaeologists and conservators*, Archetype, London.

LÓPEZ, G. 2002. *Guía de los árboles y arbustos de la Península Ibérica*. Ed. Mundi-Prensa, Madrid.

MARGUERIE, D., HUNOT, J.-Y. 2007. Charcoal analysis and dendrology: data from archaeological sites in north-western France. *Journal of Archaeological Science*, 34, 1417-1433.

MARTÍN SEIJO, M. 2005. A xestión dos recursos forestais no Castro de Zoñán (Mondoñedo, Lugo): os resultados da antracoloxía. *Gallaecia*, nº24, 169-179.

MARTÍN SEIJO, M. 2006. A xestión dos recursos leñosos no castro de Montealegre. In Aboal, R., Castro, V. (coord.) *O Castro de Montealegre (Moaña, Pontevedra)*, 301-322. Ed. ToxosOutos, Noia.

MARTÍN SEIJO, M. 2007. Estudo dos carbóns arqueolóxicos do xacemento de Os Remedios, R. Fábregas, A. Bonilla, M. César, *Monte dos Remedios (Moaña, Pontevedra). Un asentamiento de la prehistoria reciente*, 107-114. Ed. Tórculo, Santiago de Compostela.

MARTÍN SEIJO, M. 2008a. Estudo de carbóns e madeiras: paleoambiente, xestión forestal e explotación de combustibles, X.M. Ayán (coord.) *Os Castros de Neixón II*, 249-267. Ed. ToxosOutos, Noia.

MARTÍN SEIJO, M. 2008b. Escavación e consolidación no castro de Navás, Priegue, Nigrán (Pontevedra), *Actuacións Arqueolóxicas: ano 2006*, 57. Ed. Xunta de Galicia, Santiago de Compostela.

MARTÍNEZ CORTIZAS, A., FRANCO, S., FÁBREGAS, R. 2000. Evolución del paisaje y actividad humana en el área de Monte Penide (Redondela, Pontevedra): una aproximación metodológica, *Trabajos de prehistoria*, 57, nº1, 173-184.

MARTÍNEZ CORTIZAS, A., GARCÍA-RODEJA, E., PONTEVEDRA, X., NOVOA, J.C., WEISS, D., CHEBURKIN, A. 2002. Atmospheric Pb deposition in Spain during the last 4600 years recorded by two ombrotrophic peat bogs and implications for the use of peat as archive. *The Science of the Total Environment*, 292, 33-44.

MARTÍNEZ CORTIZAS, A., NOVOA, J.C., PONTEVEDRA, X., GARCÍA-RODEJA, E., LLANA, C. 1997b. Paleocontaminación. Evidencias de contaminación atmosférica antrópica en Galicia durante los últimos 4000 años, *Gallaecia*, 16, 7-22.

MARTÍNEZ CORTIZAS, A., PONTEVEDRA, X., NÓVOA, J.C., GARCÍA-RODEJA, E. 1997a. Four thousand years of atmospheric Pb, Cd and Zn deposition recorded by the ombrotrophic Peat Bog of Penido Vello (Northwestern Spain). Water, Air and Soil Pollution, 100, 387-403.

PRECIOSO, M.L. 2004. Estudio de los restos paleobotánicos en Murcia. 2. Las plantas recolectadas y sinantrópicas. *Revista Arqueomurcia*, nº2, 28pp.

REY, J., ABAD, E., CALO, N., MARTÍN SEIJO, M., QUINDIMIL, L., RICO, A., RODRÍGUEZ CALVIÑO, M., TEIRA, A. in press. Metodoloxía e criterios para o estudo dos materiais arqueolóxicos: o proxecto do Castro da Punta do Muíño. *Gallaecia*.

SÁNCHEZ-PALENCIA, F.J. (Ed.) 2000. *Las Médulas (León). Un paisaje cultural en la Asturia Augustana*. Instituto Leonés de Cultura, Dip. Provincial de León, León.

SCHÖCH, W., HELLER, I., SCHWEINGRUBER, F.H., KIENAST, F. 2004. *Wood Anatomy of Central European Species*, Online version, [www.woodanatomy.ch].

SCHWEINGRUBER, F.H. 1978. *Mikroskopische Holzanatomie*. Zurcher a.g. Zug.

SCHWEINGRUBER, F.H. 1990. *Anatomie Europäischer Hölzer*. Bern und Stuttgart.

VIGO, A. 2007. *O Castro de Zoñán (Mondoñedo-Lugo): escavacións 2002-2004*. Concello de Mondoñedo.

# ANTHRACOLOGICAL ANALYSIS FROM A BRONZE AGE NECROPOLIS AT KOKOTÓW (POLAND)

Magdalena MOSKAL-DEL HOYO
Department of Prehistory and Archaeology, University of Valencia, Av. Blasco Ibáñez 28, 46010 Valencia, Spain
magdalena.moskal@uv.es

Andrzej MATOGA and Małgorzata ZYSNARSKA
Archaeological Museum of Kraków ul. Senacka 3, 31-002 Kraków, Poland
Andrzej.Matoga@ma.krakow.pl

**Abstract**: *The Kokotów necropolis is located south-east of Kraków (Poland) and is dated to the last phases of the Bronze Age Lusatian culture (*ca. *700-600 BC). This work is based on charcoal remains coming from 112 graves. The study has two main aspects: firstly, to determine if the wood used in cremation ceremonies corresponds to a special taxonomic selection that will provide mostly palaeoethnographic information; and secondly, if the gathered wood has palaeoecological significance.*

*Within more than 4600 charcoal fragments 15 taxa were found, in which four ones appeared more frequently and were ubiquitous: oak (*Quercus *sp. deciduous), common pine* (Pinus sylvestris)*, birch* (Betula *sp.) and heather* (Calluna vulgaris)*. Besides, it was observed that the taxonomic richness in the necropolis was related to the number of graves and fragments analyzed. Even if no quantitative interpretation of the vegetation can be performed, qualitative analysis may reveal the woody plants that appeared in the proximity of the site. Furthermore, three different vegetation communities may be observed as catchment's area: pine forest developed on sandy soils, broadleaf or mixed deciduous forest dominated by oaks and riverine forest.*

**Key words**: *charcoal analysis, wood exploitation, palaeoenvironment, Bronze Age necropolis, Poland*

The necropolis of Kokotów is situated in the northern part of the Kokotów village, south-east of Kraków (Poland). It is located on sandy soils, in the proximity of the valley of the Vistula river (Figure 1). This multicultural site was discovered in 2004 and was excavated as Kokotów-site 19 during the period of 2005-2007, as part of the planned construction of motorway A4, by the Kraków Archaeological Team for Excavation of the Motorways (Institute of Archaeology and Ethnology of Polish Academy of Science, Jagiellonian University, and Archaeological Museum in Krakow). In the excavated area of almost 2 ha, 310 archaeological structures were discovered, and most of them represent burial's remains of both incineration and inhumation type. Moreover, the 247 graves found in the site constitute one of the biggest Late Bronze Age necropolises in the Kraków region. The cemetery is dated to the last phases of the Bronze Age and first period of the Iron Age, and it corresponds to a Lusatian culture that belongs to the Urn Field complex (Gedl 1975).

In the anthracological analysis the wood charcoal fragments come from 112 cremation graves, in which 40 belong to urn graves and 72 were found in pit structures (Figure 2). In comparison with other necropolises from the same period, the Kokotów cemetery demonstrated some peculiarities such as the presence of a rather big and oval pit situated on the N-S axis, in the bottom of which mainly burnt bones were located. The bones were found in pottery urns or deposited directly in the soil; nonetheless, an observation of the bone's concentrations only in one part of the pit may suggest a possible use of another kind of containers made of an organic matter that

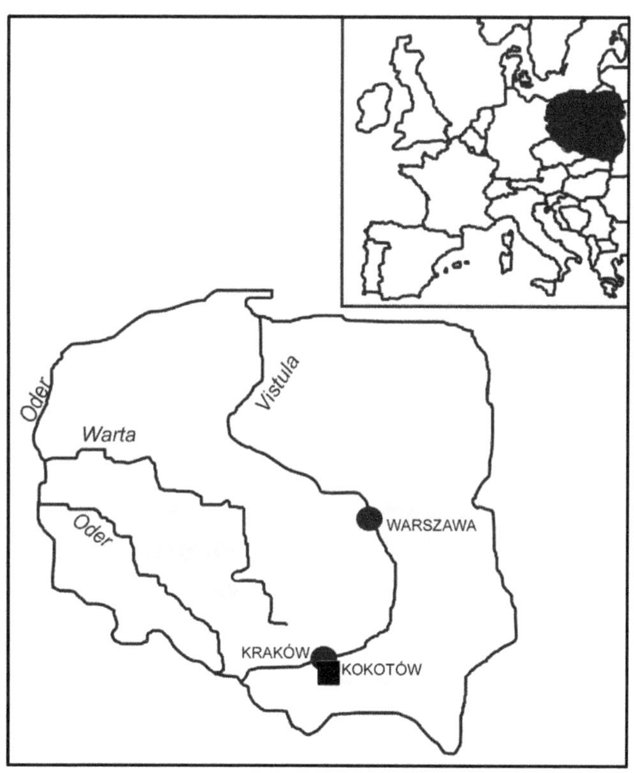


Figure 1. Location of the Kokotów necropolis

has not been preserved in the sandy soil. As a consequence, the present division of the grave types into urn graves and pit graves is rather artificial and should be taken with some caution. In the proximity of the bones, some pottery and metal offerings were also placed. The

| **Kokotow** | **Urn graves (UG; 40)** | | | | | | | | **Pit graves (PG; 72)** | | | | **All graves** | |
|---|---|---|---|---|---|---|---|---|---|---|---|---|---|---|
| **Grave structures (Number)** | **Urn (U) + pit (P) (16)** | | | | **Urn (U) or pit (P) (24)** | | | | **1 sample (43)** | | **Various samples (29)** | | | |
| **Taxa** | **U** | **%** | **P** | **%** | **U** | **%** | **P** | **%** | **P** | **%** | **P** | **%** | **Σ** | **%** |
| *Abies alba* | – | – | 1 | 0,1 | – | – | 1 | 0,3 | – | – | 1 | 0,1 | **3** | **0,1** |
| *Calluna vulgaris* | 14 | 5,1 | 90 | 11,1 | 22 | 12,6 | 29 | 8,9 | 35 | 3,2 | 132 | 6,8 | **322** | **6,9** |
| *Carpinus betulus* | – | – | – | – | – | – | – | – | 66 | 6 | 17 | 0,9 | **83** | **1,8** |
| *Corylus avellana* | 2 | 0,7 | 1 | 0,1 | – | – | – | – | – | – | – | – | **3** | **0,1** |
| *Fagus sylvatica* | – | – | 1 | 0,1 | – | – | – | – | – | – | – | – | **1** | **0,02** |
| *Fraxinus excelsior* | – | – | 1 | 0,1 | – | – | – | – | – | – | – | – | **1** | **0,02** |
| *Pinus sylvestris* | 39 | 14,2 | 193 | 23,8 | 51 | 29,3 | 99 | 30,5 | 254 | 23 | 385 | 19,7 | **1021** | **22** |
| *Alnus* sp. | 8 | 2,9 | 34 | 4,2 | – | – | 34 | 10,5 | 7 | 0,6 | 6 | 0,3 | **89** | **1,9** |
| *Betula* sp. | 95 | 34,7 | 198 | 24,4 | 46 | 26,4 | 47 | 14,5 | 38 | 3,4 | 117 | 6 | **541** | **11,7** |
| *Populus* sp. | – | – | – | – | – | – | – | – | 6 | 0,5 | – | – | **6** | **0,1** |
| *Prunus sp.* | 2 | 0,7 | – | – | 3 | 1,7 | – | – | – | – | 1 | 0,1 | **6** | **0,1** |
| *Quercus* sp. decid. | 107 | 39,1 | 252 | 31,1 | 48 | 27,6 | 101 | 31,1 | 654 | 59,2 | 1221 | 62,6 | **2383** | **51,4** |
| *Salix* sp. | 4 | 1,5 | – | – | – | – | – | – | – | – | 3 | 0,2 | **7** | **0,2** |
| *Salix* sp./*Populus* sp. | – | – | – | – | – | – | – | – | 8 | 0,7 | 1 | 0,1 | **9** | **0,2** |
| *Ulmus* sp. | – | – | – | – | – | – | 1 | 0,3 | – | – | – | – | **1** | **0,02** |
| Betulacea | – | – | 8 | 1 | – | – | – | – | – | – | – | – | **8** | **0,2** |
| Betulacea/Corylaceae | – | – | – | – | – | – | – | – | – | – | 1 | 0,1 | **1** | **0,02** |
| Corylaceae | – | – | – | – | – | – | – | – | – | – | 1 | 0,1 | **1** | **0,02** |
| Maloideae | – | – | – | – | – | – | – | – | – | – | 3 | 0,2 | **3** | **0,1** |
| Gymnosperms | – | – | 5 | 0,6 | – | – | – | – | 8 | 0,7 | 7 | 0,4 | **20** | **0,4** |
| Gymnosperms: Cones | – | – | 5 | 0,6 | 2 | 1,1 | 4 | 1,2 | 9 | 0,8 | 21 | 1,1 | **41** | **0,9** |
| Angiosperms | – | – | 7 | 0,9 | 2 | 1,1 | 7 | 2,2 | 8 | 0,7 | 19 | 1 | **43** | **0,9** |
| Indeterminable | 1 | 0,4 | 9 | 1,1 | – | – | – | – | 11 | 1 | 14 | 0,7 | **35** | **0,8** |
| Indeterminable: bark | 2 | 0,7 | 6 | 0,7 | – | – | – | – | – | – | 2 | 0,1 | **10** | **0,2** |
| **Analyzed charcoal (total: Σ)** | **274** | **100** | **811** | **100** | **174** | **100** | **325** | **100** | **1104** | **100** | **1952** | **100** | **4640** | **100** |
| **Minimum number of taxa** | **8** | | **9** | | **5** | | **7** | | **8** | | **10** | | **15** | |

Figure 2. Absolute and relative frequency of taxa documented in the Kokotów necropolis

bottom of the burial was usually covered by soil sediments, above which the rest of the funerary pyre was deposited. In addition, all the graves represent secondary deposits as no signs of cremation of the bodies above the pit (*bustum*) were detected. The remnants of the incineration were transported from a cremation place (*ustrinum*) and put together with uncharred offerings into the burial pit. The quantity of the pyre remains varies throughout the different graves, and therefore a selection of the material placed in the burial zone should be also emphasized. Furthermore, the exact area of the *ustrinum* has not been discovered.

The archaeological context of the charcoal assemblage is of major importance since it may show a possible way of its interpretation providing palaeoenvironmental and/or palaeoethnographic information. Usually the charcoal fragments associated directly to the archaeological structures, indicate a special taxonomic selection and thus give mainly palaeoethnographic information. In addition, this kind of structures is characterized by a low diversity of species since it is related to a short-span or one-time usage (Asouti and Austin 2005; Badal 1992; Carrión 2005; Chabal 1997; Heinz and Thièbault 1998; Chabal *et al.* 1999; Ntinou 2002; Smart and Hoffman 1988). For that reason, the anthracological study from the necropolis of Kokotów aims firstly to determine the wood used in cremation ceremonies in order to detect the pattern of wood selection for funerary rituals. The other interesting question is if the charcoal assemblage coming from archaeological structures such as graves may also serve as a palaeoecological source of information. In this sense, one previous anthracological study (Ntinou 2002), although concerning another kind of archaeological structures that represent a different region and climatic period, has served as inspiration and thus may provide a basis for this hypothesis. Ntinou (2002) in her analysis of hearths has demonstrated that even though a simple hearth gives a low diversity of taxa, the sum of many different structures may reflect a longer span of wood recovery and therefore may provide the palaeoenvironmental meaning. In conclusion, the anthracological analysis from the necropolis might expand the knowledge of the past flora reflected in the charcoal assemblage and offer an interesting approach to the understanding of funerary practises in the Lusatian culture, especially in the context of limited anthracological data from similar archaeological sites.

## METHODOLOGY

During the field work, soil samples of ten litres each were taken from the graves. Usually one soil sample was recovered from all kinds of funerary pits but in various graves more than one was gathered in order to make sure that the limited number of samples does not influence the results of the analysis. In addition, in the case of urn graves, the content of the urn constituted a separated sample. Charcoal fragments were obtained by the flotation method using meshes of 1,2 and 0,5 mm. In the laboratory, charcoal samples were identified using a reflected light microscope with magnifications of up to 500x. Each charcoal fragment was observed along three anatomical wood sections: transverse section, longitudinal tangential section and longitudinal radial section (Figure 2). Identifications were made by comparison with anatomical atlases (Greguss 1955, 1959; Jacquiot 1955; Jacquiot *et al.* 1973; Schweingruber 1982, 1990) and specimens coming from a modern reference collection available at the University of Valencia. Plants were classified following "Flora Europaea" (Tutin *et al.* 1993). Micrographs were taken using scanning electron microscopy Hitachi S-4100 with the EMIP 3.0 software.

The aim of the anthracological study was to determine the species of all charcoal fragments represented in the archaeological assemblage; nonetheless, this may confront some problems. First of all, many species are not differentiable within theirs genera on the basis of wood anatomy and it should be emphasized that all determinations to species level in the present analysis were based on the existence of only one species in the studied region (Lityńska-Zając 1997, 2005). Another difficulty was the bad preservation state of the charcoal pieces that as a consequence resulted in broader groups (gymnosperms, angiosperms or indeterminable fragments). A curl wood showed very irregular anatomical structure and together with bark fragments they constituted the indeterminable pieces (Schweingruber 1982). Another kind of observed alterations was due to damage caused by insects and microorganisms, nonetheless, in that case more precise botanical identifications were still possible.

Each charcoal fragment, independently of its size, is an observation unit (Badal 1992; Chabal 1997; Chabal *et al.* 1999). The quantification method was based on a counting of all charcoal fragments since the existence of a statistic law explains that all species show the same tendency of fragmentation (Chabal 1997). Besides, some experiments have indicated that the results of the fragments' count and weight measurements are highly correlated (Chabal 1997). Also, in order to quantify taxonomic diversity found in charcoal assemblages, the term "minimum number of taxa" (MNT) is used. In this sense, the taxa identified in higher taxonomic classification (e.g. angiosperm, Corylaceae) are not counted if documented as genus or species (e.g. *Carpinus*, *Corylus avellana*) in the same assemblage.

## RESULTS

The absolute number of charcoal fragments together with the relative frequency of all structures is presented in the Figure 2. 15 taxa that represent the MNT were found within more than 4600 charcoal fragments. Among the gymnosperms two taxa were determined to the species level: *Pinus sylvestris* and *Abies alba*, whereas among the angiosperms five species (*Calluna vulgaris*, *Carpinus betulus*, *Corylus avellana*, *Fagus sylvatica* and *Fraxinus excelsior*), seven genera (*Alnus* sp., *Betula* sp., *Quercus* sp. deciduous, *Populus* sp., *Prunus* sp., *Salix* sp. and *Ulmus* sp.) and one subfamily Maloideae were found.

Four taxa dominated in the charcoal assemblage and together constituted 92% of all examined fragments: oak (*Quercus* sp. deciduous), common pine (*Pinus sylvestris*), birch (*Betula* sp.) and heather (*Calluna vulgaris*) (Figure 2 and 4). The most abundant taxon was oak which reached 51% of all identified charcoal pieces. Nevertheless, taking into account the risk of an over-representation of taxa in the structures just once used, the presence and ubiquity of them in the graves should be taken into consideration. However, in Kokotów necropolis case, the absolute number of charcoal fragments may be highly correlated with the analysis of ubiquity since all above mentioned taxa were the most widespread in the site: oak appeared in 94 of 112 graves (84%), pine in 98 structures (87%), birch in 58 burials (52%), and heather in 64 graves (57%). The rest of the taxa were mainly represented by single specimens that were found sporadically in 1 to 3 graves, with the exception of alder (*Alnus* sp.) and hornbeam (*Carpinus betulus*) which were more frequent: appearing in 12 and 6 structures respectively (Figure 3).

Among 40 urn graves, 16 graves included the analysis of charcoal fragments from both the content of the urn and from the surrounding pit (Figure 2; URN GRAVES: URN + PIT). The number of charcoal fragments from urns is scarcer than the one found in the pits. This suggests that burnt bones were especially selected for the burial deposition inside the urn. Moreover, this may also be confirmed by the occurrence of urns for which no charcoal fragments were found. In general, the MNT in both categories is similar; 8 species were present in the urns while 9 were in the pits. Altogether they provided 10 different species (Figure 2). However, in 12 single graves

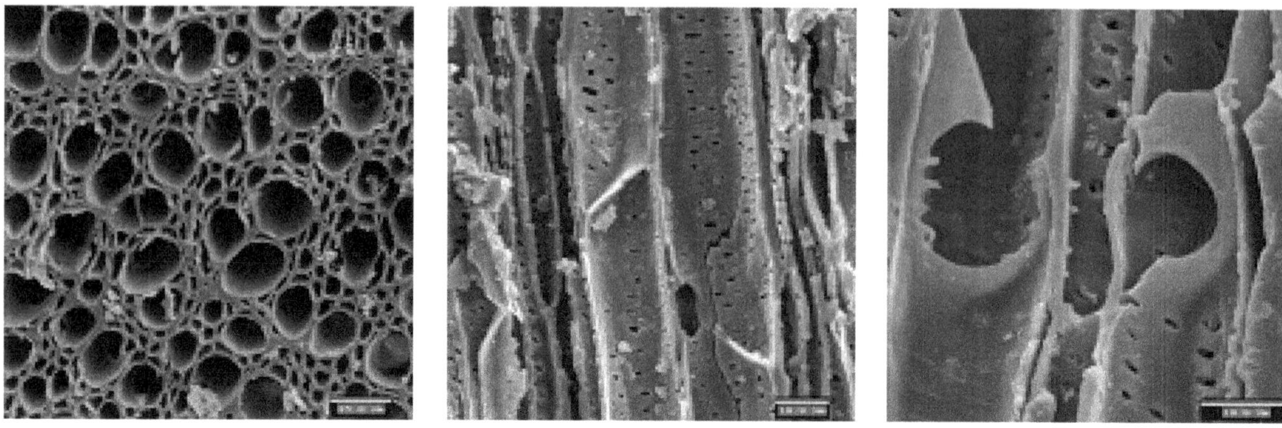

Figure 3. Relative frequency of taxa from Kokotów necropolis

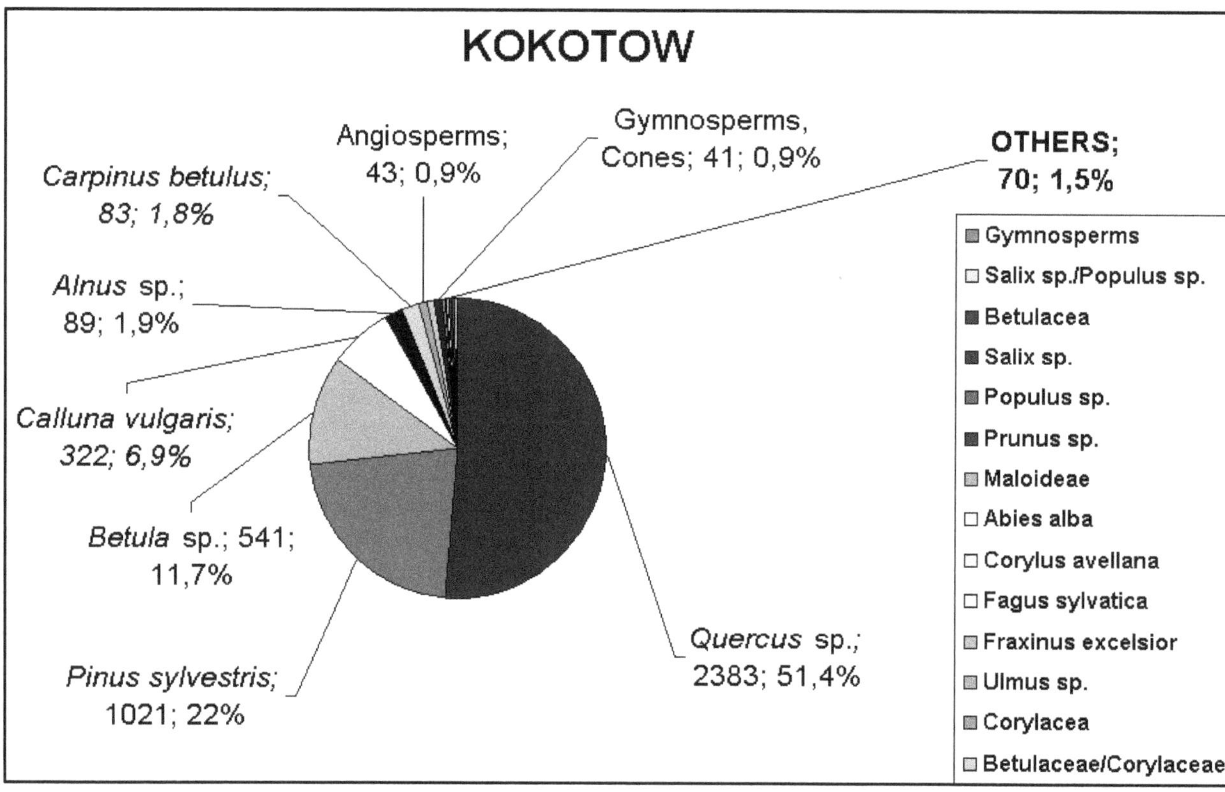


Figure 4. *Calluna vulgaris* in three anatomical sections: transversal, longitudinal tangential, and longitudinal radial. SEM micrographs: M. Moskal-del Hoyo

the MNT was higher in the adjacent pit than in the urn, and in the majority of them (75%) the MNT was between 3 and 4 with some of them reaching 5 and even 7 species, whereas the urn were characterized by a lower MNT (Figure 5; UG – URN GRAVES). It is probable that one of the factors that may influence these results is the aforementioned lower absolute number of charcoal fragments. However, there were other cases where the absolute frequency was not related directly to the diversity of taxa since in some graves the MNT is the same in both the pit with more charcoal fragments and the urn content with very limited frequency. The MNT can also be higher in the content of the urn with lower amount of charcoal fragments than in the surrounding pit exhibiting higher number of fragments.

The discrepancies between the urn and pit content may also be due to different ways of recovering the samples as charcoal fragments from the pit come from bigger samples, and thus cover a larger area of the structure. The absolute frequency of charcoal fragments is also associated with their concentration rate per sample, which in the Kokotów necropolis was generally low. The number of charcoal fragments was more elevated in the graves with more soil samples taken. Therefore, it should be expected that in graves with larger area sampled and higher charcoal abundance, the taxonomic diversity would be higher. The results nevertheless do not show this relationship undoubtedly as for graves with more than 100 charcoal pieces analyzed, coming from various samples, the MNT was similar as for others where only

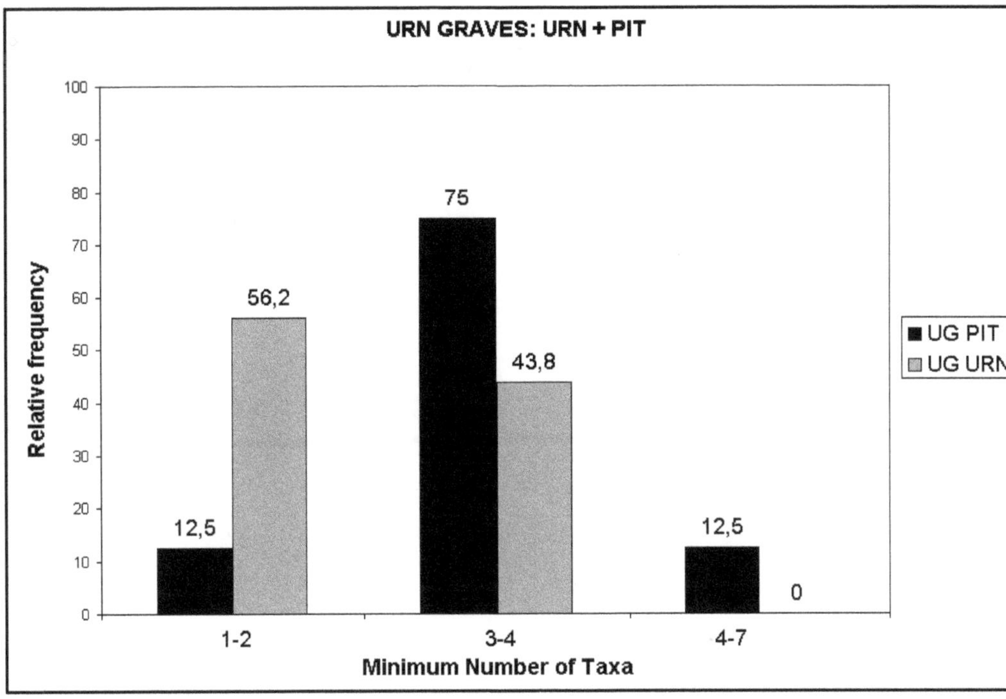


Figure 5. Relative frequency of the minimum number of taxa in urn graves.
UG P – Urn graves, pit; UG U – urn graves, urn content

one soil sample was taken and a lower quantity of charcoal fragments were recovered.

The results obtained for 24 urn graves (Figure 2; URN GRAVES: URN or PIT) that had only one area sampled – either urn content or surrounding pit – show, after comparing with the results coming from both areas sampled, lower absolute frequency of charcoal fragments as well as lower taxonomic diversity. Both results seem to be in agreement.

Anthracological results from 72 pit graves were divided into 2 groups concerning the number of soil samples taken (Figure 2; PIT GRAVES: 1 SAMPLE, VARIOUS SAMPLE). In the first group which contains 43 graves with only one sample, the MNT for 1104 charcoal fragments analyzed was 8. In detail, the concentration of charcoal fragments per sample was not very high and in the majority of the graves the number of charcoal pieces did not exceed 40. In this case, the diversity of taxa may be connected with the absolute frequency as only in the graves with 40 or more charcoal pieces the MNT. was greater or equal to four species. Moreover, in more than a half of the graves (58%) there were only 1 to 2 different species (Figure 6; PG 1S/43: PIT GRAVES with 1 sample).

On the other hand, in the graves with different sampled areas the absolute frequency (1952) and the MNT is the highest (10). Also the same pattern in the charcoal concentration can be indicated as in some of the graves, even the ones with two or three samples, the number of charcoal pieces was smaller than 40. In the majority of graves (62%), the MNT is between 3 and 4 (Figure 6; PG VS/29: PIT GRAVES with various samples). In addition, graves with higher number of charcoal fragments, which is between 120 and 230, exhibited the MNT values which were between 5 and 6. This may as well confirm the correlation between the absolute frequency and the diversity of taxa, however, it should be said that in these structures the taxonomic curves were stabilized after analyzing between 20 and 60 charcoal fragments coming from 1 to 3 samples.

Furthermore, in 2 pit graves, an additional rectangular wooden construction that surrounded the pit was found. The remnants of the construction were also burned and presented only 1 taxon: *Quercus* sp. deciduous. In both cases, the wood belonged to branches that in the preserved fragments demonstrated a minimum number of annual rings of 22 and 36 and radii of about 25-30 mm.

Another matter worth to be mentioned is the presence of different parts of the tree in the remnants of funerary pyres. The majority of charcoal fragments included trunk and many of them were small enough to make the observation of the ring curvature possible and thus detect the plant's part from which it came from. Nonetheless, in many cases the fragments showed the characteristics of the branch wood such as ring curvature (Figure 7A) or presence of the reaction wood that is especially visible in the gymnosperms as a compression wood (Figure 7B) (Schweingruber *et al.* 2006). There were 78 fragments (1,7%), except of the heather's remnants (Figure 7C), mainly of *Pinus sylvestris* that clearly came from small branches (Figure 7A) or even very young shoots (Figure 7D). This may suggest the use of branches and leaves for funerary rituals. Perhaps the occurrence of fragments of gymnosperms' cones (scales) may also confirm this

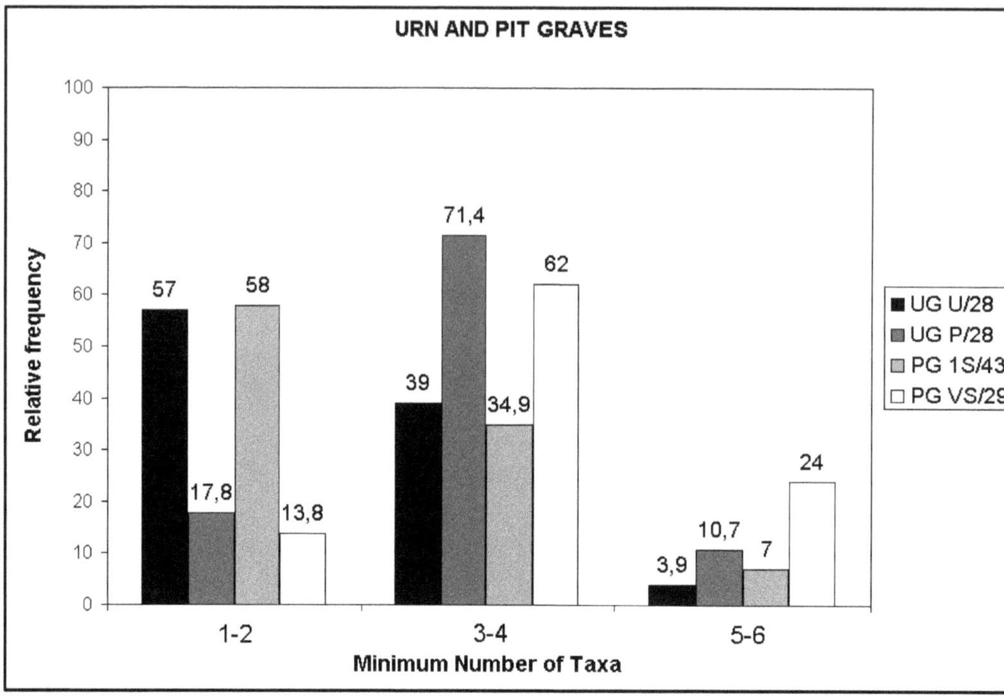


Figure 6. Relative frequency of the minimum number of taxa in all graves. UG U/28 – 28 urn graves, urn content; UG U/28 – 28 urn graves, pit content; PG 1S/43 – 43 pit graves with one sample; PG VS/29 – 29 pit graves with various samples

practice (Figure 7E). Also, other parts of trees such as bark were represented in the charcoal assemblage. In addition, different actions of organisms that attacked the wood before being burned in the pyre were detected in some fragments. Most of them correspond to microorganisms such as fungi noticed mainly in the form of hyphae or they may be related to invertebrates such as the nematodes (Figure 7F).

In summary, in all graves 15 taxa were found and the taxonomic richness in the anthracological analysis of wood charcoal from the necropolis was related to the number of graves and fragments analyzed. The taxonomic curve (Figure 8) demonstrates that the most common taxa such as pine, oak, birch and heather appeared in first two graves examined and after analyzing 50 structures 10 species were found, while in order to detect less frequent taxa it was necessary to analyze more graves.

## DISCUSSION

In the necropolis of Kokotów, the predominance of oak was clearly noticed in the absolute frequency as well as in the ubiquity analysis (Figure 2 and 4). It is followed in these both categories by common pine, birch, and heather. The high proportion of these taxa in the charcoal assemblage may indicate theirs special selection for rituals of funerary cremation. On the other hand, their dominance may also be related to their real abundance nearby the site and thus may suggest a different possibility of interpretation in which the selection could have derived from simple availability.

In present-day vegetation, oak species together with *Pinus sylvestris* constitute the major trees in the proximity of the site and their distribution may be associated with soil requirements. Pines prevailed on sandy soil where the necropolis is situated, whereas in the closest area with more fertile soil the broadleaf forest dominated by oaks occurs. Three species of oaks are considered native in Poland: *Q. robur* (pedunculate oak), *Q. petraea* (sessile oak) and *Q. pubescens* (pubescent oak), nevertheless the latter is rather restricted to the area of xerothermic communities. The two possible oak species may occur in many different forest communities from deciduous and mixed coniferous to riverine. They appear on diverse types of soils but *Q. robur* is most widespread on richer and moisture ones than *Q. petraea.* Besides, oaks form the woodland with quite broken canopy that permits the development of many kinds of trees in the undergrowth, nonetheless almost without any shrubs (Ellenberg 1988; Milecka *et al.* 2004; Seneta and Dolatowski 2004). *Pinus sylvestris* is one of the less demanding species regarding climatic conditions and is very well adapted to a climatic continentality. This explains its real ubiquity in many different woodland communities. However due to the competition it usually loses the territory that is more suitable for growth's conditions. Common pine appears mainly on acid soils and especially on sandy ones (Latałowa *et al.* 2004; Polunin and Walters 1985). Birch may also be present in both vegetation types (deciduous and mixed coniferous) as this genus has very wide habitat requirements (Seneta and Dolatowski 2004).

Nearby the site, sandy soils form a possible habitat for *Calluna vulgaris* and this species may constitute part of

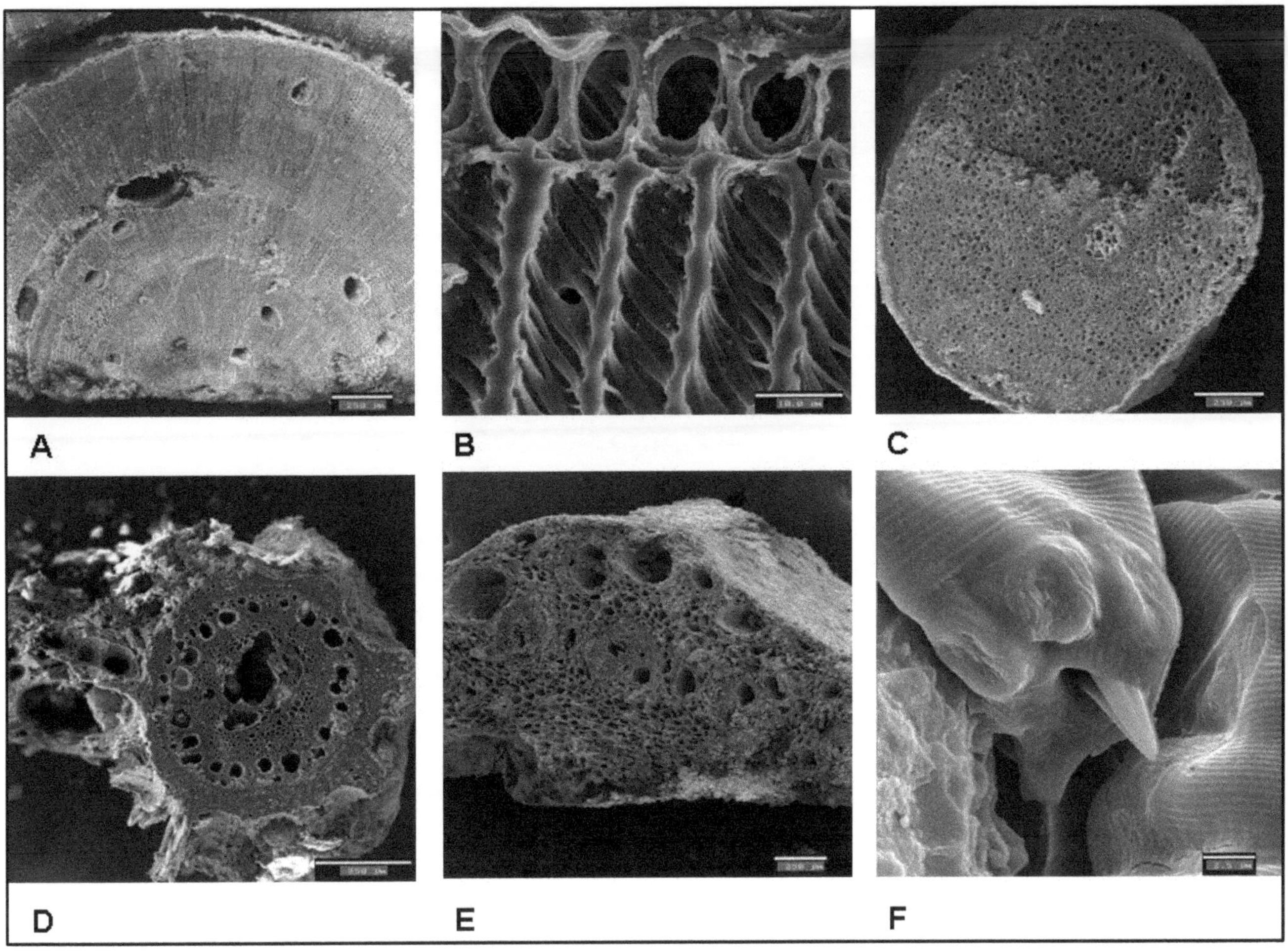


Figure 7. Charcoal fragments from the Kokotów necropolis. Branch wood: A. and B. *Pinus sylvestris* (bar: 250 µm and 10 µm); C. *Calluna vulgaris* (bar: 250 µm); D. Gymnosperm: young shoot (bar: 250 µm); E. Gymnosperm: cone (scale) (bar: 250 µm); F. Wood attacked of *Pinus* sylvestris by the nematodes (bar: 5 µm). SEM micrographs: M. Moskal-del Hoyo

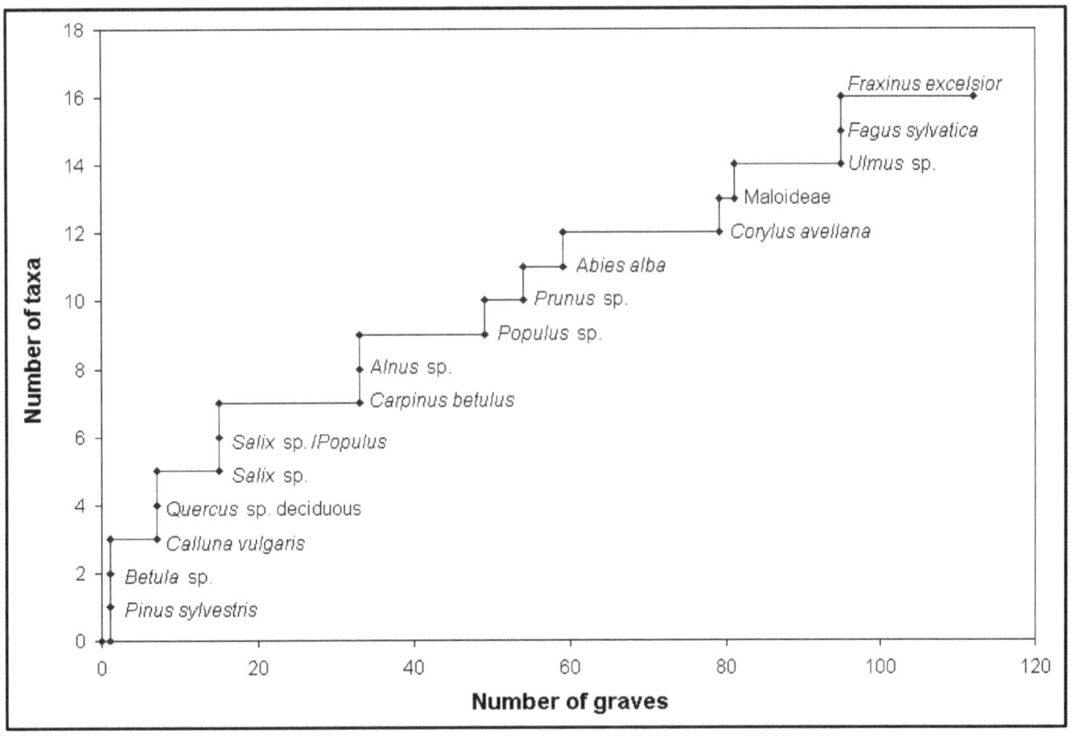


Figure 8. Taxonomic curve from the necropolis of Kokotów

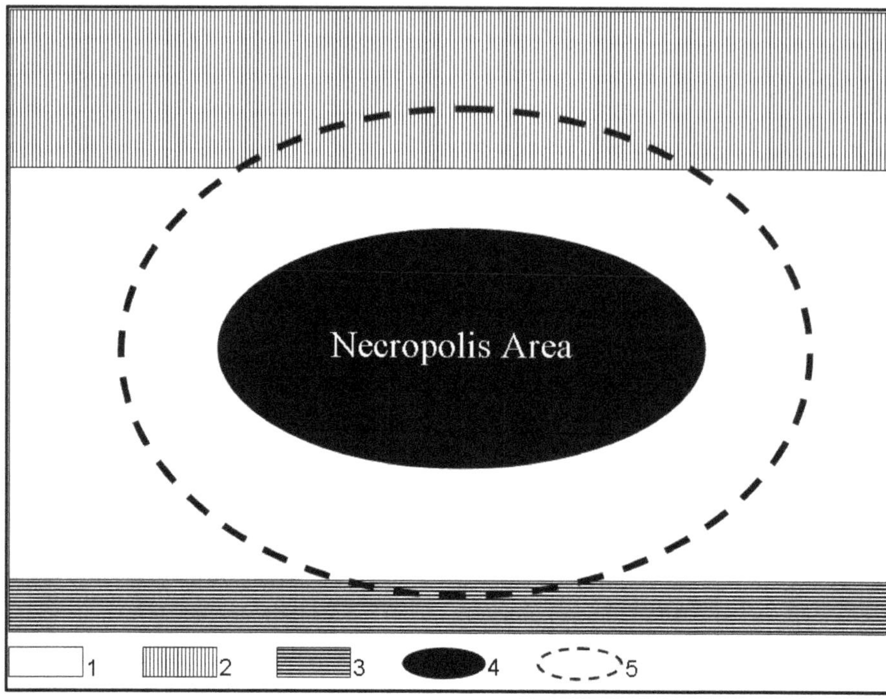


Figure 9. Different forest communities as a wood capture area for the funerary pyres.
1. Pine wood on sand soil, 2. Oak forest, 3. Riverine forest, 4. Necropolis area, 5. Wood capture area

the field layer in the mixed coniferous wood (Polunin and Walters 1988; Seneta and Dolatowski 2004; Summers *et al.* 1999). This taxon is very unusual in all anthracological records from Poland and its presence in such a significant quantity and ubiquity in the necropolis is really uncommon. Its special selection for the funerary pyre might have many different possible interpretations in which the selection for ignition purposes, its odour or its decorative aspect could be included. Conversely, it could have also covered the ground on which the *ustrinum* might have been located as the fire may favour the growing of the heather (Whittaker and Gimingham 1962). This may imply its more natural appearance in the charcoal assemblage, nevertheless it is only one of many possible explanations. Furthermore, some scarcer remains of heather were also found in the funerary pyres in other necropolis from the Iron Age from Poland (Korytnica) that are as well situated on sands (Moskal-del Hoyo, unpublished data).

Moreover, the presence of other 11 taxa of the woody flora (Figure 2) in the necropolis may also suggest that the species selection was based on availability. The taxa such as *Carpinus betulus*, *Fagus sylvatica* together with *Corylus avellana* and different species of Rosaceae family in the shrub layer could have also occurred in aforementioned oakwoods, whereas *Abies alba* could have formed part of pinewood. Besides, some of them such as *Alnus* sp., *Populus* sp., *Salix* sp., *Ulmus* sp. or *Fraxinus excelsior* might have formed part of the riverine forest. Perhaps theirs low frequency in the charcoal assemblages is related directly to a longer distance of those communities to the necropolis area (Figure 9). In summary, three different plant communities may be indicated nearby the site: deciduous forest dominated by oaks, mixed coniferous one with the predominance of common pines and the riparian woodland.

The MNT varies in diverse grave types showing differences that may result from the absolute frequency among other factors as generally the structures with more charcoal abundance present more MNT. This correlation is noticed above all while comparing the results from the urn graves for which the contents of the urns (UG U) with rather limited number of charcoal fragments has showed lower taxonomic diversity (Figure 6; UG U/28) than presented in the pits coming from all graves structures (Figure 6): pits related to 28 urn graves (UG P/28), 43 pits from the pit graves with only 1 sample taken (PG 1S/43), and 29 pits with different burial zones samples (PG VS/29). The study of the pyre remains from the pits demonstrated higher taxa diversity and in this group the influence of the amount of samples from the graves on the MNT was showed. Therefore, it is recommended in methodology to gather various samples from both areas such as pit and urn's content of the urn graves.

The presence of different parts of trees may indicate the use of some trunks together with the branches, and the latter were also possibly placed into the pyre with the leaves as some young shoots were found. The appearance in the charcoal assemblage of such tiny parts of plants may be related to their particular use such as ignition of the fire or to produce a special smoke and aroma. In addition, all the charcoal pieces that belonged to *Calluna vulgaris* were of very small dimensions as the radii usually did not exceed 2 mm. Moreover, fragments of gymnosperms cones were also found and they may indicate the use of branches with leaves or a special usage due to theirs odour while burning. The occurrence of fungi infected charcoal pieces may indicate furthermore the use of some deadwood gathered for funeral purposes (Théry-Parisot 2001), especially when it comes from the branch wood.

It is difficult to demonstrate the pattern of wood recovery for funerary pyres on the basis of anthracological analysis, however, the presence of the diversity of taxa along with different parts of trees may indicate an opportunistic behaviour. This may be due to the necessity of great amount of fuel wood as modern ethnographic observations show (Sikandar *et al.* 1985). Therefore, we may suggest that all kind of wood was possibly utilized as stored wood, freshly cut trees, branches, young shoots and dead wood recovered from forest litter.

The study of the material culture together with the anthropological analysis from the necropolis of Kokotów is still in progress. This explains why some aspects of the archaeological context such as differences between the graves resulting from a relative chronology or people's age and sex determinations have not been included in the interpretations of the anthracological analysis.

In Poland, when comparing the great amount of archaeological excavations in necropolises, there are relatively very few anthracological analyses performed, and their insufficiency is observed not only in Bronze Age sites but also in the structures representing other periods. However, some wood charcoal remains from the cemetery of Opatów (Kłobuck) that belongs also to the Lusatian culture demonstrate different taxa utilized as a fuel the funerary pyres in which the most frequent were oak and common pine (Lityńska-Zając 2004). The results from the necropolis of Pawłowice (Michałow) of the same culture show an analogous pattern (Moskal-del Hoyo, unpublished). The lack of studies on this subject does not permit to make any general conclusions about the uses of the wood for funerary purposes in the Lusatian culture.

## CONCLUSION

The study of the charcoal assemblage from the necropolis of Kokotów has provided new data related to the usage of woody plants for the funerary rituals of cremation. The results concerning their different significance together with methodological problems may be resumed in the following statements:

- Different wood species were utilized for cremation rituals as 15 taxa were documented in the prehistoric flora. The most common and ubiquitous were oak, common pine, birch, and heather.

- Perhaps the wood selection corresponds to availability of different species in the site's vicinity.

- No main taxonomic differences were observed between urn graves and pit graves as in both categories dominated the same taxa and the minimum number of taxa was between 11 and 12.

- In the majority of the graves 3 or 4 taxa were present, nevertheless in some of them only 1 taxon was found. The maximum number of taxa per grave was 7.

- The increasing species diversity in the necropolis is related to the number of graves analyzed.

- In the case of urn graves, the content of the urn demonstrates the lowest minimum number of taxa that may be a result of the selection of the remains deposited in the urns and thus may originate from lower absolute frequency.

- The wood used for funerary rituals represents different parts of the trees such as trunks, branches, bark fragments, or cones. In addition, the use of a deadwood may be indicated.

- There is a possibility of palaeoecological significance of the charcoal assemblages coming from funerary pyres' remnants based on different species listed, but no quantitative interpretations of forest composition can be done.

- Three diverse communities of vegetation may be traced as catchment's area.

### Acknowledgements

The authors would like to thank to Dr Ernestina Badal from the University of Valencia for her immense help and to the SEM Laboratory (University of Valencia) for providing access and technical support with the scanning electron microscope. This work was possible due to financial support of "Cinc Segles" grant program from University of Valencia. We are also grateful to the Kraków Archaeological Team for Excavation of the Motorways for their permission to present the results from the anthracological analysis from Kokotów-site 19.

### References

ASOUTI, E. and AUSTIN, P. 2005. Reconstructing Woodland Vegetation and its Exploitation by Past Societies, base on the Analysis and Interpretation of Archaeological Wood Charcoal Macro-Remains. *Environmental Archaeology* 10, 1-18.

BADAL GARCÍA, E. 1992. L'anthracologie préhistorique: à propos de certains problèmes méthodologiques. *Bulletin de la Société botanique de France, Actualités Botaniques* 139 (2/3/4), 167-189.

CARRIÓN MARCO, Y. 2005. *La vegetación mediterránea y atlántica de la Península Ibérica. Nuevas secuencias atracológicas.* Valencia, Diputación Provincial de Valencia.

CHABAL, L. 1997. *Forêts et sociétés en Languedos (Néolithic final, Antiquité tardive). L'athracologie, méthode et paléoécologie.* Paris, Maison des Sciences de l'Homme.

CHABAL, L., FABRE, L, TERRAL, J.-F. and THÉRY-PARISOT, I. 1999. L'Anthracologie. In A. Ferdière, Ch. Bourquin-Mignot, J.-E. Brochier, L. Chabal, S. Crozat, L. Fabre, J.-F. Terral and I. Théry-Parisot (eds.), *La Botanique*, 43-104. Paris, Errance.

GEDL, M. 1975. Kultura łużycka, Kraków, Uniwersytet Jagielloński.

GREGUSS, P. 1955. *Xylotomische Bestimmung der heute lebenden Gymnospermen*. Budapest, Académiai Kiadó.

GREGUSS, P. 1959. *Holtzanatomie der europaischen Laubhölzer und Straücher*. Budapest, Académiai Kiadó.

ELLENBERG, H. 1988. *Vegetation ecology of Central Europe*. Cambridge, Cambridge University Press.

HEINZ, C. and THIÈBAULT, S. 1998. Characterization and palaeoecological significance of archaeological assemblages during Late and Post-Glacial phases in Southern France. *Quaternary Research* 50, 56-68.

JACQUIOT, C. 1955. *Atlas d'Anatomie des bois des Conifères*. Paris, Centre techniques de bois.

JACQUIOT, C., TRENARD, Y. and DIROL, D. 1973. *Atlas d'Anatomie des bois des Angiospermes*. Paris, Centre techniques de bois.

LITYŃSKA-ZAJĄC, M. 1997. *Roślinność i gospodarka rolna w okresie rzymskim. Studium Archeobotaniczne, Vegetation and agriculture in the Roman Iron Age. An archaeobotanical study*. Kraków, Instytut Archeologii i Etnologii PAN.

LITYŃSKA-ZAJĄC, M. 2004. Drewno i węgle drzewne z cmentarzyska kultury łużyckiej w Opatowie, pow. Kłobuck, woj. Śląskie. *Sprawozdania Archeologiczne* 56, 459-464.

LITYŃSKA-ZAJĄC, M. 2005 Rekonstrukcja zbiorowisk roślinnych. In M. Lityńska-Zając and K. Wasylikowa (eds.), *Przewodnik do badań archeobotanicznych*, 437-469. Poznań, Sorus.

MILECKA, K., KUPRYJANOWICZ, M., MAKOHONIENKO, M., OKUNIEWSKA-NOWACZYK, I. and NALAPKA, D. 2004. *Quercus* L. – Oak. In M. Ralska-Jasiewiczowa, M. Latałowa, K. Wasylikowa, K. Tobolski, E. Madeyska, H.E. Wright Jr. and Ch. Turner (eds.), *Late Glacial and Holocene history of vegetation in Poland based on isopollen maps*, 189-197. Kraków, W. Szafer Institute of Botany.

NTINOU, M. 2002. *La Paleovegetación en el Norte de Grecia desde el Tardiglaciar hasta el Atlántico. Formaciones Vegetales, Recursos y Usos*. Oxford, BAR Publishing.

POLUNIN, O. and WALTERS, M. 1985. *A guide to the Vegetation of Britain and Europe*. Oxford, Oxford University Press.

SCHWEINGRUBER, F.H. 1982. *Mikroskopische Holzanatomie*. Birmensdorf, Swiss Federal Institute for Forest, Snow and Landscape Research.

SCHWEINGRUBER, F.H. 1982. 1990. *Anatomie Europäischer Hölzer*. Bern-Stuttgart, Paul Haupt Berne und Stuttgart Publishers.

SCHWEINGRUBER, F.H., BÖRNER, A. and SCHULTZE, E.-D. 2006. *Atlas of Woody Plants Stems. Evolution, Structure, and Environmental Modifications*. Berlin, Springer.

SENETA, W. and DOLATOWSKI, J. 2004. *Dendrologia*. Warszawa, Wydawnictwo Naukowe PWN.

SIKANDAR, M., TRIPATHI, B.D. and SHUKLA, S.C. 1989. Possible ecological effects of dead body cremation on the bank of river Ganga at Varanasi. *Journal of Tropical Ecology* 30, 90-95.

SMART, T.L. and HOFFMAN, E.S. 1988. Environmental Interpretation of Archaeological Charcoal, In Ch.A. Hastorf and V.S. Popper (eds.), *Current Palaeoethnobotany. Analytical Methods and Cultural Interpretations of Archaeological Plant Remains*, 167-205. Chicago University Press, Chicago.

SUMMERS, R.W., MAVOR, R.A., MACLENNAN, A.M. and REBECCA, G.W. 1999. The structures of ancient native pinewood and other woodlands in the Highlands of Scotland. *Forest Ecology and Management* 119, 231-245.

THÉRY-PARISOT, I. 2001. *Économie du combustible au Paléolithique. Anthracologie, Expérimentation, Taphonomie*, ***Centre National de la Recherche Scientifique***, Dossier de Documentation Archéologique, Paris.

TUTIN, T.H., BURGES, N.A., CHATER, A.O., EDMONDSON, J.R., HEYWOOD. V.H., MOORE. D.M., VALENTINE. D.H., WALTERS. S.M., WEBB., D.A. 1993. Flora Europaea, Volume 1. Cambridge University Press, Cambridge.

WHITTAKER, E. and GIMINGHAM, C.H. 1962. The Effects of Fire on Regeneration of *Calluna vulgaris* (L.) Hull from Seed. *The Journal of Ecology* 50, (3), 815-822.

# CHARCOALS FROM IRON SMELTING FURNACES – FUEL SUPPLY AND ENVIRONMENT OF A MEDIEVAL IRON SMELTING SITE NEAR PEPPANGE/LUXEMBOURG

Oliver NELLE and Doris JANSEN
Institute for Ecosystem Research, Christian-Albrechts-University Kiel, Olshausenstr. 40, D-24098 Kiel, Germany
Graduate School "Human Development in Landscapes", Christian-Albrechts-University Kiel, Olshausenstr. 40, D-24098 Kiel, Germany
onelle@ecology.uni-kiel.de
djansen@ecology.uni-kiel.de

Michael OVERBECK
Department for Prehistoric Archaeology, Westfälische Wilhelms-University Münster, Robert-Koch-Str. 29, D-48149 Münster, Germany
michaeloverbeck@gmx.de

**Abstract**: *Analysis of charcoals from the medieval ironworks "Genoeserbusch" near Peppange in the south of Luxembourg gives insight into fuel supply and management of woodland. Mainly charcoal of beech and hornbeam was used. In addition, charcoal of oak, hazel, willow, poplar, maple and of the Maloideae type was found in the several sampled features. In addition to the taxonomic identification of the charcoals, measurements of the diameters and calculations of the mean growth rates were made. For the diameter measurements, the bending of the annual growth rings was fitted into a circular template which is divided into five diameter-classes. We assume that woodland stands dominated by beech and hornbeam which grew in the vicinity of the ironworks were used for charcoal production, though the actual production sites are not known yet. The results of the diameter measurements show the use of small dimensioned wood implying coppice-wood-use. An alternative hypothesis could be that the trunks were used as timber and only the branches and twigs were charcoaled. However, the growth ring analysis supports the hypothesis of the use of coppiced woodland for fuel supply of the medieval iron smelting site.*

**Key words**: *charcoal analysis; diameter measurements; growth ring analysis; dendroanthracology; coppiced woodland; medieval ironworks; bloomery furnace; archaeometallurgy*

## INTRODUCTION

The medieval period in Europe was a time of intensive iron-working. During this time the demand for iron was steadily increasing. The clearance of new land and increase in population resulted in the emergence of urban centres. Furthermore, a lot of iron was needed for the manufacturing of nearly all tools and instruments, as well as for military equipment. There were different regions which were centres of this iron industry, like for example the Märkisches Sauerland in North Rhine-Westphalia (Jockenhövel *et al.* 1998) and Lahn-Dill area in Hesse (Jockenhövel and Willms 2005). Another traditional iron landscape is the southwestern part of Luxembourg (Overbeck *et al.* 2007) related to the northern and central part of Lorraine (Leroy 1997).

The iron production needed vast quantities of charcoal, produced in pits or upright kilns. The charcoal was not only needed for fuelling different processes, it was also essential as an agent for the reduction of iron oxide in the ore. The exploitation of woodlands for charcoal production might have led to overexploitation, which might have resulted in wood shortage in some areas and transportation of charcoal from still wooded regions. To tackle these questions, the usually abundant left-over charcoal at the charcoal consumption sites has to be analysed, and attempts should be made to link these sites with the charcoal production sites via charcoal spectra. The approach is a double-tracked: first, field survey with mapping of charcoal production sites or kiln sites identifies areas of charcoal production and subsequent sampling of the left-over charcoal and its analysis enables us to reconstruct woodland species composition on a fine spatial scale (e.g. Hillebrecht 1982; Bonhote and Vernet 1988; Ludemann 1996, 2003; Davasse 2000; Ludemann and Nelle 2002; Ludemann *et al.* 2004; Nelle 2002, 2003, 2008). Second, taking samples from charcoal consumption sites when archaeological excavations are carried out at iron working sites shows the use of charcoal for different technological processes (Marziani and Citterio 1999). The direct fine spatial relationship is not given at consumption sites, since charcoal might be transported from more distant charcoal producing woodlands. However, we can assume that most charcoal came from the wider vicinity of smelting sites, especially when kiln sites are known in the surroundings.

In Luxembourg, archaeological investigations revealed a huge medieval iron smelting site. Since 2003 the Department for Prehistoric Archaeology of the Westfälische Wilhelms-University Münster is working in cooperation with the Musée National Luxembourg and the Musée rural à Peppange on the smelting site

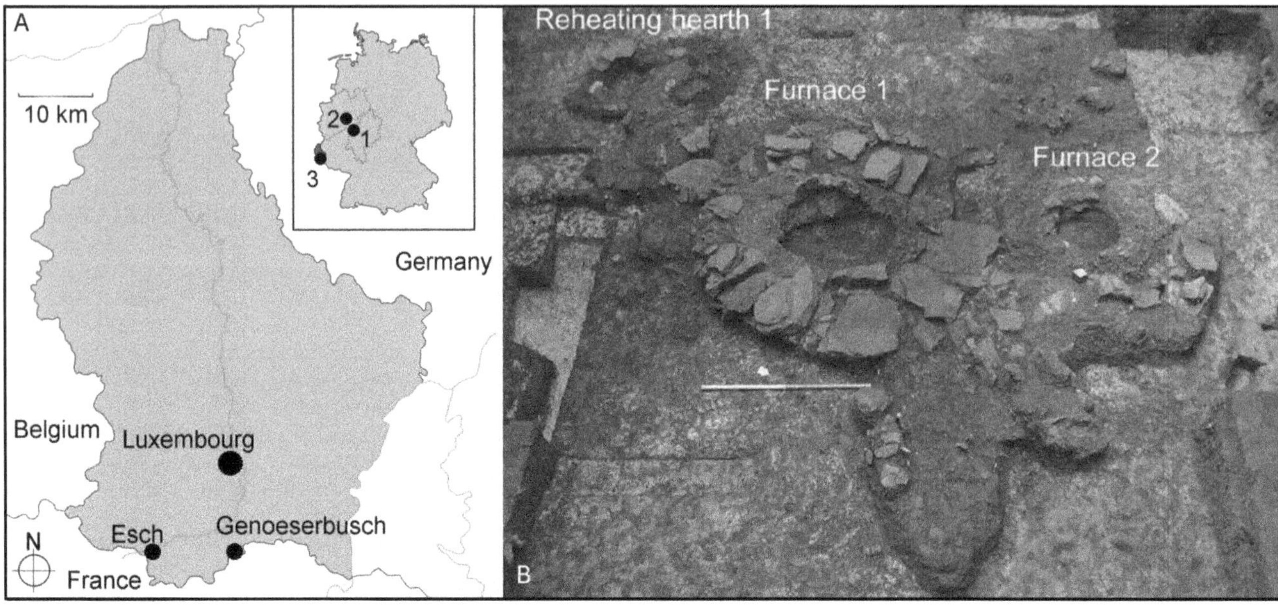


Figure 1. A Location of study site at "Genoeserbusch"/Peppange, Luxembourg (small inset: Locations of archaeometallurgical research projects mentioned in the text: 1. Dietzhölzetal (Hesse); 2. Iron and Steel Production in the Märkische Sauerland. A Process of Production from the Bloomery Furnace to the Early Osemundfinery (North Rhine-Westphalia); 3. Early iron production in Luxembourg – The medieval ironworks of "Genoeserbusch" (Luxembourg). B Furnace features of "Genoeserbusch": furnace 1, the largest furnace (outer diameter 210 cm); furnace 2, small bloomery furnace on the right side (outer diameter 120 cm) and reheating hearth 1 (outer diameter 80 cm) in the upper left corner

"Genoeserbusch" close to Peppange (Figure 1). The interdisciplinary cooperation consolidating archaeology-cal, metallurgical, botanical and geological data led to producing a significant amount of evidence that proves close connection between technological development, utilization of resources, and socioeconomic needs. Here, we present the results of the anthracological investigations, which so far concentrated on those samples which can be attributed to furnaces, roasting beds and reheating hearths.

By using a combination of wood charcoal species identification, diameter analysis and growth ring analysis, we approach the following questions:

- Which wood species and wood diameters were used?
- Did the smelters use different charcoal for different technological usages, such as roasting, smelting, reheating?
- How did the charcoal producing woodland look like? How was it managed?

## STUDY SITE, MATERIAL AND METHODS

### Medieval iron smelting furnaces near Peppange

The well kept and unusually rich ironwork from the High to Late Middle Ages of about 2,000 m$^2$ in size is situated in "Genoeserbusch", a forest near the villages of Peppange and Hellange in Luxembourg, directly on the border to France. The excavations brought to light five small bloomery furnaces (furnaces 2-6) and one much bigger furnace (furnace 1) (Figure 1, 2).

Two reheating hearths with parts of forged blooms, parts of a tuyere (nozzle) made of clay, reheating/smithing slags and hammer-scale has been found. One of them (reheating hearth 1) was almost completely conserved with an outer diameter of 80 centimetres. In addition, nineteen roasting places, two storages for wasted ore, a storage for roasted ore, four associated slag heaps of different size and several tapping channels have been located. Normally medieval ironworking sites show only a few archaeological finds with regard to pottery or iron tools. The site of "Genoeserbusch" is an exception in this respect. More than 2500 pottery shards were spread all over the medieval living floor belonging to more than 35 different vessels. The pottery gave a date range from the second half of the 13th century AD up to the start of the 14th century AD.

### Analysis of charcoal

During the excavation of the site, many hundred different charcoal samples (in total 19.9 kg of charcoal) were taken (Figure 2). The squares that were excavated were 1.25 x 1.25 metres. Charcoal was carefully collected by hand from each layer (depth of each layer: 10 cm). Charcoals pieces from 7 features separated in 16 samples were selected for analysis. For determination wood anatomical criteria like distribution and dimension of pores, resin canals, ray width and height, spiral thickenings of the pores etc. can be distinguished in different cutting sections (transversal, tangential and radial). The pieces

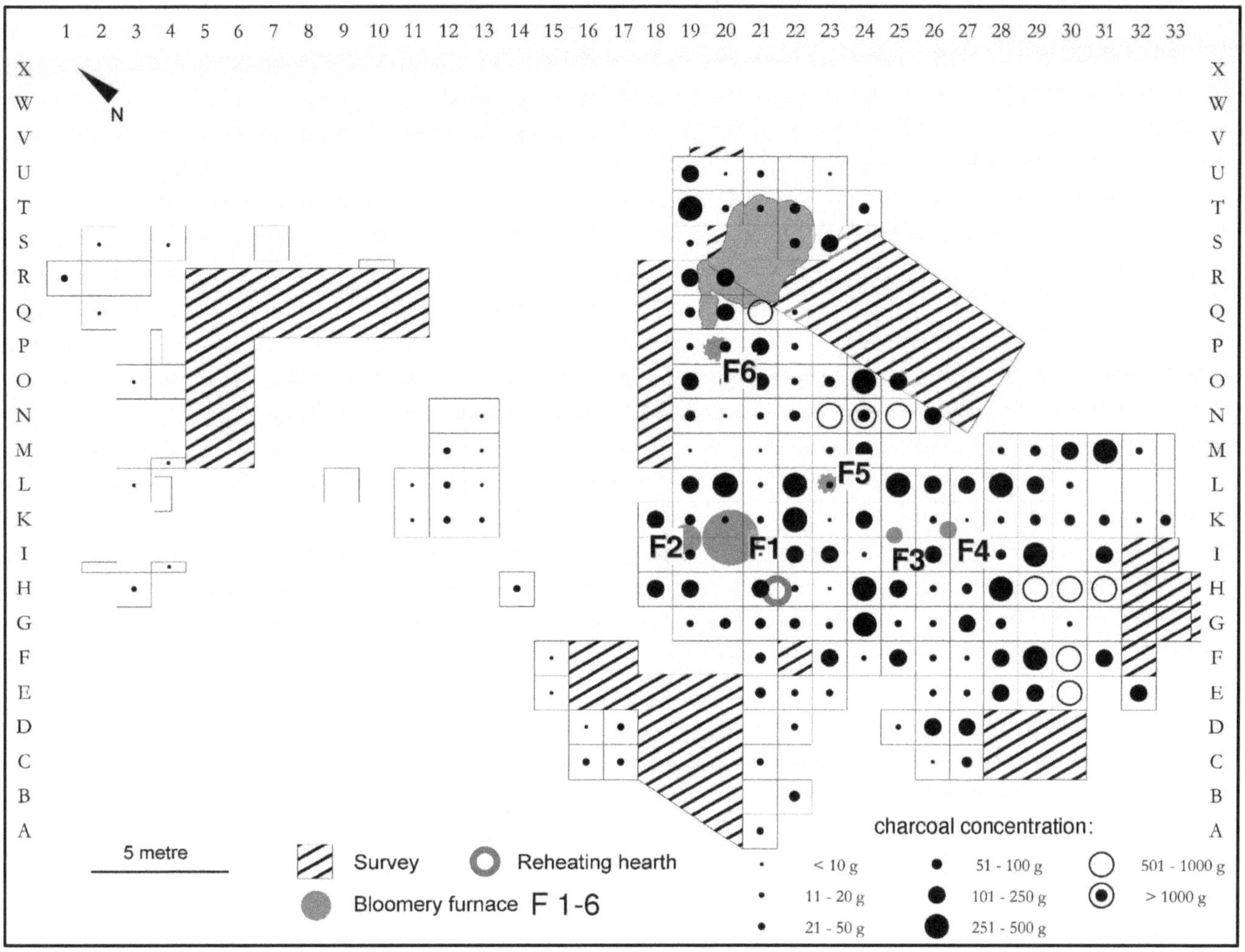


Figure 2. "Genoeserbusch". Map showing the main area of the medieval ironworks, with positions of bloomery furnaces and reheating hearths, and the distribution of charcoal per 1.25 x 1.25 m squares

were identified using the keys by Schweingruber (1990a, 1990b) and a charcoal sample collection of the Palaeoecology working group of the Institute for Ecosystem Research, University of Kiel. Charcoal fragments were weighed after identification and the overall weight per sample and species was divided by the number of fragments. This yields a "W/N" value, which represents a mean fragment size.

For estimation of the charcoals minimum diameter a circular template – divided into five diameter-classes – was used (Figure 3). On the basis of the bending of the annual growth rings and the angle of the rays it is possible to measure the minimum diameter if the charcoal pieces are big enough (Ludemann and Nelle 2002; Nelle 2002, 2003). If bark is absent (which is the case for most of the material), it is not possible to decide whether the charcoal piece derived from a branch or from the inner part of a stem. Therefore the minimum diameter is measured. During carbonization wood shrinks radially around 15-20%, but the shrinking is assumed to be proportional (Schläpfer and Brown 1948). A mean diameter (mD) was calculated for each sample using the following formula:

$$mD\ [cm] = [n(0\text{-}2\ cm)*1 + n(2\text{-}3\ cm)*2.5 + n(3\text{-}5\ cm)*4 + n(5\text{-}10\ cm)*7.5 + n(>10\ cm)*15]\ /\ n\ total$$

Mean diameter theoretically ranges between one and 15 cm. In fact, values vary usually from 3 to 13 cm. The mean diameter is an arbitrary value that does not tell us about the actual size of the wood which was used, but gives a relative value when comparing different sites. Applying this method to samples where the size of the charred wood was known, the resulting mD-values are: 4.3 cm for a campfire sample, 8.3 for a sample of charcoal produced from coppiced woods, and 10.6 for charcoal from a contemporary kiln site with timber diameters (*Fagus sylvatica*) between 10 and 30 cm (Figure 3).

Minimum age and mean growth rates were calculated for *Fagus* and *Carpinus*, applying the method of Marguerie and Hunot (2007). For this purpose the fragment width between the first and the last countable growth ring was measured and the number of growth rings was counted for each fragment having five or more growth rings. By dividing the measured width by the number of growth rings we got the mean growth rate. To compare our

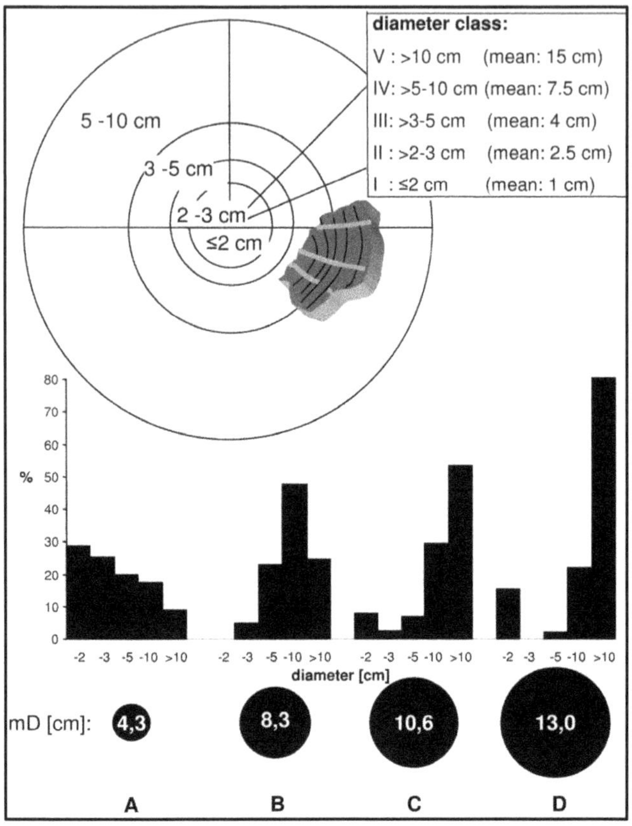


Figure 3. Diameter stencil with charcoal put in place according to growth ring bending and ray orientation, diameter classes, and reference histograms with mean diameters (mD, in cm) of selected reference samples. Charcoal samples of known wood diameter were analysed: A: camp site fire; B: charcoal production from a formerly coppiced stand; C: charcoal production *Fagus* 30-40 cm stem diameter. D: charcoal production with huge stem wood (modified from Nelle 2002)

material from "Genoeserbusch" to other samples, we analysed *Fagus* and *Carpinus* charcoals of comparable diameter class distributions and comparable fragment size (W/N) from historic and contemporary charcoal production sites of different regions (Figure 6).

## RESULTS

Sixteen samples with overall 506 charcoal fragments (399 g) were analysed. Taxonomic identification showed that *Carpinus* (*C. betulus*, hornbeam), *Fagus* (*F. sylvatica*, beech), *Quercus* (*Q. robur*, common oak or *Q. petraea*, sessile oak), *Corylus avellana* (hazel), *Salix* sp. (genus willow), *Populus* sp. (genus poplar), Maloideae (e.g. *Malus*, *Pirus, Crataegus*, *Sorbus*) and *Acer* (maple) were present (Figure 4, 5). Species spectra of the different features show all high proportions of *Fagus* and *Carpinus*. The post-hole sample contained only charcoal from *Fagus*. The mean fragment weights („W/N") range between 0.2 g and 1.9 g, with an average around 0.8 g ± 0.5 g. The biggest fragments were found in the post-hole sample, with 1.9 g. 478 of the 506 analysed charcoals were allocated to one of the five diameter classes (Figure 5). The mean diameter values are small in all samples, ranging between 3.4 cm (furnace 3) und 5.2 cm (post-hole). The distribution among diameter classes for the wood species (Figure 5) shows that oak when summed up has a slightly higher value.

Minimum age of fragments was analysed by counting the growth rings, on the Peppange material as well as material from different sites in Europe (Figure 6). For that, fragments with a minimum of 5 growth rings were used. Minimum ages range from this up to 118 years, with medians of the samples ranging from 10 to 43 years. The Peppange samples are similar to each other, and differ from most other samples because their highest minimum age is 28 years. Mean yearly increment (growth ring width) for *Fagus* and *Carpinus* is between 0.6 and 1.2 mm in the charcoal, with 0.1 mm to over 3 mm as extremes. Taking into account the 15-20%-shrinkage of the material, the resulting mean increment lies around 0.8 and 1.5 mm per year.

## DISCUSSION

### Charcoal used for iron smelting and woodland composition

At the "Genoeserbusch" iron works mainly charcoal of *Fagus* and *Carpinus* was used. In addition, oak, hazel, willow, poplar, maple and of the Maloideae subfamily was utilised. The spectra of the different features appear to be very similar. Only furnace 1 is different to the other spectra, with a nearly 50%-proportion of *Fagus* compared to 19-33% in the other features, and less *Carpinus* but more *Quercus*. Reheating hearth, roasting place and furnaces show a similar charcoal composition. This can be explained with a common provenance of the charcoal, presumably produced in the vicinity of the iron works, in charcoal kilns. However, we cannot relate the charcoal findings to certain woodlands, since at the charcoal consumption sites the charcoal might be transported from different directions, as opposed to the charcoal production sites, where the wood for charring was cut within approximately a few hundred meters around the site. The roasting place might also be worked with charcoal, coming from the same charcoal production sites. Alternatively, people used the wood offering of the vicinity, may it be to gain fire wood for the ore roasting, or may it be to produce charcoal in pits or in piles. There is no evidence for selection of certain species between the different technical features. The actual woodland composition in the vicinity of the smelting site is dominated by *Fagus*, *Carpinus* and *Quercus*. Signs of former coppicing are abundant, like multiple stemmed trees. Additionally, today there are young spruce plantations. Having not found a single charcoal of spruce shows us at least that there is no obvious contamination of our samples, since spruce was surely introduced into the area during Modern Times.

Both species which dominate the spectra immigrated late into the area after the last glaciation. *Fagus* reached the

| Samples and charcoal values | | | | | | | Taxon in % no. of fragments | | | | | | | | | | Taxon in % weight | | | | | | | | | |
|---|---|---|---|---|---|---|---|---|---|---|---|---|---|---|---|---|---|---|---|---|---|---|---|---|---|---|
| Context | Sample | N: n charcoals | W: weight [g] | W / N | n diameter | mD [cm] | *Fagus* | *Carpinus* | *Quercus* | Maloideae | *Corylus* | *Populus* | *Salix* | *Populus/Salix* | *Acer* | hard wood id. | *Fagus* | *Carpinus* | *Quercus* | Maloideae | *Corylus* | *Populus* | *Salix* | *Populus/Salix* | *Acer* | hard wood id. |
| Furnace 1 | *Of 1-3* | 40 | 44.2 | 1.1 | 40 | 3.6 | 53 | 33 | 10 | . | 3 | . | 3 | . | . | . | 51 | 31 | 9 | . | 1 | . | 8 | . | . | . |
| | *Of 1-4* | 40 | 9.8 | 0.2 | 39 | 3.6 | 38 | 40 | 23 | . | . | . | . | . | . | . | 39 | 37 | 24 | . | . | . | . | . | . | . |
| | *Of 1-5* | 32 | 13.7 | 0.4 | 31 | 3.5 | 56 | 34 | 6 | 3 | . | . | . | . | . | . | 83 | 14 | 2 | 1 | . | . | . | . | . | . |
| Furnace 2 | *Of 2-1* | 9 | 1.7 | 0.2 | 8 | 2.6 | 11 | 56 | 11 | . | . | . | . | . | . | 22 | 7 | 75 | 9 | . | . | . | . | . | . | 9 |
| | *Of 2-2* | 28 | 15.3 | 0.5 | 25 | 3.4 | 50 | 46 | . | 4 | . | . | . | . | . | . | 54 | 45 | . | 1 | . | . | . | . | . | . |
| | *Of 2-3* | 21 | 33.7 | 1.6 | 21 | 4.3 | 14 | 67 | 5 | 5 | . | . | 5 | . | 5 | . | 13 | 37 | 6 | 2 | . | . | 39 | . | 3 | . |
| Furnace 3 | | 36 | 25.7 | 0.7 | 35 | 3.4 | 19 | 53 | 8 | 14 | 3 | . | 3 | . | . | . | 53 | 27 | 7 | 7 | 4 | . | 2 | . | . | . |
| Furnace 4 | *Of 4-6a* | 40 | 49.2 | 1.2 | 40 | 3.9 | 33 | 60 | 8 | . | . | . | . | . | . | . | 25 | 70 | 5 | . | . | . | . | . | . | . |
| | *Of 4-6b* | 40 | 37.1 | 0.9 | 40 | 4.3 | 30 | 63 | . | 5 | . | 3 | . | . | . | . | 38 | 57 | . | 1 | . | 4 | . | . | . | . |
| | *Of 4-6c* | 40 | 31.4 | 0.8 | 40 | 3.7 | 33 | 58 | . | 8 | 3 | . | . | . | . | . | 37 | 57 | . | 5 | 1 | . | . | . | . | . |
| | *Of 4-7* | 24 | 25.9 | 1.1 | 24 | 4.9 | 8 | 88 | 4 | . | . | . | . | . | . | . | 23 | 73 | 4 | . | . | . | . | . | . | . |
| Roasting place | *R 14-b* | 39 | 11.9 | 0.3 | 33 | 4.3 | 28 | 56 | 8 | . | . | . | . | . | . | 8 | 32 | 59 | 4 | . | . | . | . | . | . | 5 |
| | *R 14-d* | 42 | 8.5 | 0.2 | 35 | 5.0 | 38 | 52 | 7 | . | . | . | . | . | . | 2 | 42 | 54 | 4 | . | . | . | . | . | . | <1 |
| | *R 15-c* | 18 | 4.7 | 0.3 | 10 | 5.1 | 33 | 39 | 17 | . | . | . | . | . | . | 11 | 20 | 48 | 25 | . | . | . | . | . | . | 7 |
| Reheating hearth | | 40 | 53.9 | 1.3 | 40 | 3.6 | 38 | 53 | 3 | 5 | . | . | . | 3 | . | . | 49 | 44 | 1 | 5 | . | . | . | 1 | . | . |
| Post hole | | 17 | 32.5 | 1.9 | 17 | 5.2 | 100 | . | . | . | . | . | . | . | . | . | 100 | . | . | . | . | . | . | . | . | . |
| *Sum / mean* | | *506* | *399* | *0.8* | *478* | *4.0* | *36* | *51* | *7* | *2* | *<1* | *<1* | *<1* | *<1* | *<1* | *2* | *36* | *51* | *7* | *2* | *<1* | *<1* | *<1* | *<1* | *<1* | *2* |

Figure 4. Results of charcoal analysis, ironworks "Genoeserbusch"/Peppange

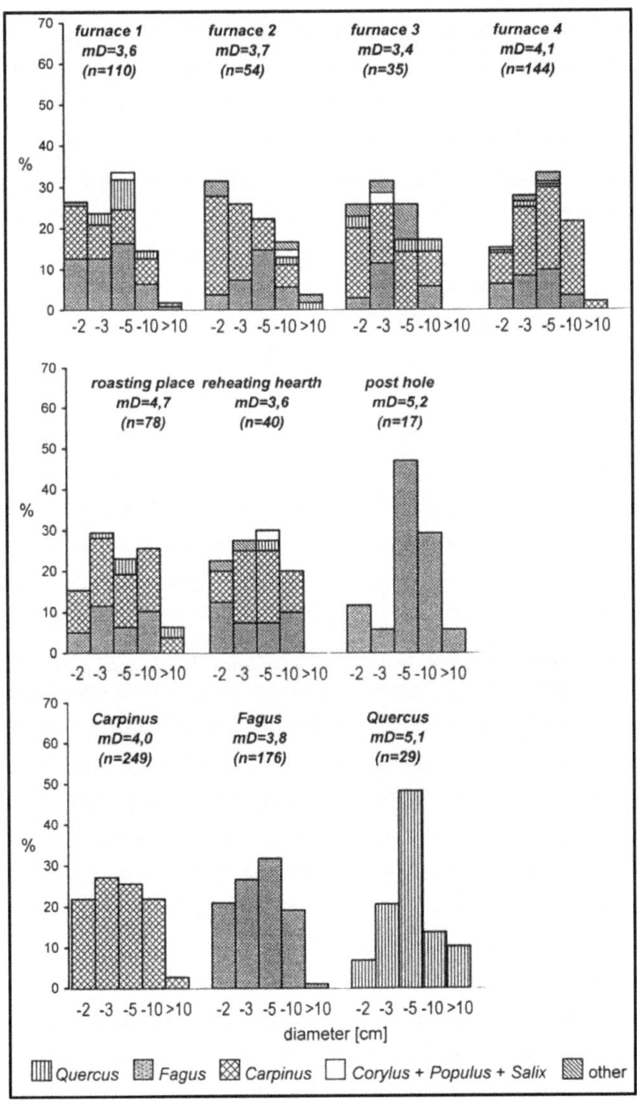


Figure 5. Results of charcoal analysis and diameter class distribution, ironworks "Genoeserbusch"/Peppange. The histograms indicate small diameter wood usage comparable to the histogram A in Figure 3. mD = mean diameter (see text), n = no. of diameter-analysed charcoals

investigation area between 5000 and 4000 cal BP, while *Carpinus* appeared not before 3000 and 2000 cal BP (Lang 1994, pp. 154, 160). In the Eifel area, *Fagus* mass dispersal happened around 3700 cal BP. *Carpinus* pollen values are considerable around 2000 cal BP indicating the presence of the taxon in the area (Kubitz 2000).

**Evidence for coppicing?**

The histograms of diameter size analysis show the use of small sized wood, compared to the campsite reference histogram in Figure 3. From this, it is possible that wood from coppiced stands where used. Alternatively, the small diameter wood distribution might be explained with the use of branch wood of bigger trees, where the stems where used for e.g. construction wood. To test the hypothesis of coppicing, we performed more detailed analysis of growth ring numbers and patterns.

The dendroanthracological analysis shows that we failed to find fragments with more than 28 growth rings, while samples of 16th-19th century charcoal production context from different sites in Germany as well as one contemporary charcoal kiln show that within comparable fragment sizes and diameter size classes, growth ring numbers of up to 118 can be found, indicating the charring of branch wood of old trees (Figure 5). The mean growth rates of the Peppange samples (samples 13-18 in Figure 6) are higher than material where branches were used for charcoal production (e.g. samples 2, 3 and 7 in Figure 6). Therefore, we can conclude that the used wood derived not from branches of older trees. This supports the hypothesis of coppiced woodland usage for fuel supply of the medieval iron smelting site. Other shown samples, e.g. 8-10 in Figure 6, potentially derived from a combination of coppicing and high forest, because they show a broad variation in the plots and might be a combination of coppiced wood and branches from old trees.

Looking on the mean growth ring widths, the Peppange material shows slightly higher values than samples 2, 3,

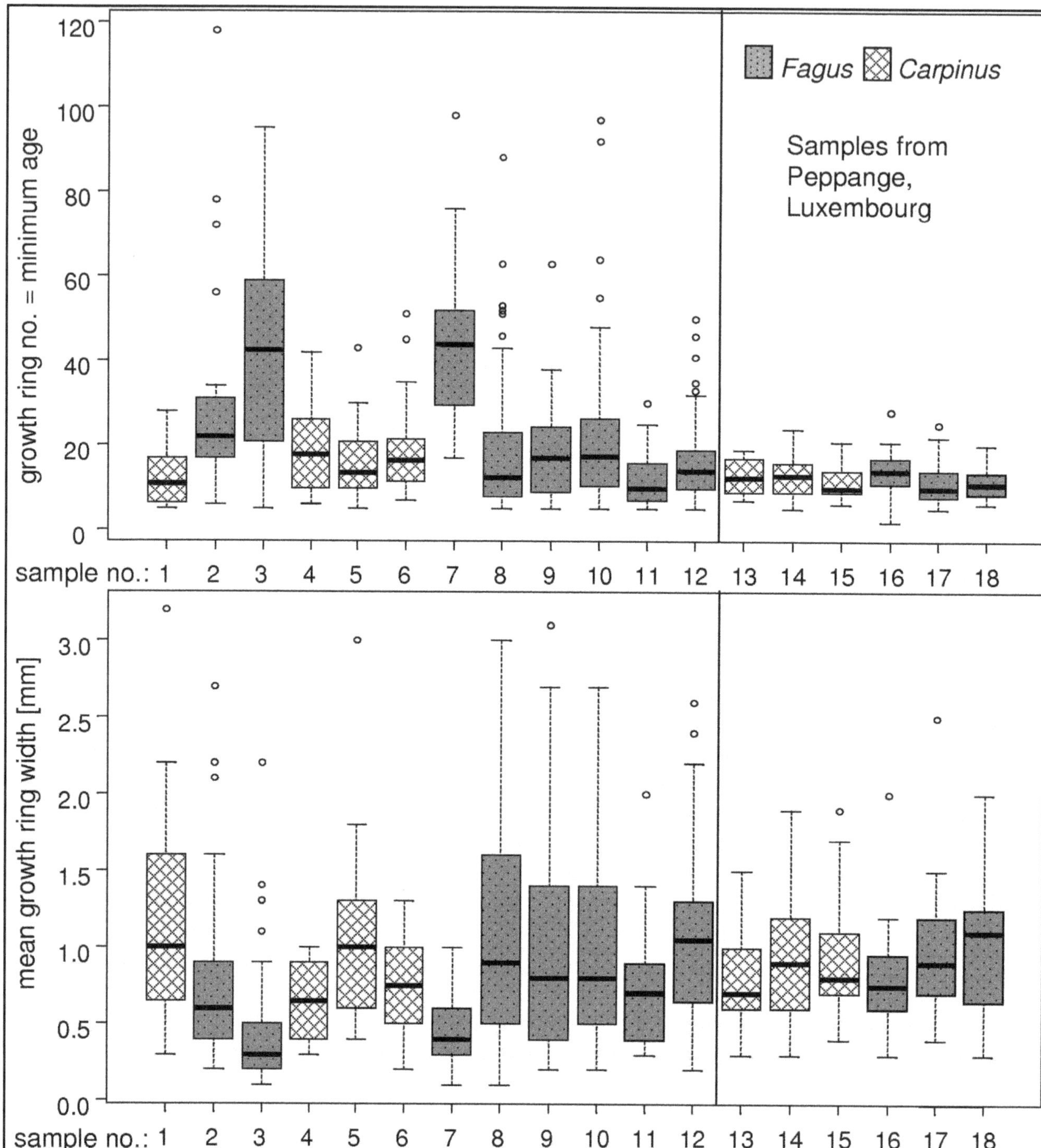


| | sample no. | region | geographical position N | E | m a.s.l. | context | W/N [g] | mD [cm] | n charcoals *Carpinus* | *Fagus* |
|---|---|---|---|---|---|---|---|---|---|---|
| BAR 1u2 | 1, 2 | Schleswig-Holstein | 54°10'01" | 09°46'27" | 20 | | 1.9 | 5.8 | 16 | 33 |
| BAR 4 | 3 | Schleswig-Holstein | 54°10'07" | 09°45'56" | 27 | | 0.7 | 5.4 | | 30 |
| K 44 | 4 | Bavarian Forest | 48°58'58" | 12°46'53" | 590 | | 0.4 | 6.9 | 14 | |
| K 45 | 5 | Bavarian Forest | 48°58'44" | 12°46'43" | 525 | | 0.6 | 6.2 | 26 | |
| K 100 | 6 | Bavarian Forest | 48°58'48" | 12°46'59" | 540 | historical charcoal prod. | 0.3 | 6.2 | 24 | |
| K 50 | 7 | Bavarian Forest | 48°57'07" | 12°51'40" | 885 | | 2.2 | 8.4 | | 35 |
| K 89 | 8 | Bavarian Forest | 48°58'10" | 12°51'40" | 1030 | | 0.6 | 8.6 | | 106 |
| KS 1 | 9 | North Hessen | 51°10'57" | 09°35'15" | 335 | | 0.5 | 5.7 | | 59 |
| KS 2 | 10 | North Hessen | 51°10'48" | 09°35'36" | 358 | | 0.8 | 6.6 | | 60 |
| GW 15 | 11 | Meckl.-Vorpommern | 54°03'56" | 13°28'25" | 31 | | 0.2 | 8.3 | | 22 |
| KoG04 | 12 | South Bavaria | 47°39'45" | 11°03'35" | 829 | contemporary charcoal prod. | 0.5 | 10.8 | | 100 |
| | 13, 16 | Peppange | | | | furnace 1 | 0.7 | 4.1 | 14 | 20 |
| | 14, 17 | Peppange | 49°29'32" | 06°08'09" | 270 | furnace 4 | 0.6 | 3.8 | 85 | 38 |
| | 15, 18 | Peppange | | | | reheating hearth | 1.1 | 2.9 | 21 | 15 |

Figure 6. Box plots of measured growth rings (upper part) and of growth ring width (lower part). Sample origin and details are given in the table below. Box plots indicate median (line in boxes), interquartile distance (= IQD, box), extreme values (o)

4, 7 and 11, while they are comparable to samples 1, 5, 8-10 and 12. This can only partly be explained with differences in environmental conditions like altitude (e.g. sample 7 from 885 m a.s.l.). The influence of former woodland management on tree ring growth is a considerable factor, but cannot put forward for explaining mean growth differences since we do not know enough of the woodland situation around the 16th-19th century kilns of which material was used for the comparison. References for better interpreting growth rates are currently established by collecting modern wood material. Thus at the moment we cannot interpret mean growth ring values in more detail.

Though the use of young trees cannot completely excluded, archaeobotanical evidence of coppicing can be drawn from a combination of taxonomic identification, diameter size distribution and dendroanthracological analysis on material of charcoal production and charcoal consumption, the latter in the case of the iron furnaces in the south of Luxembourg. Future research on this material should test whether it is possible to differentiate branch from stem wood by looking on anatomical features, like vessel or ray size.

## CONCLUSION

Wood species analysis, diameter analysis and growth ring analysis in combined interpretation point to a wood source from coppiced stands at the "Genoeserbusch" iron smelting site. At the moment, it seems very probable that the charcoal was produced in the vicinity of the site, where today's woodland show a very similar species composition of the medieval charcoal. Thus, the medieval woodland around the smelting site was managed by coppicing to provide the necessary wood for iron production. The analysis of charcoal from charcoal production and consumption sites can prove directly the existence of coppiced stands. The remaining task is now to find the charcoal production sites at "Genoeserbusch". First signs of them were discovered close to the site, and will be followed up in future field campaigns.

### References


BONHOTE, J. and VERNET, J.L. 1988. La memoire des charbonnieres. Essai de reconstitution des milieux forestiers dans une vallee marquee par la metallurgie (Aston, Haute-Ariege). *Revue Forestiere Francaise* 40, 197-212.

DAVASSE, B. 2000. *Forêts charbonniers et paysans dans les Pyrénées de l'est du moyen âge à nos jours. Une approche géographique de l'histoire de l'environment.* Toulouse (GEODE (Géographie de l'environment)).

HILLEBRECHT, M.-L. 1982. *Die Relikte der Holzkohlewirtschaft als Indikatoren für Waldnutzung und Waldentwicklung. Untersuchungen an Beispielen aus Südniedersachsen.* Göttingen, Erich Goltze.

JOCKENHÖVEL, A., WILLMS, C., ABDINGHOFF, T. and OVERBECK, M. 1998. Archaeological Investigations on the Beginning of Blast Furnace-Technology in Central Europe. In: Extended Abstracts. The Fourth Intern. Conference on the Beginning of the Use of Metals and Alloys in Shimane, May 25–27, 1998 (The beginnings of the use of metals and alloys = BUMA IV 1998) 65-69.

JOCKENHÖVEL, A. and WILLMS, C. 2005. Das Dietzhölzetal-Projekt. Archäometallurgische Untersuchungen zur Geschichte und Struktur der mittelalterlichen Eisengewinnung im Lahn-Dill-Gebiet (Hessen). *Münsterische Beiträge zur ur- und frühgeschichtlichen Archäologie* 1.

KUBITZ, B. 2000. *Die holozäne Vegetations- und Siedlungsgeschichte in der Westeifel am Beispiel eines hochauflösenden Pollendiagrammes aus dem Meerfelder Maar.* Berlin, Stuttgart, J. Cramer.

LANG, G. 1994. *Quartäre Vegetationsgeschichte Europas: Methoden und Ergebnisse.* Jena, Stuttgart, New York, G. Fischer.

LEROY, M. 1997. La sidérurgie en Lorraine avant le haut fourneau: L'utilisation du minerai de fer oolithique en réduction directe. *CNRS Editions.*

LUDEMANN, T. 1996. Die Wälder im Sulzbachtal (Südwest-Schwarzwald) und ihre Nutzung durch Bergbau und Köhlerei. *Mitteilungen Verein Forstliche Standortskunde und Forstpflanzenzüchtung* 38, 87-118.

LUDEMANN, T. and NELLE, O. 2002. Die Wälder am Schauinsland und ihre Nutzung durch Bergbau und Köhlerei. *Freiburger Forstliche Forschung* 15, 1-139.

LUDEMANN, T. 2003. Large-scale reconstruction of ancient forest vegetation by anthracology – a contribution from the Black Forest. *Phytocoenologia* 33, 645-666.

LUDEMANN, T., MICHIELS, H.-G. and NÖLKEN, W. 2004. Spatial patterns of past wood exploitation, natural wood supply and growth conditions: indications of natural tree species distribution by anthracological studies of charcoal-burning remains. *European Journal of Forest Research* 123, 283-292.

MARGUERIE, D. and HUNOT, J.-Y. 2007. Charcoal analysis and dendrology: data from archaeological sites in north-western France. *Journal of Archaeological Science* 34, 1417-1433.

MARZIANI, G. and CITTERIO, S. 1999. The effects of human impact on the arboreal vegetation near ancient iron smelting sites in Val Gabbia, northern Italy. *Vegetation History and Archaeobotany* 8, 225-229.

NELLE, O. 2002. Zur holozänen Vegetations- und Waldnutzungsgeschichte des Vorderen Bayerischen Waldes anhand von Pollen- und Holzkohleanalysen. *Hoppea, Denkschriften Regensburgische Botanische Gesellschaft* 63, 161-361.

NELLE, O. 2003. Woodland history of the last 500 years revealed by anthracological studies of charcoal kiln sites in the Bavarian Forest, Germany. *Phytocoenologia* 33, 667-682.

NELLE, O. 2008. Combining charcoal and pollen analysis: Holocene vegetation dynamics, tree species composition and woodland use in the Bavarian Forest. –. In Fiorentino, G. and Magri, D. (eds.), *Charcoals from the Past: Cultural and Palaeoenvironmental Implications. Proceedings of the Third International Meeting of Anthracology, Cavallino – Lecce (Italy), June 28th – July 1st 2004*, 183-191. Oxford, BAR Publishing.

OVERBECK, M., KRONZ, A. and NELLE, O. 2007. Medieval iron production in Luxembourg (13th-14th century AD). Archaeological evidence of the transition from bloomery furnace to high bloomery furnace (Stuckofen) in Europe. In Crew, P. and Crew, S. (eds.), *Early Ironworking in Europe II. Archaeology, technology and experiment. Plas Tan y Bwlch occasional paper No 4, Abstracts of the second International Conference at Plas Tan y Bwlch 17th-21th Sept. 2007*, 37-40. Plas Tan y Bwlch.

SCHLÄPFER, P. and BROWN, R.C. 1948. *Über die Struktur der Holzkohlen*. Zürich, EMPA.

SCHWEINGRUBER, F.H. 1990a. *Mikroskopische Holzanatomie. Formenspektren mitteleuropäischer Stamm- und Zweighölzer zur Bestimmung von rezentem und subfossilem Material.* Birmensdorf/Schweiz, Eidgenössische Anstalt für das forstliche Versuchswesen.

SCHWEINGRUBER, F.H. 1990b. *Anatomie europäischer Hölzer*. Bern, Stuttgart, Haupt.

# RADIUS OF CURVATURE MEASUREMENTS AND WOOD DIAMETER: A COMPARISON OF DIFFERENT IMAGE ANALYSIS TECHNIQUES

Sandrine PARADIS

PhD student, GEOLAB, UMR-CNRS 6042, 39[E] rue Camille Guérin 87036 Limoges, France

Alexa DUFRAISSE

CR CNRS, Archéozoologie, archéobotanique: sociétés, pratiques et environnements. Muséum national d'Histoire Naturelle, Département EGB, UMS 303-UMR 7209, CP 56, 55 rue Buffon, 75005 Paris, France

Philippe ALLÉE

PR, GEOLAB, UMR-CNRS 6042, 39[E] rue Camille Guérin 87036 Limoges, France

**Abstract**: *The study of wood diameter is needed for the understanding of wood supply management. This type of approach, which was launched in the 1990s, has improved considerably as radius of curvature measurement has gained in precision through the introduction of new equipment: stereomicroscope equipped with a camera, and image analysis software. This paper focuses on the first step of the wood diameter study: radius of curvature measurement. This measurement is unusable as it stands, but the data set can be integrated into models to provide better information on wood diameter. In addition to the circle tool technique based on tree-ring curvature, three new techniques are presented here, based on wood rays: Thales' theorem, trigonometry in a right-angled triangle, and trigonometry in an isosceles triangle. This study compares results on perfectly graduated targets and on wood samples for four radii of curvature measurement techniques, to assess their reliability and limits. Three parameters are examined: angle between two wood rays, distance between two wood rays, and radius of curvature.*

**Key words**: *wood charcoal, wood diameter, image analysis software, radius of curvature*

## INTRODUCTION

Charcoal analyses allow firewood management and past environments to be studied by means of wood diameter, tree-ring width, wood physiology, and identification of wood taxa (Vernet, 1992; Thiébault, 2002; Fiorentino & Magri, 2008). However, it is still necessary to improve analytical methods, especially for wood diameter, which is a selection criterion at least as important as species or wood humidity (Chabal *et al.* 1999). This type of approach was launched in the 1970s (Willerding, 1971) and developed and used by other authors (Hillebrecht, 1982; Marguerie, 1992; Marguerie and Hunot, 2007; Ludemann and Nelle, 2002; Dufraisse, 2002, 2006).

The determination of burnt wood diameter is composed of two steps, the first of which being radius of curvature measurement. However, this measurement indicates only the position of the charcoal fragment in a log and not the diameter. The second step is data treatment and interpretation which can be tackled by different models (see the medium wood diameter of Ludemann and Nelle (2002) or the three-dimensional wood model proposed by Dufraisse (2002, 2006), the object of ongoing research.

After a brief review of radius of curvature measurement techniques, mainly based on ring morphology, this paper will focus on new approaches based on wood rays. Measurement techniques using image analysis will be compared, and reliability and limits established, both on perfect targets and on wood samples.

## BRIEF HISTORY OF RADIUS OF CURVATURE ANALYSIS

Until now, various methods have been developed to evaluate tree-ring curvature, some of which are qualitative and others quantitative.

### A qualitative approach to tree-ring curvature

According to Marguerie (1992) and Marguerie and Hunot (2007), ring curvature estimation depends on standard classification, using constant magnification and a transparent test card. The authors have proposed three groups depending on the degree of the curvature:

- strongly curved rings;
- moderately curved rings;
- weakly curved rings.

Charcoal samples with indeterminate curvature are placed in a fourth group.

The predominance of weakly curved rings in the archaeological sample suggests the use of large diameter wood, e.g. trunks or large branches; the predominance of strongly curved rings indicates the use of small diameter wood, e.g. young trees or small branches.

This approach is not a quantitative measurement of wood diameter but merely a characterisation of ring morphology.

### Quantitative approaches based on tree-ring morphology

In the first quantitative approach, the visual estimate is guided by a graduated target (i.e. different diameters printed on a transparency) (Willerding, 1971; Lundström-Baudais, 1986; Ludemann and Nelle, 2002; Dufraisse, 2002, 2006). The target is placed on top of the charcoal fragment under a stereomicroscope to compare the curvature and angles of both the target and the charcoal fragment. This approach is a visual estimate: observer bias and measurement error are difficult to quantify, i.e. several values may be found by different observers for one charcoal fragment. This method implies the use of very broad diameter classes, so interpretation will also lack precision (Paradis, 2007).

Another approach based on tree-ring morphology uses the "circle tool" found in image analysis software (Chrzavzez, 2006). A few characteristic points are placed along the last visible ring and the software calculates an extrapolation, draws the corresponding circle and gives the diameter of the ring. Even if observer bias is less than for the use of a target printed on a transparency, this method is still influenced by ring morphology variation.

### Quantitative approaches based on wood rays

We have therefore developed another approach based on wood rays, using image analysis software. This method was tested with three geometric formulas (Figure 1):

- Thales' theorem;
- Trigonometry in a right-angled triangle;
- Trigonometry in an isosceles triangle (Paradis, 2007).

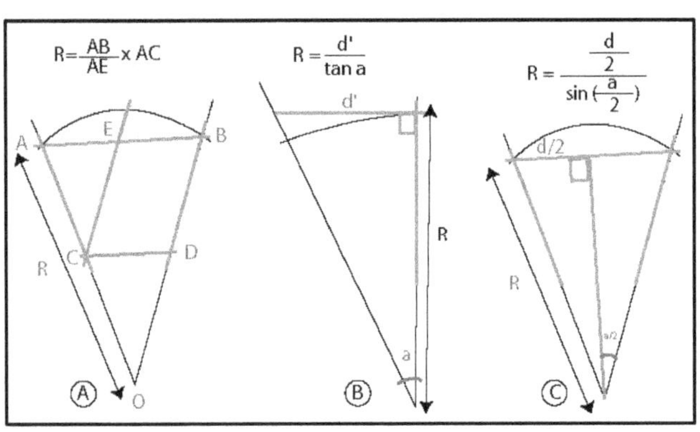


Figure 1. Radius of curvature measurement techniques based on the angle between two notional rays. A: Thales' theorem; B: Trigonometry in a right-angled triangle; C: Trigonometry in an isosceles triangle

In this work, only the four approaches which use image analysis software (Lucia Nikon) were tested and compared: the circle tool, the Thales' theorem, trigonometry in a right-angled triangle, and trigonometry in an isosceles triangle.

## MATERIAL AND METHODS

These four techniques were compared, first on perfect material (graduated targets printed on paper) in order to exclude the variations of natural material, and then on non-carbonised wood in order to test the reliability of these methods before the effects of carbonisation. Practical applications and assessment on charcoal fragments are underway (Paradis, ongoing PhD).

For each radius of curvature measurement, reliability and limits were tested for three parameters clearly identifiable on wood:

- radius of curvature which corresponds to the distance between the last visible ring and the pith of the wood;
- angle between two wood rays;
- distance between two wood rays.

Wood anatomical vocabulary is used throughout this study, but for perfect target paper printouts, the following terms are used: angle between two notional rays and distance between two notional rays.

### Test on a perfect graduated target

Perfect targets (graduated targets printed on paper) were measured with image analysis software (Lucia, Nikon). They present a large range of values for radius of curvature (0.1 cm to 15 cm), angle between two notional rays (0.5° to 70°) and distance between two notional rays (0.1 cm to 2.5 cm) (Figure 2). In total, 236 targets were tested with the four measurement techniques and the measurement was repeated three times (2832 measurements).

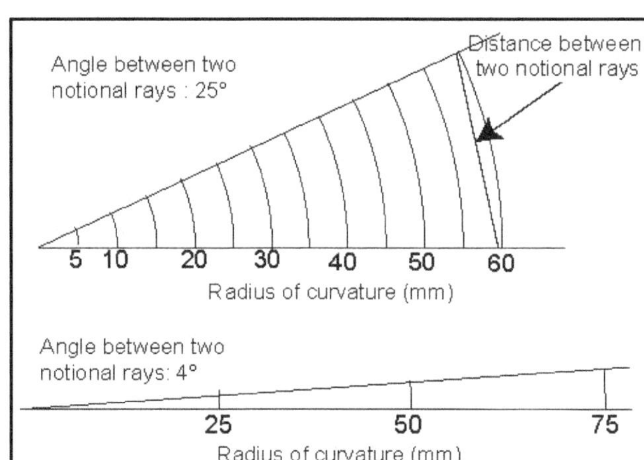


Figure 2. Perfect targets on paper printout used for the comparative study

For each measurement, the true values of radius of curvature, angle and distance between two notional rays are known, allowing the reliability of each technique to be tested.

### Test on non-carbonised wood

To measure radius of curvature on wood, small and large pieces of oak were used, with radii of curvature up to 20 cm.

Freshly cut transverse sections were used to test the four measurement techniques. Nevertheless, only two parameters, radius of curvature and distance between two wood rays, can be controlled (consequently, angles measured depend on both these parameters).

To obtain results in coherence with those from perfect targets, a large range of radii (from 0.3 cm to 20 cm) and distances between two rays (from 1 mm to 23 mm) were chosen, allowing a range of angles between 0.5° and 62°.

In total, 100 freshly cut sections of wood were tested, as in the comparative study on perfect targets, i.e. on all three parameters (angle, distance between two rays, and radius of curvature). Each measurement was repeated three times. However, as the true value of the angle is unknown, the six angle values obtained by trigonometry were averaged and used as a reference value.

### Data treatment

For each sample (perfect targets and freshly cut sections of wood), the radius of curvature measurement was repeated three times to calculate an average and then a percentage of error, which is indispensable to compare the error margin (Delta value) for large and small radius of curvature values.

Delta = average of measured radius of curvature – true radius of curvature value

Percentage of error = (delta x 100) / true radius of curvature value

A graphic representation of the percentage of error in a cumulate diagram (Figures 3 and 5) is used to compare the different measurement techniques. This graph allows two thresholds of measurement reliability to be defined. Three reliability classes are determined:

- good reliability;
- medium reliability; (this percentage of error is still acceptable for a radius of curvature measurement);
- poor reliability; (unacceptable for radius of curvature measurement).

These classes allow the reliability and limits of each measurement technique to be evaluated for the three parameters: angle between two wood rays, distance between two wood rays and radius of curvature.

A bar graph is used to present the reliability of each measurement technique for each parameter (Figures 4 and 6). The interest of the bar graph is that measurement reliability can be assessed for the three parameters. Each parameter was divided into unequal classes depending on the evolution of the percentage of error. For example, for the angle parameter, the percentage of error varies very quickly on small angles and stabilises for angles greater than 8°. The classes for angles between two rays are thus: ]0-1°], ]1-2°], ]2-4°], ]4-8°], ]8-16°], ]16-32°] and above 32°.

For distance between two rays, the classes were established in relation to the most frequent sizes of archaeological charcoal fragments: ]0-2 mm], ]2-4 mm], ]4-6 mm], ]6-10 mm], ]10-15 mm] and ]15-30 mm]. There was a specific focus on small distances between two rays to evaluate the minimal transverse section of charcoal fragment required.

For the radius, the classes are: ]0-0.5 cm], ]0.5-1 cm], ]1-1.5 cm], ]1.5-2 cm], ]2-4 cm], ]4-6 cm], ]6-10 cm], and ]10-15 cm], with an additional class for wood sections: above 15 cm.

Thus, for each class of each parameter, the percentage of the measurement per reliability class is shown, with good reliability in white, medium in grey and poor in black (Figure 4).

## RESULTS: RELIABILITY AND LIMITS OF EACH MEASUREMENT TECHNIQUE

### On perfect targets

The results of the 236 perfect targets for the four techniques are represented on the cumulate diagram (Figure 3), with the percentage of error on the x axis and the number of measurements on the y axis.

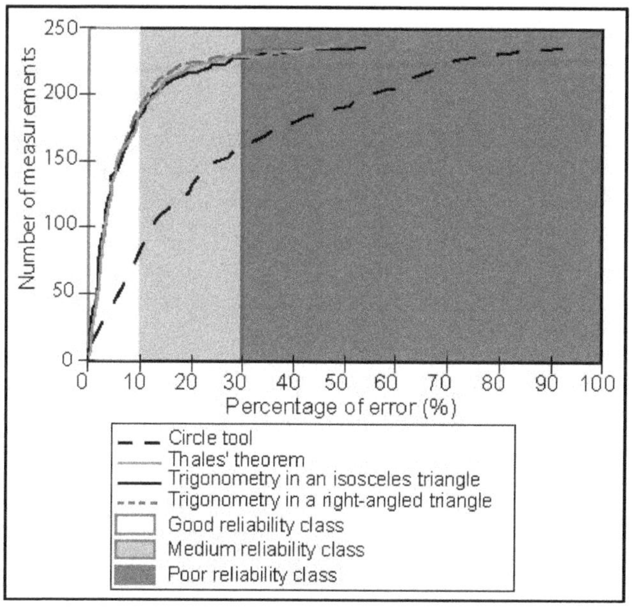


Figure 3. Graphic representation of the margin of error in a cumulate diagram for the four methods on perfect targets

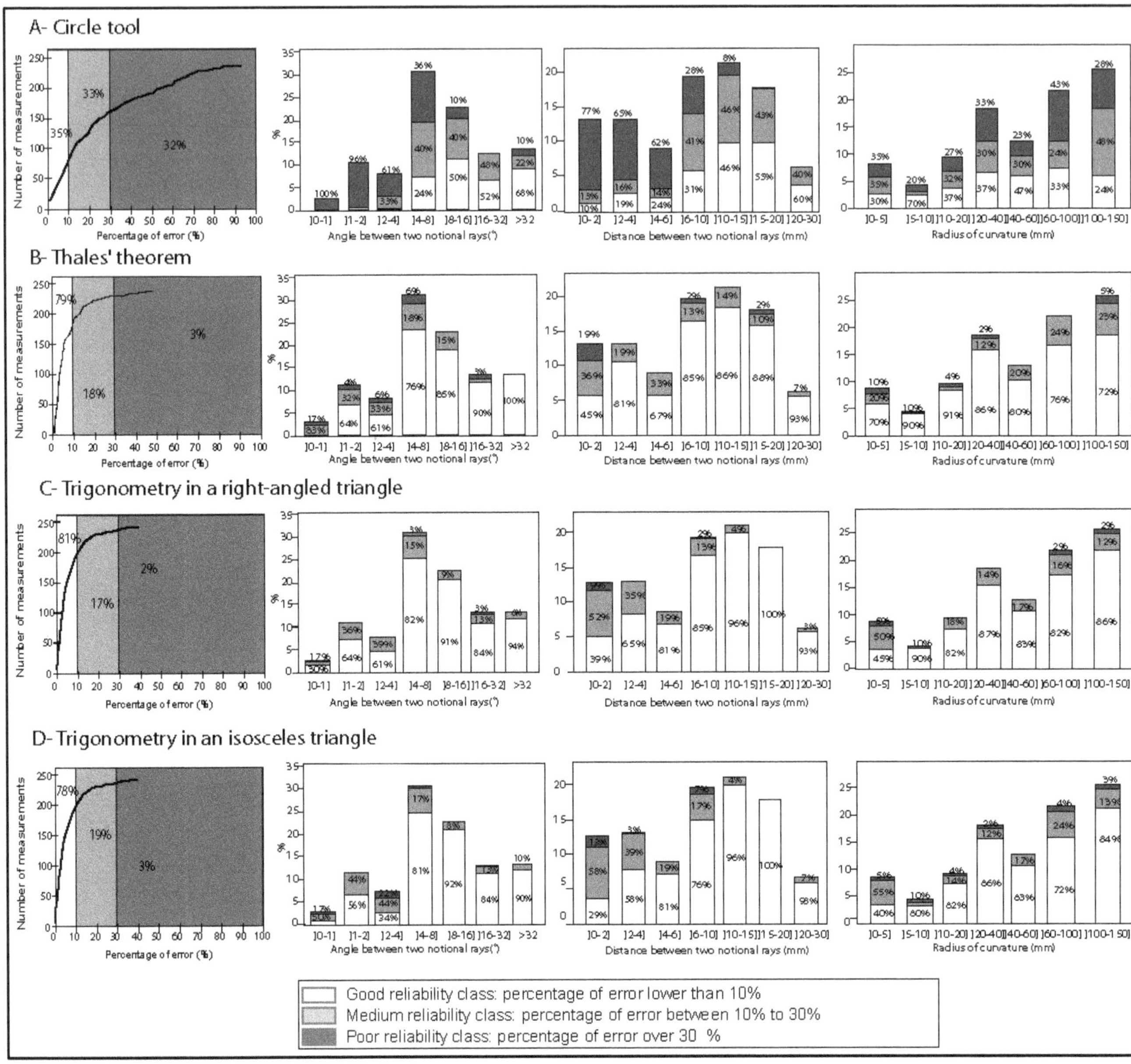


Figure 4. Reliability of the different measurement techniques on each parameter (angle between two notional rays, distance between two notional rays and radius of curvature) on perfect targets

Using this graph, the two thresholds for perfect targets were defined at 10 % and at 30%, corresponding to a 95% confidence level for most of the techniques. The following reliability classes are defined:

- good reliability, lower than 10%;
- medium reliability, between 10 and 30%;
- poor reliability, over 30%.

***The circle tool***

The circle tool presents a very high percentage of error, up to 90% (Figure 3). Unlike the other techniques there are very few measurements in the first two reliability classes as can be seen on the graph (the curve corresponding to the circle tool is below those of the other techniques).

The distribution of measurements in the percentage of error diagram indicates that 32% are in the poor reliability class and 33% in the medium class (Figure 4 A). Only 35% are in the good reliability class, which is not sufficient for radius of curvature measurement.

The study of the bar graph for each parameter (angle between two notional rays, distances between two notional rays and radius of curvature) underlines the poor reliability of this tool. In fact, more than half of the measurements with angles lower than 8° have a margin of error of over 30%, which is not suitable for radius of curvature measurement. In addition, it is not relevant to consider only the angles over 8°, due to the correlation between angle and radius of curvature ($r = 0.45$ with $p<0.0001$). In other words, if the angles lower than 8° are removed, many larger radius of curvature measurements are also removed.

For the distances between two notional rays, the measurement is less reliable when the distance is smaller than 1 cm. Yet, that is the most frequent size of charcoal fragments in archaeological contexts.

This technique is thus very attractive at first due to the ease and the speed of use, and also it is present in most image analysis software. Nevertheless, this method is unreliable for medium and large radii of curvature.

***Thales' theorem***

With this technique, only 3% of the measurements are in the poor reliability class and 79% are in the good reliability class (Figure 4 B).

The distribution of reliability for the different parameters gives very good results.

For small angles, all the measurements belong to the medium and poor classes (respectively 83% and 17%). The measurement of angles of less than 1° is technically more difficult. It is harder to draw notional rays for small angles, thus the measurement of the angle could be wrong; there is also the question of observer bias, which leads to a higher percentage of error.

The graph for distance between two notional rays underlines the greater unreliability of distances lower than 2 mm (percentage of error of 45%). This inaccuracy means that greater care is required for charcoal fragments with a small transverse section. All radius of curvature classes provide good reliability with only 0% to 10% of measurements in the poor class.

This method is reliable, but it is time-consuming (two notional rays and two series of parallel lines must be drawn and three segments must be measured (AB, AE, AC, cf. Figure 1 A). This technique is consequently not adapted for routine analysis.

***Trigonometry in a right-angled triangle***

This method is highly reliable because 81% of the measurements are in the good reliability class (lower than 10%) (Figure 4 C). The distribution of reliability for the different parameters gives almost the same result as the Thales Theorem, i.e. with lower reliability on small angles (up to 1°, only 30% of the data are in the good reliability class) and small distances between notional rays (up to 2 mm, only 39% of the data belong to the good reliability class). When distances and angles between two notional rays are small (less than 2 mm and 1°), it is advisable to obtain a representative average from repeated series of measurements.

This technique is more preferable to the Thales' theorem technique owing to its speed of use. However, automatic drawing of a right-angled triangle is not always available in the software package.

***Trigonometry in an isosceles triangle***

For this measurement technique, 78% of the measurements are in the good reliability class (Figure 4 D).

This technique appears as good as the previous one on perfect targets. Indeed, for the angle between two notional rays, angles of more than 4° indicate very good reliability (between 81 and 92%) and good reliability for angles from 1° to 4° with only 20% of the measurements in the poor class. The reliability for small angles (lower than 1°) is worse, and much greater care is required to obtain satisfactory results. For distance between two notional rays, the results provide very good data from 4 mm (between 76% and 100%), but reliability is also good for distances from 2 to 4 mm (with only 3% of the measurements in the poor class). All radius of curvature classes give good reliability, with only 0 to 10% of the measurements in the poor class.

Finally, compared with other techniques, isosceles triangle trigonometry is the fastest method because it requires only two measurements: angle and distance between two rays which reduces the potential for error.

**First results on non-carbonised wood**

Working only on perfect targets is not representative of the potential of each measurement technique, owing to the variability of natural material. Thus, all four measurement techniques were tested on 100 freshly cut transverse sections. For all techniques, it was very difficult to measure the smallest diameter (less than 1 cm) due to the wavy outline of the tree-ring close to the pith.

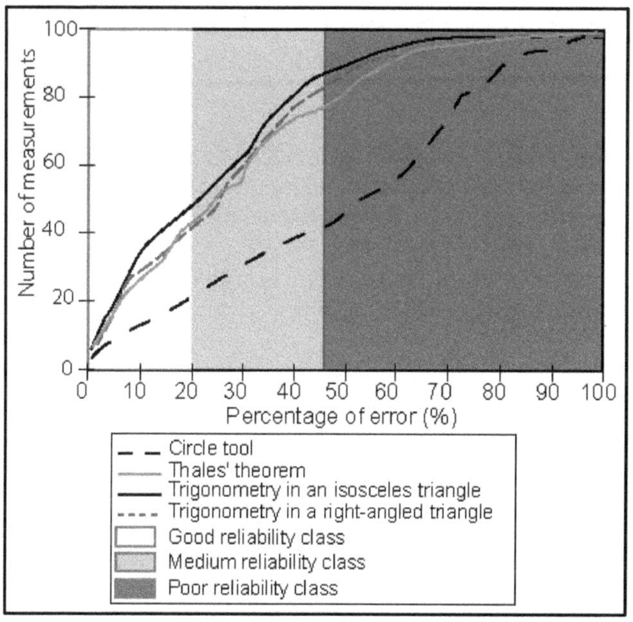


Figure 5. Graphic representation of the margin of error in a cumulate diagram for the four methods on freshly cut transverse sections

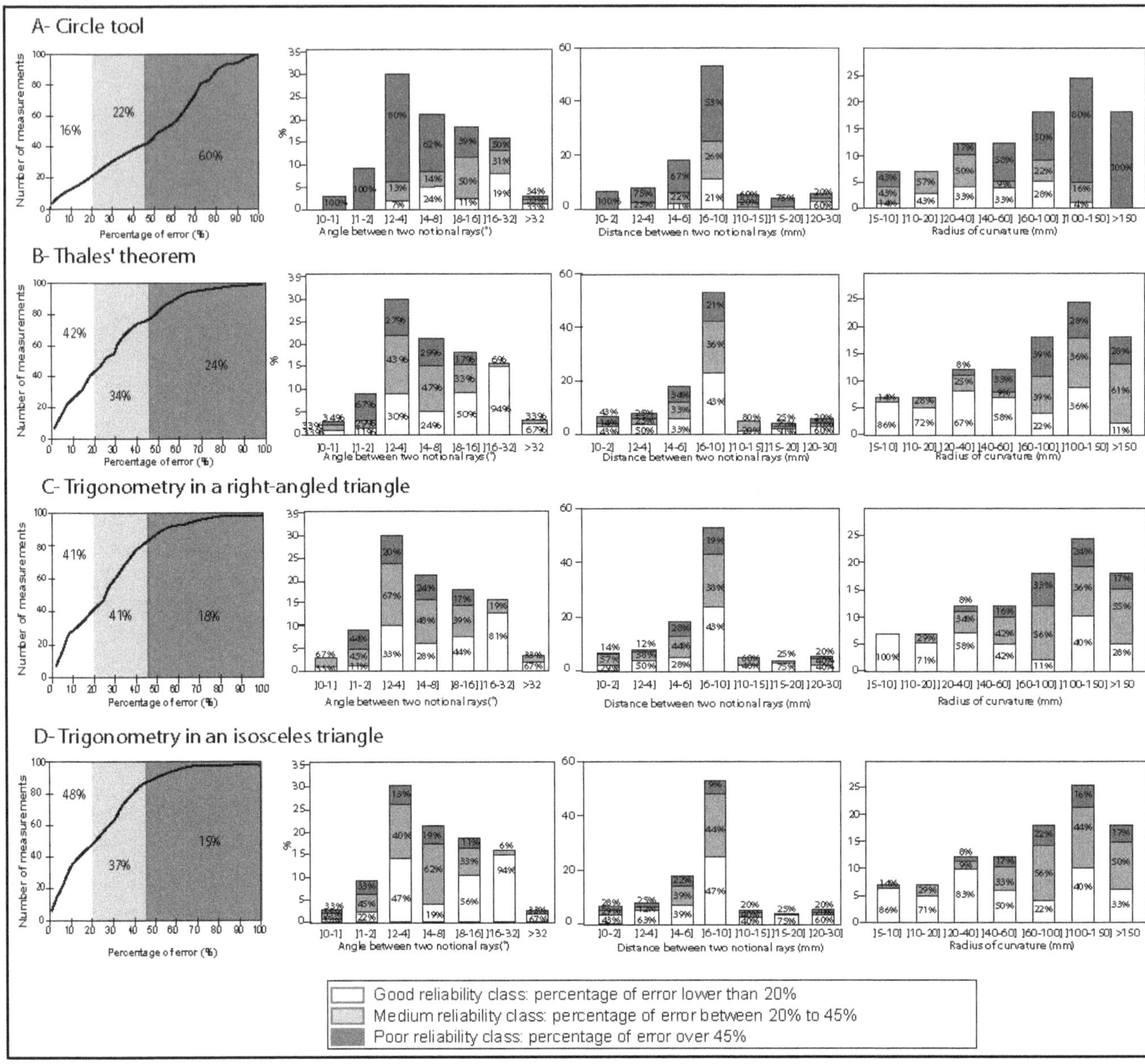


Figure 6. Reliability of the different measurement techniques on each parameter (angle between two notional rays, distance between two notional rays and radius of curvature) on freshly cut transverse sections

The stabilisation of the percentage of error is around 70% for a 95% confidence level (Figure 5). However, such a threshold is hardly usable in a study of wood diameter. Consequently, a compromise has been found with a threshold of 45% which corresponds to an 80% confidence level for most of the techniques.

The three reliability classes have been adapted to correspond to the evolution of the margin of error. Classes are defined as the following (Figure 5):

– good is lower than 20%;

– medium is between 20% and 45%;

– poor is above 45%.

***The circle tool***

On the cumulate diagram of the percentage of error, only 18% of the measurements are in the good reliability class and 60% are in the poor reliability class defined above (Figure 6 A).

For angle between two wood rays, the reliability for small angles is very poor: 100% of the measurements with angles below 2° are in the poor reliability class and more than 60% for angles from 2° to 8°; the results are better for angles over 8°. But as there is a correlation between angle and radius of curvature (cf. §3.1.1.), it is not acceptable to work only on large angles. For distance between two wood rays, the results are also very poor for all distances under 2 cm (up to 100% of the measurements are wrong) instead of 1 cm on perfect targets. An increase in the proportions of wrong measurements is undeniable from 4 cm (i.e. 8 cm of diameter) (from 58% to 100% of the measurements are in the poor reliability class).

All these observations underline the limits of the circle tool proposed in the software. In other words, with this measurement technique it is possible to measure only small radii of curvature with large angles (> 8°) on large transverse sections (>2 cm); which are not common in archaeological charcoal samples.

### *Thales' theorem*

The technique of the Thales' theorem is characterised by 76% of reliable measurements (42% in the good reliability class and 34% in the medium class) which is better for a study of radius of curvature (Figure 6 B). The distribution of reliability for the three parameters reinforces the results obtained on perfect targets: there from 34% to 67% of wrong measurements for angles below 2°, and 43% for small distances between two wood rays (less than 2 mm). It is thus necessary to be careful with such values, as previously mentioned for perfect targets.

### *Trigonometry in a right-angled triangle*

For this method, only 18% of the measurements are in the poor reliability class (Figure 6 C). In detail, from 17 to 24% of the measurements are in the poor reliability class for angles over 2°, 44% belong to the poor reliability class for angles below 2°. For distances between two wood rays of less than 1 cm, from 12 to 28% of the measurements are in the poor reliability class. Surprisingly, the smallest distances present the best results (14% and 12% of wrong measurements respectively for ]0-2 mm] and ]2-4 mm] classes) which may be due to the variability of wood samples. For radius of curvature, the poor reliability class appears from 2 cm with a percentage of 8% to 33%, which is acceptable for radius of curvature measurements.

Finally, the distribution of reliability on the three parameters shows better results than for the Thales' theorem. Thus, the technique of trigonometry in a right-angled triangle is well adapted to radius of curvature measurements, taking into account the natural variability of wood.

### *Trigonometry in an isosceles triangle*

In the trigonometry in an isosceles triangle technique, only 15% of the measurements are in the poor reliability class and 48% are in the good reliability class (Figure 6 D).

Furthermore, the results of the distribution of reliability on the bar graph also show the good reliability of each parameter, whatever the angles (angle <2°, 33% of wrong measurements), distances between two wood rays (distance <6 mm, from 22 to 28% of wrong measurements) and radius of curvature (length >4 cm, 17% to 22% of wrong measurements).

This technique provides the best results on wood sections. In addition, it is a quick and easy technique to apply, which is compatible with a routine analysis.

## DISCUSSION

Overall, the results of the percentage of error for wood are worse than for perfect targets, which is not surprising considering the natural variability of wood.

In order to establish the limits of each measurement technique on notional targets and on wood transverse sections, each parameter was divided into three classes based on characteristic values (8° for angles and 4 mm for distances between two wood rays (notional rays for perfect targets) (Figure 7):

- for angles: ]0-8°], ]8-30°] and ]30-70°];
- for distance between two wood rays: ]0-0.4 cm], ]0.4-1 cm] and ]1-3 cm];
- and for radius of curvature: ]0-2.5 cm], ]2.5-7.5 cm] and ]7.5-20 cm].

Next, a notation was attributed to each parameter as follows to allow discussion of the general reliability of each measurement technique (Figure 7):

- "- -" for more than 60% of wrong measurements;
- "-" for 45 to 60% of wrong measurements;
- "+" for 20 to 45% of wrong measurements;
- "+ +" less than 20% of wrong measurements.

As shown in figures 7 and 8, the circle tool is not recommended. In fact, its reliability is poor for almost all parameters and the value dispersion does not follow the trend of true radius of curvature values (Figure 8). Thus, it becomes impossible to measure radius of curvature above 4 cm. In addition, the absence of correlation between measured value and real values makes it almost impossible to establish a potential corrective factor.

The three other techniques based on wood rays (and not on tree-ring morphology) provide relevant results with good correlation between the measured value and true radius of curvature values.

The reliability of freshly cut transverse sections of wood is good, with more than 80% of the measurements included in the first two reliability classes. If dispersion increases clearly for large radii of curvature, all values follow the trend of the true values, unlike results for the circle tool. Thus, with these observations, measurement techniques based on wood rays are good for radius of curvature measurement. However, the Thales' theorem is not adapted for routine analysis, as it is too time-consuming. Trigonometry in a right-angled triangle requires software with an automatic right-angle feature and it is difficult to use with very large angles, which required a wide camera field to draw the straight lines for the right angle.

The most efficient technique for routine analysis is trigonometry in an isosceles triangle, which is the fastest technique and the most reliable on wood transverse

| A: targets | Angle between rays | | | Distance between rays (cm) | | | Radius of curvature (cm) | | |
|---|---|---|---|---|---|---|---|---|---|
| | ]0-8°] | ]8-30°] | ]30-70°] | ]0-0,4] | ]0,4-1] | ]1-3] | ]0-2] | ]2-6] | ]6-20] |
| Circle tool | - | ++ | ++ | - | - | + | - | + | + |
| Thales' theorem | ++ | ++ | ++ | + | ++ | ++ | + | ++ | ++ |
| Trigonometry in a right-angled triangle | ++ | ++ | ++ | ++ | ++ | ++ | ++ | ++ | ++ |
| Trigonometry in an isosceles triangle | ++ | ++ | ++ | + | ++ | ++ | + | ++ | ++ |

| B: wood sections | Angle between rays | | | Distance between rays (cm) | | | Radius of curvature (cm) | | |
|---|---|---|---|---|---|---|---|---|---|
| | ]0-8°] | ]8-30°] | ]30-70°] | ]0-0,4] | ]0,4-1] | ]1-3] | ]0-2] | ]2-6] | ]6-20] |
| Circle tool | -- | - | - | -- | -- | -- | - | - | -- |
| Thales' theorem | - | + | ++ | - | - | + | ++ | + | - |
| Trigonometry in a right-angled triangle | - | + | ++ | + | + | + | + | + | + |
| Trigonometry in an isosceles triangle | + | ++ | ++ | + | + | ++ | + | ++ | + |

Figure 7. Synthesis table. A: on perfect targets; B: on freshly cut transverse sections.
In grey, the best techniques for routine analysis

sections. In addition, taking into account the care required with small angles and distances between two wood rays, this technique is compatible with the necessity of increasing the number of measurements in order to limit observer bias.

We must also recall that for measurements on wood sections, the chosen threshold was established at 45% which provides an 80% confidence level (instead of 95% for the perfect target). Even if this percentage of error can be balanced by the measurement of numerous fragments, this choice also allows relevant diameter classes to be established for the interpretation of archaeological samples (which was not possible with a threshold of 70%: diameter classes would be too broad to be used for interpretation). As a first application, the minimal diameter classes which can be used for wood diameter study are:

]0-1 cm], ]1-2 cm], ]2-4 cm], ]4-8 cm],
]8-16 cm]; ]16-28 cm] and >28 cm

As regards the large dispersion of the measured values on large radii of curvature (Figure 8), the creation for broad classes such as ]16-28 cm] and >28 cm is indispensable to improve interpretation.

In addition, this study allows a minimal size of charcoal fragment to be estimated. Analyses show that results are acceptable for distances of about 4 mm using trigonometry in an isosceles triangle (but 6 mm for the two other techniques based on the wood rays). It is however possible to measure the radius of curvature on a small section, but it still fiddly for the manipulator.

These analyses used only oak sections, for reasons of wood ray size and tree-ring homogeneity. However, the question of other wood species is still pending. First observations on beech indicate that the technique can be used with large rays; nevertheless, for wood with small rays (like chestnut), it will be more difficult. Thus, to estimate the potential of the radius of curvature measurement on each type of wood (for large and small wood rays), further analyses are necessary.

## CONCLUSION

This work is the first which compares different measurement techniques, and is the first step towards mapping the limits and reliability of each measurement technique for a better interpretation of the data and their use models to determine wood diameter. This study shows that radius of curvature measurement based on tree-ring morphology alone must be avoided. The trigonometry in an isosceles triangle technique is the best method, with good reliability and ease of use.

This comparative study has shown the advantages and drawbacks of the techniques using image analysis software. Now it is indispensable to test these methods on charcoal fragments (Paradis, ongoing PhD). The first results show that measurement on charcoal is easier due to the better visibility of wood rays and the ease with which a fresh transverse section can be obtain by a simple split. But these analyses are still in progress.

Thanks to Farid Boumediene and Guillaume Grenouillet for their help on the statistical aspects and to Carmela Chateau, Emanuelle Grimaud and Amy Wells for English corrections. Thanks also to Romain Rouaud for discussion which was very helpful in the construction of new measurement techniques.

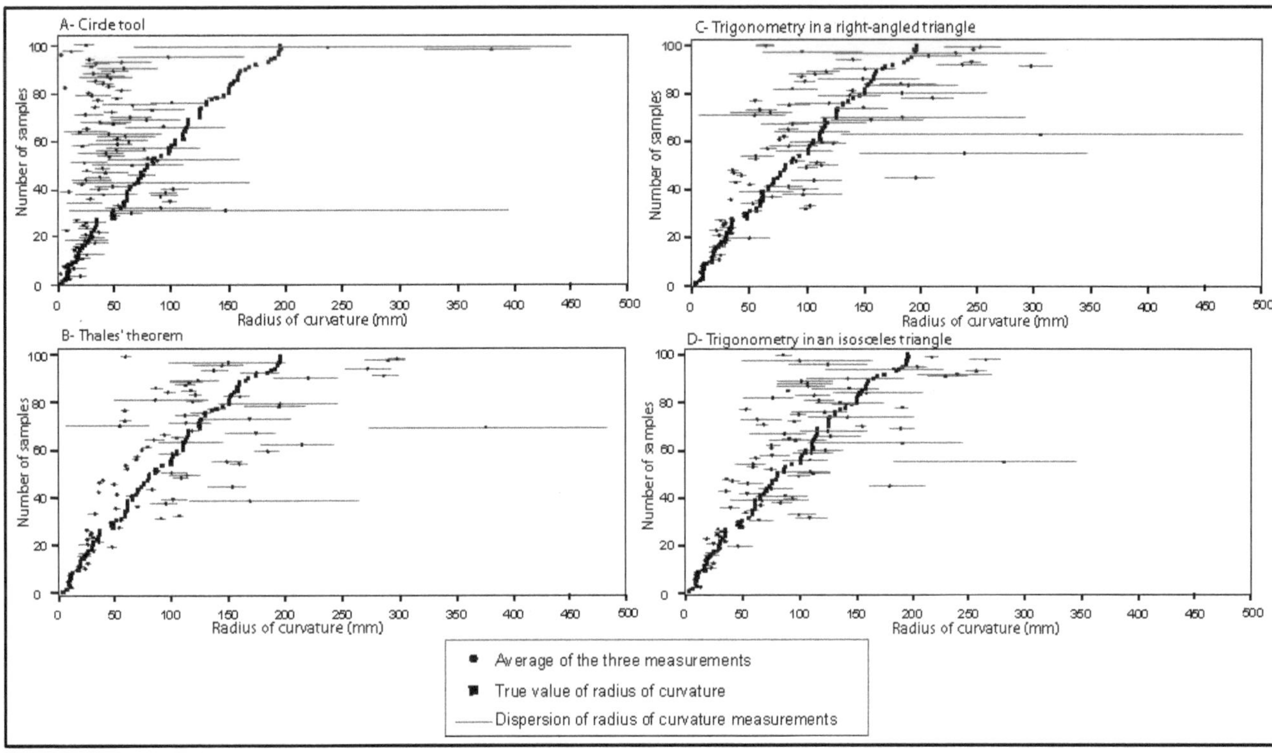


Figure 8. Dispersion of measurements around the radius of curvature on freshly cut transverse section

## References


CHABAL, L., FABRE, L., TERRAL, J.-F., and THÉRY-PARISOT, I. 1999. L'anthracologie. In J.E. Brochier C. Bourquin-Mignot, L. Chabal, S. Crozat, L. Fabre, F. Guibal, P. Marinval, H. Richard, J.F. Terral, I. Théry (éds.), *La Botanique*, 43-104. Paris, Errance (Collection "Archéologiques").

CHRZAVZEZ, J., 2006. *Apport méthodologique et étude de cas: la grotte de Fumane dans les Pré-Alpes italiennes*. Master II Environnement et Archéologie, Paris I.

DUFRAISSE, A. 2002. *Les habitats littoraux néolithiques des lacs de Chalain et Clairvaux (Jura, France): collecte du bois de feu, gestion de l'espace forestier et impact sur le couvert arboréen entre 3700 et 2500 av. J.-C. Analyses anthracologiques*. PhD Thesis, Université de Franche-Comté, Besançon.

DUFRAISSE, A. 2006. Charcoal anatomy potential, wood diameter and radial growth. In A. Dufraisse (ed.), *Charcoal analysis: new analytical tools and methods for archaeology. Papers from the Table-Ronde held in Basel 2004*, 47-59. Oxford, BAR Publishing (British Archaeological Reports International Series 1483).

FIORENTINO, G., and MAGRI, D. (eds), 2008. *Charcoals from the past: cultural and palaeoenvironmental implications. Proceedings of the third international meeting of anthracology (Lecce, Italy, June 28th – July 1st 2004)*. Oxford, BAR Publishing (British Archaeological Reports International Series S1807).

HILLEBRECHT, M.L 1982. *Die Relikte der Holzkohlewirtschaft als Indikatoren für Waldnutzung und Waldentwicklung. Untersuchungen an Beispielen aus Südniedersachsen*. Göttinger Geographische Abhandlungen 79.

LUDEMANN, T., and NELLE, O. 2002. *Die Wälder am Schauinsland und ihre Nutzung durch Bergbau und Köhlerei*. Freiburg, Forstwissenschatliche Fakultät der Universität Freiburg und Forstlische Versuchs- und Forschungsanstalt Baden-Württemberg.

LUNDSTRÖM-BAUDAIS, K. 1986. Etude paléoethnobotanique de la station III de Clairvaux. In Pierre Pétrequin (éd.), *Les sites littoraux néolithiques de Clairvaux-les-Lacs (Jura). I, Problématique générale, l'exemple de la station III*, 311-391. Paris, Maison des Sciences de l'Homme.

MARGUERIE, D. (ed.), 1992. *Evolution de la végétation sous l'impact anthropique en Armorique du Mésolithique au Moyen Age: études palynologiques et anthracologiques des sites archéologiques et des tourbières associées*. Rennes, U.P.R. n°403 du C.N.R.S.

MARGUERIE, D., and HUNOT, J.-Y. 2007. Charcoal analysis and dendrology: data from archaeological sites in North-western France. *Journal of Archaeological Science* 34, 1417-1433.

NELLE, O. 2002. *Charcoal burning remains and forest stand structure – Examples from the Black Forest (suth-west Germany) and the Bavarian Forest (south-east Germany)*. BAR International Series, 1063. Oxford: BAR Publishing. 201-207.

PARADIS, S. 2007. *Etude dendro-anthracologique: une approche méthodologique pour l'étude des calibres de bois. Application sur des charbonnières médiévales du Mont-Lozère (France, 48)*. Master Pro Archéosciences, Université de Dijon.

THIÉBAULT, S. (ed.), 2002. *Charcoal analysis. Methodological approaches, palaeoecological results and wood uses. Proceedings of the Second International Meeting of Anthracology, Paris, september 2000*. Oxford, BAR Publishing (British Archaeological Reports International Series 1063).

VERNET, J.-L. (ed.), 1992. *Les Charbons de bois, les anciens écosystèmes et le rôle de l'homme*. Colloque organisé à Montpellier en décembre 1991, Paris, Bulletin de la Société Botanique Française, Actualités Botaniques, 2/3/4.

WILLERDING, U. 1971, *Methodische Probleme bei der Untersuchung und Auswertung von Pflanzenfunden.* Vor- und frühgeschichtlichen Siedlungen. – Nachr. aus Niedersachsens Urgesch. 40: 180-198.

# ARCHAEOBOTANICAL RESEARCH IN THE PREHISTORIC BALEARIC ISLANDS: LANDSCAPE CHANGES AND CULTURAL PLANT USES

Llorenç PICORNELL GELABERT
Seminary of Prehistoric Studies and Research, University of Barcelona, Spain
Arqueobalear Research Team, University of the Balearic Islands, Spain

Gabriel SERVERA VIVES
Seminary of Prehistoric Studies and Research, University of Barcelona, Spain
Arqueobalear Research Team, University of the Balearic Islands, Spain
Geolab. CNRS, UMR 6042, University of Limoges, France

Santiago RIERA MORA
Seminary of Prehistoric Studies and Research, University of Barcelona, Spain

Ethel ALLUÉ MARTÍ
Institut Català de Paleoecología Humana i Evolució Social (Unitat asociada al CSIC),
University Rovira i Virgili, Tarragona, Spain

**Abstract**: *In this paper we present the results from charcoal and pollen analyses carried out in two prehistoric funerary sites in the Balearic Islands (Son Ferrer funerary mound in Mallorca and Cova des Pas in Menorca) with the aim to discriminate cultural uses of plants among the prehistoric societies and their relation with the shaping of the Mediterranean landscape of these islands. With the aim of developing a holistic approach, the results are analyzed and discussed in comparison with archaeobotanical data from contemporary domestic and funerary sites and with natural pollen records from the islands. The integration of archaeobotanical data obtained in diverse archaeological contexts (domestic, ceremonial and funerary) with pollen records from natural sediments let us identify different plant uses associated with different social spheres (economic, funerary), from which cultural uses and meanings of plants can be suggested. The results obtained with the data currently available in the Balearic Islands evidences the potential of this contextual and holistic approach to palaeo- and archaeobotanical data, which will be increased with new analysis in both fields (palaeoenvironmental and archaeobotanical analysis).*

**Key words**: *Anthracology/charcoal analysis, pollen analysis, Mediterranean landscape, Balearic Islands, Prehistory, human-landscape interactions, cultural uses of plants*

## INTRODUCTION

The aim of this research is to identify cultural uses of plants among prehistoric societies of the Balearic Islands (Figure 1). To achieve this goal, results from charcoal and pollen analyses carried out at the Cova des Pas cave (Menorca) and Son Ferrer funerary mound (Mallorca) will be discussed. Anthracological and palynological results will be integrated and confronted firstly with data obtained in contemporary domestic and funerary archaeological sites, and secondly with pollen records obtained in natural sediments from the islands. This holistic approach (e.g. Newman *et al.* 2007) will allow us to discuss the cultural patterns of plant uses among Balearic prehistoric communities.

Moreover, the confrontation of archaeobotanical datasets and pollen records will permit to improve the understanding of the shaping of a Mediterranean landscape in the Balearic Islands. Pollen records obtained in the Balearic Islands suggest that rapid, wide-ranging changes occurred in the islands' vegetation during the Holocene with a major change in plant communities occurring at *c.* 4000/3000 cal BC (Burjachs *et al.* 1994; Burjachs 2006; Pérez-Obiol *et al.* 2000; Pérez-Obiol 2007; Yll *et al.* 1994; 1995; 1999). It has been suggested that this rapid landscape change was caused by the impact of the first sporadic human presence in the islands, during the so-called period of "earliest visits", archaeologically dated at *c.* 3700/3000 cal BC (Calvo *et al.* 2002). Human impact intensified during a second phase of island colonisation, when human occupation stabilised in all biotopes throughout the largest Balearic Islands (Mallorca and Menorca) between *c.* 2870/2500 cal BC (Guerero *et al.* 2007; 2008) and 2350-2150 cal BC (Alcover 2008).

However, these changes in vegetation have also been attributed to a change in climate towards drier conditions and an increase in seasonality. Some scholars have suggested that these landscape changes could have been the combined result of insularity, climate variability and human impact (Pérez-Obiol 2007; Yll *et al.* 1999).

In order to have a better understanding of the islands' landscape shaping during the Balearic Prehistory,

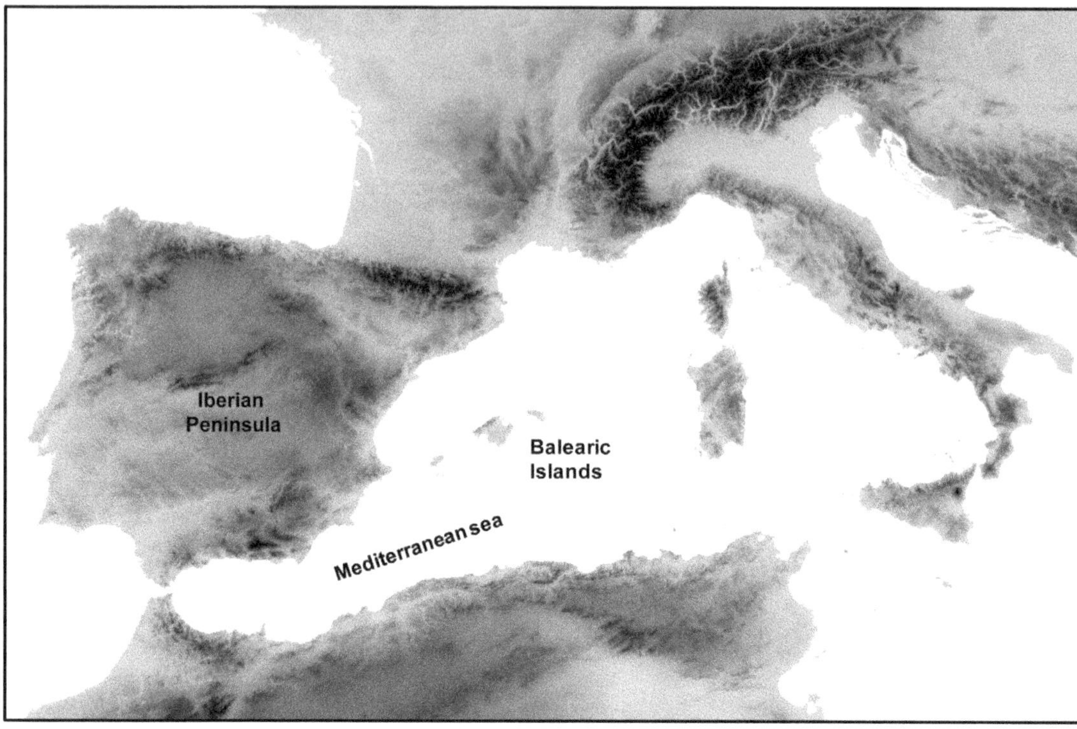


Figure 1. Map of the Balearic Islands

additional factors should be analyzed. It must be considered that the human-environment relationships have a dual nature, both material-physical and cultural-symbolic. In this sense, anthropological studies suggest that human cultures interpret nature in order to be able to manage its resources successfully. In consequence, the cultural perception of the environment could not be interpreted as exclusively epiphenomenal to human adaptations to the environment (Descola 2005). Therefore, the cultural perception of spaces, plants and animals also take part in the regulation of the interactions between people, space and nature (Ingold 2006) and, in consequence, the perception of the environment constitutes a relevant factor in the shaping of cultural landscapes. Recent archaeological theories and approaches have introduced these new perspectives in the interpretation of past landscapes (Thomas 2002; Ejarque and Orengo 2009). In the Mediterranean region, human-plant interactions have largely influenced the shaping of the landscape and, as such, the Mediterranean environments must be broadly considered cultural landscapes (Roberts *et al.* 2001).

Ethnography (Rival 1998), ethnoarchaeology (Guiot 2002, 2003; Picornell *et al.* 2011) and archaeobotany (Newman *et al.* 2007) have demonstrated that the social significance of plants influences their uses and, in consequence, contributes to landscape configuration. This paper proposes an approach to the cultural meanings and uses of plants by prehistoric groups and the role that these cultural significances play in the shaping of cultural landscapes in a Mediterranean insular environment. The proposed approach analyses two complementary aspects: firstly, the material relation between humans and plants and the way in which human activities transformed the Balearic landscapes; and, secondly, the cultural uses of plants and their relation with the cultural perception of these environments.

We propose that the study of the cultural uses given to of plants can be approached through the integration of anthracological and palynological analyses carried out in prehistoric funerary archaeological sites. The results of these analyses can then be compared with those obtained in archaeobotanical data from other domestic and funerary sites in the Balearic islands (Figure 2). This comparison of records from different social contexts should enable us to analyze a wide array of plants selected for domestic and funerary practices.

## GEOGRAPHICAL SETTING

The Balearic archipelago is located in the western Mediterranean (Figure 1). It is composed of five islands, Mallorca and Menorca being the larger ones. Mallorca is the most mountainous island, with the Tramuntana range reaching an altitude of 1453 m.a.s.l. Mediterranean maritime climate regime predominates in the islands. The climate regime is characterised by hot summers and mild winters with the highest precipitations occurring in autumn and early spring (Bolós and Vigo 1984). The bioclimate of the islands corresponds to a Mediterranean thermoclime. In Mallorca, however, a meso-Mediterranean bioclimate regime is widely represented, while small areas of the mountain summits are characterised by a supra-Mediterranean climate. The islands' ombroclimates are more diverse, ranging from a sub-humid type in Menorca and in the highest areas of Mallorca, to a dry type on the lower plains of southern Mallorca (Rivas Martínez 1995, 1997).

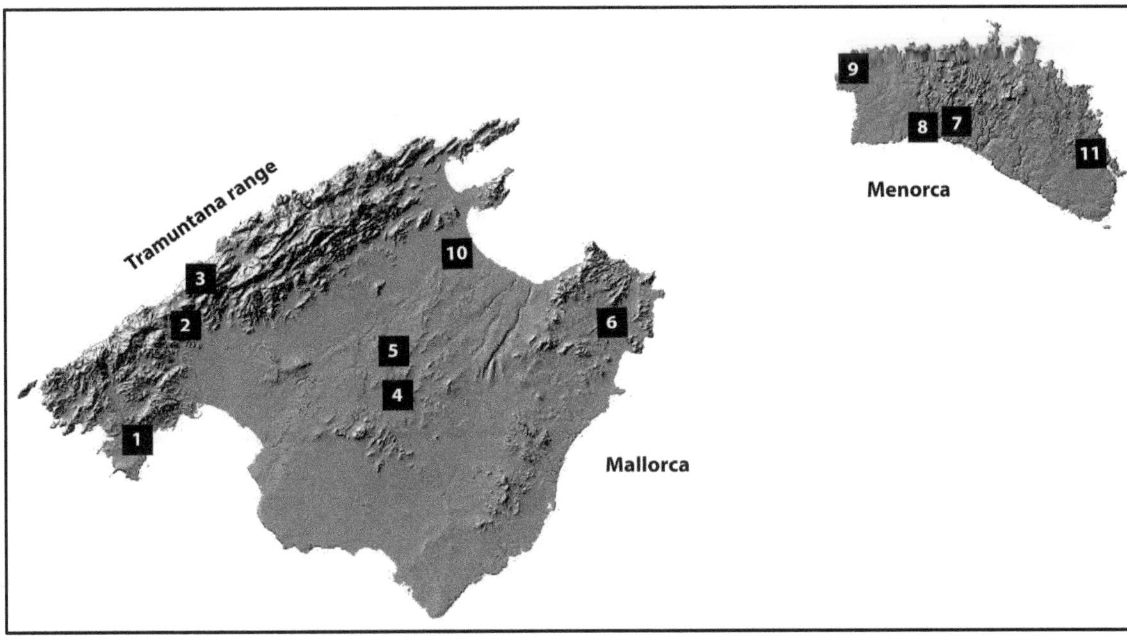


Figure 2. Archaeological sites with anthracological and/or archaeopalinological analysis in Mallorca

Evergreen oak forests (*Quercus ilex*) are spread over the inland while littoral sectors are dominated by a Mediterranean macchia formed by *Olea europaea* var. *sylvestris*, *Pistacia lentiscus*, *Pinus halepensis* and diverse species of *Juniperus*. Relict communities of *Buxus balearica* and deciduous trees, such as *Acer opalus*, can be found in Mallorca's Tramuntana range. Zones above 1200 m are characterized by an oromediterranean belt formed by shrubs and grasses with a wide diversity of endemic flora (Rita 1998).

## SITE DESCRIPTION

The Son Ferrer funerary mound, located in North-Western Mallorca (Figure 2) is a ritual and funerary archaeological site used by prehistoric people over a period of 2,000 years. The site was established at c. 2000 BC as a hypogeum excavated in the bedrock. This space was reused as a collective burial site between 400 and 200 cal BC. At c. 1100-850 cal BC, new ritual areas were added to the hypogeum and in 800 cal BC a stepped tumulus was erected over it. Finally, between 200 BC and 0 cal BC/AD, child corpses were buried in the external part of the mound and at the entrance of the hypogeum (Calvo *et al.* 2005; Picornell *et al.* 2012).

Cova des Pas is a karstic cave located in a rock shelter of the Trebaluger ravine, in Southern Menorca (Figure 3). The cave was used as a collective burial site at the end of the Late Bronze Age and Early Iron Age, at 900-800 cal yr BC (Van Strydonck *et al.* 2010). A total of 66 individuals were deposited into the cave using wooden stretchers after the corpses were flexed, covered with leather shrouds and tied up using ropes (Fullola *et al.* 2007). The exceptional preservation of the organic remains and materials in this cave (including wood, hair, organ tissues, rope, vegetal fibres, leather shrouds and coprolites), makes it one of the most important archaeological sites in the Balearic Islands. The preservation of this organic material could probably be the result of specific environmental conditions inside the cave. In southern Menorca, other funerary caves, such as Mussol and Càrritx (Figure 3) have also been excavated (Lull *et al.* 1999).

## MATERIALS AND METHODS

In Son Ferrer funerary mound, a dual strategy was used to recover charcoal assemblages inside the hypogeum (Stratigraphical Unit 9): manual recovery of charcoal pieces during the excavation process and flotation of all the sediment contained in the stratigraphical unit. Through manual recovery, 293 charcoal fragments were collected and identified. The charcoal assemblage resulting from flotation technique was sub-sampled and 476 charcoal fragments > 2 mm were analysed. Moreover, the complete assemblage of 79 charcoal fragments contained in the ceremonial fire related with infant inhumations (Stratigraphical ***Unit*** 64) were analysed. Besides, in Cova des Pas, 40 charcoal fragments from a ceremonial fire related with the first inhumations into the cave were identified.

Dried sediment samples from both sites were treated for pollen analyses following standard extraction procedures (Faegri *et al.* 1989) and 300 pollen grains were identified per sample. In Cova des Pas, pollen percentages have been calculated on the basis of the total pollen sum while in Son Ferrer mound, Cichorioideae has been excluded from pollen sum due to its overrepresentation in some archaeological layers. Pollen identifications are based on atlas and morphological keys (Beug, 2004; Faegri *et al.* 1989; Reille, 1992, 1995, 1998; Valdés *et al.* 1987), and a specific pollen atlas of the Balearic flora (Boi and Llorens, 2007).

| | Son Ferrer Funerary Mound | | | Cova des Pas |
|---|---|---|---|---|
| | SU 9 Nº frags | % | SU 64 Nº frags | Fireplace Nº frags |
| *Acer* sp. | 1 | 0,13 | – | – |
| *Arbutus unedo* | 1 | 0,13 | – | – |
| *Cistus* sp. | 71 | 9,23 | – | – |
| *Hedera* sp. | 1 | 0,13 | – | – |
| *Lamiaceae* | 20 | 2,6 | – | 4 |
| *Olea europaea* | 54 | 7,02 | 3 | – |
| *Pinus* cf. *halepensis* | 228 | 29,64 | – | – |
| *Pistacia* cf. *terebinthus* | 4 | 0,52 | – | – |
| *Pistacia lentiscus* | 229 | 29,77 | 65 | 32 |
| *Quercus* deciduous | 1 | 0,13 | – | – |
| *Rhamnus alaternus/Phyllirea* sp. | 17 | 2,21 | – | – |
| *Rhamnus* t. *lycioides/oleoides* | 2 | 0,26 | – | – |
| *Rosmarinus officinalis* | 18 | 2,34 | – | 3 |
| cf. *Arbutus unedo* | 1 | 0,13 | – | – |
| cf. *Cistus* sp. | 2 | 0,26 | – | – |
| cf. *Erica* sp. | 1 | 0,13 | – | – |
| cf. *Olea europaea* | – | – | 1 | – |
| cf. *Lamiaceae* | 4 | 0,52 | – | – |
| cf. *Pistacia lentiscus* | 13 | 1,69 | 1 | – |
| cf. *Rosmarinus* | 1 | 0,13 | – | – |
| Undiferenciated Angiosperm | 65 | 8,45 | 9 | 1 |
| Undiferenciated Conifer | 35 | 4,55 | – | – |
| **Total nb of fragments** | **769** | **100** | **79** | **40** |

Figure 3. Archaeological sites with anthracological and/or archaeopalinological analysis in Menorca

Pollen analyses have been carried out in four sediment samples from constructional and funerary contexts of the Son Ferrer funerary mound. The constructional levels correspond to the stratigraphical units 65 and 56 dated at 800-700 BC and 200-0 BC, respectively. The ritual levels include stratigraphical unit 17, an external ritual space dated at 1100-850 BC, and stratigraphical unit 9, which corresponds to the collective burial inside the hypogeum, dated at 400-200 BC. In the Cova des Pas Cave, 70 sediment samples were treated for pollen analyses. These samples were obtained from different parts of the bodies, such as heads, thorax, coccyx, arms and legs.

## RESULTS

At the SU 9 of the Son Ferrer funerary mound, 13 plant taxa have been identified in an assemblage of 769 charcoal fragments, including *Acer* sp., *Arbutus unedo, Cistus* sp., *Hedera* sp., Lamiaceae, *Olea europaea, Pinus* cf. *halepensis, Pistacia* cf. *terebinthus, Pistacia lentiscus,* deciduous *Quercus, Rhamnus alaternus/Phyllirea* sp., *Rhamnus* t. *lycioides/oleoides* y *Rosmarinus officinalis* (Figure 3). Two taxa, *Pinus* cf. *halepensis* and *Pistacia lentiscus,* represent more than half of the assemblage, with each taxon reaching values of 30%. In addition, *Cistus* sp. and *Olea europaea* are present with values between 10% and 5%. The rest of the taxa record values lower than 3% (Figure 3). At the same site, the 79 charcoal fragments recovered at the fireplace related with the infant burials (SU 64), correspond to *Pistacia lentiscus* with an occurrence of *Olea europaea* (Figure 3).

In Cova des Pas, the assemblage of 40 charcoal fragments from the fireplace related with the first inhumations corresponds to *Pistacia lentiscus* – the most abundant taxon – with the presence of *Rosmarinus officinalis* and Lamiaceae (Figure 3).

Pollen assemblages from Son Ferrer mound show a scarce representation of Arboreal Pollen -recording values between 6% and 15%- and shrubs –with percentages between 3% and 25% (Figure 4). Pollen assemblages obtained in ritual and burial contexts (SUs

| | UE 9 | UE 56 | UE 65 | UE 17 |
|---|---|---|---|---|
| *Pinus* | 6,82 | 1,55 | 0,98 | 6,22 |
| *Quercus ilex* type | 1,70 | 2,33 | 4,39 | 1,91 |
| *Tamarix* type | – | – | – | 0,96 |
| *Quercus* sp | – | 1,55 | – | 0,96 |
| Olea | 1,70 | – | 0,49 | 2,39 |
| *Olea/Phillyrea* type | – | – | – | 0,48 |
| *Corylus* | – | – | 0,98 | – |
| Deciduous *Quercus* | 0,57 | 1,55 | – | – |
| *Tilia* | 0,57 | – | – | – |
| *Alnus* | 1,70 | 1,55 | – | – |
| *Salix* type | 1,70 | 0,78 | – | 1,44 |
| *Pistacia* | – | – | 0,98 | – |
| Ericaceae | – | – | 0,98 | – |
| *Juniperus* | – | 0,78 | 1,46 | – |
| *Ephedra fragilis* | 13,64 | 2,33 | 6,34 | 22,49 |
| *Ephedra altissima* | – | – | – | 1,44 |
| *Chamaerops humilis* | – | 0,78 | – | – |
| Poaceae | 20,45 | 32,56 | 21,95 | 17,22 |
| *Phragmites* | – | 1,55 | 0,49 | 0,48 |
| Urticaceae | – | 0,78 | – | – |
| *Plantago major/media* type | – | – | 0,49 | – |
| *Plantago* sp | 3,98 | 7,75 | 7,32 | – |
| *Plantago lanceolata* type | 2,27 | 3,88 | 6,83 | – |
| *Rumex* sp | – | – | 0,98 | 0,96 |
| *Chelidonium/Glaucium* type | 0,57 | 0,78 | – | – |
| Chenopodiaceae | 11,36 | 10,08 | 0,98 | 10,53 |
| *Lotus* type | 1,70 | – | – | – |
| *Humulus* type | – | – | – | 0,48 |
| *Cannabis* type | 1,14 | – | 1,46 | – |
| *Carduus* type | 4,55 | 5,43 | 13,17 | – |
| *Xanthium* type | 4,55 | 1,55 | 9,27 | 0,48 |
| Asteoideae | 3,41 | 13,18 | 4,39 | 1,44 |
| Cichoriodeae * | 51,70 | 103,88 | 34,15 | 31,58 |
| Brassicaceae | 2,84 | 0,78 | 0,98 | 3,35 |
| *Lamium* type | 5,68 | 1,55 | 0,49 | 10,53 |
| *Helianthemum* | – | 1,55 | 11,71 | 0,48 |
| Apiaceae | 0,57 | 1,55 | – | 0,48 |
| Cyperaceae | – | – | – | 0,48 |
| *Potentilla* | – | – | 0,49 | – |
| Caryophyllaceae | – | – | 0,49 | – |
| *Ranunculaceae* | 1,70 | 1,55 | – | – |
| *Saxifraga* | 1,14 | 1,55 | – | 0,48 |
| *Sedum* | 0,57 | – | – | – |
| *Euphorbia* | – | – | 1,46 | – |
| *Verbascum* type | – | – | – | 2,39 |

| | UE 9 | UE 56 | UE 65 | UE 17 |
|---|---|---|---|---|
| *Scrophularia* type | – | – | – | 0,48 |
| *Linaria* type | 0,57 | 0,57 | 0,57 | 0,57 |
| Scrophulariaceae undiff. | 4,55 | 0,78 | 0,49 | 11,48 |
| Indeterminables | 22,16 | 41,86 | 39,51 | 14,35 |
| AP | 14,77 | 10,08 | 5,85 | 12,92 |
| Shrubs | 14,20 | 3,10 | 9,76 | 24,88 |
| Herbs | 71,02 | 86,82 | 83,41 | 61,72 |
| Ruderal+Nitrophilous | 32,95 | 43,41 | 45,37 | 16,75 |

Figure 4. Results of the archaeopalinological analysis of Cova des Pas (Servera *et al.* 2008)

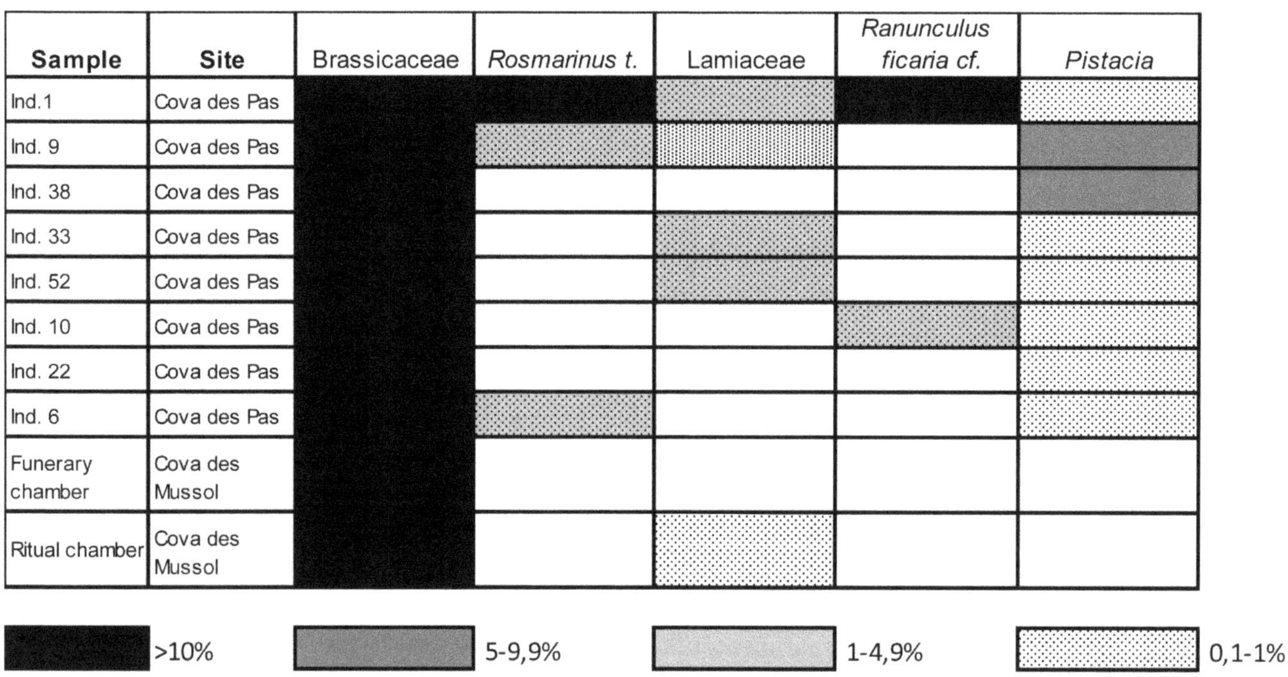


Figure 5. Histogram showing the percentage pollen data from Son Ferrer funerary mound (from Picornell *et al.* 2008).
The black columns correspond to S.U. 17, ritual area added to the hypogeous 1100-850 cal yr BC.
The grey columns correspond to S.U. 9, funerary use of the hypogeous 500-200 cal yrs BC

17 and 9) show high percentages of herb taxa such as *Lamium* type, *Ephedra* sp, Brassicaceae, Scrophulariaceae, Poaceae, Cichorioideae, Asteroideae and Chenopodiaceae, as well as remarkable presences of *Pinus* sp. and *Olea europaea.* Occurrences of *Quercus ilex* type, *Salix* sp., *Juniperus* sp., *Tamarix*, *Plantago* sp. and *Lotus* type are reported. The most frequent pollen grains in constructive layers (SUs 65 and 56) are Poaceae, Asteroideae, Cichrioideae and *Plantago* sp. with a remarkable presence of *Helianthemum*, *Quercus ilex* type and Chenopodiaceae (Figura 4).

Pollen assemblages of Cova des Pas also show low values of Arboreal Pollen and high diversity of herbs pollen taxa. These assemblages record high values of entomophilous plants, mainly Brassicaceae *Ranunculus ficaria* cf. and *Rosmarinus*. Other represented taxa are: *Plantago* sp., Poaceae, Asteroideae, Cichoriodeae, Apiaceae and Ericaceae as well as cultivated taxa such as *Ficus* and Cerealia (Figure 5).

## DISCUSSION

### Palaeoenvironmental background: the shaping of the Mediterranean landscape in the Balearic Islands

Pollen records obtained in natural sediments indicate a change in the islands' vegetation between 4000 and 3000 cal BC (Yll *et al.* 1999) coinciding with the stable human occupation of every biotope in the islands between c. 3000/2500 and 1700 cal BC (Calvo *et al.* 2002). Pollen records show that before this change in vegetation, deciduous trees such as *Corylus, Acer* or deciduous *Quercus*, along with *Buxus*, *Taxus baccata* and *Ilex aquifolium,* dominated the landscape. Since 4000/3000

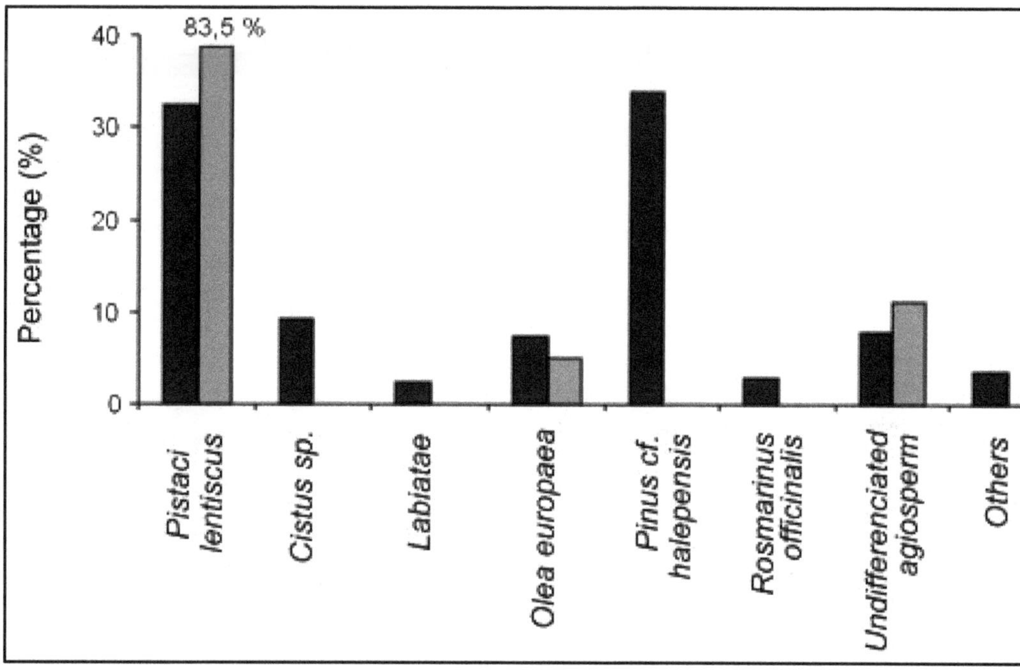


Figure 6. Histogram showing the percentage charcoal data from Son Ferrer funerary mound (from Picornell *et al.* 2008). The black columns correspond to S.U. 9, funerary use of the hypogeous 500-200 cal yrs BC. The grey columns correspond to S.U. 64, ritual fire related with child burials 200-o cal yrs BC/AD

cal BC, the presence of these plants decreased to the extent that *Corylus* disappeared from islands and *Buxus* only remained as a relict plant in Tramuntana mountain range in the North of Mallorca (Yll *et al.* 1999). Similarly, other trees, such as *Acer* and *Taxus*, have survived in small relicts in the mountains of Mallorca (Pérez-Obiol *et al.* 2003). At this time, *Olea* and evergreen *Quercus* expanded and were established as the dominant woody plants in both islands, while the presence of *Pinus* expanded, specially in Mallorca. Thus, pollen records indicate the establishment at this time of the evergreen Mediterranean forests and *macchias,* which have dominated the Balearic landscape over the last four millennia (Yll, *et al.* 1999).

Anthracological analyses carried out at archaeological sites should provide additional information for a better understanding of this process of vegetation change and their causes. Currently, little information is available on the islands' earlier settlements, but charcoal analyses have been carried out at the sheep and goat pen layers of the Son Matge rock shelter (Tramuntana range, Mallorca) (Figure 2), dating to the beginning of the Third millennium (Bergadà *et al.* 2005). In these layers, *Buxus* dominates but the presence of more xeric taxa, such as *Olea europaea, Pistacia lentiscus* and *Rhamnus/Phillyrea* is also reported (Picornell *et al.* 2009-2010). It has been suggested that some of these plants could be used as tree fodder by pastoral human groups using the northern mountains of Mallorca for grazing. Moreover, charcoal analyses from the nearby (and contemporaneous) rock shelter site of Son Gallard (Tramuntana range, Mallorca) (Figure 2) report the presence *Juniperus* and *Olea europaea* being used as fuelwood by these earlier pastoral groups (Picornell *et al.* 2009-2010).

On the other hand, the use of *Buxus* wood to manufacture artefacts has also been documented during the Bronze Age (c. 1700-1000/900 BC) in Mallorca and Menorca islands (Piqué and Noguera 2002), but *Buxus* charcoal has not been yet reported in archaeological layers of this period. With the available data, it could be suggested that prehistoric peoples exploited the relicts of *Buxus* in the mountain areas of Mallorca and exported the wood as raw material or as manufactured artefacts to Menorca (Piqué 1999), as it could be suggested by the fact that pollen records from Menorca evidence that *Buxus* disappeared from this island around 3000 cal BC (Burjachs 2006).

Deciduous trees have been identified in the anthracological record of the Late Iron Age in Mallorca. At the Son Ferrer site, deciduous *Quercus* and *Acer* charcoal fragments were present in the funerary layer dated at c. 500-200 cal BC. Today, *Acer opalus* is present as a relict in the mountains of northern Mallorca, but currently there are no deciduous oaks on the Balearic Islands (Rita coord.). In relation to the noted presence of *Corylus* in the pollen records from natural sediments, it should be noted that archaeobotanical evidences of this taxon has not been found in the anthracological analysis. Moreover, any significant palynological evidences have been documented in the studied archaeological sites. In this case, archaeobotany does not furnish us data to properly decipher the role of human in the reduction of these deciduous trees.

The archaeobotanical studies conducted at two domestic sites on the central plain of Mallorca (archaeological sites of Son Fornés and Son Ferragut) (Figure 2) reveal a change in the composition of anthracological assemblages during the Iron Age (Piqué and Noguera

2002). In both villages, *Olea europaea* was almost the sole source of fuelwood for domestic fires prior to *c.*500 cal BC, but after that date, a diversification in the fuelwood supply and the use of *Pinus halepensis* and shrubs, such as *Cistus*, *Erica*, *Pistacia* and *Rhamnus/Phillyrea* are reported.

Pollen studies from the constructive layers of Son Ferrer (SUs 56 and 65), provide additional information to better understand the changes in charcoal records. High levels of NAP (mainly Poaceae) indicate that the mound was located in an open landscape, dominated by herbs and shrubs during the Bronze Age. This open environment was probably the result of long-term human activity in the area before the mound was built c. 1650/1500 – 1300 BC. However, the pollen record from the Alcúdia lagoon, located at the opposite side of Mallorca island (Figure 2), does not provide evidences of major landscape changes at *c.*500 cal BC, although a decrease in deciduous trees is recorded when Mediterranean shrubs expand as evidenced by the increase of pollen value in *Pistacia* and Ericaceae (Burjachs *et al.* 1994). Changes in the management of forest resources occurred at the time of economic and cultural transformations in the Balearic Prehistory, and were characterized by an increase in social and political hierarchy and an intensification of external contacts with Mediterranean colonial peoples (Guerrero *et al.* 2002; Lull *et al.* 1999). These facts should have had influence in the organization of the forest resources management and, consequently, on the landscape, too.

### The cultural dimensions of plants

Charcoal analysis of the ceremonial fire associated with the first burials in Cova des Pas indicates that *Pistacia lentiscus*, *Rosmarinus officinalis* and other Lamiaceae were used as fuelwood in this funerary context. In addition, the archaeopalynological analysis of this site has enabled us to identify plant offerings and other plant uses in funerary practices (Servera *et al.* 2008; Riera *et al.* 2008). The most prevalent taxa reported in the corpses are Brassicaceae, together with other taxa, such as *Rosmarinus* t. *Ranunculus* cf. *ficaria*, *Plantago sp.*, *Pistacia* and Poaceae (Figure 4). The presence of pollen clumps and tetrads indicate that flowers were probably deposited as funerary offerings (Bui-Thi-Mai and Girard, 2002). Variability in the presence of flowers and plant offerings is reported and it seem to be associated with different body parts. This fact suggests a complex pattern of plant uses in burial practices and a similar complexity in the meanings attributed to plants by Balearic prehistoric peoples.

In the hypogeum of Son Ferrer funerary mound (SU 9, 500-200 BC), charcoal analysis shows the presence of resinous taxa such as *Pinus halepensis*, *Pistacia lentiscus* and *Cistus*, as well as that of aromatic plants such as *Rosmarinus officinalis* and undifferentiated Lamiaceae (Figure 3). These evidences suggest that funerary fires were probably carried out inside the hypogeum with aromatic purposes. *Pistacia lentiscus* was almost the only wood found in the ceremonial fire associated with child burials (SU 64, 200 BC-0 BC/AD) (Figure 3) suggesting a cultural preference towards this plant in funerary ceremonies. This interpretation is strengthened by the fact that high percentages of *Pistacia lentiscus* charcoal have also been reported at other funerary sites, such as that of Cova des Càrritx in Menorca (Piqué 1999), whereas in domestic fires, the presence of this plant is scarce (Piqué & Noguera, 2002). Similarly, *Rosmarinus officinalis* and Lamiaceae charcoal are testimonial in domestic archaeological contexts, but they have been recorded in funerary caves such as Cova des Càrritx (Piqué 1999) and Cova des Pas in Menorca, as well as in the Son Ferrer mound in Mallorca (Figure 3). The fact that the presence of both resinous and aromatic plants has been reported in caves containing corpses suggests that the aromatic properties of these plants might have played an important role in funerary rituals of the islands (Javaloyas *et al.* 2009).

In the Son Ferrer funerary mound, pollen analyses carried out in the ceremonial (SU 17, 1100-800 cal BC) and burial (SU 9, 500-200 cal BC) contexts (Figure 4) reveal a higher presence of *Lamium* type, *Ephedra*, Scrphulariaceae and Brassicaceae than in constructive layers. These data could corroborate the evidence that some plants could be selected and used with ceremonial purposes.

The selective use of plants during the Prehistory of the Balearic Islands can also be observed in other species. *Quercus ilex/coccifera*, for example, has not been reported in the anthracological record (domestic or ceremonial) or as a raw material used in wooden artefacts (Piqué 1999; Piqué and Noguera 2002). However, pollen studies conducted in natural sediments show that this taxon has been one of the most widespread plant in Mallorca and Menorca since 4000-3000 cal yr BC (Burjachs 2006; Burjachs *et al.* 1994; Yll *et al.* 1999). Moreover, *Quercus ilex/coccifera* type pollen has also been reported in archaeological sediments although this taxon does not appear as charcoal in the same sites, as in the village of Ses Païsses (Burjachs 2005) and the funerary sites of Son Ferrer and Cova des Pas (Servera *et al.* 2008). It would seem, therefore, that prehistoric societies avoided using the evergreen oak as fuelwood. However, more data will be needed to identify the cultural, economic and/or ecological factors that explain this pattern of use.

## CONCLUSION

Anthracological and archaeopalynological spectra obtained from ceremonial and funerary sites differ from those documented in domestic spaces. Thus, *Rosmarinus officinalis* and Lamiaceae charcoal, together with high percentages of *Pistacia lentiscus*, appear only in ritual fires related to funerary practices. This suggests that these species could be selected for their aromatic properties for funerary fires. On the other hand, *Quercus ilex/coccifera* has not been documented in the anthracological spectra from prehistoric sites, although the taxon was present in the landscape as shown by the pollen diagrams. Finally, we have identified the important role played by flowers as part of the grave goods and offerings in funerary rites.

By integrating archaeobotanical data obtained in different archaeological contexts (domestic, ceremonial and funerary) with pollen records from natural sediments, it is possible to propose specific contextual uses for plants. Thus, we have identified different plant uses associated with different social spheres (economic and funerary), from which cultural uses and meanings of plants can be pointed out.

Therefore, the case of the Balearic Islands suggests that cultural plant uses could derive from their meanings and, as such, this cultural meaning of plants and places filters human impact on the environment. Accordingly, although we are still working with preliminary data from a broad interdisciplinary project, this research aims at analysing the formation of insular landscapes as the result of centuries of socio-environmental interactions.

Although it has been claimed that human influence has had an influential role in the shaping of the Balearic landscapes, it is difficult to establish a direct correlation between the changes identified in the palaeoenvironmental record and the cultural meanings of plants. At present, we need to increase the number of archaeobotanical studies being conducted at archaeological sites of different chronology and functionality, and of paleobotanical analyses undertaken in well-dated natural sediments.

The potential of this contextual and holistic and cultural approach to palaeo- and archaeobotanical data is self-evident if we consider that these preliminary conclusions have been drawn from the few data available for the Balearic Islands. New studies will certainly provide further arguments to reinforce the relevance of cultural perceptions of the environment as key factors in the shaping of the cultural landscape.

## Acknowledgments

The archaeobotanical research has been carried out as part of an interdisciplinary project: Producing, Consuming, Exchanging. Exploitation of Resources and External Interaction of the Balearic Communities during the Late Prehistory, HAR2008-00708 (Ministry of Science and Innovation, Spanish Government). Llorenç Picornell is the holder of a doctoral fellowship from the Ministry of Science and Innovation, Spanish Government.

## References

ALCOVER, A. 2008. The First Mallorcans: Prehistoric Colonization in the Western Mediterranean. *World Prehistory* 21, 19-84.

BERGADA, M., GUERRERO, V. and ENSENYAT, J. 2005. Primeras evidencias de estabulación en el yacimiento de Son Matge (Serra de Tramuntana, Mallorca) a través del registro sedimentario. *Mayurqa* 30, 153-180.

BEUG, H-J., 2004. *Leitfaden der Pollenbestimmung für Mitteleuropa und angrenzende Gebiete.* Verlag Friedrich Pfeil, Munich.

BOI, M., LLORENS, L., 2007. *Atlas polínico de las Baleares*. Flora endémica. Govern de les Illes Balears. Conselleria de Media Ambient. Palma.

BOLOS, O. and VIGO, J. 1984. *Flora dels Països Catalans*. Barcelona, Barcino.

BURJACHS, F., PEREZ-OBIOL, R., ROURE, J.M. and JULIA, R. 1994. Dinámica de la vegetación durante el Holocenos en la isla de Mallorca. In *Trabajos de palinología básica y aplicada,* X Simposio de Palinologia (APLE), collective volume. Valencia, Universitat de València.

BURJACHS, F. 2005. Informe de l'anàlisi palinològica del jaciment arqueològic de Ses Païsses (Artà, Mallorca, Illes Balears). In Aramburu, J. and Hernández, J. (eds.) *Memoria de las excavaciones arqueológicas en el poblado talayótico de Ses Païsses (Artà, Mallorca).* Campañas 1999-2000, CD-ROM, Palma de Mallorca.

BURJACHS, F. 2006. Palinología y restitución paleoecológica. *Ecosistemas* 15 (1), 7-16.

CALVO, M. and GUERRERO, V. 2002. *Los inicios de la metalurgia en Baleares. El calcolítico (c. 2500-1700 cal BC*). Palma de Mallorca, El Tall Editor.

CALVO, M., FORNES, J., GARCIA, J., IGLESIAS, M.A. and JUNCOSA, E. 2005. Condicionantes espaciales en la construcción del turriforme escalonado de Son Ferrer (Calvià, Mallorca). *Mayurqa* 30 (1), 485-508.

DESCOLA, P. 2005. *Par-delà nature et culture*. Paris: Editions Gallimard.

EJARQUE, A. and ORENGO, H.A. 2009. Legacies of Change: the shaping of cultural landscapes in a marginal mediterranean mountain range, the Garraf massif, north-eastern Spain. *Oxford Journal of Archaeology* 28(4): 425–440.

FAEGRI, K., IVERSEN, J., 1989. *Textbook of pollen analysis*, fourth ed. John Wiley and Sons, Chichester.

FULLOLA, J.M. *et al.* 2007. La Cova des Pas (Ferreries, Menorca): un avanç. In *L'Arqueologia a Menorca: eina per al coneixement del passat.* Collective volume, 95-110, Maó, Consell Insular de Menorca.

GUERRERO, V., CALVO, M. and SALVÀ, B. 2002. La cultura talayòtica. Una sociedad de la Edad del Hierro en la periferia de la colonización fenicia. *Complutum* 13, 221-258.

GUERRERO, V., CALVO, M., GARCIA, J. and GORNES, S. 2007. Prehistoria de las Islas Baleares. Registro arqueológico y evolución social antes de la Edad del Hierro, Oxford, *British Archaeological Reports International Series* 1690.

GUERRERO, V. and CALVO, M. 2008. Resolviendo incertidumbres. Nuevos datos sobre las primeras ocupaciones humanas de las Baleares. In Hernández, M, Soler, J and López, J. (eds.) IV Congreso el Neolítico peninsular, Alicante, Museo Arqueológico de Alicante.

GUIOT, H. 2002. They plant what they will burn. The impact of traditional burning on the landsacapes of Wallis and Futuma (western Polynesia). Proposals for an ethnoarchaeological and palaeoenvironmental study. In S. Thiébault (ed.) *Charcoal analysis. Methodological Approaches, Paleoecological results and Wood Uses*, 251-254. Oxford, BAR Inernational Series.

GUIOT, H. 2003. Organisation du territoire et répresentations culturelles sur le fonctionement de l'écosystème insulare: quelques cas polynésiens ('Uvea, Futuma, Nuku Hiva). In C. Orliac (dir.) *Archéologie en Océanie insulaire. Peuplement, sociétés et paysages.* Paris, Editons Artcom.

INGOLD, T. 2006. *The perception of the environment. Essays in livelihood, dweling and slill.* New York, Routledge.

JAVALOYAS, D., PICORNELL, Ll. and SERVEA, G. 2009. Plantas y fenomenología de la muerte durante el Bronce Medio y Final en Menorca. In I Jornadas de Jóvenes en Investigación Arqueológica: *Dialogando con la Cultura Material*. Collective volume. Madrid.

LULL, V., MICO, R., RIHUETE, C. and RISCH, R. 1999. *Ideología y sociedad en la prehistoria de Menorca. La Cova des Càrritx y la Cova des Mussol.* Barcelona, Consell Insular de Menorca.

NEWMAN, C., O'CONNELL, M. and DILLON, M. 2007. Interpretation of charcoal and pollen data relating to a late Iron Age ritual site in eastern Ireland: a holistic approach. *Vegetation history and Archaeobotany* 16, 349-365.

PEREZ-OBIOL, R. 2007. Palynological evidence for climatic change along the eastern Iberian Peninsula and Balearic Islands. *Contributions to Science* 3 (3), 415-419.

PÉREZ-OBIOL, R, YLL, E., PANTALEÓN-CANO, J. and ROURE, J.M. 2000. Evaluación de los impactos antrópicos y los cambios climáticos en el paisaje vegetal de las Islas Baleares durante los últimos 8000 años. In V. Guerrero and S. Gornés (eds.) *Colonización humana en ambientes insulares. Interacción con el medio y adaptación cultural*, 444-454. Palma de Mallorca, Universitat de les Illes Balears.

PEREZ-OBIOL, R., SAEZ, Ll. and YLL, E. 2003. Vestigis florístics postglacials a les Illes Balears I dinàmica de la vegetació holocènica. *Osiris* 18, 77-94.

PICORNELL, Ll., GUERRERO, V. and CALVO, M. 2009-2010. Anàlisis antracològiques a Son Matge i Son Gallard (Valldemossa, Mallorca). Algunes hipòtesis sobre la dinàmica de la vegetació i l'explotació forestal durant el Calcolític a Mallorca. *Mayurqa* 33: 315-332.

PICORNELL, Ll., ASOUTI, E. and ALLUE, E. 2011. The ethnoarchaeology of firewood management in the Fang villages of Equatorial Guinea, central Africa: Implications for the interpretation of wood fuel remains from archaeological sites. *Journal of Anthropological Archaeology* 30 (3): 375-384.

PICORNELL, Ll., SERVERA, G., CALVO, M., RIERA, S. and ALLUE, E. 2012. The study of prehistoric sacred sites and sacred plants. A case study of the Son Ferrer funerary mound (Majorca, Balearic Islands). In Pungetti, G., Oviedo, G. and Hooke, D. (eds.) *Sacred species and sites. Guardians of biocultural diversity*, 428-434. Cambridge, Cambridge University Press.

PIQUE, R. 1999. La gestión de los recursos leñosos en Cova des Càrritx. In Lull, V., Micó, R., Rihuete, C. and Risch, R. *Ideología y sociedad en la prehistoria de Menorca. La Cova des Càrritx y la Cova des Mussol*, 489-520. Barcelona, Consell Insular de Menorca.

PIQUÉ, R. and NOGUERA, M. 2002. Landscape and Management of Forest Resources in the Balearic Islands During the II-I Millennium BC. In W.H. Waldren and J.A. Ensenyat (eds.) *World Islands in Prehistory. International Insular Investigations,* 292-300. Oxford, British Archaeological Reports International Series 1095.

REILLE, M. 1992. *Pollen et spores d'Europe et d'Afrique du Nord.* Lab. de Bot. Hist. et Paly., URA CNRS 1152, Marseille.

REILLE, M. 1995: *Pollen et spores d'Europe et d'Afrique du Nord.* Suppl. 1, Lab. de Bot. Hist. et Paly., URA CNRS 1152, Marseille.

REILLE, M. 1998. *Pollen et spores d'Europe et d'Afrique du Nord.* Suppl. 2, Lab.de Bot. Hist. et Paly., 1152, Marseille.

RIERA, S., MYRAS, Y. and SERVERA, G. 2008. Anàlisis pol·líniques a la Cova des Pas: l'ús de plantes en les pràctiques funeràries. *UNICUM* 7: 26-29.

RITA, J. 1998. Els pisos de la vegetación de la Serra de Tramuntana. In *La Serra de Tramuntana, aportacions per a un debat*, collective volume. Palma de Mallorca, Sa Nostra/GEM.

RITA, J. (coord.). *Virtual Herbarium of Western Mediterranean (Balearic Islands),* http://herbarivirtual.uib.es. Palma de Mallorca, Universitat de les Illes Balears.

RIVAL, L. (ed.) 1998. *The social life of trees. Anthropological perspectives on tree symbolism.* Oxford, Berg.

RIVAS MARTINEZ, S. 1995. Bioclimatic Classification System of the World. *Folia Botanica Madritensis* 16: 1-25.

RIVAS MARTINEZ, S. 1997. Syntaxonomical Synopsis of the potential natural plant communities of North-America I. *Itinera Geobotanica* 10: 5-148.

ROBERTS, N., MEADOWS, M.E. and DODSON, J.R. 2001. The history of Mediterranean-type environments: climate, culture and landscape. *The Holocene* 11 (6), 631-634.

SERVERA, G., RIERA, S., MIRAS, Y., PICORNELL, Ll., ARMENTANO, N., ALLUE, E., ESTEVE, X. and BOI, M. 2008. Plants and death in the Late Bronze Age of Minorca Island: plant uses, funerary offerings and symbolism in the "Cova des Pas" collective burial cave. *Terra Nostra* 2008/2, 254-255.

THOMAS, J. 2002. Archaeologies of Place and Landscape. In I. Hodder (ed.) *Archaeological Theory Today*. Cambridge, Polity Press.

VALDES, B., DIEZ, M.J., FERNADEZ, I. 1987. *Atlas polínico de Andalucía Occidental.* Instituto de desarrollo regional, Universidad de Sevilla, Excma. Diputación de Cádiz.

Van STRYDONCK, M., BOUDIN, M., GUERRERO, V.M., CALVO, M., FULLOLA, J.M. and PETIT, M.A. 2010. The necessity of sample quality assessment in $C^{14}$ AMS dating: the case of Cova des Pas (Menorca – Spain). *Nucl. Instrum. Methods Phys. Res.* B, 268 (2010), pp. 990-994.

YLL, E., PEREZ-OBIOL, R. and JULIA, R. 1994. Vegetational change in the Balearic Islands (Spain) during the Holcene. *Historical Biology* 9, 83-89.

YLL, E., PEREZ-OBIOL, R., PANTALEON-CANO, J. and ROURE, J.M. 1995. Dinámica del paisaje vegetal en la vertiente mediterránea de la Península Ibérica i las Islas Baleares desde el Tardiglaciar hasta el presente. In T: Alexandre and A. Pérez (eds.) *Reconstrucción de paleoambientes y cambios climáticos durante el Cuaternario*, 319-328. Madrid, Centro de Ciencias Medioambientales CSIC.

YLL, E., PANTALEÓN-CANO, J., PÉREZ-OBIOL, R. and ROURE, J.M. 1999. Cambio climático y transformación del medio durante el Holoceno en las islas Baleares. *Saguntum-Plav* Extra 2, 44-51.

# SIGNATURE OF FOREST FIRES IN PRAIRIE SOILS

Elena PONOMARENKO
Ecosystem Archaeology Services, 1139 Agincourt Rd., Ottawa ON, K2C 2H8, Canada
eponomarenko@yahoo.com

Darwin W. ANDERSON
Department of Soil Science, University of Saskatchewan, Saskatoon, Saskatchewan, S7N 5A8, Canada
darwin.anderson@usask.ca.

**Abstract**: *The discovery of charcoal-cored iron-manganese concretions in prairie soils allows reconsideration of the generally accepted in situ growth theory. Such concretions not only indicate a particular set of conditions at the time of deposition, i.e., forest landscape but are also indicative of deforestation events such as large-scale fires followed by heavy rain. Initial results of radiocarbon dating of concretions from two concretionary layers in five sites within Canadian prairies yielded ages ranging from 10000 years BP in the lower concretionary layer of the southernmost site to 1300 years BP in the upper concretionary layer of the northernmost site. Given the commonality of closed depressions in the grasslands landscapes, this discovery has the potential to improve paleo-ecological reconstructions of the Holocene for grassland regions in such aspects as determination of the past species composition in a particular site, spatial distribution of tree-covered areas within the grasslands, and the time of deforestation events.*

**Key words**: *charcoal-cored concretions, prairie soils, deforestation, fire history, Holocene, paleoecology, anthracology, charcoal*

## INTRODUCTION

Millimeter-sized, reddish brown to black, sub-rounded to ovoid pellet-like bodies generally known as iron-manganese concretions occur widely from acidic forest soils to alkaline prairie soils. Such concretions are particularly common in soil profiles of depressions that have (or could have in the past) impeded drainage and received exaggerated water influx (Joffe 1936). Iron-manganese concretions may contribute up to 8% of the bulk soil weight and up to 98% of the weight in > 2 mm size fraction in gleyed eluviated (albic) horizons of Aqualf soils (Arshad and St. Arnaud 1980). Concretions can contain up to 60 times more Mn than the bulk soil, several times more Fe, and have a marked concentration of P (Drosdoff and Nikiforoff 1940). They are thought to form authigenically in soils with marked contrasts in redox potential, from strongly reduced during periods of saturation, to be oxidized as the soil drains. Reduced forms of Fe and Mn migrate to foci where they form Fe and Mn oxides that cement together the particles that are being enveloped. Soil concretions can be ordinated into two morphological groups. One is ochre-colored concretions of varying, but generally oblong to tubular shape, often hollow in the center. The process of growth of such concretions along plant roots was modeled in vitro (Bidwell *et al.* 1968). The other is brown to black colored concretions of round to oblong ("pebbly") shape. The formation of such concretions was not reproduced experimentally. The presence of round concretions is sporadic; finding them in the soils of one depression does not mean that they will appear in a neighboring depression with similar hydrological parameters (Personal observation, DWA). Below we will be discussing this second type of iron concretions, referring to them as IC.

## PREVIOUS WORK

The globular growth of Fe-bacteria colonies, well described in aquatic and wetland (peat) environments, does not provide an explanation for the growth of round concretions in the soil due to a high density of mineral soils. Under a substrate density > 1.5 g/cm$^3$ it is expected that the growth will at least partially follow channels, cavities or pore spaces that are plentiful in living soils (as it happens with calcareous concretions), rather than distending the compacted soil groundmass. Several features of IC correspond to secondary surficial deposition of iron oxides on the surface of a large round core rather than to a primary build-up of a concretion due to the iron deposition:

- Removal of iron oxides with either ditionite or diluted HCl does not destroy the core of the IC: it is preserved as a "large black fragile particle" (Pawluk and Dumanski 1973; Arshad and St-Arnaud 1980; Cescas *et al.* 1970).
- The proportion of iron oxides in some of IC is lower than that of the encasing soil (Childs 1975).
- IC of different sizes, sampled from the same layer have significantly different Fe:Mn ratio, with smaller IC containing a greater proportion of Fe oxides (Drozdoff and Nikiforoff 1940).

If iron has been deposited on the surface of a large core, then the rounded shape of the resulting concretions should be associated with the roundness of the particle on which iron has been deposited. Apart from the unlikely mechanism of particles rounding within the soil, particles can be rounded due to their transport along the soil surface, and then be deposited/incorporated in the soil profile. Such transport is also short in time, as in soils it is

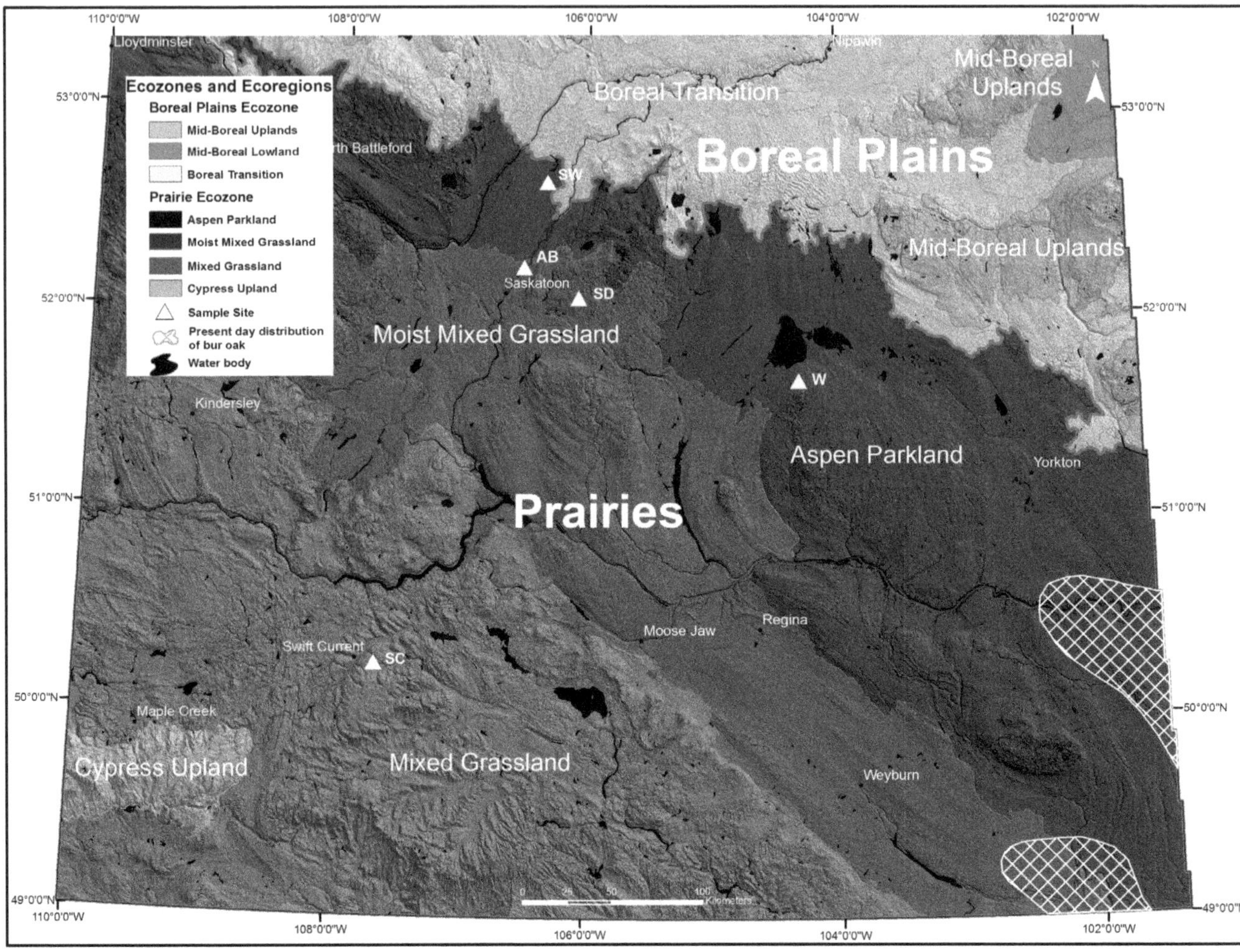


Figure 1. Locations of study sites within prairie ecoregions

correlated to a period when the land surface is not vegetated. Therefore, to become rounded over a short distance and time, the material must be soft. Common materials of mineral origin capable of fast rounding are composed of soft minerals of friable nature (carbonates, shales, poorly cemented sandstones, etc.); the most common organic material capable of even faster rounding, is charcoal.

A review of published data indicates that a number of soil parameters, chemistry, physical properties and paleo-ecology support the hypothesis of charcoal core in IC:

- Pore space in IC is described as chambered pores and stalactite-like structures (Pawluk and Dumanski 1973, Cescas *et al.* 1970), which is typical for plant material and unusual for the pore space between mineral grains in the soil matrix.
- A carbon content of IC is higher than that of encasing soil (Joffe 1936), which points to the presence of plant-originated material.
- An endothermic peak around 500 $C^0$ diagnostic for coal and charcoal (MacKenzie 1970) was recorded on thermograms of IC (Gallaher *et al.* 1973): This peak was interpreted as caolinite, in spite of the absence of the second diagnostic peak for caolinite around 950 $C^0$.

## OBJECTS AND METHODS

Grasslands are the deeply altered ecosystems. More than 90% of Canadian grasslands, or prairies, are farmlands, and the remaining land was probably used for some agricultural purposes in the past. Based on the modern distribution of plant associations, the Prairie ecozone is divided into several ecoregions (north to south):

- Aspen Parkland: vegetation cover is represented by aspen groves on the moister north-facing slopes and low-lying depressions and grasslands on the drier south-facing slopes and hillocks.
- Moist Mixed Grassland: trees and shrubs occur along permanent sloughs and streams.
- Mixed Grassland: dry grassland, no trees and shrubs grow in the open prairies, shrubs occur mostly along rivers.

Within the glaciated region of the prairies, stretching to the southern margin of glaciation in Montana, the Dakotas and Minnesota, typical components of relief are wetland depressions (Carter *et al.* 2005). Millions of depressions (known locally as potholes or sloughs) formed at the time of deglaciation in areas with hummocky or disintegration moraine. Hummocky

| Sites | Concretionary horizon, depth | Maximum size of concretions, mm | Number of concretions analyzed, identified | Laboratory, sample code | $^{14}C$ age, years BP | Calibrated age, yrs (CALIB REV5.0.2) |
|---|---|---|---|---|---|---|
| Swift Current | Bh 117-123 cm | 3 | 5 grams of unsorted concretions (deciduous species) | CAMS 84158 | 9710 ± 170 | 9107 ± 249 BC |
| Saint-Denis | Bh 106-120 cm | 2 | 5 grams of unsorted concretions (deciduous species) | CAMS 84157 | 10060 ± 220 | 9771 ± 383 BC |
| Wynyard | Ae 60-70 cm | 8 | 1 large concretion (*Populus sp.*) | UCIAMS 6048 | 7865 ± 20 | 6673 ± 20 BC |
| Marten Slough, Aberdeen | Ae 63-70 cm | 7 | 1 large concretion (*Populus sp.*) | UCIAMS 6049 | 3160 ± 40 | 1432 ± 18 BC |
| Seager Wheeler Historic Farm, Rosthern | AeB 27-30 cm | 7 | 1 large concretion (*Quercus sp.*) | CAMS 84581 | 1300 ± 40 | AD 714 ± 42 |

Figure 2. The bedding depth, size range and radiocarbon age of the concretions studied

moraine with enclosed depressions covers about 60% of the Prairie Ecozone, including all ecoregions. Some of the depressions are filled with water year-round (permanent wetlands), whereas others hold water only in spring/early summer: for several months (seasonal wetlands) or weeks (ephemeral wetlands). Their catchment areas rarely exceed several hundred meters in diameter, some of the depressions being connected by drainage channels. Level rather than rounded bottoms of the sloughs suggest in-filling with local sediments, probably episodically at times of climatic stress or following disturbances in the adjacent areas such as fire, grazing and ploughing. The wetlands are delineated by a boundary in soil, vegetation or topography. The vegetation within the wetland includes sedges, rushes and other hydrophytic plants. Rings of willow and aspen often grow adjacent to the wetland, especially in more moist regions such as the aspen parkland. Soils within the depressions are mainly Gleysolic by the Canadian Soil Classification, Aquoll and Aqualf suborders by Soil Taxonomy.

IC were collected from Luvic Gleysols in closed depressions at five sites in the prairies of central and southern Saskatchewan, Canada (Figure 1). Two of the sites (Seager Wheeler Farm and Wynyard) are located in the Aspen Parkland ecoregion, two more (Aberdeen and St. Denis) in Moist Mixed Grasslands, and one (Swift Current) in Mixed Grasslands.

In the prairie ecozone of Saskatchewan, two concretionary layers have been recorded during soil surveys of prairie wetlands-"sloughs". IC are present within sublayers in Ae horizon (depth 20 to 40 cm), or at depth – at the contact of soil with the parent material, in Bh hori-zons (depth 60 to 120 cm). IC from the two layers also occur in lower diffuse concentrations within transitional AB horizons, in this location it is interpreted to have been remobilized by pedoturbation following the initial deposition.

Of the five sites studied, three sites have an "upper" concretionary layer and two sites have a "lower" concretionary layer (Figure 2). The sites with the upper concretionary layer were located within large depressions with a rim of aspen tree cover. The sites with the lower concretionary layer were located within small, closed depressions (seasonal sloughs) in treeless prairies. Only solitary trees (willow and/or aspen) grow at the edges of some larger sloughs in the area.

IC were separated from the groundmass by wet-sieving through a nest of three sieves with 0.4 mm, 2 mm and 5 mm-openings. In each assemblage, length and width of 20 to 30 largest IC were measured. Internal structure of the IC was studied using reflected-light microscopy and SEM. An SEM-EDS (Energy Dispersive Spectrometry) analysis of morphological zones within the IC was completed. Bulk unsorted samples of IC from the sites with a Bh concretion layer and a single large charcoal-cored IC from the sites with the Ae concretionary layer were submitted for AMS radiocarbon dating. Calibration of radiocarbon dates was carried out using radiocarbon calibration program CALIB REV5.0.2 (Stuiver and Reimer 2005).

## RESULTS AND DISCUSSION

### *Size and shape of concretions*

In the Ae layer, IC ranged in diameter up to 8 mm (Figure 2), and constituted up to 6% of total soil mass. In the lower Bh layer, IC were up to 3 mm in diameter and made up about 2% of the soil mass. IC have an asymmetrical (oblong or pebble-like) shape (Figure 3) and a length to width ratio of 1.2 to 1.6.

### Internal structure of concretions

Most of the IC had two morphological zones with a distinct and sharp contact between them: black to brownish-black "core" and an outer layer of much lighter, yellowish-brown to reddish-brown hue (Figure 3).

Figure 3. A pebble-shaped concretion from an upper concretionary layer. The central zone (a charcoal fragment) is more porous than the outer zone which is more dense (mineral coating)

Soaking the IC in dilute HCl solubilized the external coating, whereas the dark-colored internal part remained as a fragile brownish-black, to black, particle more elongated than the whole concretion. Solubilized components included iron oxides and carbonates, indicating that the coating had mixed iron-calcareous cement. Insoluble components of the coating were fine sand- to silt-sized silicate grains, partially aggregated in larger particles with siliceous cement.

The dark core consisted of several domains (Figure 4): 1) degraded charcoal, 2) well-preserved charcoal impregnated with amorphous silica, 3) deposits/pseudomorphosis of manganese oxides over initial charcoal/plant structure, and 4) a proportion of mineral grains, deposited within tubular hollows.

In the Bh layer, anatomical plant structures were recognizable only in some domains of the concretion cores (Figure 5). Charcoal in the concretions from the upper Ae layer was much better preserved and allowed for wood identification.

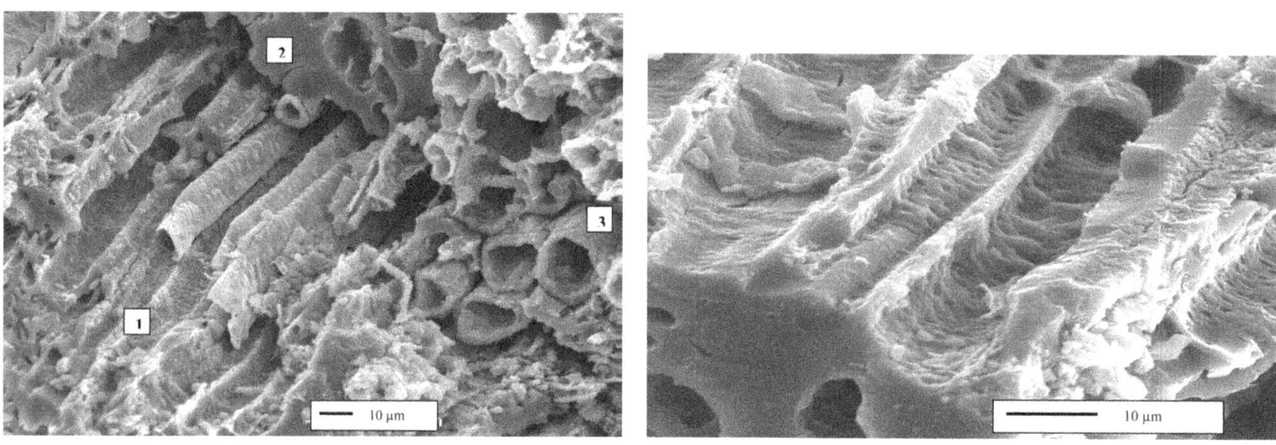


Figure 4. Top: degraded charcoal (1), well-preserved charcoal impregnated with amorphous silica (2), tubular structures are manganese oxides that have replaced carbon in the charcoalified anatomical structures (3). Bottom: close-up of a silica-impregnated charcoal

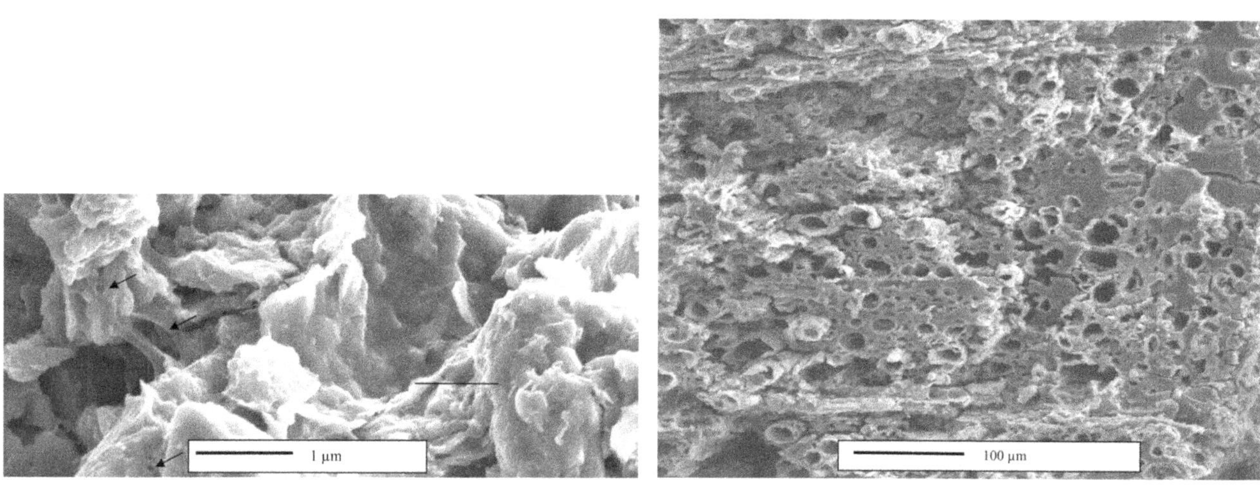


Figure 5. Charcoal preservation in the lower (top photograph) and upper (bottom photograph) concretionary layers. In the lower concretionary layer, plant anatomical structures (such as pits and fragments of perforation plates indicated by arrows) are recognizable only in some domains of the concretion cores. In the upper layer, the degree of charcoal preservation is often sufficient for botanical identification (here: *Quercus* sp.)

Figure 6. Formation of the rounded shape of concretions. Top: a mud cap is added to an abraded, somewhat rounded charcoal fragment. Bottom: close-up of a transversal section through a concretion core showing the abrasion of charcoal fragment at the contact with the mud coating

Figure 7. Imprints of wood structure at the contact of the mud coating with the charcoal core

In all sites, IC assemblages contained charcoal of deciduous species, mainly *Populus* sp. or *Salix* sp., with a proportion of *Quercus* sp. (Leucobalanus group) in the northernmost site. The first two genera are represented in the prairies near the sampling sites, whereas the present range for bur oak (*Quercus macrocarpa* Michx) extends only into eastern Saskatchewan (Fig. 1), about 400 to 500 km to the east (Hosie, 1969).

The charcoal fragments in IC are encased by mineral coating – "mud cap". In the concretions with an initially cylindrical shaped charcoal core, abrasion of edges of the charcoal fragments at their contact with the mud cap can be noticed in transversal sections (Figure 6). The overall pebbly shape of the concretions is achieved due to addition of a mud cap to the abraded/slightly rounded charcoal fragment. The mud caps have clear imprints of wood structure at their contact with the charcoal core, pointing to the wet state of the mineral mass during coating (Figure 7).

Silica cement is deposited along cavities in the mud coating, as well as within tubular hollows and vessels in the charcoal core. Plant opals (phytoliths) are a recognizable part of the silica deposits in the cavities within the mud coating of IC (Figure 8).

**Suggested mechanism of IC formation**

We postulate that formation of charcoal-cored IC was the result of multiple events that proceed from i) presence of forest cover, ii) large-scale fire, iii) lateral transport by water, iv) deposition within sloughs v) precipitation of Fe and Mn oxides.

i. Forest cover. To generate the charcoal core of the concretions, shrubs and trees had to be present in the landscape as a potential fuel. To create accumulations of charcoal in depressions of relief, the forest cover had to be more extensive than currently present. On the basis of charcoal identification, the forest cover included species of *Quercus*, *Populus* and/or *Salix*.

ii. Fire. The origin of charcoal-cored IC must necessarily involve large fires that formed a large amount of char particles and ash, and destroyed the vegetation cover at least within the catchment's area.

iii. Transport. The ovoid, pebble-like flattening of the coated char fragments suggests their lateral transport prior to deposition. The transport distances are limited to several hundred meters due to the small catchment areas of prairie sloughs. Transport was probably as overland flows or sheet flows. Within such thin flows charcoal buoyancy would not remove charcoal particles from their interaction with substrate, leading to a consequent abrasion and shaping of charcoal fragments. Destruction of surface vegetation by fire would have provided an opportunity for mineral substrate to be liquefied during intense precipitation events providing slurry that may have been absorbed by charcoal fragments during transport. Formation of the slurry would be followed by down slope mass-transport during which charcoal particles became coated with mud and

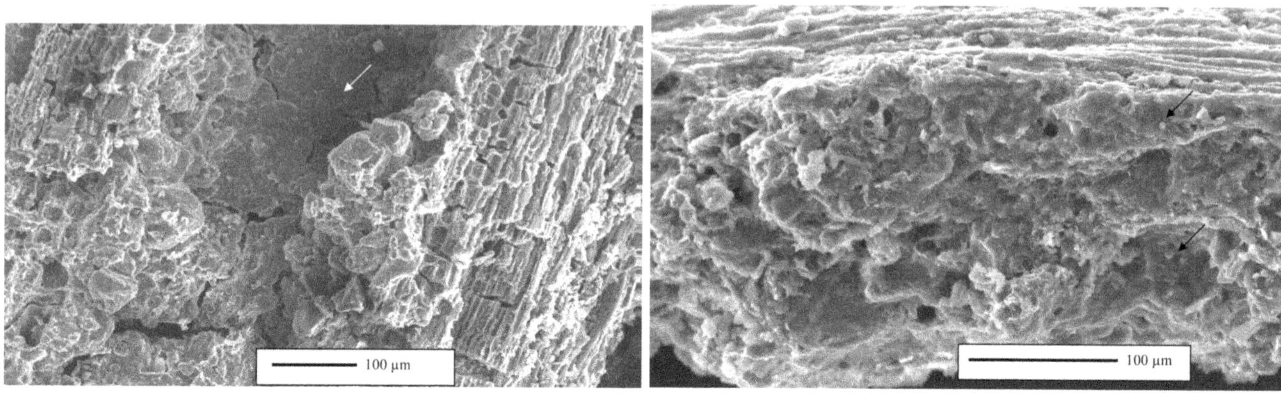


Figure 8. The top photo shows the amorphous silica impregnating the tubular hollows in the charcoal core. Bottom: silica deposited along cavities in the mud coating of the concretion. The bottom photo shows amorphous silica deposited in the cavities in the mud coating of a concretion. Arrows indicate round silt-sized plant opals (phytoliths) in the mud coating

carbonates (perhaps ash-derived), and their deposition in "sloughs". To allow for long-distance transport of comparatively large charcoal particles, such flow (e.g., heavy rain) would have to occur shortly after fire, before the soil surface became re-vegetated.

iv. Deposition. Due to high buoyancy of charcoal, it tends to be deposited in closed depressions. Depending on the flow velocity, charcoal fragments can be carried to depressions of a different order. According to the studies of IC distribution in the relief, the concentration of IC is maximal at the perimeter of closed depressions, declining in both directions from the rim (Zaidelman 1991). According to our observations, IC were more abundant within the band adjacent to the current boundary of the sloughs from outside then at the centers of the sloughs. Such distribution corresponds with the deposition of charcoal fragments from a hypersaturated flow, whereas lower load would probably result in a more uniform distribution of charcoal fragments within the repositories. The presence of silica coatings over charcoal points on alkaline conditions in the repository following the transport/deposition, probably due to influx of soluble ash from the flow.

v. Post-depositional transformation. Deposition of Fe and Mn on the exterior coating is a secondary event that occurred under wet and reduced conditions within the soil. Although the foregoing could result in the build-up of iron and manganese deposits on char particles in any concave area of relief, the actual formation of concretions would be favored in those seasonally wet depressions with alternating conditions of reduction and oxidation. During wet periods and under anaerobic conditions, Fe and Mn are reduced, solubilized and able to move in the soil to centers of concentration. The highly adsorptive surface of char particles may provide those centers, with more oxidizing conditions possible because of the pore spaces within char and their generally large size in comparison to the soil matrix. It is likely as well that some Fe-Mn microorganisms may grow preferentially on the char particles. Some fungi specialize in growing on charcoal, and are known to produce fungal manganese peroxidase, capable of destroying various aromatic compounds (Hofrichter *et al.* 1998). We observed elongated, tubular structures shown by electron microprobe to consist mainly of Mn oxides in the concretions studied (Figure 4), and considered them to be a result of the replacement of charcoal within the former plant anatomical structures by Mn oxides, perhaps by a process involving fungi. The deposition of Mn and Fe oxides can start at any time following charcoal deposition, and may occur at various depths. If mud-coated charcoal fragments have been overlaid by younger deposits, with Mn and Fe oxides precipitating on these fragments either prior or following their burial, it would appear as the concretionary layer at depth.

**Paleoecological implications of charcoal cores in concretions**

The presence of charcoal in the IC has some important paleoecological implications. Finding charcoal-cored IC in currently treeless prairies marks an episode (or episodes) of deforestation of a previously forested or shrub-covered landscape. Similarly, finding charcoal-cored IC in currently forested areas points on an episode of deforestation in the past. Were such a combination of events, i.e., forested landscape — large-scale fire — heavy rain, a frequent occurrence in the Northern Great Plains, we would expect the presence of numerous concretionary layers in prairie sloughs. However, in prairie soils of the former Late Wisconsin glaciated North America no more than two distinctly different stratigraphic concretionary layers have been reported in Holocene soils. To the south beyond the extent of Late Wisconsinian glaciation at least three concretionary layers occur (Philippe *et al.* 1972.). The concretionary layers are divided by non-concretionary deposits of varying grain size composition, with silt deposits being more common.

The thickness of the overlaid non-concretionary layers is several times greater than the thickness of concretionary layers. The limited number of the concretionary layers in prairie soils may indicate that the formation of charcoal-cored concretions was a rare event. Contrarily, the concretions within the same layer could result from multiple deforestation events that coincided with low-intensity sedimentation, whereas much more intense sedimentation followed in time leading to the formation of a buried layer of uneven-aged concretions. Our work has been exploratory, recognizing that more systematic studies are needed to fully make use of our findings.

### Age of concretions

Although ferruginous concretions described as "alternating vitreous and metallic lustres where iron oxides were infused into charred wood fragments and phytoliths" were noted years ago (Pawluk and Dumanski 1973), the concretions have never been radiocarbon dated. Here we present the first data on the age of iron-manganese concretions from prairie soils. Due to the small size and low carbon content of concretions from the lower Bh layer in our study sites, bulk unsorted samples of IC were submitted for radiocarbon dating. The ages of unsorted concretion samples from the lower concretionary layer in two sites were 10060 ± 220 BP and 9710 ± 170 BP (Figure 2), coinciding within the error of the method. The age values correspond with a major and abrupt shift from mainly forested to prairie vegetation in the Northern Plains *c.* 10000 BP (Yansa and Basinger 1999). In the upper Ae layer, a large size and higher carbon content of IC allowed for radiocarbon dating of single concretions. We find radiocarbon dating of single concretions much more reliable/interpretable, as it is unknown *a priori* whether the concretions from the same layer originate from one or more deforestation events. Age of single large (5 to 7 mm in diameter) charcoal-cored IC from the upper Ae layer varied in our three study sites, with the calendar age increasing from AD 714 ± 42 in the northernmost site to 1432 ± 18 BC (intermediate site) and to 6673 ± 20 BC in the southernmost location (Figure 1). The data show that IC from upper Ae concretionary layers in our study sites have a younger age than IC from the lower concretionary layers. These dates show as well that it takes less than 1300 years to form the iron-manganese concretions in the prairie soils: further radiocarbon dating of IC may demonstrate even a shorter formation time. The presence of charcoal-rich layers at depth indicates that the deposition of the charcoal cores of the concretions was followed by the deposition of sediments. Therefore, the age of concretions provides a time marker for the development of the horizons bedded above the concretionary layers, such as Ae and Ah horizons: apparently, formation of bleached eluviated horizons may take as short as 1300 years.

Much more extensive radiocarbon dating of concretions from the same layers/sites, as well as from the variety of geographical locations is needed for any further interpretation of the dates. A regional systematic sampling of IC in depressions across the ecoregions has the potential to contribute a more complete record of paleo-ecological conditions throughout the Holocene.

## CONCLUSION

Information on forest cover of the prairies in the middle and early Holocene is limited due to the absence of preserved macroscopic debris. Charcoal cores of Fe-Mn concretions provide a medium that can yield information on forest species composition and timing of fires. It may also provide information on the transition from forest cover to grassland and hence transient ecoregion conditions, possibly related to paleoclimatic changes. Finding radiocarbon datable objects in grasslands is always a challenge. The charcoal-cored concretions have a potential as a new source of paleo-ecological information, deserving a detailed and systematic study.

### References

ARSHAD, M.A., and St. ARNAUD, R.J. 1980. Occurrence and characteristics of ferromanganiferous concretions in some Saskatchewan soils. *Canadian Journal of Soil Science*. 60, 685.

BIDWELL, O.W., GIER D.A., and CIPRA, J.E. 1968. Ferromanganese pedotubules on roots of *Bromus inermis* and *Andropogon gerardii*. *Cong. Soil Sci., 9th*, Adelaide, Australia, 1968, Trans., V.4.

CESCAS, M.P., RYNER, E.H., and HARMER, R.S. 1970. Ferro-manganiferous soil concretions: a scanning electron microscope study of their micropore sructures. *Soil Science Society of America Proceedings* 34: 642-644. Childs, C.W. 1975. Composition of iron-manganese concretions from some New Zealand soils. *Geoderma* 13: 141-152.

DROSDOFF, M. and NIKIFOROFF, C.C. 1940. Iron manganese concretions in Dayton soils. *Soil Science* 49, p. 33-40.

GALLAHER, R.N., PERKINS, H.F., TAN, K.H. and RADCLIFFE, D. 1973. Soil concretions. II. Mineralogical analysis. *Soil Science Society of America Proceedings* 37: 469-472.

HOFRICHTER, M., SCHEIBNER K., SCHEEGASS I., and FRITSCHE, W. 1998 Enzymatic combustion of aromatic and aliphatic compounds by manganese peroxidase from *Nematoloma frowardii*. *Applied Environmental Microbiology*, vol. 64:2, p. 399-404.

HOSIE, R.C. 1969. Native trees of Canada. Queen's Printer for Canada, Ottawa, ON. p. 186.

JOFFE, J.S. 1936. *Pedology*. New Brunswick, N.J.

MACKENZIE, R.C. 1970. Solid fuels, in R.C. Mackenzie, ed. *Differential thermal analysis*. Vol. 1, 708-715, London, Academic Press.

PAWLUK, S. and DUMANSKI, J. 1973. Ferrugineous concretions in poorly drained soils of Alberta. *Soil Science Society of America Proceedings* 37, 124-127.

PHILIPPE, W.R., BLEVINS, R.L., BARNHISEL R.I. and BAILEY, H.H. 1972. Distribution of concretions from selected soils of the inner blue grass region of Kentucky. *Soil Science Society of America Proceedings* 36: 171-173.

YANSA, C.H. and BASINGER, J.F. 1999. A postglacial plant macrofossil record of vegetation and climate change in southern Saskatchewan, in D.S. Lemmen and R.E. Vance (eds.) *Holocene Climate and Environmental Change in the Palliser Triangle: a geoscientific context for evaluating the impacts of climate change on the southern Canadian Prairies*, Geological Survey of Canada, Bulletin, 534: 139-172.

# RECONSTRUCTING SPECIES COMPOSITION AT THE TIME OF LAND CLEARANCE: TWO APPROACHES COMPARED

Elena PONOMARENKO
Ecosystem Archaeology Services, 1139 Agincourt Rd., Ottawa ON, K2C2H8, Canada
eponomarenko@yahoo.com

Donna CROSSLAND
Kejimkujik National Park and National Historic Site, Maitland Bridge, Annapolis County, NS, B0T 1B0, Canada
Donna.Crossland@pc.gc.ca

Judy LOO
Atlantic Forestry Centre, 1350 Regent Street South, Fredericton, New Brunswick, Canada

**Abstract**: *Botanical identification of charcoal from plough horizons in abandoned farmlands was applied to reconstruct pre-agricultural species composition of Acadian forest in Kouchibouguac National Park (Canadian Maritime Lowlands). Results of the reconstruction were compared to archival descriptions of the forest* ca *1800, just prior to intensive anthropogenic modification. Both methods detected a high diversity of tree species including species that are currently not common on the landscape, such as* Ulmus *and* Ostrya. *Both methods indicated that forests in the area were mixed and dominated by mid- to late successional species* (Picea spp., Tsuga canadensis, Betula spp., Acer spp., Fagus grandifolia, Abies balsamea *and* Pinus strobus), *whereas early successional species, such as* Populus spp., Pinus resinosa, Larix laricina *and* Quercus rubra *were present at low levels. According to both methods,* Pinus banksiana *– a fire-dependant species that is common and widespread in the modern landscape, was absent at the time of land clearance. Reconstructed occurrences of species were generally similar using either method, but anthracological research indicated lower occurrences of some coniferous canopy species, such as* Tsuga canadensis, Abies balsamifera *and* Picea sp., *and higher occurrences of species that fill-in subordinate canopy positions, particularly* Thuja, *in charcoal assemblages in relation to witness trees results. Comparison of the two methods confirmed applicability of the analysis of charcoal assemblages for the site-oriented reconstruction of tree species composition at the time of land clearance.*

**Key words**: *charcoal botanical analysis, historical forest composition, ecosystem dynamics, disturbance return interval, vegetation management, paleoecology, anthracology, witness trees*

## INTRODUCTION

Many Canadian national parks and national historic sites feature former farmlands that are slowly returning to a forested state. Restoration of forest in such areas requires knowledge of the historical species composition, i.e. species composition prior to land clearance. European colonization of the Atlantic coast of Canada began in the 1700s, and during the last 300 years coastal forests have been progressively modified by burning, logging, and introduced forest pests. As a result, modern forests in the area do not necessarily reflect either a potential species composition or species composition at the time of land clearance. In some areas of the Maritimes, the composition of early forests at the time of agricultural colonization can be reconstructed using surveyor plan drawings and descriptions from diaries of the pioneers and early travelers. In other areas, clearing occurred within several years, before any systematic descriptions of the initial forest cover were documented. According to the historical descriptions of the methods of forest clearing for farming in the Maritimes, trees were first logged; their trunks and large branches collected for a variety of uses, and the logging slash burned (Johnston 1851). Logging slash was often piled around stumps of cut trees to hasten disintegration. Burning of the logging slash produced ash and charcoal associated with the site clearance event. Once ploughing began, the charcoal was incorporated into the soil and re-distributed within the plough layer (Johnston 1851). This study explored the possibility of reconstructing pre-agricultural forest cover using botanical identification of soil charcoal from plough layers of modern and abandoned farmlands. Results of the charcoal analysis on the forest composition at the time of land clearance were compared with the analysis of archival data, (using witness trees and early forest descriptions), to validate both approaches.

## OBJECTS AND METHODS OF THE COMPARATIVE STUDY

### Study area

The study area is situated on the east coast of New Brunswick, Canada, within the 'Eastern Lowlands' ecoregion (Rowe 1972; DNRE 1996) (Figure 1). This region is defined by low elevation (approximately three-quarters of the land is less than 15 m above sea level) and limited topographic variation. Gentle moraines, a few eskers and kames, and short, steep banks of streams and meandering rivers dissect the landscape in an east-west

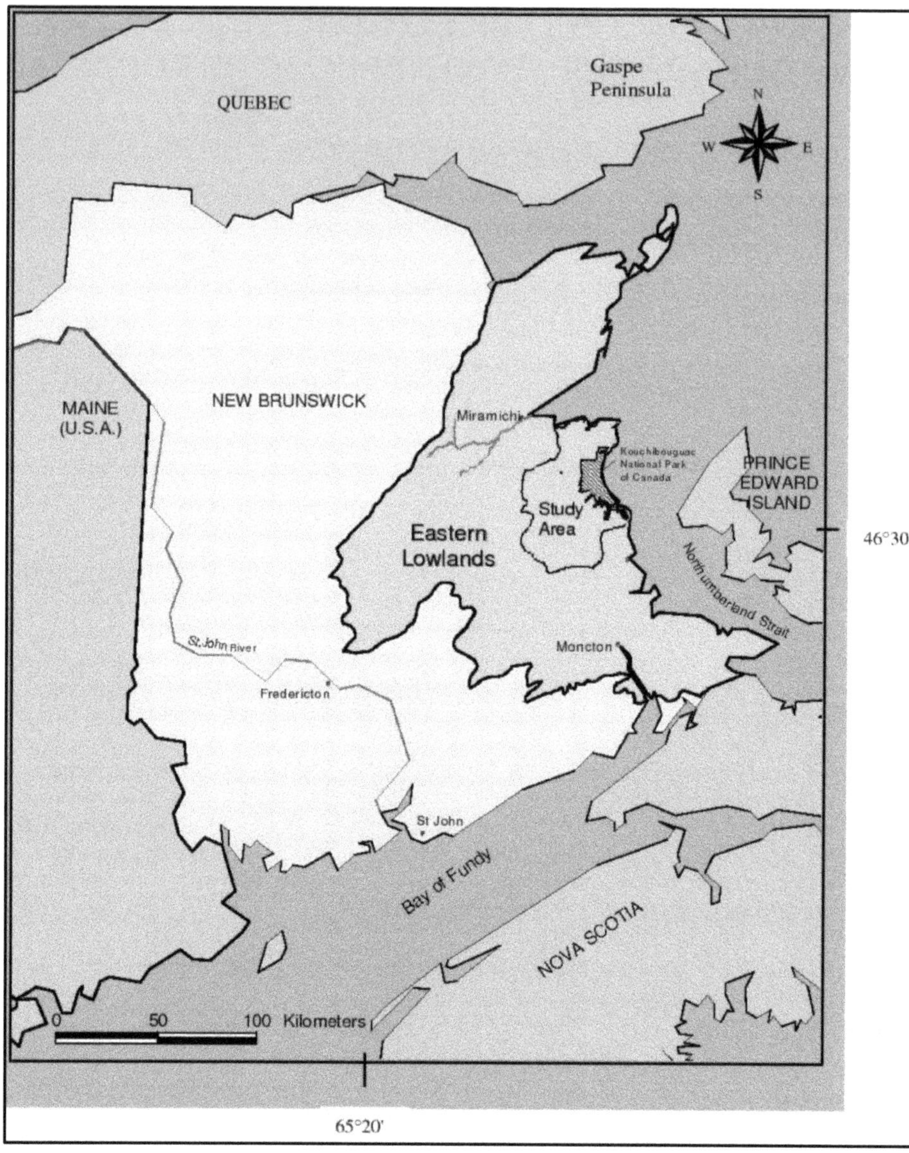


Figure 1. Location of the Eastern Lowlands Ecoregion, archival study area, and Kouchibouguac National Park (anthracological study area)

pattern to empty into the Northumberland Strait (Wang and Rees 1983; Beach 1988). Poor drainage is a prominent feature, resulting in large raised peat bogs over approximately 1/5 of park surface (Beach 1988). Local drainage patterns are discontinuous, often changing rapidly over distances of only a few meters. Kouchibouguac National Park (KNP) is a 239 km$^2$ protected area of the Eastern Lowlands that conserves over 125 km$^2$ of Acadian forest.

The Acadian Forest Region covers most of the area of the Maritime Provinces of Canada (Loucks 1962; Rowe 1972), and comprises approximately 35 tree species forming a high variety of associations and community types. Red and black spruce (*Picea rubens* Sarg., *P. mariana* (Mill.) BSP.) and their hybrids, balsam fir (*Abies balsamea* (L.) Mill.), red maple (*Acer rubrum* L.), Eastern hemlock (*Tsuga canadensis* (L.) Carr.) and white pine (*Pinus strobus* L.) comprise a distinctive association on the landscape (Loucks 1962). Another prominent species, eastern white cedar (*Thuja occidentalis* L.), is considered an edaphic climax species on heavy soils with impeded drainage (Dryade 1979). KNP features the largest protected eastern cedar stand east of Ontario (Desloges 1980). Eastern larch (*Larix laricina* (DuRoi) K. Koch) and black spruce predominate on expansive bog lands (Loucks 1962).

The Acadian Forest Region was the earliest Canadian forest type to be exploited for timber resources. European settlement began in earnest in the study area in 1820-1830, and local forests were subjected to increased resource harvesting from that time onwards. Contemporary forests of the study area exhibit extensive stands of intolerant pioneer species with trembling aspen (*Populus tremuloides* Michx.), white birch (*Betula papyrifera* Marsh.), grey birch (*B. populifolia* Marsh.), white spruce (*Picea glauca* (Moench) Voss), larch, and green alder (*Alnus viridis subsp. crispa* (Aiton) Turrill) (Loucks 1962; DNRE 1996). Widespread fires appear to have favored coniferous species, particularly black spruce and jack pine (*Pinus banksiana* Lamb.) (Rowe 1972).

Park vegetation currently comprises 26 tree species, forming 37 forest community types based on tree species composition, understory vegetation and site condition (Beach 1988). Black spruce, red spruce, balsam fir, trembling aspen, and white cedar, in various associations, account for approximately 70% of species on forested land (Desloges 1980). Narrow bands of better drained soil feature tolerant hardwood growth, including sugar maple (*Acer saccharum* Marsh.), beech (*Fagus grandifolia* Ehrh.), and small pockets of white ash (*Fraxinus americana* L.). These species, along with yellow birch (*Betula alleghaniensis* Britt.), are minor components, and have probably been greatly reduced in frequency, as the better-drained landscapes were favored for human settlement and agriculture (Desloges 1980).

**Anthracological research of historical forests**

Prior to European colonization, the inland area of Kouchibouguac National Park was used mainly as wintering grounds by the Mi'kmaq, a non-agricultural First Nations people (Leonard 1996). Starting from the late 1700s, the area has been logged and most of the moderately to well-drained sites were cleared for farmsteads. Once cleared, fields were continuously cultivated until 1969 when the area was acquired for the national park. Many of the abandoned fields were still open during the time of research, after 50 years of abandonment, due to a slow rate of natural re-forestation.

Thirty abandoned fields were examined in KNP over a mosaic of sites with varied drainage conditions: coastal plain, watershed bogs, swales, sand ridges, and well to moderately drained river banks. During land clearing, mostly well- to moderately drained sites were selected for cultivation, but due to the high complexity of the landscape each field included wetter and dryer spots, with wetter sites requiring drainage into dugout ditches prior to cultivation.

A single field was used as the sampling unit, with the size of individual fields varying from 3 to 20 ha. For purposes of this study, we assumed that the long-term ploughing averaged out the initial variations in the species composition within the forest patch that has been cleared for farming. Ploughing causes not only vertical mixing and incorporation of charcoal into the ground mass of the plough layer, but also charcoal fragmentation and long-distance lateral transportation of fragments as suggested by the studies of potsherds in archaeological sites. A single plowing action moves a particle only for 0.5-2.0 meters (Van Oost *et al.* 2000), but repetition of such action over many years leads to a significant re-distribution of charcoal particles. The distance and extent of such movement can be estimated by research on the distribution of ceramic fragments in archaeological sites: long-term field experiments and surficial archaeological surveys show that the dispersal may reach ~100 meters after 200 years of plowing (Yorston *et al.* 1990). According to our observations on fields that were recently cleared from forests, piles of logging slash could be distributed approximately 20 m from one another. We assumed that if several logging slash burn sites were located within a field during land clearance, they would provide a sufficient mixing of combustion products throughout the field.

Local soils are developed on late Pleistocene-early Holocene deposits. Gray Luvisol and Luvic Gleysol are formed in minor areas of marine clay deposits, and Humo-Ferric Podzols are common on sand deposits of various genesis and thickness (Wang and Rees 1983). In most of our study sites, loamy glacial deposits are overlain by a veneer of fluvioglacial or aeolian sand from 0.3 m to over 2 meters deep. As a result, plough layers are predominantly sand, which allowed for extraction of charcoal by dry sieving in field.

Two sampling approaches were used:

- In each site, charcoal was collected from a trench wall (7 x 1.5 m) so that the whole perimeter of the trench was represented in the charcoal assemblage.
- Charcoal was sampled from test pits in catenas running across each field to ensure that data was representative for each field. The test pits were of a uniform size, 50 x 50 x 25 cm, and were spaced at 20 m intervals. The soil mass from the plough layer within the 50 $cm^2$ squares was sifted through a sieve with 2 mm openings, and all charcoal fragments larger than 2 mm were collected for additional identification. Number of the sampled charcoal fragments and their size varied substantially within the catenas: in some test pits, we were able to collect only several charcoal fragments less than 4 mm in diameter, whereas in others, charcoal was abundant and up to 1 cm in diameter. It was realized later that those test pits rich in charcoal were located in close proximity to logging slash fireplaces. We were not able to process all the charcoal sampled within the catenas, as the amount of material by far exceeded the available time and budget. At this stage, identifications were limited to 5 test pits within each field.

The number of charcoal fragments identified for each field site had to meet statistical requirements for representativeness, but was constrained by the fact that charcoal identification is a time consuming procedure: it was not feasible to do more than 50 identifications per sample unit due to time and budget constraints. We identified a minimum of 30 fragments for sites with a low species composition (1-3 taxa), and 50 fragments for sites with a higher species diversity. A total of approximately 1300 charcoal fragments were identified.

Charcoal fragments were broken manually and analyzed in reflected light. For botanical identification, existing manuals (e.g. Barefoot and Jenkins 1982) were consulted along with reference materials collected from the park. Charcoal was identified with confidence to the genus level, in some cases to a group of species within a genus. Other taxa were narrowed to the species level based on the species geographical ranges.

## EXPECTED ERRORS

Besides the charcoal associated with land clearance, it was possible that plough layers could contain both older and younger charcoal from other events. The former could originate from pre-agricultural forest fires and the latter could be charcoal from furnaces and woodstoves, sometimes added to the fields by local farmers as fertilizer. Although both components could be represented in the charcoal pool, we assumed that the charcoal associated with the land clearance would prevail due to the immense amount of logging slash produced during land clearing.

Charcoal fragments larger than 2 mm are rarely preserved in the surficial (0-25 cm) layer of soil following a forest fire for a longer duration than one, or at most two, tree generations. Shortly after a fire, charcoal fragments become buried deeper in the soil profile by pedoturbation (e.g., uprootings and animal activity) (Ponomarenko 1999), while the rest become progressively reduced in size and disaggregated by ploughing. Therefore, we assumed there would be a negligibly small proportion of old charcoal fragments found in the plough layers that exceeded 2 mm in diameter. However, if a fire had occurred shortly prior to the land clearance the plough layer could be contaminated by pre-clearance charcoal. Some morphological criteria were used to eliminate charcoal from furnaces from the analyzed assemblages: we removed the fragments that had a high percentage of soluble ash within charcoal fragments and no clay coating on the surface of the charcoal fragments, as charcoal within the plough layers is becomes progressively coated/impregnated by clay (Ponomarenko and Anderson 2000).

### Archival records of historical forest composition

Archival documentation on early forest types has been analyzed for approximately the same area (Crossland 2006). Archival information consisted of both witness trees and early forest descriptions located at the Provincial Archives of New Brunswick. In the early 1800s, land surveyors divided the landscape into various-sized parcels, marking trees on the bounds of each plot. Location and species of such "witness trees" were recorded on surveyor plan drawings of land grants to private citizens from years 1805 to 1909. Pre-settlement forest descriptions were also reconstructed using 46 survey sketches, to produce 306 records of forest descriptions.

The analysis extended to lands beyond KNP in order to ensure an adequate witness tree sample size The study area was approximately 10 times greater than the area of Kouchibouguac National Park (2,420 $km^2$ and 239 $km^2$ correspondingly), extending both along the coast and into inland (Fig. 1). Witness trees were widely distributed throughout the study area, but were more concentrated on riparian zones. Witness tree densities varied from 0.3 to 2.6 trees/$km^2$, averaging 1.1 tree/$km^2$. The resulting witness tree database consisted of 2537 trees.

A total of ~2500 witness trees were used for the subsequent analysis and its comparison with results of botanical identification of ~1300 charcoal fragments. Information derived from both methods characterizes the Acadian forest ca 1800, just prior to heavy anthropogenic modification.

## RESULTS AND THEIR INTERPRETATION

### Age of identified charcoal fragments

Since most of the charcoal fragments originated from small branches and twigs, we assumed that the absolute age of the radiocarbon dated charcoal did not exceed several decades. However, some of charcoal fragments of shade-tolerant species, such as *Fagus* and *Tsuga* could be over 80 years old as estimated by the number of annual rings in the fragments that had an initial cylindrical shape preserved.

According to historical descriptions, the trees logged at the time of the land clearing greatly varied in size and, hence, in age (Crossland 2006). The age of the cleared forest patches also varied from young stands to old-growth forests. Ages of shade-tolerant species, such as Eastern hemlock and red spruce, can achieve 400 years, whereas, long-lived, light-demanding species, such as white pine and red oak can attain 250 years (Burns and Honkala 1990). Species composition at the time of land clearance reflects the conditions that existed at the time of establishing the forest communities, in some loci up to 400 years ago.

Direct radiocarbon dating of single charcoal fragments from the plough layer in three sites yielded $^{14}C$ ages 264 ± 55 BP (BGS2525), 230 ± 40 BP (CAMS 86042), and 195 ± 35 BP (Beta -190069), the dates being consistent with the time of agricultural colonization of the area.

### Taxa reconstructed

Charcoal analysis allowed for reconstruction of 20 taxa (Figure 2). The number of taxa recovered per site varied from 2 in drained cedar bogs to 10 in a well-drained riparian zone, revealing a high complexity of forest cover at the time of land clearance. More than 6 taxa were recorded in charcoal assemblages from ~50% of sites.

The results are presented in two forms, as species frequencies and species abundance.

**Mean tree species frequencies** were calculated as a percent of sites where a species occurred. Eight genera dominated the area at the time of land clearance, occurring in more than 50% of the sites (in descending order): *Picea* (90%) > *Thuja, Betula* (~80%) > *Acer* (65%) > *Tsuga, Pinus* (Strobus type) (~50%), *Abies, Fagus* (~40%) (Figure 3). Less common were such taxa as *Populus, Quercus (*Erythrobalanus type*), Pinus* (Resinosa type), *Larix, Ulmus, Alnus, Fraxinus, Taxus, Ostrya*, and *Prunus.*

| Taxa identified by the witness tree study | | Taxa identified by the charcoal study | |
|---|---|---|---|
| **Survey terms** | **Scientific Names** | **Taxa identified by the charcoal anatomy** | **Probable species according to geographical ranges** |
| **maple** | *Acer* spp. | *Acer* | *Acer rubrum, Acer saccharum, Acer pensylvanicum* |
| red maple, swamp maple, white maple | *Acer rubrum* | | |
| moosewood | *Acer pensylvanicum* | | |
| rock maple | *Acer saccharum* | | |
| **fir** | *Abies balsamea* | *Abies* | *Abies balsamea* |
| **alder** | *Alnus incana* | *Alnus* | *Alnus incana, Alnus viridis* |
| **birch** | *Betula* spp. | *Betula* | *Betula alleghaniensis, Betula papyrifera* |
| yellow birch, black birch | *Betula alleghaniensis* | | |
| white birch | *Betula papyrifera* | | |
| | | *Corylus* | *Corylus cornuta* |
| **beech** | *Fagus grandifolia* | *Fagus* | *Fagus grandifolia* |
| **ash** | *Fraxinus* spp. | *Fraxinus* | *Fraxinus americana, Fraxinus nigra* |
| white ash | *Fraxinus americana* | | |
| **ironwood** | *Ostrya virginiana* | *Ostrya* | *Ostrya virginiana* |
| **larch**, juniper, tamarack | *Larix laricina* | *Larix* | *Larix laricina* |
| **cherry** | *Prunus* spp. | *Prunus* | *Prunus virginiana* |
| **poplar** | *Populus spp.* | *Populus* | *Populus tremuloides* |
| **pine** | *Pinus* spp. | | |
| white pine | *Pinus strobus* | *Pinus Strobus type* | *Pinus strobus* |
| pitch pine, red pine | *Pinus resinosa* | *Pinus Resinosa type* | *Pinus resinosa* |
| **spruce** | *Picea* spp. | *Picea* | *Picea rubra, Picea mariana* |
| white spruce | *Picea glauca* | | |
| **oak** | *Quercus rubra* | *Quercus* Erythrobalanus type | *Quercus rubra* |
| | | *Salix* | *Salix spp* |
| | | *Taxus* | *Taxus canadensis* |
| **cedar** | *Thuja occidentalis* | *Thuja* | *Thuja occidentalis* |
| **hemlock** | *Tsuga canadensis* | *Tsuga* | *Tsuga canadensis* |
| **elm** | *Ulmus americana* | *Ulmus* | *Ulmus americana* |

Figure 2. List of taxa reconstructed using two approaches: charcoal study and witness tree study

Based on the geographical ranges of tree species possibly existing in the lowlands (Hosie 1990), each of these genera can be represented by one to several species (Figure 2).

It is possible that the occurrence of *Picea* and *Betula* was more frequent because these genera are represented by several species associated with contrasting hydrological conditions. The frequent occurrence of *Acer rubrum* is consistent with a broad range of habitats of this species (Burns and Honkala 1990). Eastern white cedar *(Thuja occidentalis)* is represented in charcoal assemblages from a much broader range of sites than the sites where it is present currently. In the present day Kouchibouguac forest, this species occupies mostly wet depressions and transitional bogs. However, a broader geographical analysis suggests that Eastern white cedar has a very high plasticity in a wide range of soil conditions (Hofmeyer, *et al.* 2009). It can grow on acidic soils of transitional bogs and on calcareous soils of uplands where it forms pure stands of "upland cedar" (Fernald 1919). It pioneers sand dunes (Scott and Murphy 1986) and abandoned pastures (de Blois and Bouchard 1995), but also is very shade-tolerant and commonly grows in subordinate canopy positions in mixed-species stands (Burns and Honkala 1990). One of the factors controlling regeneration of Eastern white cedar in the modern forest cover is browsing by white-tailed deer *(Odocoileus virginianus),* a species introduced to the area following the agricultural colonization (Squires 1946).

The high occurrence of Eastern white cedar in charcoal assemblages may indicate that prior to the land clearing wet depressions existed throughout the forest. However,

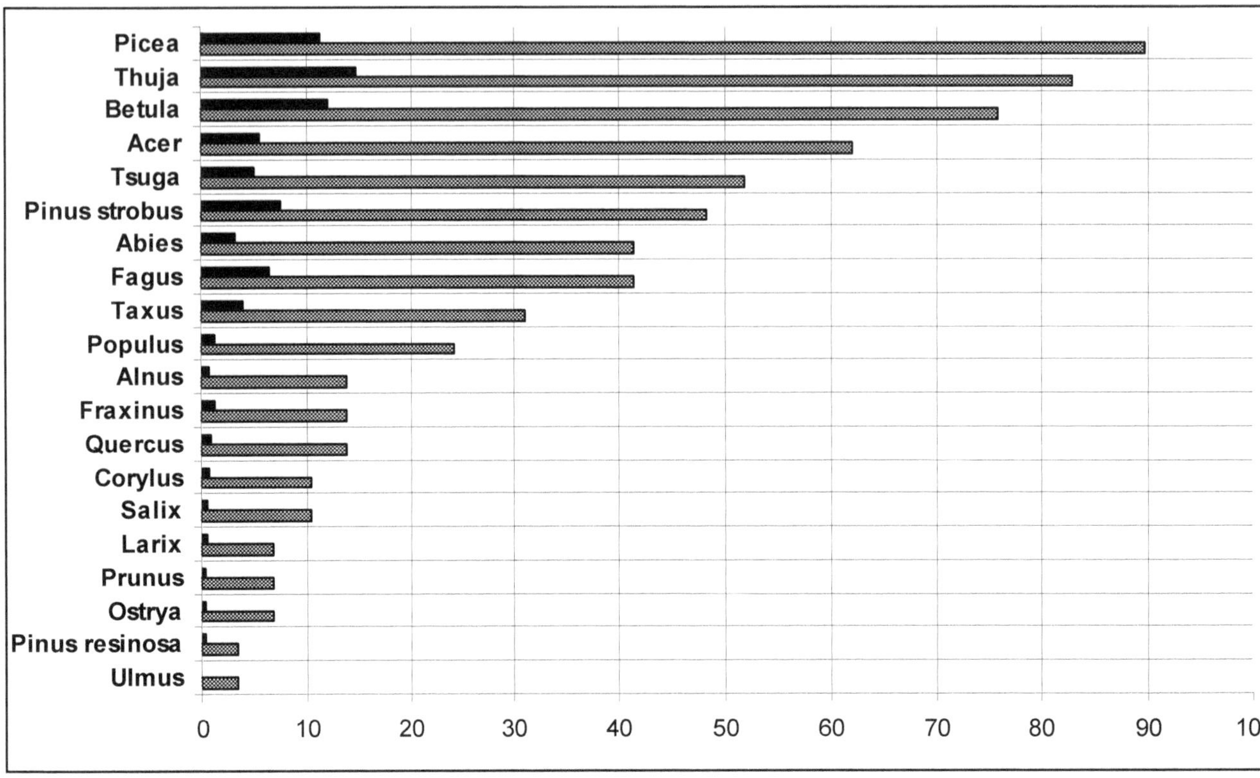


Figure 3. Frequency (in gray) and abundance (in black) of various species in charcoal assemblages

it may also indicate that prior to the land clearance Eastern white cedar occupied a much wider range of edaphotops than currently.

**Species abundance** has been estimated as a percentage of charcoal fragments belonging to species in assemblages from a particular site, and also as a mean percentage calculated for all study sites. To estimate the proportions of various arboreal species in each site, total percentage of identifiable components in each assemblage was assumed to be 100%; non-identifiable components excluded.

It must be noted that in the sites with the highest species diversity recorded in charcoal assemblages, identification of even 50 fragments may be insufficient for the appropriate characterization of the percentage of each species. Moreover, our estimates of species abundance, (that were dependant on the estimates of species frequencies), are in turn reliant on less well known parameters, such as ashing coefficients (ash:charcoal ratio) of each species during initial combustion and differential preservation of charcoal from various species (different rates of resistance to mechanical weathering and grinding in the soil). Furthermore, proportions of charcoal represented between arboreal species may reflect varying proportions of crown wood versus trunk wood, as trunk wood does not greatly contribute to the charcoal assemblages associated with land clearance according to the descriptions of the land clearing process (Johnston 1851).

The mean percentage of fragments belonging to each species did not exceed 19%. This can be interpreted as the absence of strongly dominant species over the studied range. Mean percentage of nine dominant taxa in decreasing order is as follows (Figure 4): *Thuja* (18%) > *Betula, Picea* (14%) > *Pinus* (Strobus type) (13%) > *Tsuga, Acer, Fagus* (7%) > *Taxus* (6%)> > *Abies* (4%).

In the sites where the maximum abundance of a species was recorded, the percentage of the species in charcoal assemblages was much higher than the mean abundance, reaching 70% for *Thuja*, 53% for *Pinus* (Strobus type), ~44% for *Betula* and *Fagus*, 38% for *Picea*, and ~32% for *Tsuga and Acer*. Other species such as *Salix, Ulmus, Alnus, Prunus, Ostrya* did not have high abundance on any studied sites.

It appears that the most common species (9 species recorded in most of the sites) were also the most abundant ones, whereas the uncommon species contributed a fairly low number of charcoal fragments in sites where they occurred (Figure 3). However, there are some inversions: e.g., *Acer* being a highly common species produced a relatively low proportion of charcoal fragments in the sites where it occurred. Contrarily, *Pinus* (Strobus type), being a less frequent species, contributed a relatively large proportion of fragments to the charcoal assemblages.

**Spatial distribution and species associations**

Tree species associations were highly variable over the landscape. In many sites (56%), the charcoal assemblages reflected a mixed forest composed of both deciduous and coniferous species. Coniferous species dominated

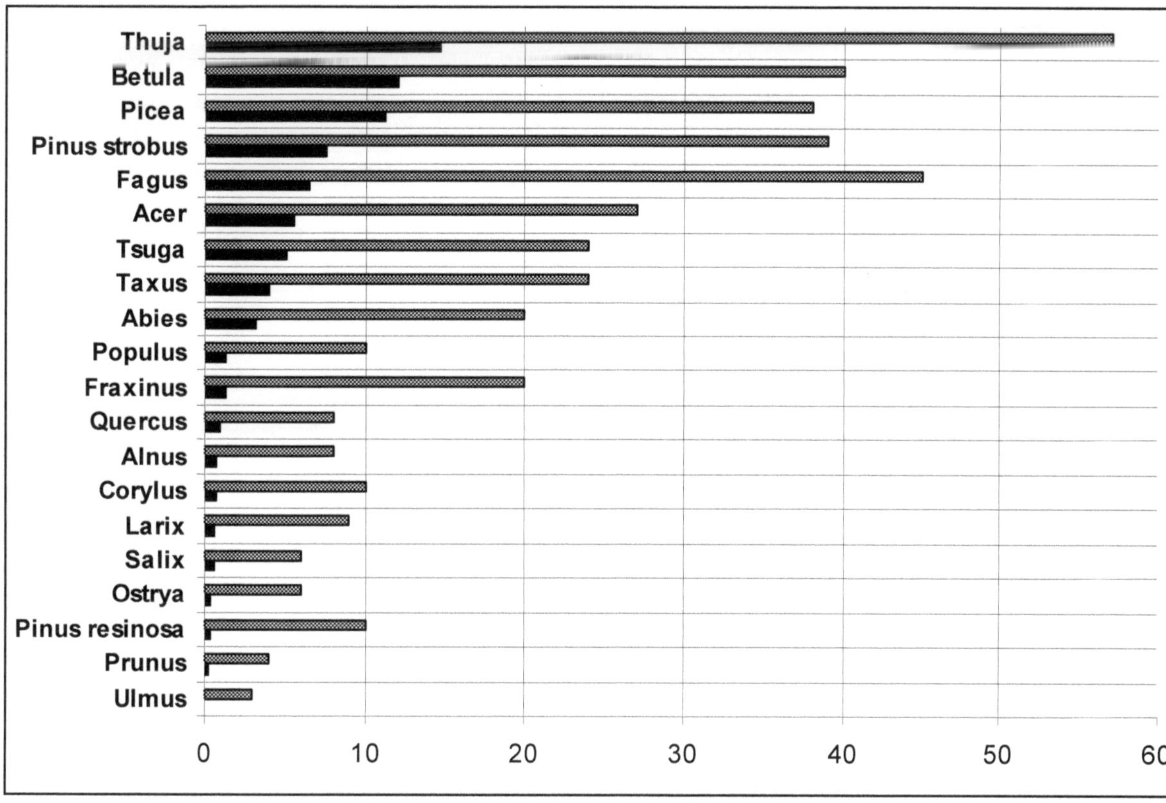


Figure 4. Mean abundance (shown in black color) and maximum abundance (gray color) of reconstructed taxa in charcoal assemblages from 30 sites

charcoal assemblages on 36% of the sites, comprising over 75% of identified fragments. In 20% of the sites, charcoal assemblages were composed almost entirely of coniferous species (>90% of fragments). In 11% of the sites, deciduous species dominated the charcoal assemblages, contributing over 75% of identified fragments. In approximately 4% of sites, charcoal assemblages were composed entirely of deciduous species (100% of fragments).

The number of reconstructed taxa and the species composition varied within geomorphological components of the landscape, such as watersheds, riparian zones, and coastal areas (Figure 5). Of 20 taxa identified, 13 were represented on watershed sites and 14 in the coastal sites. Riverine sites appeared slightly more diverse accounting for 15 taxa.

Species frequencies were distributed among the three geomorphological units as follows:

**On watersheds**, *Picea* was the most common taxon appearing in all sites (100%). Less common were *Thuja* (67% of sites), and *Acer, Betula* and *Tsuga* (50%). *Fraxinus, Pinus, Abies*, *Salix* and *Taxus* occurred in approximately 30% of sites. The least common taxa were *Fagus, Alnus, Populus*, *Larix* (~16%) and *Corylus* (8%).

**In coastal sites**, *Acer*, *Betula* and *Picea* were ubiquitous, appearing in charcoal assemblages from all study sites (100%). Very common were *Thuja* (80%), and *Pinus* (Strobus type) (60%). *Quercus, Alnus* and *Abies* were found in 40% of sites, and such species as *Prunus, Ostrya* and *Corylus* in 20% of sites. *Tsuga* was notably absent in charcoal assemblages from the coastal sites, whereas *Fagus* was found in only 20% of the sites.

**In riparian zones**, *Thuja* and *Picea* were almost equally common, appearing in 100% and 90% of sites correspondingly. *Pinus* and *Tsuga* occurred in 70% of sites. Slightly less common were *Betula* (60%); *Acer, Fagus*, and *Abies* (50%); *Populus* and *Taxus* (40%). *Quercus* charcoal was found in 20% of sites. The least common taxa were *Prunus, Ulmus, Ostrya* and *Corylus* (10%).

Several taxa were restricted to certain geomorphological units. *Ulmus* was found only in riverine sites, *Alnus* only in seashore sites, and *Fraxinus*, *Larix* and Salix only in watershed sites. *Quercus* was found mainly in seashore sites (40%) and, less frequently in riverine sites (20%), completely lacking in watershed sites. *Populus* occurred in 40% of riverine sites and 16% of watershed sites, lacking in the seashore sites.

**Interpretation of species composition**

Comparison of species composition in the three geomorphological units with the ecology of these species (Burns and Honkala 1990) allows for the following interpretation of the forest cover in these units at the time of land clearance.

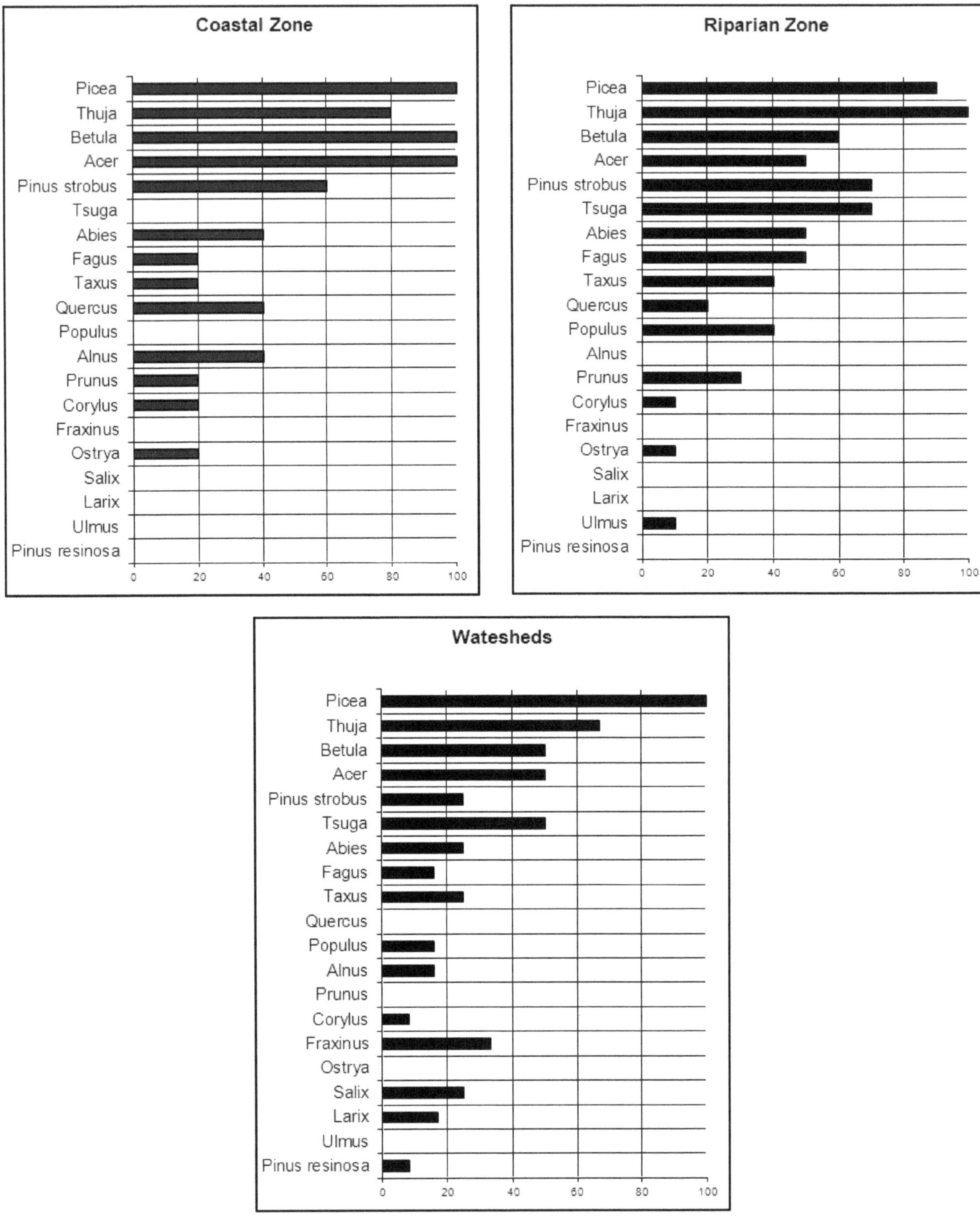


Figure 5. Tree species frequencies of tree species in charcoal assemblages from three geomorphological zones of Kouchibouguac National Park

**On watersheds**, the canopy layer was commonly composed of spruce (most probably, red spruce in well-drained sites and black spruce in sites of poorer drainage) with a proportion of Eastern hemlock, balsam fir, and deciduous species: red maple and birch (presumably mainly yellow birch, given its association with late-successional forest complexes). White pine was part of the canopy layer in more localized stands, and American beech occurred only in well-drained sites. Canada yew was a common understory shrub in moderately to well-drained sites, and beaked hazel grew in more open, light-canopy forests on well-drained soils.

The currently drained bogs that existed at the time of land clearance were occupied by tamarack and Eastern white cedar, whereas margins of bogs, small creeks and beaver ponds were a suitable habitat for white pine, quaking aspen and gray alder. Wet depressions provided a niche for black ash and Eastern white cedar. Finally, various willow species could inhabit any of the above-mentioned wetlands.

Given poor drainage conditions on watersheds, the frequent occurrence of *Fraxinus* could be at least partially due to presence of black ash (*Fraxinus nigra*). However, its occurrence in watershed sites is higher than the proportions of any other wetland species, therefore it is possible that both black and white ash (*Fraxinus americana*) were present in the landscape at the time of the land clearance.

**In coastal sites** with sufficient drainage, the canopy layer could be composed of white pine, red oak, red maple, American beech, balsam fir, and species of birch and spruce. Some of these forests must have been early successional (or recently disturbed) to allow for reproduction of such understory species as pin cherry *(Prunus pensylvanica),* beaked hazel (*Corylus cornuta*) and possibly green alder (*Alnus viridis*). Eastern white cedar and possibly gray alder *(Alnus incana)* inhabited more poorly drained coastal sites.

**In riparian sites**, canopy layer could be composed of white pine, Eastern hemlock, red oak, red maple, American beech, balsam fir, and species of spruce (most probably red spruce *(Picea rubens)* and birch (both yellow and white birch). Some riparian forests had to be early successional (or recently disturbed) to allow for reproduction of such species as quaking aspen (*Populus tremuloides*), pin cherry, American elm, and beaked hazel. Although the riparian zone is generally well-drained, there were numerous loci of poorer drainage inhabited by Eastern white cedar.

**Frequency of natural disturbances**

The composition and spatial distribution of reconstructed taxa allow for some interpretation of natural disturbances at the time of land clearance.

The species composition at the time of land clearance was dominated by mid- to late successional, shade-tolerant species. Reproduction of such species is governed by small-scale disturbances or gap dynamics (Pickett and White 1985). Supporting this observation was the high frequency and high abundance of Eastern white cedar in most of the study sites. This species is usually absent in recently disturbed sites.

In a few sites, light-demanding species were a part of the forest cover, appearing either in the canopy (white pine, red oak, quaking aspen, American elm, and species of ash) or in understorey – e.g., pin cherry, species of alder, and beaked hazel. Reproduction of light-demanding, early successional species requires periodic disturbances of a larger scale, such as fires, windstorms, flooding, and insect infestations. Therefore, the presence of light-demanding species in the charcoal assemblages associated with the time of land clearance indicates the occurrence of such disturbances, and the proportion of sites where such species have been found reflects frequency of the disturbances. Such disturbances had to closely precede the establishment of the light-demanding species in the site. Hence, the maximum longevity of the species provides us with an estimate of the maximum time since the last disturbances in the site.

Long-lived, light-demanding canopy species, such as red pine, white pine and red oak can achieve age of 250-300 years in the area, therefore their presence in the landscape/site may indicate the occurrence of some disturbances (mainly fire) as much as ~250-300 years prior to the land clearance event. Red pine, a species directly associated with fire for reproduction, was found in only one site. White pine was recorded in 45% of study sites, while both pine and oak species were recorded in 20% of the sites. Presumably, these numbers correlate with the frequency of large-scale disturbances during 250-300 years preceding the land clearance.

It was assumed that the sites lacking long-lived, light-demanding canopy species were not affected by disturbances within the maximum lifetime of these species; whereas those sites that featured these species were affected by disturbances within the species longevity interval. Therefore, it was possible to calculate the disturbance return interval as a division of the indicative species longevity on the fraction of sites where those species were present.

The maximal disturbance return interval calculated using the frequency of red oak – the species that requires canopy replacing disturbances, such as fire and wind, is 1250 years (=250 / 0.2, where 250 is species longevity and 0.2 is the fraction of sites with this species).

Red oak was always found in association with white pine, but white pine occupied a wider range of sites in association with a wide range of species. It may indicate that reproduction of white pine in the area is governed by a wider spectrum of factors than the reproduction of red oak. Both red oak and white pine (where they dominate forest stands) are considered to be largely dependent on fires for their reproduction (Burns and Honkala 1990). At the same time, our observations on white pine reproduction in the area indicate that white pine can also be established on margins of bogs, creeks and beaver ponds, replacing deciduous vegetation that is selectively consumed by beaver. The width of the "feeding zone" cleared by beaver foraging can reach 60 meters (Rossell *et al.* 2005, Johnston and Naiman 1990). Therefore, calculating the canopy replacing disturbance return interval based on white pine frequency would probably underestimate the real disturbance return interval and can be used as a minimal assessment. Thus, minimal estimate is 555 years (=250 / 0.45, where 250 is species longevity and 0.45 is a fraction of sites with this species). Combining these two estimates, one can conclude that during the period between 450 and 200 years ago the canopy replacing disturbance return interval for the studied area was between approximately 1250 and 550 years.

Such estimates were calculated on the assumption that all sites are equally prone to disturbances.

However, a proportion of the "disturbance-dependant species" varied among geomorphological units. Red oak and white pine were much more common in riparian and coastal sites than in watersheds, with red oak completely absent on watersheds. Red oak was found in ~70% of coastal sites, but only 30 % of riverine sites. Assuming that all sites within the same geomorphological unit are equally prone to disturbances, the minimal and maximal estimates will be 555/833 years in riparian zone, 417/625 years in coastal sites, and 1000 to over 3012 years on watersheds.

Short-lived species, such as quaking aspen, American elm, gray alder, willow species, and black ash require the appearance of openings in the canopy at least once every 50-80 years. All these species are associated with wetlands that are very numerous in the area. Periodic (decadal to centennial) pulsation of wetland margins is typical for the area and can be caused by a number of factors, such as changes in atmospheric precipitation, changes in ground water regimes, and beaver activity. Such pulsation can provide new habitats for reproduction of the short-lived wetland species with a periodicity sufficient for their maintenance in the landscape.

**Comparison with historical records**

Results of the charcoal analysis have been compared with the analysis of archival descriptions of the forest composition (Crossland 2006). Results of botanical identification of ~1300 charcoal fragments from 30 sites were used for comparison with results of the analysis of ~2500 witness trees. The archival data represented a much higher sample of trees, as each of the analyzed charcoal fragments did not represent a single tree. Also, the archival data represented a much larger study area, as the majority of witness trees were located outside of KNP.

Each research method detected a high diversity of species in 19th century forests (Figure 2), including tree species that are generally not common on the landscape, such as *Ulmus americana* and *Ostrya virginiana*.

Archival analyses revealed presence of 20 taxa in pre-clearance forests: 18 of them were assigned to certain species, whereas 2 could be determined only to genus level.

Charcoal study captured 20 taxa: it recorded all tree genera that were represented in the witness tree record, and the method allowed some genera to be consistently identified to the species level, e.g. *Pinus strobus* was separated from *P. resinosa* that was found in only one study site. Witness tree results also confirmed minor amounts of *Pinus resinosa* (a frequency of 0.6% in the landscape). According to both methods, *Pinus banksiana*, a fire-dependant species that is common and widespread in the modern landscape, was absent at the time of land clearance. Species of *Picea, Betula,* and *Acer* were not distinguished through charcoal study, nor were they consistently discerned in the witness tree record, so further quantitative and comparative analyses were confined to the genus level.

Both methods revealed the presence of shrub or understorey species, such as *Alnus* and *Prunus*. Charcoal analysis identified three additional shrub genera that were not detected in the witness tree record: *Taxus canadensis, Corylus cornuta,* and *Salix spp*. An understorey tree species discerned in the witness tree record, but not through charcoal analysis was *Acer pensylvanicum*.

Both methods indicated that historic forests in the area were mixed and comprised of mid- to late successional species (*Picea sp, Tsuga canadensis, Betula sp, Acer sp, Fagus grandifolia, Abies balsamea* and *Pinus strobus*) in the majority of sites. Early successional and shade-intolerant species, such as *Populus sp., Pinus resinosa* and *Quercus rubra* were present at very low levels according to both methods.

Occurrence of taxa reconstructed by the analysis of charcoal assemblages have been compared with the occurrence of taxa calculated using witness trees records. For this comparison we excluded shrub species (*Salix, Corylus* and *Taxus*) from the analysis since they were not generally used as witness trees and compared only 16 tree genera common for both data sets. To make the two sets of data comparable, we calculated the occurrence of each species in charcoal assemblages as $n/(N_1 + N_2 +...+ N_x)$, where n is the number of sites where a species occurred, and N is the total number of species recorded at each study site. In total there were 173 tree species occurrences in charcoal assemblages. These data were compared with a total 306 occurrence records of witness trees, also expressed as a percent of total tree species occurrences.

The occurrences of species calculated by these two methods were remarkably similar (Figure 6), given the differences in sampling distribution and sample sizes. Both methods revealed the same eight most frequent taxa, with *Picea* being the most frequent taxon. Occurrences of *Betula, Acer, Fagus, Pinus* (Strobus type) and *Fraxinus* were very similar in either method. Smaller trees and shrubby species, such as *Thuja occidentalis, Ostrya virginiana, Prunus pensylvanica, Populus spp.*, and *Alnus spp.* were represented in the witness tree data at much lower occurrences than indicated by charcoal study. Results differed, however, among coniferous species such as *Tsuga canadensis, Abies balsamea* and *Picea sp.*, which had much higher occurrences in witness tree

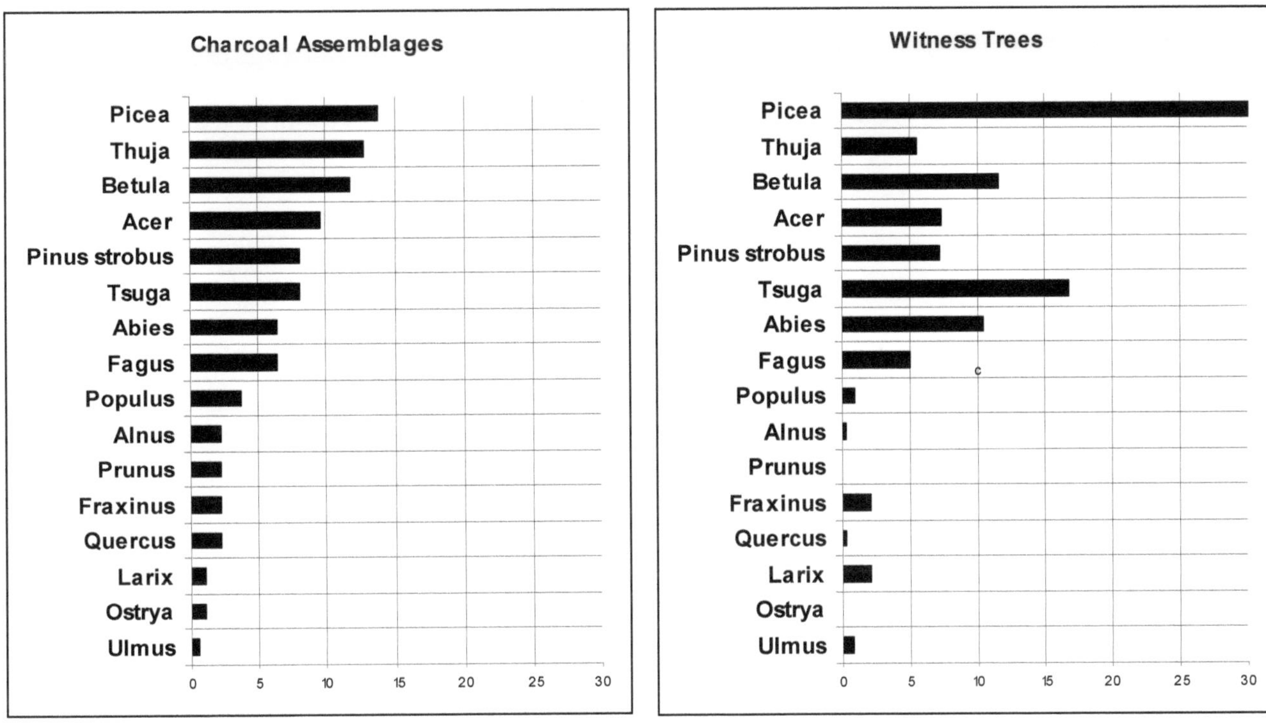


Figure 6. Comparison of tree species occurrences as estimated by the studies of charcoal assemblages and witness tree records

results in relation to charcoal assemblages. Under-representation of smaller understorey tree species and over-representation of coniferous canopy species in archival records may be explained by a bias towards selecting larger trees during surveying.

Some of the discrepancies between the two methods may also reflect the differences in sample site locations: e.g., *Quercus* is over-represented and *Tsuga* underrepresented in charcoal assemblages in relation to witness tree data. The former species is most frequent and the latter is least frequent in coastal sites where a smaller portion of witness trees were present to compare to the charcoal research sites. Finally, some of the discrepancies may reflect differences in preservation of charcoal of different species, which requires additional study.

## CONCLUSION

Comparison between the two methods confirmed applicability of the analysis of charcoal assemblages from plough layers for the reconstruction of tree species composition at the time of land clearance. This presents an opportunity of applying the proposed anthracological approach for reconstructing historical forest composition in agricultural areas that were previously forested.

### References

BEACH, H. (Ed.) 1988. *The Resources of Kouchibouguac National Park: Resource Description and Analysis.* Kouchibouguac National Park, Envir. Canada.

BURNS, R.M. and HONKALA, B.H. tech. coords. 1990. *Silvics of North America: 1. Conifers; 2. Hardwoods. Agriculture Handbook 654.* U.S. Department of Agriculture, Forest Service, Washington, DC. 2, 877 pp.

CROSSLAND, D. 2006. *Defining a forest reference condition for Kouchibouguac National Park and adjacent landscape in Eastern New Brunswick using four reconstructive approaches.* Unpublished MScF thesis, University of New Brunswick.

De BLOIS, S. and BOUCHARD, A. 1995. Dynamics of *Thuja occidentalis* in an agricultural landscape of southern Quebec. *J. Veg. Sci.* 6(4), 531-542.

DESLOGES, C. 1980. *Les ressources naturelles du Parc national Kouchibouguac. Parcs Canada, région de l'Atlantique.* Halifax, Nova Scotia. 137 pp.

DNRE. 1996. *An ecological land classification system of New Brunswick.* Ecosystem Classification Working Group, Department of Natural Resources and Energy, Fredericton, NB.

DRYADE (Le Groupe) 1979. *Kouchibouguac National Park forest vegetation. Parks Canada.* Unpublished report. 186 pp.

DUCHESNE, L.C. and HAWKES, B.C. 2000. Fire in northern ecosystems. In Brown, J.K. and Smith, J.K. (Eds.) *Wildland fire in ecosystems: Effects of fire on flora.* USDA Forest Service Gen. Tech. Rep. RMRS-GTR-42, 2: 35-51.

FERNALD, M.L. 1919. Lithological factors limiting the ranges of *Pinus banksiana* and*Thuja occidentalis. Rhodora* 21 41-67.

HOFMEYER, P.V. 2008. Ecology and silviculture of northern white-cedar (*Thuja occidentalis L.*) in Maine. Ph.D. dissertation. Univ. of Maine, Orono, ME. 160 pp.

HOSIE, R.C. 1990. *Native trees of Canada.* Fitzhenry and Whiteside Ltd., Ont. 380 pp.

JOHNSTON, J.F.W. 1851. *Notes on North America.* Vol. I and II., C.C. Little and J. Brown, Boston. 512 pp.

LEONARD, K.J. 1996. Mi'kmaq *culture during the Late Woodland and Early Historic periods.* Thesis, U. Toronto, Dept. of Anthropology, 329 pp.

LOUCKS, O.L. 1962. A Forest Classification for the Maritime Provinces. *Proceedings of the Nova Scotian Institute of Science.* Vol. 25, Part 2.

PICKETT, S.T.A. and WHITE, P.S. 1985. *The ecology of natural disturbance and patch dynamics.* Academic Press, Inc. Toronto, ON. 472 pp.

PONOMARENKO E.V. and ANDERSON, D.W. 2000. The importance of charred organic matter in Black Chernozemic soils. *Can.J.Soil Sci.* 81, 285-297.

PONOMARENKO E.V. 1999. Methodological approaches to the analysis of the succession process in soil cover. In Smirnova, O.V., Shaposhnikov, E.S. (eds.). *Forest succession in protected areas of Russia and problems of biodiversity conservation.* Russian Botanical Society. St-Petersburg. p. 34-57. (In Russian).

ROSSELL F., BOZSERT, O., COLLEN, P. and PARKER, H. 2005. Ecological impact of beavers *Castor fiber* and *Castor canadensis* and their ability to modify ecosystems. *Mammal Review,* 35, 3-4, 248-276.

ROWE, J.S. 1972. *Forest Regions of Canada.* Publication No. 1300. Ottawa: Canadian Forestry Service, Department of the Environment.

SCOTT, M.L., and MURPHY, P.G. 1986. Regeneration patterns of northern white-cedar, an old-growth forest dominant. *Am. Midl. Nat.* 117, 10-16.

SQUIRES, W.A. 1946. Changes in mammal populations in New Brunswick. *Acadian Naturalist,* 2:26-44.

Van OOST, K., GOVERS, G., Van MUYSEN, W. and QUINE, T.A. 2000. Modeling translocation and dispersion of soil constituents by tillage on sloping land. *Soil Society of America Journal,* 64, 1733-1739.

WANG, C. and REES, H.W. 1983. *Soils of the Rogersville-Richibucto Region of New Brunswick.* Agriculture Canada, 239 pp.

YORSTON, R.M., GAFFRNEY, V.L. and REYNOLDS, P.G. 1990. Simulation of artifact movement due to cultivation. *Journal of Archaeological Science,* 17, 67-83.

# THE USE OF WOOD IN ARGARIC SETTLEMENTS OF THE SOUTH-EASTERN IBERIAN PENINSULA

Maria Oliva RODRÍGUEZ-ARIZA
Centro Andaluz de Arqueología Ibérica, Edif. B-1 Universidad de Jaén, 23071 Jaén, Spain
moliva@ujaen.es

**Abstract**: *The settlements of the Argaric culture that developed in the south-eastern Iberian Peninsula during the Middle Bronze Age had in most cases one or two levels of generalized fire. The largest of these fires, whether provoked or chance, surprised the inhabitants of the village, who were forced to flee from their houses. Some of the utensils and materials went directly from a systemic context of use to an archaeological context without other depositional processes intervening, as in the case of wastes and abandonment. Therefore the use inferences are direct and the economic conclusions are important. All the varieties of trees, shrubs, and herbaceous species used for the different activities of daily life and that the environment offered indicate an intimate knowledge of and relationship with the plant environment. The species most used in everyday work came from the immediate surroundings, while for more special items, the radius of collecting the plant materials covered a larger territory.*

**Key words**: *Anthracology, Argaric Culture, Bronze Age, south-eastern Iberian Peninsula, Wood, Plant Use*

## THE ARGARIC CULTURE AND ANTHRACOLOGICAL DATA

Between 2000 and 1800 BC, in the south-eastern Iberian Peninsula, the Argaric Culture began to emerge, spreading westwards to the current provinces of Granada and Jaén, and north-eastwards to Murcia and Alicante. The end of the this culture is dated at 1400-1200 BC.

The settlements belonging to the Argaric Culture are situated over more or less steep slopes, in river valleys, dominating an expansive territory. The dwellings, rectangular in floor plan and composed often of several rooms, were aligned along narrow terraces stepped into the hillslopes or they formed more compact nuclei with narrow, twisted streets in the high parts of the villages. The economy of these settlements was primarily agricultural, although with a notable amount of herding. The society was highly stratified and complex, as demonstrated by the funerary ritual and the difference in the wealth of the possessions found (Contreras *et al.* 1997).

The Argaric excavated sites that have been submitted to anthracological analyses on which our study is based, are: Cerro de la Virgen, Castellón Alto, Fuente Amarga, Loma de la Balunca, and Terrera del Reloj of the province of Granada (Rodríguez-Ariza, 1991, 1992a, 1992b; Rodríguez-Ariza and Ruiz, 1995; Rodríguez-Ariza *et al.* 1996a; Rodriguez-Ariza, *et al.* 1996b; Contreras *et al.* 1997; Ruiz and Rodríguez-Ariza, 2003); Peñalosa (Rodríguez-Ariza and Contreras, 1992; Contreras *et al.* 1995; Contreras *et al.* 1997; Rodríguez-Ariza, 2000) and Cerro del Alcázar of the province of Jaén (Rodríguez-Ariza, 2011); El Castillejo de Gádor (Rodríguez-Ariza, 2001a), Fuente Álamo (Schoch y Schweingruber, 1982; Carrión, 2004), and Gatas (Gale, 1999; Rodríguez-Ariza *et al.* 1998; Rodríguez-Ariza, 2000) of Almería; and Terlinques (Machado *et al.* 2004) of the province of Alicante (Figure 1).

## ARGARIC PALAEOVEGETATION ACCORDING TO ANTHRACOLOGY

Biogeographically, the south-east of the Iberian Peninsular belongs to the Mediterranean region and, within the biogeographic or chorological typology, includes mainly the Luso-Extremadurense and Castellano-Maestrazgo-Manchega provinces (Rivas Martinez, 1988). The altitudinal bioclimatic belts of this zone indicates, by the presence of the different plant communities present, five successive bells of vegetation (thermo-, meso-, supra-, oro- and cryoromediterranean), although most settlements of those referred to here belong to the thermo- and mesomediterranean levels.

Figure 2 shows the taxa determined in each of the settlements, resulting in a floral list of 59 taxa. The plant associations vary depending on the vegetation level and the biogeographic province in which each archaeological site is situated. However, in most of these, we find a degradation of the original plant formations, the settlements studied here falling within the framework of the last anthracological phase defined by Vernet and Thiébault (1987).

## ARCHAEOLOGICAL EVIDENCE FROM CHARCOAL FOR THE USE OF WOOD BY ARGARIC COMMUNITIES

The use of plant resources that the environment provides was fundamental for the survival of prehistoric peoples. Among communities of farmers and herders, such as the ones studied here, wood was one of the main resources, which was obtained from forests and woodlands of these territories. The uses in general and the choice of the wood for the construction of the dwelling, cooking, illumination, heating, etc. is beyond doubt (Fig. 2). The use of plant species as a specific function has been well determined, and in many cases long distances were

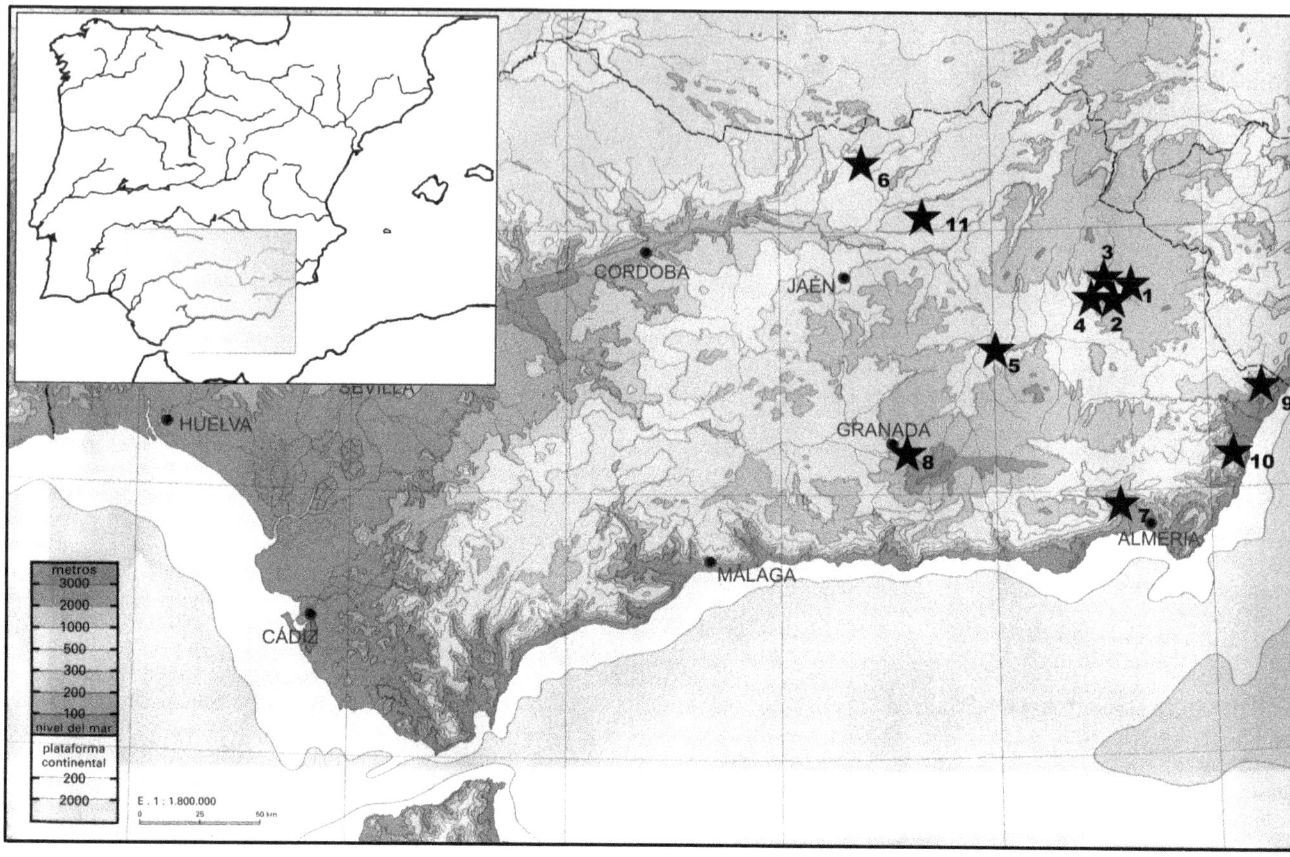


Figure 1. Map of South Iberian Peninsula with the archaeological sites. 1. Cerro de la Virgen, 2. Castellón Alto, 3. Fuente Amarga, 4. Loma de la Balunca, 5. Terrera del Reloj, 6. Peñalosa, 7. Castillejo de Gador, 8. Cerro de la Encina, 9. Fuente Álamo, 10. Gatas, 11. Cerro del Alcázar

| **Taxa / Site** | Cerro de la Virgen | Castellón Alto | Fuente Amarga | Loma de la Balunca | Terrera del Reloj | Peñalosa | Cerro del Alcázar | Castillejo de Gádor | Fuente Álamo | Gatas | Terlinques |
|---|---|---|---|---|---|---|---|---|---|---|---|
| *Acer* sp. | – | – | + | + | – | – | – | – | – | – | – |
| *Alnus glutinosa* | – | – | – | – | + | – | – | + | – | – | – |
| Angiosperma | – | – | – | – | – | – | – | – | – | – | + |
| *Arbutus unedo* | – | – | – | – | – | + | + | – | + | – | + |
| *Atriplex halimus* | – | + | – | – | – | – | – | – | + | – | – |
| Chenopodiaceae | – | – | – | – | – | – | – | – | + | – | – |
| Cistaceae | – | – | – | – | – | – | – | – | + | – | – |
| *Cistus* sp. | + | + | – | – | + | + | + | – | + | + | – |
| *Crataegus* sp. | – | – | – | – | – | – | – | + | – | – | – |
| *Dapnne* sp. | – | – | – | – | – | – | – | – | + | – | – |
| *Ephedra* sp. | – | + | – | – | – | – | – | – | + | – | – |
| *Erica* sp. | – | – | – | – | – | – | – | – | + | – | + |
| Ericaceae | – | – | – | – | – | – | – | – | + | + | – |
| *Ficus carica* | – | – | – | – | – | – | – | – | + | – | – |
| *Fraxinus* sp. | + | + | – | + | + | + | + | + | – | – | + |
| Gymnospermae | – | – | – | – | – | – | – | + | + | – | + |
| *Hedera helix* | – | – | – | – | – | – | – | – | + | – | – |
| *Juniperus* sp. | – | + | – | – | + | – | – | + | + | – | + |
| Labiatae | – | – | – | – | – | – | – | – | + | – | – |
| Leguminosae | + | + | + | + | + | + | + | + | + | + | – |
| *Myrtus* | – | – | – | – | – | – | – | – | – | + | – |

| **Taxa / Site** | Cerro de la Virgen | Castellón Alto | Fuente Amarga | Loma de la Balunça | Terrera del Reloj | Peñalosa | Cerro del Alcázar | Castillejo de Gádor | Fuente Álamo | Gatas | Terlinques |
|---|---|---|---|---|---|---|---|---|---|---|---|
| Monocotyledonae | – | + | – | – | – | – | – | + | + | – | + |
| *Nerium oleander* | – | – | – | – | + | – | – | – | – | + | – |
| Olea europaea | – | – | – | – | – | + | + | + | + | + | + |
| *Phillyrea* sp. | – | + | – | – | – | + | – | – | – | – | – |
| *Pinus halepensis* | + | + | + | + | + | – | + | – | + | – | + |
| *Pinus nigra* | – | + | + | – | – | – | – | + | – | – | – |
| *Pinus pinaster/pinea* | – | – | – | – | – | – | – | – | + | – | – |
| *Pinus sylvestris* | – | + | – | – | – | – | – | + | – | – | – |
| *Pinus nigra/sylvestris* | – | + | + | + | + | – | – | + | – | – | – |
| *Pinus* sp. | – | + | + | + | + | – | + | – | + | + | – |
| *Pistacia lentiscus* | – | – | – | – | – | + | + | + | + | – | + |
| *Pistacia terebinthus* | – | – | – | – | – | – | – | – | + | – | – |
| *Pistacia* sp. | – | – | – | – | – | + | – | – | + | – | – |
| *Prunus* sp. | – | – | – | – | – | – | – | – | – | + | – |
| Pomoideae/*Prunus* | – | – | – | – | – | – | – | – | – | + | – |
| *Populus* sp. | – | + | – | + | + | – | – | + | – | – | – |
| *Quercus ilex/coccifera* | + | + | + | + | + | + | + | + | + | – | + |
| *Quercus faginea* | + | + | + | + | + | – | – | + | – | – | – |
| *Quercus suber* | – | + | – | + | + | + | – | + | – | + | – |
| *Quercus* deciduous | + | + | + | – | + | – | – | – | – | – | – |
| *Quercus* evergreen | – | – | – | – | – | + | – | – | + | – | – |
| *Quercus* sp. | – | – | – | – | – | + | + | – | + | + | – |
| *Rhamnus* sp. | + | – | – | – | – | – | – | – | + | + | – |
| *Rhamnus-Phillyrea* | – | – | – | – | – | – | – | – | + | + | + |
| *Rhus* sp. | – | – | – | – | – | – | – | – | – | + | – |
| *Retama* sp. | – | + | – | – | – | + | + | + | – | – | – |
| Rosaceae | – | – | – | – | – | – | – | – | – | – | – |
| *Rosmarinus officinalis* | + | + | + | – | + | – | + | + | + | + | + |
| *Salix* sp. | + | + | + | + | + | – | – | + | – | + | – |
| *Salix/populus* | – | + | – | + | – | – | – | + | + | + | – |
| *Spartium junceum* | – | + | – | – | – | – | – | – | – | – | – |
| *Stipa tenacissima* | – | + | – | – | – | – | – | + | – | – | – |
| *Tamarix* sp. | + | + | – | + | + | – | – | – | + | + | + |
| *Tetraclinis articulata* | – | – | – | – | – | – | – | – | + | – | – |
| *Teucrium* | – | – | – | – | – | – | – | – | – | + | – |
| *Viburnum* sp. | – | – | – | – | – | – | – | – | – | – | + |
| *Viburnum tinus* | – | – | – | – | – | – | – | – | + | – | – |
| *Vitis* sp. | – | – | + | – | – | – | – | – | – | – | – |
| **Number of Taxa** | **12** | **26** | **12** | **15** | **20** | **13** | **12** | **21** | **33** | **22** | **15** |

Figure 2. List of taxa determinated by the anthracological analysis (+ Present, – Absent)

covered to find the type of wood or firewood sought, when not available in the surroundings of the settlement. Most of the remains have disappeared, since wood is perishable. In general, the lignin remains conserved are those that had undergone a process of partial carbonization, either intentionally or by chance.

In any case, the sites studied here in most cases had one or two levels of generalized fire. Most of them, whether by accident or intentional, surprised the inhabitants of the village, who had to hurry away from their houses. Therefore, part of the utensils and materials passed directly from a systemic context of use to an

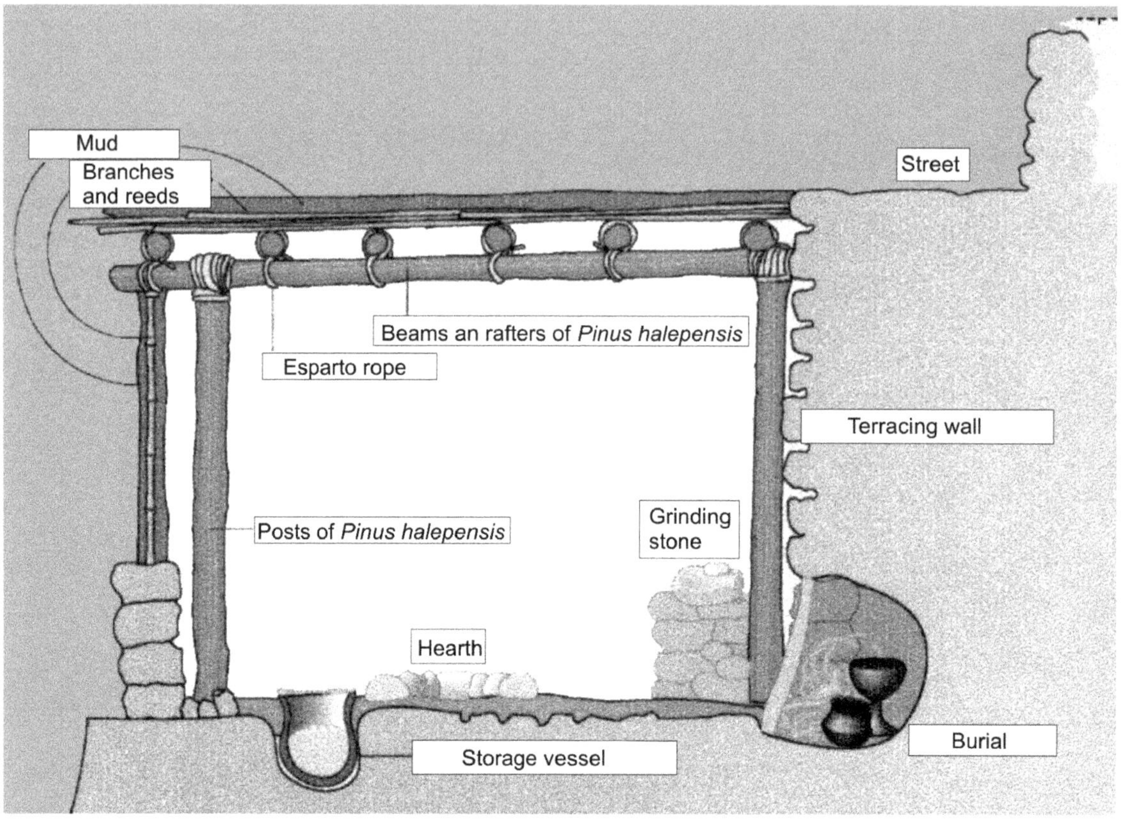


Figure 3. Schematic section of an argaric hut

archaeological context without the mediation of other depositional processes such as wastes or abandonment and thus the inferences on use are direct and the economic conclusions are important. Nevertheless, the distinction of the origin of the different charcoals is a complex process as generalized fire of all the wooden items leaves a mixture (Rodríguez-Ariza, 1993) that hampers identification.

**Elements of construction**

During the Bronze Age, there was a substantial change in urban configurations. The villages were situated primarily in high or strategic zones and the rectangular or quadrangular hut was predominant. The walls were normally built on stone plinths with the upper parts built of adobe or mud over woven cane, sometimes with the surface smoothed. The roofs were branches and mud for waterproofing, supported by wooden beams and posts. The materials used in the construction of these huts were stone, mud, cane, gypsum, wood, and esparto grass (*Stipa tenacissima*), all coming from the surroundings of the village. The wood and carbon remains from the construction of the dwellings at the Argaric sites studied here can be differentiated as two types: those used in posts and beams and cross members as opposed to those used for roofs.

At the structural level, the huts were formed by an armature of beams supported over vertical posts. Over the armature of beams, smaller cross members were laid perpendicularly (Figure 3). The posts and beams, at most of the sites, were made of *Pinus halepensis*, although cases have been documented of *Pinus nigra*, *Quercus faginea* and, in zones lacking pine, *Quercus ilex* and *Quercus suber*. In some settlements, such as Castillejo de Gádor, beams of *Alnus glutinosa* have been identified.

At some sites, fragments of beams that were square in section have been documented, as in the case of de Castellón Alto (Figure 4), where all the fragments of this type are of *Pinus nigra*, a specie that does not grow, at present, near the settlement and had probably to be brought back from the mountains over a distance of some 10 to 15 km. In Castillejo de Gádor, beams of *Alnus glutinosa*, *Populus*, *Tamarix*, and *Pinus nigra/sylvestris* have been identified. The practise of squaring beams in certain dwellings appears to have been commonplace, as such beams are appearing at the sites that we are currently studying.

In Peñalosa, there are large fragments of charcoal that have been identified as beams or posts, which invariably belong to *Quercus ilex*, *Q. faginea* or *Q. suber* or *Quercus* hybrids, in some cases presenting insect burrows, indicating that this wood before its carbonization had been used a long time after being cut.

The use of *Quercus ilex* and *Quercus suber* as posts is somewhat surprising, as the limbs of these trees are twisted and not large, in addition to being heavy and difficult to transport and work with. The use of posts of *Quercus ilex* has been found at the Valencian Bronze Age site of Lloma de Betxí (Pedro and Grau, 1991), although

Figure 4. Square beam of *Pinus nigra* of Castellon Alto

the beams were of *Pinus halepensis*. Also, at the Bronze Age site of La Mancha, la Motilla del Azuer, ten posts of *Quercus ilex* and one of *Quercus sube*r have been determined (Rodríguez-Ariza *et al.* 1999). These data appear to indicate that when no tree with straight, high trunks ais found in the surroundings of the village, such as pines, poplars, etc., *Quercus ilex* or *Quercus suber* are used.

As cross members in Castillejo de Gádor, *Populus* has been identified, with fragments slightly squared, presumably to fit more firmly over the beams. At the rest of the sites in the zones of Granada and Almería, the cross members were usually *Pinus halepensis*, while at Peñalosa they were *Quercus ilex/coccifera* or *Quercus suber*.

The plant composition of the roof remains has been defined at several sites:

- Fuente Amarga: Leguminosae, *Salix, Quercus ilex/coccifera* and *Quercus faginea.*
- Loma de la Balunca: Leguminosae, *Salix* and *Quercus ilex/coccifera.*
- Terrera del Reloj: Leguminosae, *Salix, Tamarix, Populus* and *Nerium oleander.*
- Castellón Alto: *Retama*; *Salix, Tamarix*, *Populus, Atriplex halimus* and Monocotyledonae (e.g. *Phragmites*)
- Peñalosa: bark, *Pistacia lentiscus, Cistus* and *Olea europaea.*
- Castillejo de Gádor: *Retama, Salix, Rosmarinus officinalis, Olea europaea* and *Quercus ilex/coccifera.*

Similar uses of plant species were found, with a predominance of shrubby species such as Leguminosae (e.g. *Retama*) at the mesomediterranean belt, together with *Pistacia lentiscus, Olea europaea, Cistus* and *Rosmarinus officinalis* at sites situated in the thermomediterranean. Also, the branches of tree species were used. On the one hand, there were riparian species such as *Salix, Tamarix, Populus*, and *Nerium oleander* and, on the other, the climacic vegetation such as *Quercus ilex/coccifera* together with the occasional

Figure 5. Reeds (*Phragmites*) for the construction of roofs of several huts of Castellon Alto

appearance of *Quercus faginea* and the use of bark in shingle-like pieces placed for shedding water from the roofs in Peñalosa. Reeds have been documented for the construction of roofs in several huts of Castellón Alto (Figure 5).

In general, the remains include small-diameter branches and in several huts bark has been preserved; the growth rings of the wood could be observed to estimate the general period in which the material was collected. In the Loma de la Balunca, most of the branches were between two and three years old and were cut during the autumn/winter; however, in the Terrera del Reloj the cutting appears to have been concentrated in a period of spring/summer. Therefore, the construction or repair of the roofs appears to have followed the seasons in relation to climatic change. Thus, the fact of undertaking these tasks mainly in autumn could signify prevention against winter rain and cold, and the knowledge that the wood is preserved better if cut when the plant is at its lowest vegetative activity. In addition, it coincides with the period of the year in which agricultural activities begin to diminish after the sowing of the cereals.

The posts and beams of the roof were fixed together mainly by esparto cords, for which there was a great variety of forms at sites such as Castellón Alto (Contreras *et al.* 1997). In general, tree species of large size or hardness were used for the structural elements to support greater weight, such as the posts and beams, while for the lighter elements such as the roof, herbaceous and shrubby species were used, or the thin branches of trees.

**Materials for energy and combustion**

One of the most frequent uses for wood was for burning, as energy or combustion. The structures used for combustion at the sites studied here were basically domestic fireplaces, although there were some associated with metallurgy. Because of the periodic cleaning of these structures, the charcoal remains were not usually preserved, and many of these fireplaces, delimited by several stones or marked by a simple pit, provide only ash with no charcoal. In many cases, the composition of the firewood used has to be deduced from the charcoal remains found in the levels of rooms associated with the different hearths. Thus, at most sites, as found in Terrera del Reloj and Castellón Alto, *Quercus ilex/coccifera* was the main taxon used for the fireplaces, together with other species such as rosemary, legumes and tamarish.

At Fuente Álamo two fireplaces revealed tree species represented by *Pinus* (e.g. *P. pinaster/pinea*) and *Olea europaea*, while another two hearths contained thicket species, mainly *Rosmarinus officinalis* and *Pistacia lentiscus* (Carrión, 2004).

The Structural Complex VIIe of Peñalosa at another stage (Phase IIIa) presented a floor of occupation in which a total of six charcoal fragments were collected, with a total weight of 0.08g, a meagre small of charcoal, possibly related to the periodic cleaning of the structure; this implies that the charcoal came from the last fire. The results of the anthracological analysis showed the predominance of *Quercus ilex/coccifera*, followed by *Quercus suber*. It should be highlighted that most of the charcoal fragments analysed showed a marked carbonization process, perhaps indicating their use in the process of melting minerals detected in this structure.

**Use of wood in funerary contexts**

The use of wood in rituals of these people, especially in funerary contexts, has been attested to in Castellón Alto, Fuente Amarga, and Terrera del Reloj. In the first of these settlements, the use of wood was found in the closure of sepulchral niches 11, 24, 25, and 34, in the form of slabs or posts of *Pinus halepensis*. Tomb niche 121 was sealed with three planks of black pine (*Pinus nigra*), carefully fitted and with a series of notches indicating a prior function. These planks replaced the large slabs of stone used for this function in most of the other funerary niches.Therefore, the wood was used here as a constructive element to replace boulders and stone slabs. The relationship as a symbolic element within the ritual appears not to have existed. Such a relationship did exist in niches 3 and 9 of Fuente Amarga, where the individuals were laid out with great care over a plank of *Pinus halepensis* and bark, probably of pine (Figure 6). These formed a bed that separated them from direct contact with the earth. In Terrera del Reloj, there were also the two types described above: in niche 10 covering the pit, and in niche 2 forming a sort of box that held the individual, were remains of *Pinus halepensis*.

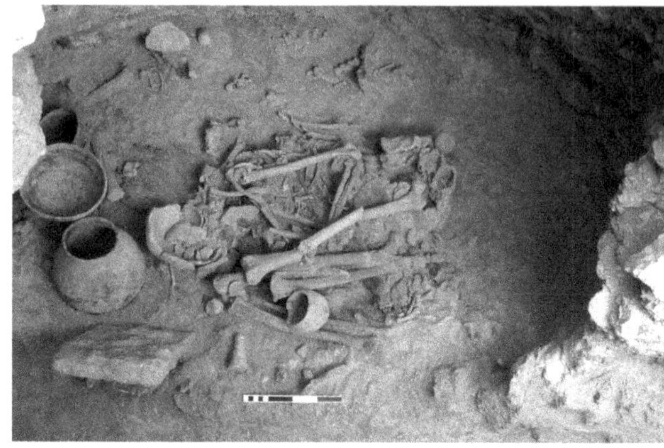

Figure 6. Individual on a table of *Pinus halepensis*. Tomb 3 of Fuente Amarga

There may be several ritual meanings to this act (Frazer, 1986:142-151) and may correspond to ideological issues within the overall community, such as personal or family identification.

**Utensils**

As indicated above, the generalized burning of the villages, apart from conserving the wood, implies a mixture of all the charcoal from the construction, from the houses, and the from the possible furnishings and utensils present in the different dwellings. Therefore, except in closed contexts such as the niches, it is difficult to recover fragments of utensils. Nevertheless, in some cases, the fastidious examination of some cylindrical fragments that appeared to be natural branches sometimes indicates whether they had been carved from a larger block and, therefore, are utensils. In one case, furthermore, the species determined (*Pinus nigra*) was not determined in the rest of the group of charcoals analysed from la Terrera del Reloj, supporting the contention that it was a utensil made with wood that was not present nearby. At this same site, a fragment that had a notch was identified as *Tamarix*.

Both at this site as well as at Castellón Alto, bronze utensils were recovered with part of the handle of wood, generally forming part of the belongings deposited together with buried individuals, although the decomposition of the wood impeded its taxonomic identification. We should highlight the copper axe with the complete handle of *Quercus ilex/coccifera*, although it was fragmented in three pieces in niche 121 of Castellón Alto. The axehead inserted into the handle perpendicularly instead of parallel to the handle, and thus would have been used as an adze (Molina *et al.* 2003).

**Basketry and crafts from other plant materials**

At all of these sites, the remains of unworked and worked *Stipa tenacissima* appear carbonized as a consequence of the aforementioned generalized fires in these villages. At the level of fibre preparation, two processes can be distinguished, one with crude esparto and the other with

Figure 7. Fragment of basket of *Typha*

the esparto split or crushed. In terms of manufacturing, there are two broad categories: cord and basketry. Each of these groups presents a great variety of typologies that correspond to a varied functionality and to a practical character for everyday life of these prehistoric communities. Thus, baskets, cords, mats, sandals, etc. have been recovered (Contreras *et al.* 1997). Also, other plants appear, such as caning (*Typha* sp.) for basket making (Figure 7) (Contreras *et al.* 1997).

A singular item was bark, indicated above in roof construction at Peñalosa. At this site, the appearance of a lid indicates the multiple uses that it could have had: bowls, hives, etc. In Castellón Alto, several fragments have also been recovered, among which three samples bear signs of manipulation with the presence of holes in two of them and sanding of another with the form of a half sphere. These fragments could be used as lids with holes serving to lash them to some sort of vessel. Their appearance at this site more than 20 km from the zones where at that time the cork oak grew naturally poses the possibility of trade or bartering of this material (Contreras *et al.* 1997).

**References**


CARRIÓN, Y. 2004. Análisis antracológico del yacimiento de Fuente Álamo (Cuevas del Almanzora, Almería): usos de la madera y paleovegetación. In L. Hernández and M.S. Hernández (eds.), *La Edad del Bronce en tierras Valencianas y zonas limítrofes,* 477-486. Villena.

CONTRERAS, F., MORALES, A., PEÑA, L., ROBLEDO, B., RODRIGUEZ-ARIZA, M.O., SANZ, J.L. and TRANCHO, G. 1995. Avance al estudio de los ecofactos del poblado de Peñalosa (Baños de la Encina, Jaén). Una aproximación a la reconstrucción medioambiental, *Anuario Arqueológico de Andalucía 1992.II*: 263-274, Cádiz.

CONTRERAS, F., RODRÍGUEZ-ARIZA, M.O., CÁMARA, J.A. and MORENO, A. 1997. *Hace 4000 años... Vida y muerte en dos poblados de la Alta Andalucía. Catálogo de la exposición.* Junta de Andalucía, Fundación Caja de Granada. Granada.

FRAZER, J.G. 1986. *La rama dorada.* Ediciones Fondo de Cultura Económica.

GALE, R. 1999. Gatas: Análisis antracológico. En P.V. Castro, R.W. Chapman, S. Gili, V. Lull, R. Micó, C. Rihuete, R. Risch y M.E. Sanahuja (eds.): *Proyecto Gatas 2. La dinámica arqueológica de la ocupación prehistórica.* Arqueología Monografías. Junta de Andalucía. Sevilla: 292-307.

MACHADO, M.C., JOVER, F.J. and LOPEZ, J.A. 2004. Primeras aportaciones antracológicas del yacimiento de Terlinques (Villena, Alicante). In In L. Hernández and M.S. Hernández (eds.), *La Edad del Bronce en tierras Valencianas y zonas limítrofes,* 363-368. Villena.

MOLINA, F., RODRÍGUEZ-ARIZA, M.O., BOTELLA, M. and JIMÉNEZ, S. 2003. La sepultura 121 del yacimiento argárico de El Castellón Alto (Galera, Granada). *Trabajos de Prehistoria* 60-1: 1-6.

PEDRO, M.J. de and GRAU, E. 1991. Técnicas de construcción en la edad del Bronce: la Lloma de Betxí (Paterna, Valencia). IInd Deia Conference of Prehistory (Deia, 1988) B.A.R. Int. S-574: 339-363.

RIVAS MARTÍNEZ, S. 1988. *Memoria del mapa de Series de vegetación de España 1:400.000.* ICONA, Madrid.

RODRÍGUEZ-ARIZA, M.O. 1991. An anthracological approach to the vegetation of the Baza-Huéscar Basin in III millennium BC. In A. Ramos, G. Martínez, G. Rios and J.A. Afonso. *Flint production and exchange in the Iberian Southeast, III millenium BC,* 105-109. Universidad de Granada e Instituto Tecnológico y Geominero de España. Granada.

RODRÍGUEZ-ARIZA, M.O. 1992a. Human-plant relationships during the Copper and Bronze Ages in the Baza and Guadix Basins (Granada, Spain). *Bulletin de la Société Botanique de France,* 139, 451-464.

RODRÍGUEZ-ARIZA, M.O. 1992b. *Las relaciones hombre-vegetación en el Sureste de la Península Ibérica durante las Edades del Cobre y Bronce a partir del análisis antracológico de siete yacimientos arqueológicos.* Tesis doctoral microfilmada. Universidad de Granada.

RODRÍGUEZ-ARIZA, M.O. 1993. Los procesos de formación y transformación del registro arqueológico en los estudios antracológicos. *Arqueología espacial,* 16-17, 371-390.

RODRÍGUEZ-ARIZA, M.O. 2000a. El análisis antracológico de Peñalosa. In F. Contreras (coord). *Proyecto Peñalosa. Análisis histórico de las comunidades de la Edad del Bronce del Piedemonte meridional de Sierra Morena y Depresión Linares-Bailén,* 257-272. Arqueología Monografías, Junta de Andalucía, Granada.

RODRÍGUEZ-ARIZA, M.O. 2000b. El paisaje vegetal de la Depresión de Vera durante la Prehistoria Reciente. Una aproximación desde la Antracología, *Trabajos de Prehistoria* 57-1, 145-156, Madrid.

RODRÍGUEZ-ARIZA, M.O. 2001. Análisis antracológico de El Castillejo de Gador (Almería). In B. Gómez, M.A. Respaldiza and M.L. Pardo (eds.) *III Congreso Nacional de Arqueometría,* 173-182. Sevilla.

RODRÍGUEZ-ARIZA, M.O. 2011. "Charcoal analysis in the district of Loma (Jaén, Spain)", E. Badal, Y. Carrión, E. Grau, M. Macías y M. Ntinou (eds.): *5th International Meeting of Charcoal as cultural and biological heritage*: 147-148, *Saguntum, Extra-11*. Departament de Prehistoria I Arqueologia, Universitat de Valéncia, Valencia.

RODRÍGUEZ-ARIZA, M.O. and CONTRERAS, F. 1992. Contrastación antracológica entre diferentes complejos estructurales del yacimiento de la Edad del Bronce de Peñalosa (Baños de la Encina, Jaén). *Actas Primeras Jornadas sobre Arqueología medioambiental a través de los macrorrestos vegetales*, Madrid 1991.

RODRÍGUEZ-ARIZA, M.O. and RUIZ SANCHEZ, V. 1993. Acción antrópica sobre el medio natural en el Sureste de Andalucía durante la Prehistoria Reciente y Época Romana. In *Investigaciones arqueológicas en Andalucía 1985-1992 (Proyectos)*, 417-428. Consejería de Cultura y Medio Ambiente de la Junta de Andalucía, Huelva.

RODRÍGUEZ-ARIZA, M.O. and RUIZ SANCHEZ, V. 1995. Antracología y palinología del yacimiento argárico de Castellón Alto (Galera, Granada*). Anuario Arqueológico de Andalucía 1992.II*, 169-176.

RODRÍGUEZ-ARIZA, M.O., RUIZ SANCHEZ, V., BUXÓ, R. and ROS MORA, T. 1996a. Palaeobotany of a Bronze Age community. Castellón Alto (Galera, Granada, Spain), *Actes du Colloque d'Archéométrie 1995 de Périgueux, Revue d'Archéometrie, Rennes, Suppl. 1996*: 191-196.

RODRIGUEZ-ARIZA, M.O., VALLE, F. and ESQUIVEL, J.A. 1996b. The vegetation from the Guadix-Baza (Granada, Spain) during the Copper and Bronze Ages based on Anthracology, *Arqueologia e calculatori* 7, 537-558.

RODRIGUEZ-ARIZA, M.O., STEVENSON A.C. and Editors 1998. Vegetation and its explotation. In P.V. Castro, R.W. Chapman, S. Gili, V. Lull, R. Micó, C. Rihuete, R. Risch and M.E. Sanahuja (eds.) *Aguas Project. Palaeoclimatic reconstruction and dynamics of human settlement and land-use in the area of middle Aguas (Almería), in the south-east of the Iberian Peninsula.* European Commission Environment and Climate Programme., 62-68. Luxembourg.

RODRÍGUEZ-ARIZA, M.O., NAJERA, T. and ROS MORA, T. 1999. Una valoración paleoecológica de la Motilla del Azuer a partir del análisis antracológico, in J. Capel Martínez (Ed.): *Arqueometría y Arqueología.* Monográfica Arte y Arqueología 47: 11-23. Universidad de Granada.

RUIZ, A. and RODRÍGUEZ-ARIZA, M.O. 2003. "Paisaje y asentamiento entre los iberos de la Cuenca del Río Guadalquivir (S. VI al III a.n.e.)", en *Ambiente e paesaggio nella Magna Gracia,* Atti del quarantaduesimo convegno di studi sulla Magna Grecia, pp. 261-278. Taranto.

SCHOCH, W. and SCHWEINGRUBER, F.H. 1982. "Holzkohlenalytische ergebnisse aus der bronzezeitlichen siedlung Fuente Alamo. Prov. Almeria. Spanien", *Archäologisches Korrespondenzblatt*, 12.

VERNET, J.L. and THIÉBAULT, S. 1987. An approach to northwestern Mediterranean recent prehistoric vegetation and ecologic implications, *Journal of Biogeography*, 14, 117-127.

# ANTHRACOLOGY IN THE UPPER DORDOGNE VALLEY: A TOOL FOR THE HISTORY OF A CHARCOAL PRODUCING FOREST

Romain ROUAUD

PhD student, Géolab UMR-6042-CNRS, 39E Rue Camille Guérin 87036 Limoges, France

Philippe ALLEE

Professor, Géolab UMR-6042-CNRS, 39E Rue Camille Guérin 87036 Limoges, France

**Abstract**: *The study of the charcoal producing forest of the upper Dordogne valley is associated with a collective research project called "geohistory and geoarchaeology of the Limousin foresters' land" led by Geolab.*

*The upper Dordogne valley is a unique massif of 1000 square kilometres of sloping forests located in the French regions of Auvergne and Limousin. This wide natural area offers a very interesting environment in which to exercise a biogeographical approach, but it has a little-known historical context. Indeed, this forest was exploited during both modern and contemporaneous times for its charcoal which was transported via the Dordogne River to the Bordeaux economical basin. Everywhere in these sloping forests, the vestiges of this activity are present. There are thousands of charcoal burning sites form the archaeobotanical* corpus. *This paper focuses on the charcoal analysis of 11 charcoal burning sites.*

*The first anthracological results indicate a change in the arboreal composition of the forested landscape during the past three or four centuries. In this case, anthracology allows us to piece together the history of the tree diversity, which was influenced by charcoal burning.*

**Key-words**: *Anthracology, charcoal burning, fire-places, woodland cover, Dordogne valley, 16th-19th centuries*

## INTRODUCTION

Nowadays we know the importance of the relationship between Man and Environment, and how it has developed over time, which has enabled an understanding of the woodland composition (Blondel, 1986; Dubois, 1993). Demographic growth and industrial development have necessitated combustibles during the last five centuries in France. Meanwhile, charcoal burning evolved in importance in many forests and has influenced the composition of the forest landscape (Corvol, 1987; Davasse, 2000; Bonhôte *et al.* 2002). In France, charcoal burning studies developed at the end of the 1980s in the Languedoc and Pyrenees Regions, but little charcoal research on forest history has been conducted in the Massif Central Region (Bonhôte, 1987, 1992; Davasse, 1992; Izard, 1992a, 1992b; Métailié, 1992).

The present study is included in a "Collective Research Program" which focuses on the ancient forests of the Limousin Region. In this paper, the territory studied is located between the Limousin region and the Auvergne region, to the northwest of the Massif Central Region. The upper Dordogne valley is known for its staves trade (planks used to build barrels) with the Bordeaux vineyards, during the modern and the contemporary periods (Cocula-Vaillières, 1981; Bombal, 1981). It's an epic story which is engraved in the local memory. However, because there are few works and archives regarding this activity, not many details are known about it. This article refers to the discovery of many remains of charcoal burning in this area, and yet the local people were unaware of this occurrence.

The aim of this work is to show the role and the importance of the charcoal burning activity on the evolution of the landscape composition of the upper Dordogne valley. Two approaches are used in this study: the use of archives and the charcoal analysis of the charcoal burning sites. Only the anthracological aspects are introduced in this work which is organised in two parts. The first presents the work from field prospecting at two scales, and then in the second part, the first results of charcoal analysis are analysed.

## MATERIAL AND METHODS

### Study area

The Dordogne River rises at the Puy de Sancy (1886 m a.s.l., near Clermont-Ferrand) and flows into the Gironde estuary near Bordeaux (Figure 1). The Upper Dordogne Valley constitutes the first 150 kilometres upstream from the river between the Auvergne volcanoes to the East and the sedimentary ground of the Périgord to the West. The study area represents a group of deep valleys which contains a large slope forest of 1000 square kilometres. The climatic conditions are Atlantic but with a mountain influence in height; annual rains are between 800 and 2000 millimetres and the annual temperatures average between 12°C and 6°C.

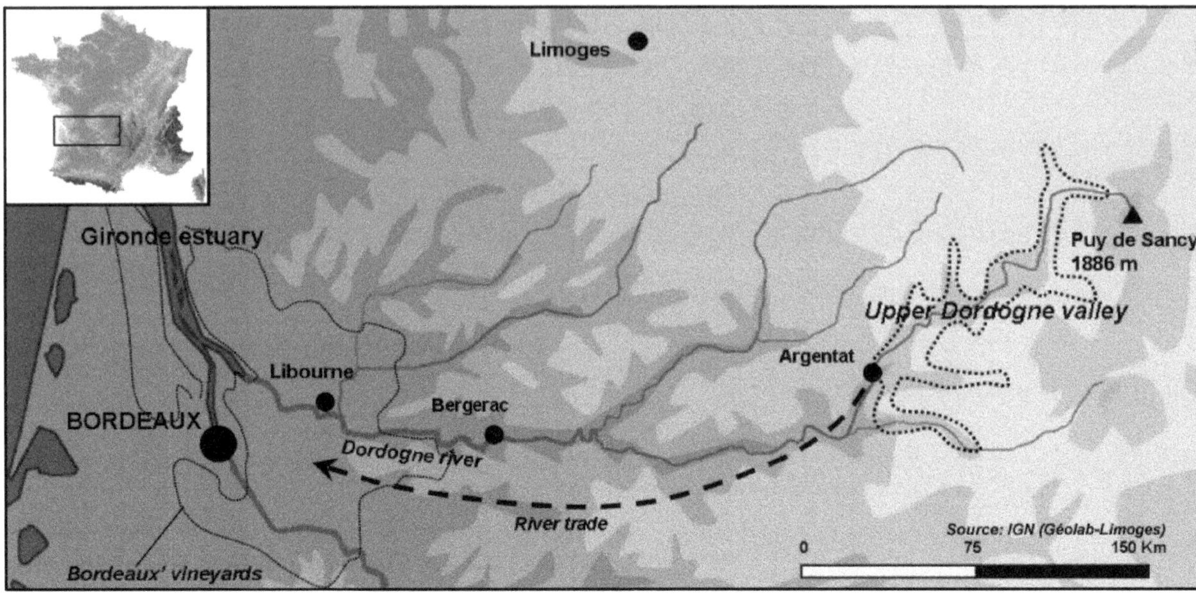


Figure 1. Location of the Upper Dordogne Basin

Today, these slope forests are composed of old coppices developed on a stony ground, with a dominance of oaks, beeches and hornbeams. The forest vegetation is organised around an altitude gradient Western to Eastern. Chestnut trees (*Castanea sativa*) disappear at about 500 metres in altitude and the hornbeam (*Carpinus betulus*) at about 600 metres. When hornbeams disappear, firs (*Abies alba*) become regular on the north-facing slopes and dominate with the beech between 1000 and 1400 metres. All over the basin, oaks dominate 200 to 800 metres in the South-facing slopes. The beech is very present in the oldest planting and in the shaded sites (in ravines for example) (Dupias and Lavergne, 1968; Dupias, 1977).

## Research and inventory of charcoal burning places

It is impossible to conduct finely-honed research in such a large area. To obtain an idea of the extent of the charcoal burning, field research has been carried out on a large scale in numerous spots all over the study sites. They represent about ten square kilometres split between 17 sites, but one zone is better-known than the others (Figure 2). In the Luzège valley, the faces explored were limited to the Lamazière-Basse village at 6 square kilometres. Every time the slopes were accessible, all faces were inspected from the bottom to the top.

For each platform found, different parameters were logged and recorded in a Geographic Information Systems database:

- The localisation by GPS.
- The actual woody vegetation by the Braun-Blanquet abundance/dominance method is logged on a surface of about 2000 square metres.
- The measurement of the place, the length, the width and the height of the excavation.
- The charcoal level depth with a sediment description.

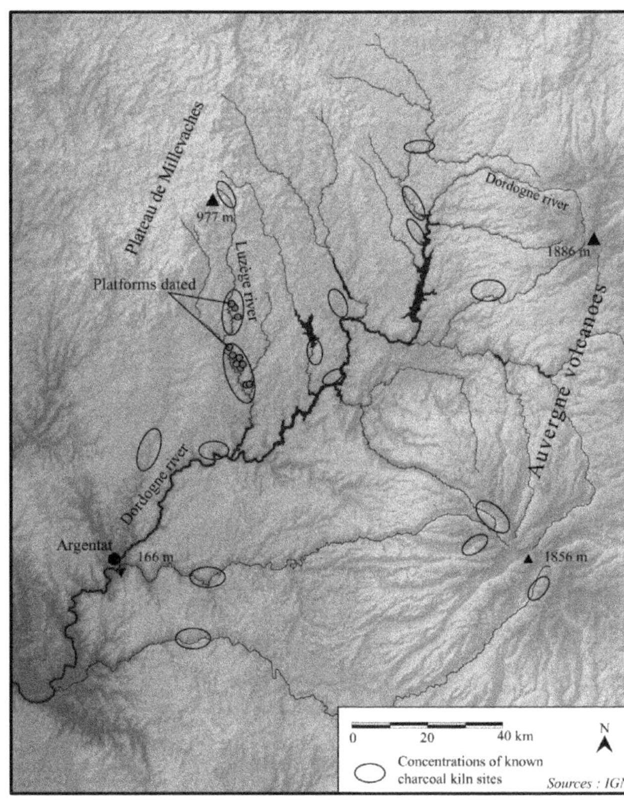


Figure 2. Topographic layout of Upper Dordogne valley

## Sampling and method of analysis

A first anthracological study of the platforms was undertaken to consider the influence of charcoal burning in local and regional vegetation. Charcoal layers are the remains of many burning kilns. The anterior works demonstrate that practices of kiln construction, burning wood, and charcoal collection harmonized the spatial deposit of distinct taxa on the surface of the platform (Fabre, 1996; Davasse, 2000; Ludemann *et al.* 2004;

Figure 3. Reusing the last charcoal sediment to cover the new kiln

Figure 4. "World War II" kiln from a traditional charcoal burning site (platform)

Euba, 2008). However, the charcoal burners reused the last charcoal sediment to build the new kiln (Figure 3). Therefore, the charcoal layers are often homogeneous for many centimetres of depth. For the moment we are unable to know the relations between this depth and the many charcoal burning events.

Consequently a standardized sampling strategy was used. Systematically, charcoal sediment was taken every three or five centimetres to obtain a minimum of three samples between the bottom and the surface. This sampling was done in a pit of twenty centimetres on the side, if possible, downhill from the burning places, between the centre and the edge. A centre sampling it is to be avoided for two reasons:

- Firstly, in many cases, the most important charcoal layer was not located in the centre of the platforms but at the periphery.
- Secondly because the centre of the burning places represents a specific event: it is the centre of the chimney kiln which can be built with special taxa. So, a sampling here may distort the representativeness of the charcoal assemblages.

The charcoal sediment is sieved at two and four millimetres, but only the fragments larger than four millimetres were analysed.

The determination of the wood taxa of charcoal fragments follows Jacquiot (Jacquiot *et al.* 1973), Schweingruber (Scweingruber, 1990) and Vernet (Vernet, 2001), using a stereoscope and an incident-light microscope (Nikon) as well as a reference collection of charred, known wood. The quantity of each taxon was determined by the number of pieces counted.

### Dating and the charcoal remains corpus

In this paper only the Luzège valley results are presented. Dates have been obtained by radiocarbon analysis from the Centre for Applied Isotope Studies of the University of Georgia (USA) and calibrated with the software "Oxcal v4.1.7". In all, 17 charcoal levels have been analysed, but only 13 dated per 10 charcoal burning places. Actually, the platforms, Ch 17.2, Ch 36, and Ch 116 have been dated once with an above ground sample and another time with a bottom layer sample. The World War II kiln has been dated by a relative chronology due to its unique contents: a closed iron kiln with the charcoal remains inside (Figure 4).

All the radiocarbon dates fall between 240BP and 50BP. Owing to the occurring of an important radiocarbon plateau, the dating is gauged at an interval of three centuries. After calibration, all the dates fall between the sixteenth and the nineteenth century (Figure 5). This period corresponds with the height of the Dordogne River trade, a time during which the forest was exploited and its produce transported to the Bordeaux basin, downstream the river.

## RESULTS

### The extent of the charcoal burning forest

Thanks to a great amount of field prospecting, out of 95 % of the prospected forests, we have identified and located about 550 charcoal burning places. All are present in very hilly landscapes, distributed all over the slopes. There are always charcoal burning places on all faces, but the density varies between each site. These first results indicate that an important charcoal producing forest could have existed. However, even if this field research is extensive, it is still not sufficient to know the real importance of this activity. That is why it is necessary to compare the field prospecting within a small area to have a better estimate of the density, the topographic situation, and the size of charcoal burning places.

In the Luzège valley 350 platforms are found within 6 square kilometres, which represents about 60 platforms per square kilometre. Platforms are present everywhere on these slopes, in all possible situations. Some are situated in ravines, while others are on the rock slides, and some have stone retaining walls. For example, the charcoal burning place "Ch. 226" has a retaining wall of

| | Radiocarbon age Years BP conv[a] | ± | Calibrated date AD[b] |
|---|---|---|---|
| Ch 107,1 -9/-11 | 50 | 40 | 1689 - 1926 |
| Ch 107,2 0/-3 | 70 | 40 | 1682 - 1933 |
| Ch 225 -20/-23 | 100 | 40 | 1680 - 1939 |
| Ch 226 fond | 150 | 40 | 1665 - 1952 |
| Ch 17,2 -9/-12 | 120 | 40 | 1675 - 1942 |
| Ch 17,2 S2 | 180 | 40 | 1648 - 1954 |
| Ch 116 S1 | 120 | 40 | 1679 - 1940 |
| Ch 116 S2 | 190 | 30 | 1644 - 1955 |
| Ch 108,2 -5/-9 | 190 | 40 | 1644 - 1955 |
| Ch 108,1 | 200 | 40 | 1641 - 1955 |
| Ch 36 S1 | 160 | 40 | 1663 - 1953 |
| Ch 36 S2 | 240 | 40 | 1492 - 1954 |
| Ch 45 | 240 | 40 | 1520 - 1955 |

[a] *Years BP conv* Years before present (1950)
[b] Calibration 2 sigma

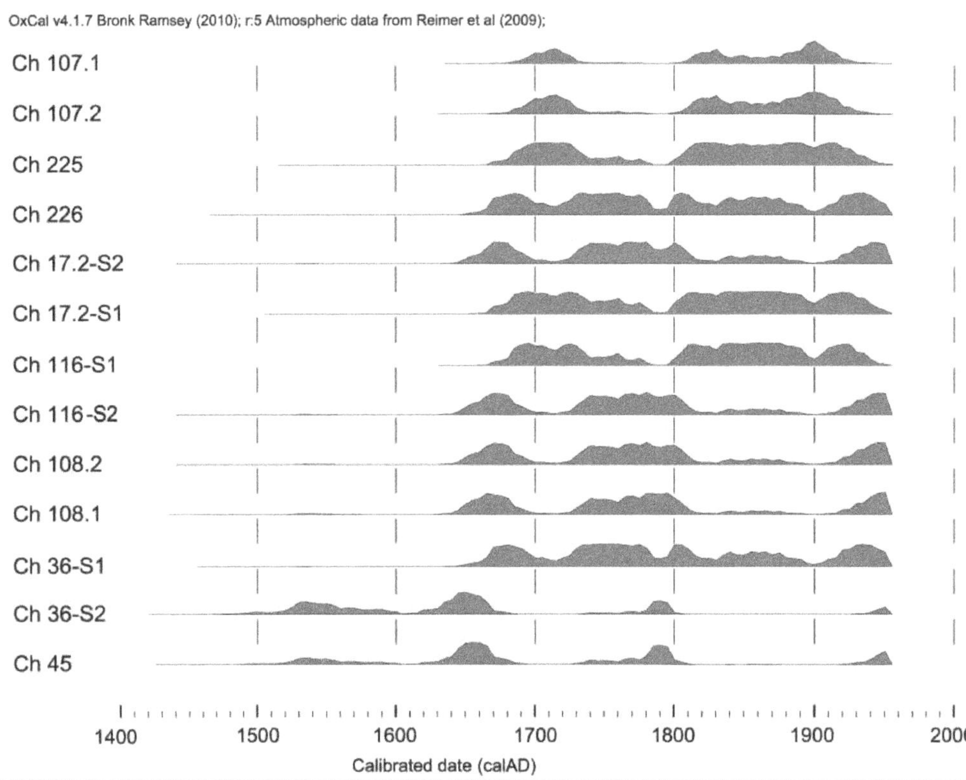


Figure 5. Radiocarbon dating in the Luzège valley (Calibrated dates)

2.60 metres in height, and the tallest retaining wall in the entire basin is 4.80 metres. The important building efforts represented by these examples offer further proof of a real will to produce charcoal.

Recent prospecting made on a finer scale than that of the Luzège valley has indicated a platform density upwards of one hundred per square kilometre. For example, in the Chavanon valley, 38 platforms have been found on 0.2 square kilometres. Also on the same surface, in a part of the Dordogne valley near the village of St-Merd-de-Lapleau, 25 platforms have been inventoried. The inventory of the Luzège valley is unbalanced and not exhaustive, and it seems that the density is similar to the other inventories. If this density is homogeneous all over the basin, thousands of platforms remain to be discovered.

**Tree vegetation of the charcoal producing forest**

We have analyzed 1275 pieces of charcoal, and they have yielded a diversity of 17 taxa: ten trees and seven shrubs (Figure 6).

- For the trees: deciduous *Quercus*, *Carpinus betulus*, *Fagus sylvatica*, *Castanea sativa*, *Corylus avellana*, *Tilia* sp., *Fraxinus* (cf. *excelsior*), *Alnus* (cf. *glutinosa*), *Betula* sp., *Populus* (cf. *tremula*).

- For the shrubs: *Salix* sp., *Juniperus* (cf. *communis*), *Sorbus* (cf. *aucuparia/torminalis*), *Sorbus* (cf. *aria*), *Evonymus europaeus*, *Frangula alnus*, *Ilex aquifolium*.

The principal taxa determined are deciduous *Quercus* with about 50% of the charcoal pieces, *Carpinus betulus*

| | Taxon | Number | % |
|---|---|---|---|
| Trees | *Quercus* | 580 | 46 |
| | *Carpinus* | 289 | 23 |
| | *Fagus* | 127 | 10 |
| | *Corylus* | 50 | 4 |
| | *Castanea* | 48 | 4 |
| | *Tilias/Acer* | 10 | <1 |
| | *Alnus* | 9 | <1 |
| | *Fraxinus* | 8 | <1 |
| | *Betula* | 2 | <1 |
| | *Populus* | 1 | <1 |
| Shrubs | *Ilex aquifolium* | 25 | 2 |
| | *Frangula* | 10 | <1 |
| | *Salix* | 9 | <1 |
| | *Evonymus* | 4 | <1 |
| | *Juniperus* | 3 | <1 |
| | *Sorbus* | 2 | <1 |
| | *Hedera helix* | 1 | <1 |
| | Pomoideae | 1 | <1 |
| | *Quercus/Castanea* | 51 | 4 |
| | Undetermined | 44 | 3.5 |
| | | 1275 | 100 % |

Figure 6. Taxa composition of 17 charcoal levels analysed

| | Charcoal pieces analysed | Taxa diversity | Orientation of the slope |
|---|---|---|---|
| Ch 226 0/-8 | 60 | 2 | Southwest |
| Ch 17,2 0/-3 | 50 | 4 | Northwest |
| Ch 17,2 -9/-12 | 50 | 3 | Northwest |
| Ch 36 S1 | 100 | 2 | West northwest |
| Ch 36 S2 | 99 | 3 | West northwest |
| Ch 225 -20/-23 | 47 | 1 | Southwest |
| Ch 226 fond | 50 | 1 | Southwest |
| Ch 116 S1 | 65 | 3 | Southwest |
| Ch 108,2 -5/-9 | 103 | 7 | North northwest |
| Ch 108,1 | 83 | 6 | North northwest |
| Ch 17,2 S2 | 73 | 3 | Northwest |
| Ch 116 S2 | 38 | 4 | Southwest |
| Ch 107,2 0/-3 | 85 | 4 | North northwest |
| Ch 107,1 0/-3 | 115 | 7 | North northwest |
| Ch 107,1 -9/-11 | 93 | 6 | North northwest |
| Ch 45 | 62 | 8 | Northwest |
| 2th war kiln | 101 | 11 | Northwest |

Figure 7. Distribution of the analysed pieces and the taxa diversity of the charcoal levels

(20%), *Fagus sylvatica* (8%), *Castanea sativa* (3%) and *Corylus avellana* (3%). Assemblages get 4 taxa on average with a maximum of eleven. (Figure 7 and 8) Shrubs represent 4.4% of the charcoal pieces analyzed but are found on 50% of assemblages. Even if their representation is low, their presence nevertheless constitutes significant data.

Among these heterogeneous results, it is possible to determinate two types of assemblages, diversified and poorly diversified:

- In the poorly diversified assemblages, deciduous *Quercus* prevails or is in co-dominance with *Fagus sylvatica* or *Carpinus betulus*.
- Among the diversified assemblages, *Carpinus betulus* rules, but is associated with other species like *Fagus sylvatica, Castanea sativa* or *Corylus avellana*.

As shown in Figure 9, despite some differences, there are similarities between the charcoal burning vegetation and the actual forest vegetation, as if there was continuity of the wood landscape. The most diversified samples are near the alluvial plain, while there is a prevalent impoverishment of the assemblages on the south facing slopes. This, lead us to believe that the natural conditions, orientation and the topography in this case, play a major rule in the composition of the charcoal burning vegetation.

## DISCUSSION

### An important activity unknown until today

The great effort of prospecting shows the importance of the charcoal burning activity during modern times on the slope forests of the High Dordogne Valley, now with thousands of known platforms. This inventory permits us to think that the number of charcoal burning places is very high, maybe even to the point of hundreds of thousands of platforms. Also, charcoal burning is an important activity, probably even more so than the staves production. In fact, staves were produced with big oaks of about 60 centimetres in diameter (Lacroix, 2002). But today, there are no oaks of this diameter on the slopes.

With this incredible number and density of charcoal burning places, it is possible to believe that the slope forests have produced much charcoal, a production which was transported to the Bordeaux basin or for other uses, such as the metallurgy in Périgord (an important production area of iron between Bordeaux and the upper Dordogne valley), or even domestic uses (Lamy, 1987). Charcoal burning has been a major forester practice in the basin which may have had an impact on the landscape. Nevertheless, this work demonstrates the existence of this heritage which was unknown until now.

### Practices, charcoal utilizations and the heritage of the landscape

The existence of diversified assemblages and the similarity between the charcoal burning vegetation and

| % | *Quercus* | *Carpinus* | *Fagus* | *Castanea* | *Corylus* | *Tilia/Acer* | *Fraxinus* | *Alnus* | *Betula* | *Populus* | *Salix* | *Juniperus* | *Hedera helix* | *Sorbus* | *Evonymus* | *Frangula* | *Ilex* |
|---|---|---|---|---|---|---|---|---|---|---|---|---|---|---|---|---|---|
| Ch 226 0/-8 | 97 | - | 3 | - | - | - | - | - | - | - | - | - | - | - | - | - | - |
| Ch 17,2 -9/-12 | 96 | - | - | - | - | - | - | - | 2 | - | - | 2 | - | - | - | - | - |
| Ch 17,2 0/-3 | 85 | - | 9 | - | 2 | - | - | - | - | - | - | 4 | - | - | - | - | - |
| Ch 36 S1 | 91 | 9 | - | - | - | - | - | - | - | - | - | - | - | - | - | - | - |
| Ch 36 S2 | 87 | - | 5 | - | - | - | - | - | - | - | - | - | - | - | - | 8 | - |
| Ch 225 -20/-23 | 100 | - | - | - | - | - | - | - | - | - | - | - | - | - | - | - | - |
| Ch 226 fond | 100 | - | - | - | - | - | - | - | - | - | - | - | - | - | - | - | - |
| Ch 116 S1 | 42 | 37 | - | - | - | - | - | - | - | - | - | - | - | - | - | - | 22 |
| Ch 17,2 S2 | 38 | - | 60 | - | - | - | - | - | - | - | - | - | - | - | 2 | - | - |
| Ch 108,2 -5/-9 | 48 | 36 | 12 | 1 | 1 | 1 | - | - | - | - | - | - | 1 | - | - | - | - |
| Ch 108,1 | 1 | 73 | 6 | 16 | 2 | - | - | - | - | - | 1 | - | - | - | - | - | - |
| Ch 116 S2 | 42 | 34 | - | 18 | 5 | - | - | - | - | - | - | - | - | - | - | - | - |
| Ch 107,2 0/-3 | - | 61 | 24 | - | 14 | - | - | - | - | 1 | - | - | - | - | - | - | - |
| Ch 107,1 0/-3 | 10 | 60 | 12 | 11 | - | 3 | - | - | - | - | 1 | - | - | - | - | 2 | - |
| Ch 107,1 -9/-11 | 1 | 49 | 18 | 21 | - | - | - | - | - | - | 9 | - | - | - | 1 | - | - |
| Ch 45 | 38 | 14 | 3 | - | 17 | 10 | 12 | 3 | - | - | - | - | - | 2 | - | - | - |
| 2th war kiln | 32 | 1 | 21 | 1 | 22 | - | 1 | 7 | 1 | - | - | - | - | 1 | 2 | - | 12 |
| *Total* | *49* | *25* | *11* | *4* | *4* | *1* | *1* | *1* | *< 1* | *< 1* | *1* | *< 1* | *< 1* | *< 1* | *< 1* | *< 1* | *2* |

Figure 8. Anthracological assemblages in the Luzège valley

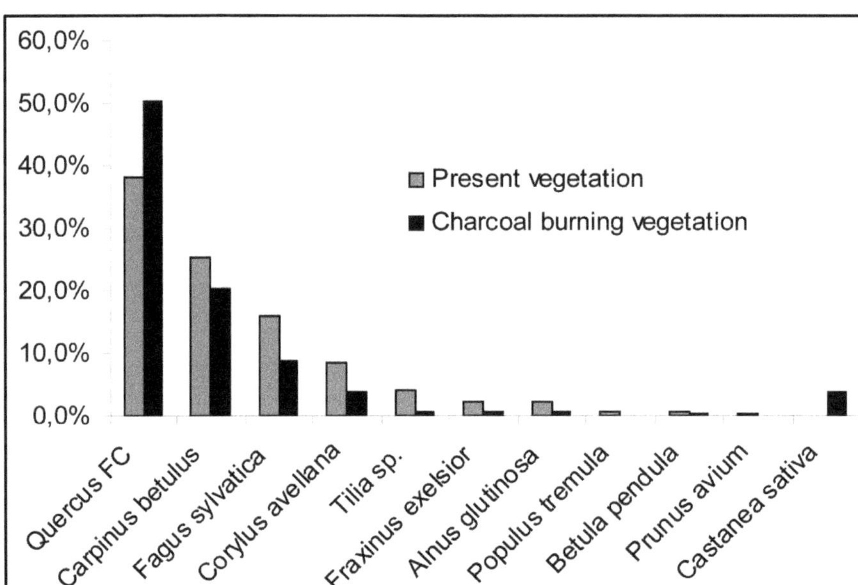


Figure 9. The Luzège valley tree composition, today and in the past

the current forest allow us to consider that the charcoal makers did not choose trees to burn (Ludemann, 2002, 2003) (Figure 8). The frequency of shrubs in the assemblages confirms this hypothesis. Shrubs like *Frangula alnus* or *Evonymus europaeus* produce wood pieces of small diameter which are not adapted for charcoal production. So, the charcoal makers used larger wood pieces around the platform and the smaller diameter woods were employed to build the chimney or to light the kiln. This refers to a forestry management by clear cutting, a well-known practice and reported by ethnological studies (Bouchey, 1970; Musset, 1996; Hanus, 2000). Thus, if clear cutting is a usual practice, the anthracology approach to charcoal burning places represents a good study tool for the historic arboreal composition of the old forests.

**Impoverishment of the vegetation during the charcoal-burning apex**

If the diversified assemblages provide proof of the usage of the vegetation near the charcoal burning places,

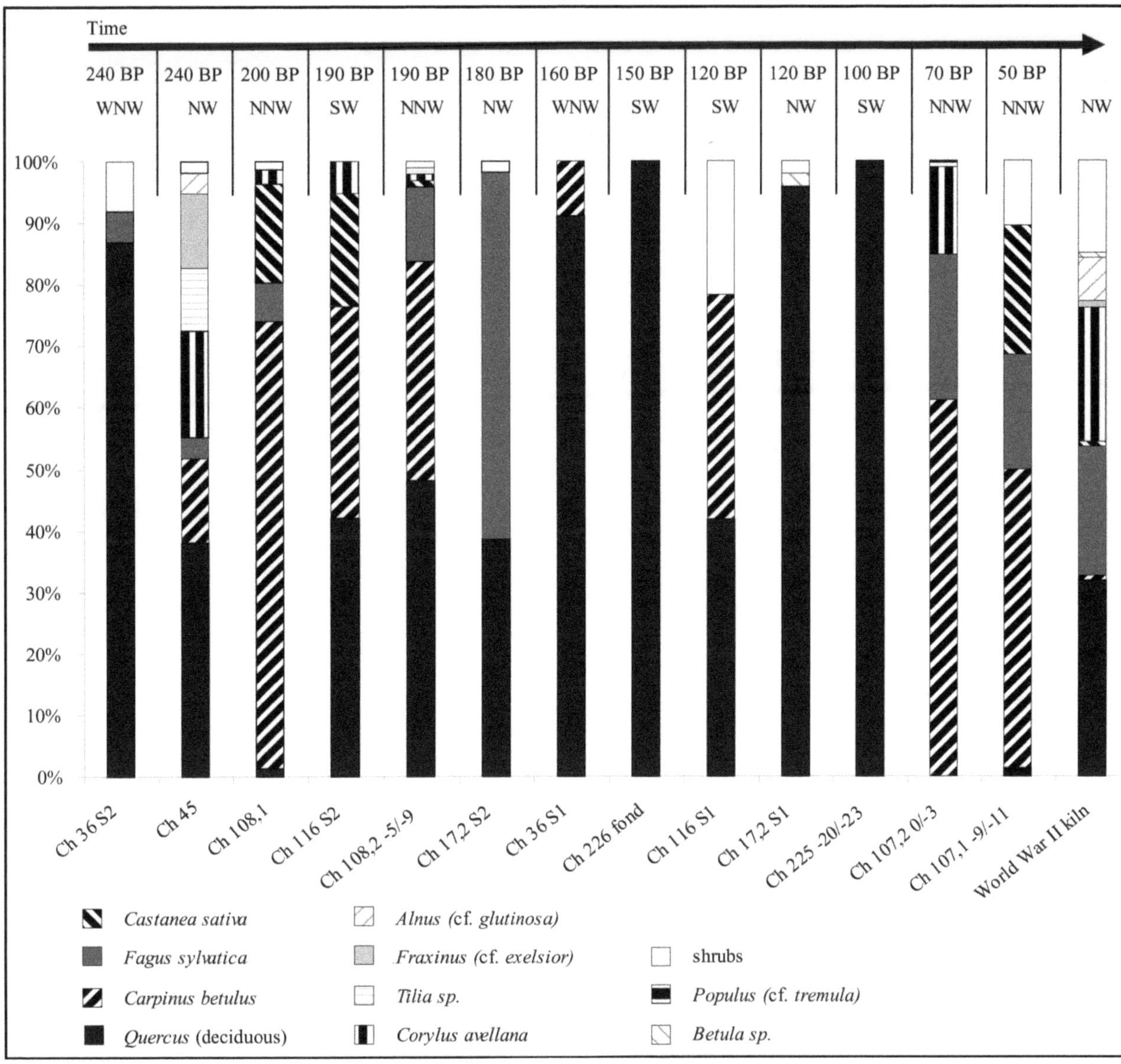


Figure 10. Anthracological chronological assemblages

nevertheless, poorly-diversified taxa assemblages exist. Moreover, it is possible to see in Figure 10 the grouping of the poorly diversified assemblages in the period included between 180 BP and 100 BP.

Despite that, it is necessary to consider radiocarbon dating with caution as it can be imprecise for the last four centuries, the period which corresponds to the probability spikes (the Dordogne river trade apex) (Coculat-Vallières, 1981), from *c.* 1750 AD to *c.* 1900 AD. The grouping of the assemblages dominated by *Quercus*, *Carpinus* and *Fagus*, during this period, allows us to think that the economical and demographical growth of the Bordeaux basin generated an important production of charcoal in these forest slopes (Poussou and Butal, 1980). The field prospecting results show the importance and the extent of the charcoal burning forest in this study area.

To conclude, the diagrams of Figure 9 show a change in the taxa proportions in relation to the taxa ecology. In fact, heliophile taxa, like the oak and the chestnut tree, are more present in the anthracological assemblages than in the actual vegetation inventory; the tendency is the reverse for the sciaphile taxa (Rameau *et al.* 1989). This point suggests a practice favouring the heliophile species at the expense of the sciaphile species, and the platform "Ch 17.2" gives a good example of this phenomenon.

Even if it does not permit us to affirm that, in some cases, there is a selection of woods by the charcoal makers, the synthesis of the different remarks does allow us to conclude that the persistence of clear cutting could induce a selection of the best-adapted species.

**Towards a diversification of the forest**

Records of vegetation taken near platforms reveal that the forest structure originates, in 80 % of the cases, from a coppice management. The average tree age does not surpass 150 years, and at the lower extremity is 50 years. These forest population plantings are the remains of a charcoal-producing forest and suggest that the management has been abandoned during the important rural depopulation of the second half of the 20th century. In the case of an exclusive management of the forest for charcoal burning, the species which do not endure the coppice management have been supplanted: the oak, the hornbeam, and the beech are the species the most adapted to this practice. Figure 8 goes along with this, because it is possible to observe that the anthracological assemblages show larger taxa diversity outside the trade apex of the Dordogne valley. Moreover, sciaphile taxa are more represented in the vegetation of today than in the assemblages. These taxa should have taken advantage of the decrease in forest management to develop themselves, like *Corylus avellana*, *Tilia sp.*, *Populus tremula*, *Prunus avium*, etc.

Thus it is possible to conclude that the enriched taxa of this older charcoal-producing forest disappeared in the 1950s due to the abandonment of the activity during the end of the nineteenth century, competition with fossil fuels, and railway development (Musset, 1996; Loddo & Mucci, 1999).

## CONCLUSION

The aim of the anthracology of the charcoal burning places is to reconstitute the composition and the physiognomy of the older fuel forests. These first results try to show the importance of the influence of charcoal burning on the diversity of the forest when intensively practised on the long term. However, to improve upon this work and these hypotheses, other radiocarbon dating and anthracological analyses must to be executed; it is a work in progress.

Finally, anthracology reaches its limits and does not resolve the socio-economical aspects of charcoal burning. In this circumstance, anthracology does not replace the necessary contribution of historic archives. Locally, the octroi records (archives of the taxed import merchandise) at the municipal archives of the port city down-stream from the upper valley should give information to the quantities of the charcoal shipments coming and going. It is the only information which could confirm that the charcoal production was traded from the Bordeaux basin. This approach would link anthracological results to the socio-economical realities of the concerned period.

### References


BLONDEL, J. 1986. *Biogéographie évolutive*. Paris, Masson.

BOMBAL, E. 1981. *La Haute Dordogne et ses gabariers*. (1903), Treignac, Les monédières.

BONHÔTE, J. 1987. La mémoire des charbonnières. Essai de reconstitution des milieux forestiers dans une vallée marquée par la métallurgie (Aston, Haute-Ariège). Revue forestière française, 40-3, 197-212.

BONHÔTE, J. 1992. La genèse de la mémoire des charbonnières. De la problématique historique à la démarche éco-biogéographique. In J-P. Métailié (ed.), *Protoindustries et histoire des forêts*, 193-206. Toulouse, Les Cahiers de l'Isard.

BONHÔTE, J., DAVASSE, B., DUBOIS, C., IZARD, V., MÉTAILIÉ, J.P. 2002. Charcoal kilns and environmental history in the eastern Pyrenees. A methodological approach. *British Archaeological Reports*, Series 1063, 219-228.

BOUCHEY, E., 1970. *Le charbonnier dans les bois*. Besançon, Folklore comtois.

COCULA-VAILLIÈRES, A-M. 1981. *Un fleuve et des hommes. Les gens de la Dordogne au XVIIIe siècle*. Paris, Tallandier.

CORVOL, A. 1987. *L'Homme aux bois, histoire des relations de l'homme et de la forêt, XVIIe-XXe siècle*. Paris, Fayard.

DAVASSE, B. 1992. Anthracologie et espaces forestiers charbonnés. Quelques exemples dans la moitié orientale des Pyrénées. *Bulletin de la Société Botanique de France* 139, 597-608.

DAVASSE, B. 2000. *Forêts, charbonniers et paysans dans les Pyrénées de l'Est du moyen âge à nos jours. Une approche géographique de l'histoire de l'environnement*. Toulouse, Geode.

DUBOIS, J.-J. 1993. L'approche de la « biogéographie historique »: concepts, méthodes, limites à l'interface de la phytodynamique et de l'histoire forestière, Phytodynamique et biogéographie historique des forêts. *Colloques Phytosociologiques*, 20, 7-13.

DUPIAS, G. 1977. *Carte de la Végétation de la France, Clermont-Ferrand n°52*. Toulouse, Centre National de la Recherche Scientifique.

DUPIAS, G. and LAVERGNE, D. 1968. *Carte de la végétation de la France, Aurillac n°58*. Toulouse, Centre National de la Recherche Scientifique.

EUBA, I. 2008. *Analysis antracologico de estructuras altimontanas en el valle de la Vans-Sierra del Cadi (Alt Urgell) y en el valle del Madriu (Andorra): explotacion de recursos forestales del Neolitico a época moderna*. Unpublished PhD thesis, Universitat Rovira y Virgili de Tarragona.

FABRE, L. 1996. *Le charbonnage historique de la chênaie à "Quercus ilex L." (Languedoc, France): conséquences écologiques*. Unpublished PhD thesis, Université de Montpellier 2.

HANUS, Ph. 2000. *Je suis né charbonnier dans le Vercors. Petite histoire des hommes dans la forêt*. Parc Naturel Régional du Vercors.

IZARD, V. 1992a. La typologie des charbonnières: méthode d'inventaire pour l'étude diachronique du charbonnage. In J-P. Métailié (ed.), *Protoindustries et histoire des forêts*, 223-235. Toulouse, Les Cahiers de l'Isard.

IZARD, V. 1992b. L'art du charbonnier: contributions ethno-botanique et géographico-historique à l'étude des paysages métallurgiques d'après l'anthracol-analyse des charbonnières. *Bulletin de la Société Botanique de France* 139, 587-596.

JACQUIOT, C., TRENARD, Y. and DIROL, D. 1973. *Atlas d'anatomie des bois des angiospermes (essences feuillues) tome 1&2.* Paris, Centre Technique du Bois.

LACROIX, J-P. 2002. L'évolution des besoins et des ressources en chêne de tonnellerie. In A. Corvol (ed), *Forêt et vigne, bois et vin,* 277-294. Paris. L'harmattan.

LAMY, Y. 1987. *Hommes de fer en Périgord au XIXe siècle.* Lyon, La Manufacture.

LODDO, D. and MUCCI, A. 1999. *Il canto della Carbonara: Charbonniers italiens du département du Tarn.* Cordes. Editions CORDAE/La Talvera.

LUDEMANN, T. 2002. Anthracology and forest sites-the contribution of charcoal analysis to our knowledge of natural forest vegetation in south-west Germany. In S. Thiébault (ed.), *Charcoal analysis, methodological approaches, paleaoeological results and wood uses,* 209-217. British Archaeological Reports, 1063. BAR Publishing, Oxford.

LUDEMANN, T. 2003. Large scale reconstruction of ancient forest vegetation by anthracology, a contribution from the Black Forest. *Phytocoenologia* 33, 645-666.

LUDEMANN, T,, MICHIELS, H-G. and NÖLKEN, W. 2004. Saptial patterns of past wood exploitation, natural wood supply and growth conditions: indications of natural tree species distribution by anthracological studies of charcoal-burning remains. *Europeans Journal Forest Research* 123, 283-292.

MÉTAILIÉ, J-P. 1992. Anthracologie, proto-industries et nouvelles perspectives en histoire des forêts. *Bulletin de la Société Botanique de France* 139, 579-586.

MUSSET, D. 1996. *De mémoire de charbonniers.* Mane, Les Alpes de Lumière.

POUSSOU, J-P. and BUTAL, P. 1980. *La fierté d'une capitale provinciale.* In R. Etienne (ed), *Histoire de Bordeaux*, 177-212. Toulouse, Privat.

RAMEAU, J-C., MANSION, D. and DUMÉ, G. 1989. *Flore forestière française, plaines et collines.* Tome 1, Paris, IDF.

SCHWEINGRUBER, F-H. 1990. *Anatomie microscopique du bois.* Birmensdorf, Editions de l'Institut fédéral de recherches sur la forêt, la neige et le paysage.

VERNET, J-L. 2001. *Guide d'identification des charbons de bois préhistoriques et récents. Sud-Ouest de l'Europe: France, Péninsule ibérique et Iles Canaries.* Paris, Centre National de la Recherche.

# LOWER PALAEOLITHIC CHARCOAL FROM IRIKAITZ-GELTOKI SECTOR (BASQUE COUNTRY, SPAIN)

Mónica RUIZ-ALONSO
Grupo de Investigación en Arqueobiología. Instituto de Historia Centro de Ciencias Humanas y Sociales (CCHS). Consejo Superior de Investigaciones Científicas (CSIC). c/ Albasanz, 26-28 28037 Madrid, Spain
monica.ruiz@cchs.csic.es

Lydia ZAPATA and Alvaro ARRIZABALAGA
Dept. Geografía, Prehistoria y Arqueología (University of the Basque Country UPV/EHU). Tomás y Valiente s/n. 01006 Vitoria-Gasteiz, Spain

**Abstract**: *We present the results of the charcoal analysis of the site of Irikaitz (Zestoa, Basque Country, Spain). This is an open air site with an archaeological sequence which includes Lower Palaeolithic, Upper Palaeolithic and post-Palaeolithic, the only site so far with Lower Palaeolithic occupations in primary position in the Basque region. Archaeobotanical studies of these early chronologies in the Iberian Peninsula are particularly scarce. Irikaitz adopted from the beginning of the research project a systematic sampling strategy and recovery methodology through flotation in order to retrieve all types of plant macro-remains. Some finds have also been hand-picked in situ. Macro-remains recovered at the site present a very good preservation. 2878 charcoal fragments have been analysed. The most important taxon in all samples is* Quercus *subgenus* Quercus *(deciduous oaks). Some species possibly linked to wet soils along streams are also very important, such as:* Corylus avellana *(hazel),* Salix *(willow) and* Fraxinus excelsior *(ash). Taxa linked to shrub and open communities have also been identified* (Ericaceae, Leguminosae). *The general results suggest species living under mild climatic conditions which might correspond to the Eemian or the Holsteinian.*

**Key words**: *charcoal analysis, carpology, macroremains, Lower Palaeolithic, Eemian, Holsteinian*

## GENERAL GEOGRAPHICAL FRAMEWORK

The site of Irikaitz, according to the classification of Rivas-Martínez (1987) is located in the Euro-Siberian Region, Atlantic Superprovince, Cántabro-Euskaldun Subprovince, Cantabro-Euskaldun Sector, Euskaldun Subsector. It belongs to the region of the Atlantic valleys, close to the coast of the Bay of Biscay. Much of the surface of the land area consists of Mesozoic materials (Lower Cretaceous). Despite the prevalence of different types of basic rocks (limestone and marl), soils are widely markedly acidic, and devoid of carbonates and active lime. The high rainfall of the whole area is the main factor of washing and acidification of soils. The present climate is very humid and warm, summer drought hardly exists. The relief is very pronounced, with a continuous succession of hills and mountains crossed by narrow valleys. Given the hilly topography and high rainfall, the entire region is crisscrossed by a variegated network of streams and small rivers. The main channels of the hydrographic network are short course rivers, with a general direction south-north (Carreras, 1980).

The site is located in northern Iberia in the province of Guipuzcoa, in the municipality of Zestoa (Figure 1). This area has a relief with modest hills and short valleys (Carreras, 1980). The current vegetation of the site consists of grassland and crops, in an environment of Cantabrian alder grove, cultivated pine and very degraded oak forest and mixed acidophilus Atlantic forests. The present potential vegetation of the area is the acidophilic oak and *Quercus robur*-Atlantic mixed forest with alder and some patches of Cantabrian evergreen oak (Aseginolaza *et al.* 1989, 1992).

## THE ARCHAEOLOGICAL SITE OF IRIKAITZ

The archaeological excavations began at Irikaitz in 1998, directed by Á. Arrizabalaga and M.J. Iriarte, in the context of the documentation and rescue of a deposit that was in danger, in view of municipal plans to enlarge the local football ground. Irikaitz is located within an area of great wealth of archaeological heritage, with significant sites nearby, such as Ekain (Altuna and Merino, dirs, 1984) or Amalda (Altuna, *et al.*, dirs, 1990). Right from the start of the first season, we were able to identify the existence of a large area with an open-air Palaeolithic deposit. Until 2002, a surface area of 48 m$^2$ was opened in the Geltoki sector, very close to the old spa resort of Zestoa, over the Urola railway line. Sedimentological studies and all analyses which can provide information on the formation of the site are extremely important to help us interpret the meaning of the plant macroremains that we have identified. From the standpoint of the sediment we have differentiated so far two sedimentary cycles in Irikaitz: a top sedimentary cycle and a bottom sedimentary cycle, made of large chunks of clay. The lower cycle is at least three metres deep and rests on the bedrock surface. It presents circular or elliptical depressions with radiuses of several meters which are over half a meter deep. In the described stratigraphy it comprises Level G. VI. The upper cycle sits on the former and clogs its depressions (Level G. V) and shows

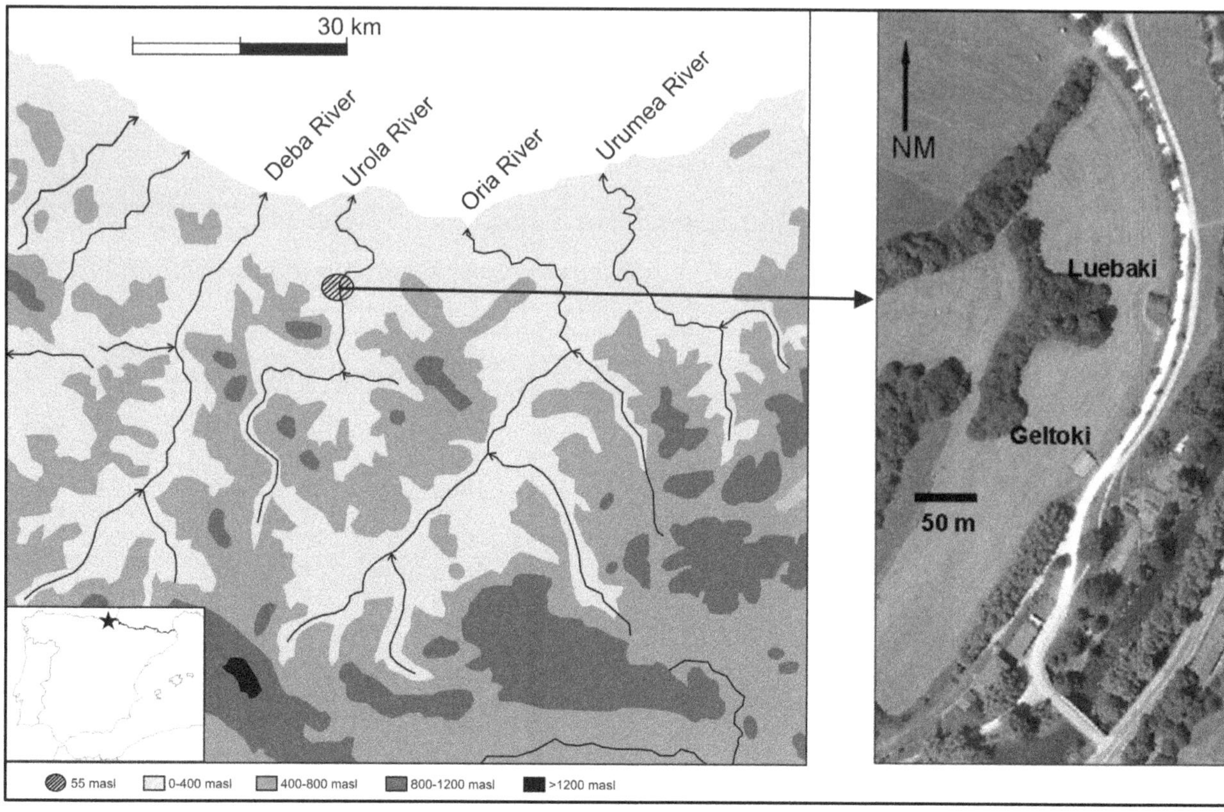


Figure 1. Location of Irikaitz

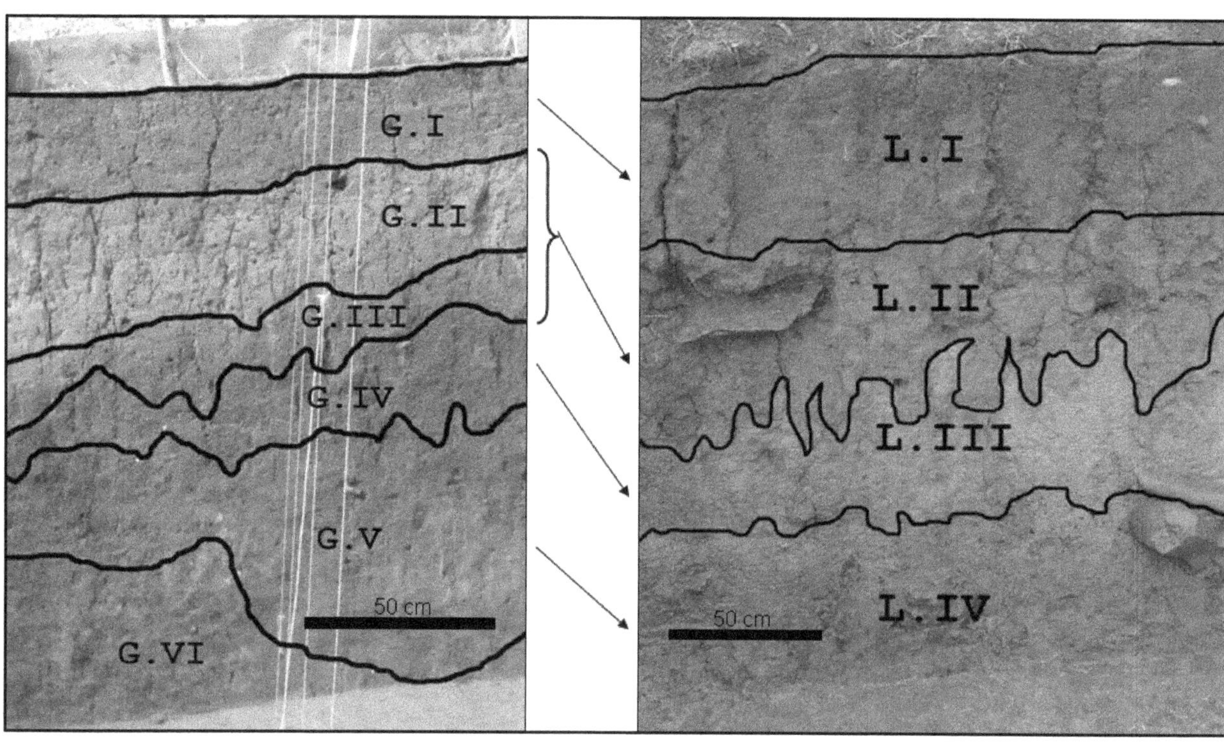


Figure 2. Stratigraphic sequence of Irikaitz in both sectors, Geltoki (G, left) and Luebaki (L, right) (Arrizabalaga and Iriarte 2011)

a stratigraphy consistent with the slope of the current topography of the site. Level G. I includes mixed materials due to modern agricultural work. Levels G. II and G. III show signs of lithics percolation from later Gravettian camp sites even if most of the sedimentary material corresponds to the Lower Palaeolithic. Both, G. IV and G. V levels appear intact on the basis of data provided by lithic assemblages and pollen analysis (Arrizabalaga and Iriarte 2002, 2003, 2005a, 2005b, 2011; Arrizabalaga *et al.* 2003) (Figure 2).

The work in this sector was provisionally concluded in 2003, when attention focused on the excavation of the Luebaki sector, which began in 2002 and has continued to the present. This large site (archaeological material has been collected over an area of some eight hectares) is exceptional because of its primary character and the presence of techno-complexes attributable to the Lower Palaeolithic. Several reports have been published on the study of the stratigraphy, archaeobotany and lithic techno-complexes found in the Geltoki sector (Arrizabalaga and Iriarte 2002, 2003, 2005a, 2005b, 2011; Arrizabalaga *et al.* 2003). The sedimentary characterization of the stratigraphic sequences of Geltoki and Luebaki are very similar (GI = LI, G. G. III II = L II, G. IV = L. III; GV = L. IV, G. VI = LV).

## MATERIALS AND METHODS

### Method of preservation of botanical macro remains

All plant macroremains identified so far in Irikaitz have been preserved by charring. The possibility of having chemicals carbonization was considered because in the remains of charcoal some fragments of lignite appeared. Therefore, in 2003 a DTA analysis (Differential Thermal Analysis) was performed in the laboratory of Inorganic Chemistry at the UPV/EHU by T. Rojo. It confirmed a carbonization by combustion with a peak temperature around 200° C (Haines, 2002).

### Sampling strategy and recovery techniques

The analysis of plant macroremains was included in the initial design of the excavation where interdisciplinary analyses of pollen, sediments and other environmental materials were proposed. Sampling has been carried out during the excavation and all stratigraphic units –as defined by the archaeologists- have been sampled following different techniques.

The five levels that have been described above where sampled: 1) Surface level (Sup), 2) Level G. II, 3) Level G. III, 4) Level G. IV and 5) Level G. V. However, the superficial lever will not be taken into account for the purpose of this work since it is disturbed due to agricultural activity. It also included very few charcoal remains Levels G. II and G. III are assigned to a moment not far from the Lower Paleolithic, although the presence of both, Upper Paleolithic lithics and Chalcolithic materials, is detected. Specifically, 60% of the archaeological material collected from Level G. II corresponds to the Lower Paleolithic while at Level G. III this percentage would reach 90%. Level G. IV and Level G. V contain materials assignes to the Lower Palaeolithic with no signs of disturbance. Level G. VI which for the moment is considered sterile and has not been sampled.

We have tried to carry out a sampling strategy that best fits our purpose and the characteristics of the site. First, the remains of charcoal in the deposit of Irikaitz were recovered by systematic hand picking with a three-dimensional record that tracks the fragments at least within each grid. This can thus provide data on the spatial distribution or dispersion of the charcoals. To this manual collection we added dry sieving with a 2 mm mesh where the charcoals were separated by hand from the sediment. This became difficult since charcoal is wrapped in clay and can be fragmented so afterwards it has been carefully cleaned in the laboratory. This is also the reason why the fragments of charred wood have not been weighted, because the amount of adhering sediment is great and masks the genuine weight of the fragment. Given the large volume of sediment excavated, only a specific area of the site has been sampled by flotation. We have chosen a column in Table F9 sector 9 where the palynological samples were taken. Thus, in future studies, data from both disciplines will be correlated.

## RESULTS

Plant macroremains other than charcoal are extremely scarce in Irikaitz. However, it is interesting to point out that two charred fragments of *Corylus avellana* pericarp have been found in Levels G. II and G. III. This paper will focus on the analysis of the wood charcoal. We examined a total of 2960 fragments > 2 mm from three excavated areas independently but the quantification of data has been carried out jointly as they relate to the same unit (Geltoki Sector).

Wood identified in Irikaitz belongs to a minimum of 18 taxa (Figures 3 and 4): *Acer* tp. *campestre*, *Alnus* sp., *Betula* sp., *Carpinus* sp., *Cornus* sp., *Corylus avellana*, *Erica* sp., *Fagus sylvatica*, *Frangula alnus*, *Fraxinus* sp., Leguminosa, *Populus* sp., *Quercus* subg. *Quercus*, *Quercus*/*Castanea*, Rosacea (Pomoidea, *Prunus* sp.), *Salix* sp. and *Ulmus* sp.

The presence of *Quercus* subg. *Quercus* (deciduous and marcescent oak) (Figure 5) stands out exceeding 67% in the three older levels. *Alnus* is another important component of the samples. In Level G. V it is 4% of the wood identified. It increases in Levels G. IV and G. III with percentages close to 13% and 21% in Level G. II. Next in importance is *Corylus avellana* (6-9% in all levels from G. V to G. II) and *Fagus sylvatica* (3-14%) (Figure 6).

Within the charcoals analysed, 60 fragments have not been identifiable due to distorted anatomies (knots) or because cells have crystallized or vitrified not allowing to recognize anatomical diagnostic patterns.

Nineteen fragments have been defined only as angiosperms due to the difficulty observing their anatomical features. Besides, three fragments from Level G. II have been identified as immature wood of possibly *Quercus*. They are small branches where only the basic anatomical characters of *Quercus* wood could be observed.

| | SUP | G. II | G. III | G. IV | G. V | Total |
|---|---|---|---|---|---|---|
| *Acer* tp *campestre* | | | | | 2 (2,1%) | **2** |
| *Acer* sp. | | 8 (1,1%) | | 1 (0,1%) | | **9** |
| *Alnus* sp. | 1 (14,3%) | 145 (20,9%) | 160 (12,7%) | 106 (12,9%) | 4 (4,2%) | **416** |
| *Betula* sp. | | 2 (0,3%) | 2 (0,1%) | 5 (0,6%) | | **9** |
| *Carpinus* sp. | | 3(0,4%) | | 1 (0,1%) | | **4** |
| *Cornus* sp. | | | | | 4 (4,2%) | **4** |
| *Corylus avellana* | | 41 (5,9%) | 101 (8%) | 74 (9%) | 7 (7,4%) | **223** |
| *Erica* sp. | 3 (42,8%) | 11 (1,6%) | 3 (0,2%) | 2 (0,2%) | | **19** |
| *Fagus sylvatica* | 3 (42,8%) | 95 (13,7%) | 81 (6,4%) | 32 (3,9%) | 3 (3,1%) | **214** |
| *Frangula alnus* | | 6 (0,9%) | 3 (0,2%) | | | **9** |
| *Fraxinus* sp | | 8 (1,1%) | 21 (1,7%) | 21 (2,5%) | | **50** |
| Leguminosa | | 4 (0,6%) | 3 (0,2%) | 1 (0,1%) | 1 (1%) | **9** |
| Pomoidea | | 5 (0,7%) | 13 (1%) | 9 (1,1%) | 3 (3,1%) | **30** |
| *Populus* sp. | | 5 (0,7%) | 9 (0,7%) | | | **14** |
| *Prunus* sp. | | 6 (0,9%) | 2 (0,1%) | | | **8** |
| *Quercus* subg. *Quercus* | | 343 (49,4%) | 854 (67,8%) | 560 (68,2%) | 69 (72,6) | **1826** |
| *Quercus/Castanea* | | 1 (0,1) | 2 (0,1%) | 3 (0,4%) | | **6** |
| *Salix* sp. | | 12 (1,7%) | 6 (0,5%) | 4 (0,5%) | 2 (2,1%) | **24** |
| *Ulmus* sp. | | | | 2 (0,2%) | | **2** |
| **Total** | **7** | **695** | **1260** | **821** | **95** | **2878** |
| Indeterminated immature wood (*Quercus*?) | | 3 | | | | **3** |
| Indet. Angiosperm | | 4 | 4 | 10 | 1 | **19** |
| Not identifiable | | 12 | 27 | 20 | 1 | **60** |
| *Corylus* pericarp fragment | | | 1 | 1 | | **2** |

Figure 3. Absolute and percentage data. Geltoki Sector (n = 2878)

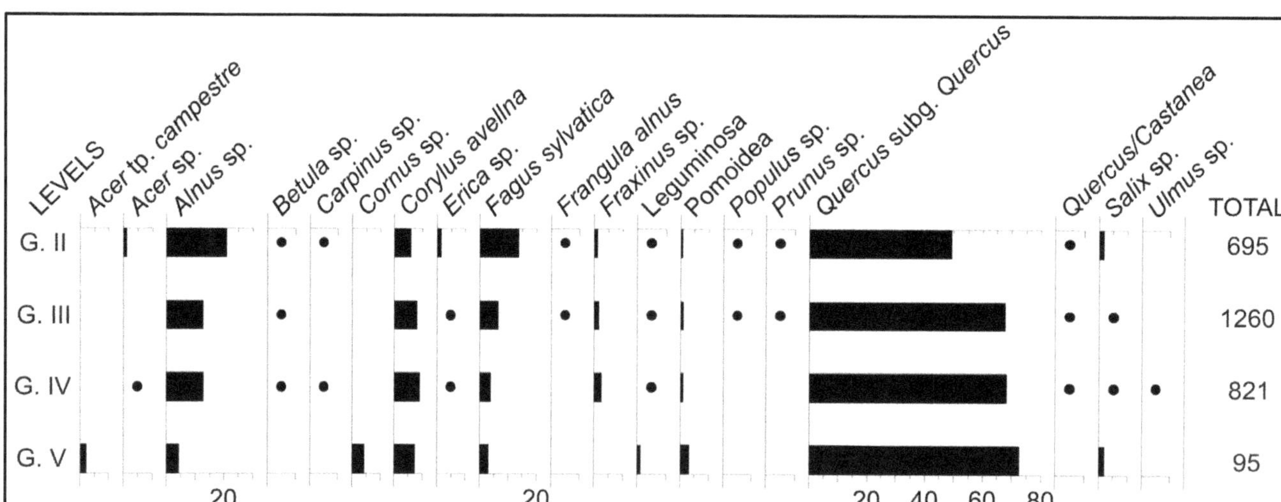


Figure 4. Charcoal histogram. The percentages less than 1% are represented by a symbol (point)

## DISCUSSION

A total of 2960 fragments of charcoal have been analyzed. We think that for Levels G. II, G. III and G. IV the identified number is sufficient to draw statistically valid conclusions. A different case is the surface level and Level G. V where the number of fragments is lower than what is routinely examined in archaeological contexts.

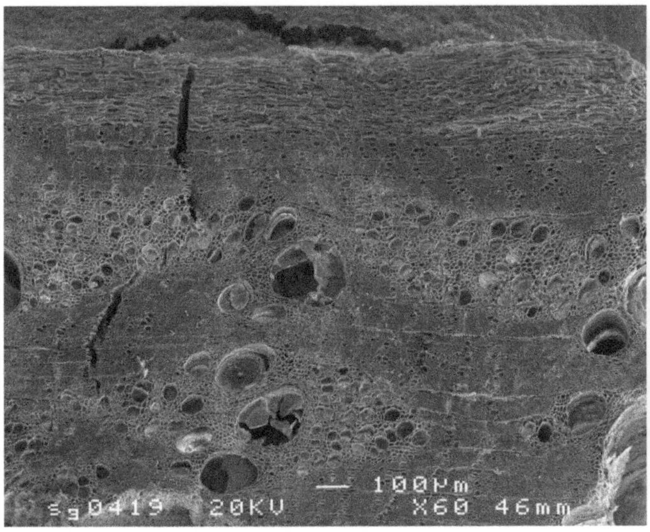


Figure 5. *Quercus* subg. *Quercus* (deciduous and marcescent oak) transversal section. Ring-porous. Rays uni- and multiseriate

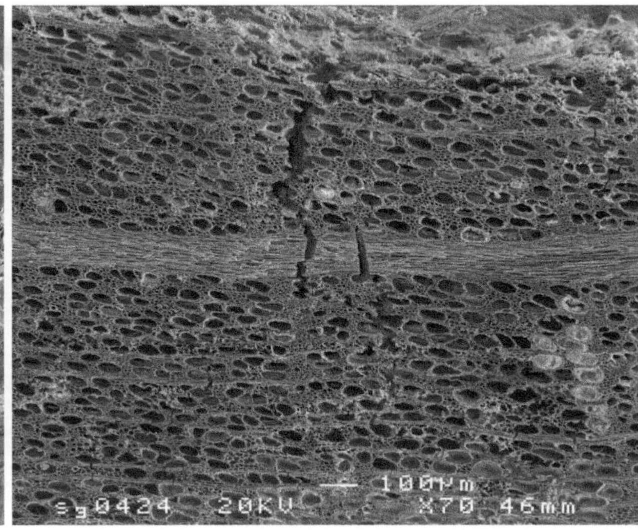


Figure 6. *Fagus sylvatica* (ash) transversal section. Diffuse-porous to semi-ring-porous. Uniseriate to multiseriate rays

**Chronology of the deposit**

The first problem of interpretation we face, since Levels G. II and G. III include materials from different chronologies, is whether charcoal located at these levels has been introduced from other contexts. We have different possibilities since wood charcoal as other debris such as small flints or ceramic fragments, may be affected by vertical movements within the sediment due to their small size: a) that all charcoal is percolated, b) that only a part of the material has moved or c) that all wood fragments are in their original position and belong to a single chronology per context. We find it difficult to discriminate which of the three possibilities is correct, so we have to be cautious about the conclusions that these materials can provide. However, the overall composition of Levels G. III and G. II is not very different from what we appreciate in Levels G. IV and G. V which have no symptoms of percolation.

A first approximation to the chronology of the site seems to lead to an interglacial time that could be 260-240 Kyr for the Holstenian or 150-90 thousand for the Eemian. The hypothesis that the chronology of the site relates to an interglacial period is supported by these elements:

1. The analysis of the recovered industry is consistent with a chronology of the Eemian or Holsteinian interglacials (Arrizabalaga *et al.* 2003).

2. Paleopalynological data (Arrizabalaga *et al.* 2003) indicate that for Level G. V a diversity of trees appears (*Carpinus*, *Quercus*, *Frangula*) with typical taxa of the riparian vegetation, influenced by the proximity of the river Urola (*Alnus* and *Corylus*). Among the herbaceous species *Ericaceae* and *Poaceae* are included and other riparian species (*Ranunculaceae* and *Cyperaceae*) are also present. There is also an important presence of spores. It can be suggested that Level G. V (Lower Palaeolithic) had favourable climatic conditions associated with an interglacial period.

3. The data obtained in our study, as discussed below, indicate that all levels analyzed may belong to a warm moment.

In conclusion, according to the available palaeoenvironmental data (Arrizabalaga *et al.* 2003) and artefactual evidence (Arrizabalaga and Iriarte 2002, 2003, 2005a, 2005b, 2011; Arrizabalaga *et al.* 2003), the chronology of the site seems to lead to an interglacial period. The presence of *Carpinus* in the charcoal samples and in the pollen record, a taxon that during the Holocene has not been present in the inner valleys of the Cantabrian region, rules out the possibility of the material to be Holocene. There are sparse references about the presence of *Carpinus* in the Western Pyrenees during the Holocene, but these are punctual (Sánchez-Goñi 1996b, Aizpuru and Catalán 1983). *Carpinus* geographical distribution was more widespread in the Iberian Peninsula during the interglacial periods we have mentioned (Arrizabalaga *et al.* 2003; Turner, 2000; Sánchez-Goñi *et al.* 1999; Peñalba, 1992)

The Eemian (OIS 5e) was the penultimate warm or interglacial period during the Quaternary, similar to the current Holocene (Foucault and Raoult, 1985). It is recognized in the sequences of vegetation as a period of tree dominance (Shackleton *et al.* 2003). The second chronological scenario for Irikaitz is the Holstenian interglacial whose chronology is not well established. An important limitation is the scarcity of paleobotanical data for this period. Climate warming occurred along with a high moisture content which allowed for expansion in Europe of large forests and the development of more thermophilic taxa such as spruce, pine, oak, etc. (Muñoz *et al.* 1999).

**Origin of charred wood**

An issue of great importance when interpreting the results of our analysis is to understand how the arrival of the wood charcoal to the site took place. Although in some places it appears to be more abundant, charcoal recovered in Irikaitz is scattered. We see at least three different possibilities to explain the origin of the wood: a) that charcoal has arrived to the sediment naturally, b) that it has been used by humans as fuel, and c) that it is the result of both. According to the information available today, in Irikaitz we have considered a set of criteria for discussing the different possibilities to explain the origin of charcoal.

There is no evidence of wood charcoal being concentrated in hearths or combustion structures, which would document a clear anthropic origin of at least part of the sample. The existence of fires in the past has been approached from different types of studies as pedoanthracological (Berger and Thiébault, 2002; Delhon, 2002; Thinon, 1992) or charcoal microscopic studies in palynological samples (Edwards, 1998; Moore, 1996; Sánchez Goñi, 1996a) although the key and the most difficult question to answer in all cases is whether they are natural, climate-induced, or anthropogenic fires. In the scientific literature there are some, though limited, references to the presence of charcoal in the Eemian paleosols (Rainho Teixeira *et al.* 2002; Van Vliet-Lanoe, 1976). According to Van Vliet (1976), some remnants of charcoal are found in the last interglacial paleosols. They note that in Belgium, northern France and Poland these charcoals are produced after a damaging type pedogenesis, a slight podzolization and the simultaneous action of ice and thawing. According to palynological and paleobotanical analysis, these events are associated with boreal forest vegetation dominated by *Pinus*. Several authors (Talon *et al.* 2005; Peterken, 1996; Rackham, 1980) have indicated that a deciduous forest is hardly affected by fire. Hardwood forests occupy mesic and wet habitats. Their leaves are high in water and the wood contains no resins. Some of these forests when grown in a dry and infertile soil burn. Mostly the shrubs will burn, but these fires rarely reach the treetops. By contrast, the coniferous forests do burn more easily (Peterken, 1996). In Irikaitz available archaeobotanical evidence suggests that the tree species present in the environment are hardwood. These are generally unaffected by the fire although we cannot completely rule out the existence of natural fires (Moore, 1996). Natural fires start when lightning lights up a tree or wood and soil debris. Most often occur in dry summers. Usually a limited area burns and then they go extinct. Charcoals might be the result of natural fires originated in situ in the immediate vicinity of Irikaitz but they could also come from relatively remote areas. Several weathering agents as water and air currents could have been responsible for spreading the charcoals (Thinon 1992) Samples of charcoal flotation in Irikaitz include small fragments, from c. 250 μ, but there are also abundant fragments > 4 mm, perhaps less susceptible to be transported by atmospheric agents such as winds.

Charcoal appears in Irikaitz along the whole sequence continuously, linked to the occurrence of archaeological material. This is an important argument to support the hypothesis of anthropogenic origin of the remains as archaeological materials are in primary position, i.e. the materials have not undergone postdepositional removal. If we accept this idea it is not unreasonable to think that charcoal has not undergone significant post-depositional movement.

Another argument is that our data do not show a clear selection of one or more fuels. Anthracological samples from Irikaitz offer a high degree of homogeneity both in terms of presence-absence of taxa as well as in trends in percentages. A wide variety of woody taxa is documented, similar to pollen. If we met a narrower spectrum, which could be best supported by comparison with–palynological data, one might think that there had been a selection of certain fuels. A further problem is that a selection of certain woods might also derive from strictly natural causes because natural fires may favour thick bark species such as oak. Moreover, the issue of human selection of wood is linked in this case to particularly poorly known issues such as cognitive and behavioural patterns of human species other than *Homo sapiens*. In the absence of specific studies, we cannot rule out that these groups engage in the opportunistic use of wood resources without any preference or any particular fuel management.

In conclusion, according to the information we have and the diversity of ideas that support the different possibilities on the origin of the charred wood at Irikaitz, we believe that our position should be based on caution and that no hypothesis can be completely ruled out until more data is available for the site.

## CONCLUSION

Despite the many different archaeological projects being developed in the Iberian Peninsula, and the efforts made by specialists, there are still many gaps related to the sampling of plant macroremains. In the case of Irikaitz this is also linked to problems typical of Lower Paleolithic deposits (poor preservation, difficulties for dating…). To date, anthracological studies for early chronologies are still scarce, so the study in Irikaitz is pioneer in this regard. The excavation is ongoing and we hope it will provide new information and throw some light upon some of the problems we have opened with this work.

Plant macroremains from Irikaitz have been preserved by charring. Wood charcoal does not come from hearths or specific structures but is scattered in the sediment. With the exception of two fragments of *Corylus avellana* pericarp from Levels G. IV and G. III, plant macroremains from Irikaitz consist of wood charcoal. The following taxa have been identified: *Acer tp. campestre*, *Acer* sp., *Alnus* sp., *Betula* sp., *Carpinus sp.*, *Cornus sp.*, *Corylus avellana*, *Erica* sp., *Fagus sylvatica*,

*Frangula alnus*, *Fraxinus* sp. *Leguminosae*, *Populus sp.*, *Quercus* subg. *Quercus*, *Quercus / Castanea*, *Rosaceae* (*Pomoideae*, *Prunus* sp.), *Salix* sp. and *Ulmus* sp. The diversity of taxa increases in the most recent samples. *Quercus* subgenus *Quercus* and *Corylus avellana* decrease their percentage progressively in the more modern levels while *Fagus sylvatica* and *Alnus* sp. increase. *Quercus* is clearly the most represented genus in the site.

With all the caution that this type of study requires, we can say that the overall structure of the vegetation that our analysis reflects is mostly composed of species inhabiting in temperate and wet environments. The highest percentages in all samples correspond to *Quercus* subgenus *Quercus*. Riparian species (*Populus*, *Alnus*, *Salix*, *Fraxinus* ...) are also an important component of the vegetation represented. Also shrub-species such as ericaceous and leguminous plants have been identified.

A first working hypothesis has been that the vegetation represented in the study shows an interglacial stage. Our data are so far consistent with the working hypothesis for the site that frames it in a Holsteinian or Eemian interglacial. This point will hopefully be confirmed in future contributions by bringing together different disciplines, such as dating by different techniques, the study of lithic assemblages, palynology, etc.

Due to the lack of hearths or structures in the site, a major problem in Irikaitz, because it links with the interpretation that we give to the samples, is the origin of wood, whether it is anthropic or not. This is an issue that remains open, especially because of the chronology in which we move for this deposit (Eemian or Holsteinian).

## Acknowledgments

The work of A. Arrizabalaga and L. Zapata is part of: 1) Research Group at the UPV / EHU IT-288-07 funded by the Basque Government, 2) Project HAR2008-03976/HIST the Ministry of Education and Science: *El medio como catalizador de los comportamientos humanos durante el Pleistoceno Superior y Holoceno en el Pirineo Occidental y proximidades: su articulación en unidades territoriales*.

## References

ALTUNA, J., BALDEÓN, A. and MARIEZKURRENA, K. (eds) 1990. *La Cueva de Amalda (Zestoa, País Vasco)*. Ocupaciones paleolíticas y postpaleolíticas, Eusko Ikaskuntza, San Sebastián.

ALTUNA, J. and MERINO, J.M. (eds) 1984. *El yacimiento prehistórico de la cueva de Ekain (Deba, Guipúzcoa), Colección Beca José Miguel de Barandiarán*, 1, 351 p. Eusko Ikaskuntza, San Sebastián.

ARRIZABALAGA, A. and IRIARTE, M.J. 2002. El yacimiento arqueológico de Irikaitz (Zestoa, Gipuzkoa). Aportación al conocimiento del Paleolítico antiguo en Euskal Herria In *XV Congreso de Estudios Vascos: Ciencia y Cultura vasca, y redes telemáticas*. Eusko Ikaskuntza. 2002, 115-122. San Sebastián.

ARRIZABALAGA, A. and IRIARTE, M.J. 2003. El yacimiento arqueológico de Irikaitz (Zestoa, País Vasco). In *Descripción del depósito y caracterización industrial de su nivel IV*. XI Reunión Nacional de Cuaternario, 205-210. Oviedo.

ARRIZABALAGA, A. and IRIARTE, M.J. 2005a. El Paleolítico antiguo en el medio pirenaico occidental, in Santonja, Pérez-González and Machado (eds.) *Geoarqueología y Patrimonio en la Península Ibérica y el entorno mediterráneo*, 219-234. Soria.

ARRIZABALAGA, A. and IRIARTE, M.J. 2005b. Irikaitz (Zestoa, País Vasco). Estudio preliminar del Paleolítico inferior en el sondeo "Geltoki", in Bicho, N. (ed.) *O Paleolítico*, 267-275, Universidade do Algarve, Faro.

ARRIZABALAGA, A. and IRIARTE, M.J. 2011. Lower and Upper Palaeolithic settlements in Irikaitz (Zestoa, Basque Country, Spain). Deconstruction of a Pleistocene archaeological site in the Eastern Cantabrian range. *Revista Cuaternario y Geomorfología* 25 (1-2), 105-119.

ARRIZABALAGA, A, IRIARTE, M.J. and RUIZ ALONSO, M. 2003. El yacimiento arqueológico de Irikaitz (Zestoa, País Vasco). Contextualización arqueobotánica y en el marco del Paleolítico Inferior vasco. In *XI reunión nacional de cuaternario*, 211-218. Oviedo, Asturias.

ASEGINOLAZA, C., GÓMEZ, D., LIZAUR, X., MONTSERRAT, G., MORANTE, G., SALAVERRIA, M.R. and URIBE-ECHEVARRIA, P.M. 1989. *Vegetación de la Comunidad Autónoma del País Vasco*. Vitoria-Gasteiz. Gobierno Vasco.

ASEGINOLAZA, C., GÓMEZ, D., LIZAUR, X., MONTSERRAT, G., MORANTE, G., SALAVERRIA, M.R. and URIBE-ECHEVARRIA, P.M. 1992. *Mapa de vegetación de la Comunidad Autónoma del País Vasco. Hoja-63-IV Azkoitia*. Madrid. Departamento de Urbanismo, Vivienda y Medio Ambiente, Viceconsejería de Medio Ambiente.

BERGER, J-F and THIÉBAULT, S. 2002. The study and significance of charcoal as an indicatot of ancient fires: an application to the middle Rhone valley (France) In S. Thiébault (ed.) *Charcoal Analysis. Methodological Approaches, Palaeoecological Results and Wood Uses. Proceedings of the Second International Meeting of Anthracology (Paris, 2000). British Archaeological Reports. International Series* 1063, 25-41. BAR Publishing, Oxford.

CARRERAS CANDI, F. 1980. Geografía general del País Vasco. Volumen III y IV, Gipuzkoa. Ed *La gran enciclopedia vasca*. Bilbao.

DELHON, C. 2002. Phytolith and pedoanthracological analysis of a Holocene pedosedimentary sequence from the middle Rhone valley (France) In S. Thiébault (ed.) *Charcoal Analysis. Methodological Approaches, Palaeoecological Results and Wood Uses. Proceedings of the Second International Meeting of Anthracology (Paris, 2000). British Archaeological Reports. International Series* 1063, 169-178. BAR Publishing, Oxford.

EDWARDS, K.J. 1998. Detection of human impact on the natural environment: palynological views In J. Bailey (ed.) *Science in Archaelogy. An agenda for the future*, 69-88. English Heritage, London.

FOUCAULT, A. and RAOULT, J-F. 1985. *Diccionario de Geología*. Masson. Barcelona.

HAINES, P.J. 2002 *Principles of thermal analysis and calorimetry*. The Royal Society of Chemistry, Cambridge.

MOORE, J. 1996. Damp Squib. How to fire a major deciduous forest in an inclement climate. In T. Pollard y A. Morrison (eds.) *The early prehistory of Scotland*, 62-73. Edinburgh University Press, Edimburg.

MUÑOZ AMILIBIA, A.M., CABRERA VALDÉS, V., FERNÁNDEZ VEGA, A., RIPOLL LÓPEZ, S., HERNANDO GRANDE, A., MENÉNDEZ FERNÁNDEZ, M. and RIPOLL PERELLÓ, E. 1999. *Prehistoria I*. Universidad Nacional de Educación a Distancia. Madrid.

PEÑALBA, M.C. 1992 Biogeografía holocena de las principales especies forestales del norte de la Península Ibérica Eusko Ikaskuntza. Cuadernos de Sección. Historia 20, 391-409. Donosita.

PETERKEN, G.F. 1996. *Natural Woodland. Ecology and Conservation in Northern Temperate Regions*. Cambridge University Press, Cambridge.

RACKHAM, O. 1980. *Ancient Woodland*. Edward Arnold, London.

RAINHO TEIXEIRA, S., DIXON, J.B., WHITE, G.N. and NEWSOM, L.A. 2002. Charcoal in solis: A preliminary view. *Soil Science Society of America. Soil Mineralogy with Environmental Applications, Soil Science Society of America. Book Series*, 7. Madison, Wisconsin.

RIVAS-MARTINEZ, S. 1987. *Memoria del Mapa de series de vegetación de España*. Ministerio de agricultura, pesca y alimentación. Madrid.

SANCHEZ GOÑI, M.F. 1996a. Image analysis of charcoal particles: application to environmental studies. *Microscopy and Analysis*. Julio 1996, 7-9.

SÁNCHEZ GÓNI, M.F. 1996b. Vegetation and sea level changes during the Holocene in the estuary of the Bidasoa (Southern part of the bay of Biscay) Quaternaire, 7, (4), 207-219.

SHACKLETON, N.J., SÁNCHEZ-GOÑI, M.F., PAILLER, D. and LANCELOT, Y. 2003. Marine Isotope Substage 5e and the Eemian Interglacial. In *Global and Planetary Change* 36, 151-155.

SÁNCHEZ- GOÑI, M.F. EYNAUD, F., TURON, J.L., SHACKLETON, N.J. 1999. High resolution palynological record off the Iberian margin: direct land–sea correlation for the Last Interglacial complex. *Earth and Planetary Science Letters* 171, 123-137.

TALON, B., SERGE PAYETTE, S., FILION, L., DELWAIDE, A. 2005 Reconstruction of the long-term fire history of an old-growth deciduous forest in Southern Québec, Canada, from charred wood in mineral soils *Quaternary Research*. 64, (1), 36-43.

THINON, M. 1992. *L'analyse Pedoanthracologique, aspects méthodologiques et applications*. Thèse de doctorat ès Sciences. Université Aix-Marseille III, 317 p.

TURNER, C. 2000 The Eemian interglacial in the North European plain and adjacent areas Geologie en Mijnbouw / Netherlands Journal of Geosciences 79 (2/3), 217-231.

Van VLIET-LANOË, B. 1976. Corrélation entre la présence de charbon de bois au sommet de Paléosols et les dégradations climatiques. *Pédologie*. 31, 97 p.

# ANTHRACOLOGICAL ANALYSIS FROM A MINING SITE IN THE EASTERN ALPS TO EVALUATE WOODLAND USES DURING THE BRONZE AGE

Anton Stefan SCHWARZ and Klaus OEGGL
Institute of Botany, University of Innsbruck, Sternwartestrasse 15, 6020 Innsbruck, Austria
Stefan.Schwarz@uibk.ac.at; Klaus.Oeggl@uibk.ac.at

Rüdiger KRAUSE
Institute of Archaeology, University of Frankfurt, Grüneburgplatz 1, 60629 Frankfurt am Main, Germany
R.Krause@em.uni-frankfurt.de

**Abstract**: *This anthracological case study from the Montafon (Vorarlberg, Austria), a valley system in the Eastern Alps, deals with woodland use in prehistory. The studies were conducted on 16 systematically collected soil samples from the cultural layers of a prehistoric settlement dated from the Early to the late Middle Bronze Age. The size classes of the studied charcoal pieces reveal that primarily branches and twigs of trees were gathered for fuel, whereas the species spectra of the charcoal assemblages reflect the composition of the former woodlands and the succession over time: during the Early Bronze Age a spruce* (Picea) *forest mixed with some beech* (Fagus sylvatica), *fir* (Abies) *and maple (**Acer**) was dominant. Later during Middle Bronze Age the constant occurrence of the pioneer species birch* (Betula), *hazel* (Corylus avellana) *and rowan* (Sorbus) *reflect the anthropogenic thinning of the forest. Complementary palynological studies of a nearby mire corroborate these results and document an immense demand on farming land or timber by the decrease of spruce* (Picea) *and fir* (Abies) *linked to an expansion of grasses* (Gramineae) *and cultural indicators during the Bronze Age.*

**Key words**: *Bronze Age – Anthracology – Woodland uses – Copper ore mining – Alps*

## INTRODUCTION

The Montafon valley (Vorarlberg, Austria) represents an almost closed inneralpine settlement chamber, which is well-suited for investigations on the course of settlement as well as the dynamics and motivations that contributed towards its habitation by prehistoric man. Until recently, only a scant number of single finds from the Bronze and Iron Ages had been reported in the Montafon area, and settlements from those periods were lacking completely (Krause 2001; Vonbank 1966). This state of knowledge has changed fundamentally through systematic field work and investigations on the vegetation history, so that today we know that this area was settled continuously since the Late and Final Neolithic period (the beginning of the third millennium BC) and that settlement increased even more so during the Bronze and Iron Ages (Krause *et al.* 2004).

Large copper and iron ore deposits in this region suggest that the exploitation of these important sources of raw material constituted a substantial motive for prehistoric colonization. The ore deposits in the Montafon are located peripheral to the vast metalliferous zones in the Eastern Alps and – perhaps for that reason – they have been poorly investigated until now. Therefore it is an important aim to compare the prehistoric (e.g. Bronze Age) mining activities and the settlement development associated with these activities in the Montafon with those known from larger prehistoric mining districts in the Inn valley, such as the Kitzbühel area and the Mitterberg, in order to study and better comprehend the similarities, as well as the differences, between these areas. In the context of a working hypothesis it is our opinion that, according to the archaeological data gained from our investigations so far, this habitat should be seen against the background of the prospecting and exploitation of copper ores (Krause 2001; Krause 2009 in press).

## AREA INVESTIGATED

Up to now diverse habitation sites constituted by one well-fortified central establishment at the 'Friaga Wald' at Bartholomäberg (Vorarlberg, Austria; 47°06'N, 9°54'E) and several peripheral open settlements are documented, which are indicative of a certain settlement hierarchy (Figure 1). Thus, the subject of settlement archaeology has the rare opportunity to investigate various aspects of settlement development and dynamics in the small closed settlement chamber of the Schruns basin. The prehistoric hilltop settlement in 'Friaga Wald' at Bartholomäberg is situated at 940 m above sea level, approximately 240 m up from the valley floor above Schruns and the confluence of the Litz River flowing out of the Silbertal and the Ill issuing from inner Montafon. Archaeological excavations, started in the years 2000-2003, brought forth impressive evidence of a strongly fortified Bronze Age settlement, dated to the 16th/15th century BC. The site was first settled already in the late Early Bronze Age, according to radiocarbon dating in the 18th/17th century BC, and greatly expanded later, during the transition to the Middle Bronze Age in the 16th

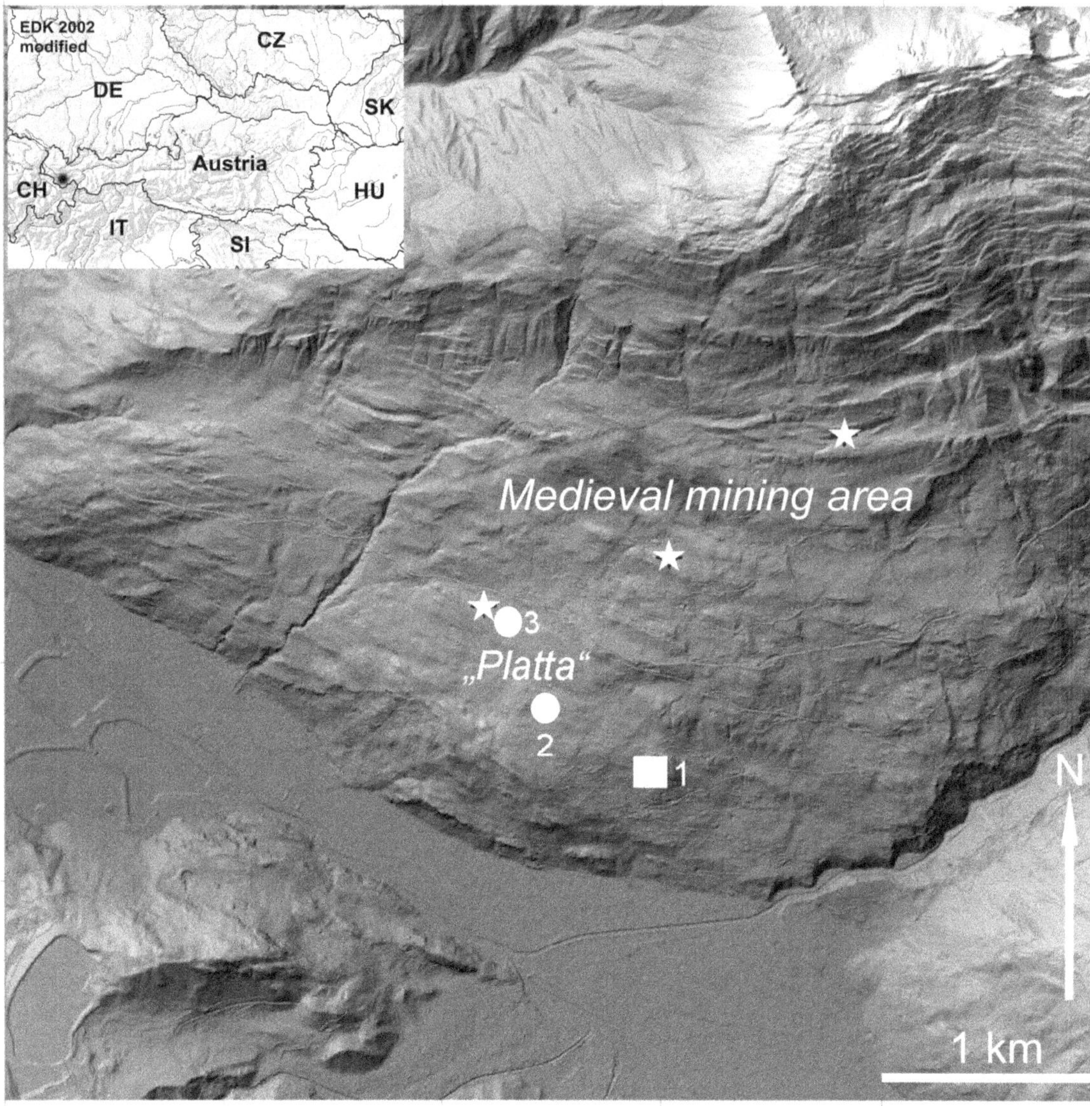


Figure 1. Montafon, Vorarlberg, Austria. Model of the terrain of Bartholomäberg with the Schruns Basin in the foreground. The prehistoric settlements, dating to phases of the Bronze and Iron Age, are situated in a favourable location on the mountain terrace Platta. At a higher location on the mountain is the Medieval mining area. Designated are three moors (with stars), where palynological investigations were conducted by Klaus Oeggl, University of Innsbruck, as well as prehistoric settlements: 1. 'Friaga Wald'; 2. 'Bodaweg'; 3. 'Buxwaldstrasse' (3D terrain model after VoGis-Data of the Land Vorarlberg, modified; EDK 2002 Schweizer Weltatlas, modified)

century BC (Krause 2005). The topography of the site is quite marked, and the fortified settlement was protected by an 80 m long and up to 3 m wide wall towards the slope and the hinterland of the mountain terrace. Moreover, the settlement hill was secured towards the hinterland by a ditch hewn into the bedrock. Within the enclosing walls and ditch artificial terraces were created for houses and for other structures (Figure 2).

An important objective of archaeological research and field studies was to locate other prehistoric settlements at the Bartholomäberg and farther in Montafon. Results from archaeobotanical research on the mires have shown that settlement density at Bartholomäberg must have been much greater in the Bronze as well as the Iron Age than evidenced by the archaeological findings thus far. The 'Platta' on Bartholomäberg is a wide open mountain terrace with fruitable soils as economical and agricultural area. The moderate elevation and the terrace's orientation towards the south are inductive for agriculture. At the centre of the large mountain terrace there are several terraces in the vicinity of the fortified settlement. During the course of prospecting the terrain by core drillings, in different locations at the 'Bodaweg' remains of black cultural layers with charcoal were found. Archaeological excavations carried out from 2005 to 2008 and radiocarbon dating have shown two settlement layers from the late Early Bronze Age (18th/17th century BC) and the Middle Bronze Age (14th/13th century BC). Typical features of the Middle Bronze Age settlement phase are at least 18 pits filled with burning remains (Figure 3), charcoal and a lot of burned stones. The question raised on the original purpose of these burning processes. Anthracological studies might help to interpret

Figure 2. Bartholomäberg, Montafon. Reconstruction of the Middle Bronze Age settlement in 'Friaga Wald', based on the digital topographic map and archaeological findings. In the foreground is the fortification wall that encloses the complex on the mountain's incline. Behind it is the settlement terrace, formed by the terrace wall, with 6-8 reconstructed block houses, each with a ground plan of c. 4 x 5 m. 3D-model by Martin Schaich, ArcTron Dokumentation, Altenthann

Figure 3. Bartholomäberg, Bodaweg. Bronze Age settlement with so called 'Feuergruben', filled with burning remains, charcoal and burned stones

and to understand this situation. We developed the idea that all the burned stones were used as heating stones for grains or all kinds of fruits (Krause 2006, 56).

The topography of the Bronze Age settlement at the 'Bodaweg' shows evidence of an open, unfortified settlement. This finding is indeed of great interest with regard to the development of the Bronze Age settlement; apparently after the abandonment of the Middle Bronze Age habitation area, a new settlement was established at another place, yet within sight of the hill fort in 'Friaga Wald'. In the meanwhile in 2008 we have found a third Bronze Age site settlement on another mountain terrace at the 'Buxwaldweg' with a well preserved cultural layer and decorated pottery from the transition from Early to Middle Bronze Age.

## MATERIAL AND METHODS

### Sampling, sample processing and identification of charred remains

At the excavation site 'Bodaweg' several pits with charcoal, stones and few artefacts were excavated. 18 Soil samples with a standard volume of 10 l (in 1 case only 5 l was possible) were collected systematically from the unearthed pits for macro remain and charcoal analysis. Standard flotation procedure (Jacomet and Kreuz 1999) was used to extract the plant and charcoal remains from the soil material. The remains were split into four fractions by using staggered sieves with mesh sizes of 2, 1, 0.5 and 0.25 mm. Diaspores were analysed using a stereo microscope (magnification 32x) and identification was made according to standard literature (Berggren 1981; Bertsch 1941; Brouwer and Stählin 1955; Cappers *et al.* 2006) in addition to the reference collection of alpine seeds and fruits at the Botanical Institute of Innsbruck University. Charcoal analysis was conducted on a sub sample of 100 randomly collected charcoal fragments and analysed using a reflected-light microscope (Zeiss Axioskop) at a magnification of 500x. Standard literature (Grosser 1977; Schweingruber 1990) and an interactive wood anatomy key (Schoch *et al.* 2004) were used for species identification. In addition some important dendrological and taphonomical features of each charcoal fragment were recorded (Bräker 1979; Heiss and Oeggl 2008). The radius of the outermost growth ring was measured by adjusting the growth ring curvature and wood ray angle of the charcoal pieces to a diameter stencil and furthermore diameter classes were built according to Ludemann 1996. The mean diameter (mD) value, an arbitrary relative value for comparing data of different sites (Ludemann 1996, 1999,

| Bartholomäberg Bodaweg | | | | | | | | | | | | | | | | | | | |
|---|---|---|---|---|---|---|---|---|---|---|---|---|---|---|---|---|---|---|---|
| Period | | Early Bronze Age | | | | | Middle Bronze Age | | | | Middle Bronze Age | | | | | | | | |
| Type of pit | | type 'eba' | | | | | type 'a' | | | | type 'b' | | | | | | | | |
| Sample number | Sum | 8 | 9 | 10 | 13 | 39 | 2 | 6 | 4 | 12 | 7 | 5 | 1 | 11 | 16 | 18 | 23 | 24 | 28 |
| Sample volume [l] | 175 | 10 | 10 | 10 | 10 | 5 | 10 | 10 | 10 | 10 | 10 | 10 | 10 | 10 | 10 | 10 | 10 | 10 | 10 |
| Sample weight [kg] | 33,87 | 1,96 | 1,35 | 1,55 | 0,68 | 0,87 | 1,78 | 1,68 | 1,79 | 1,68 | 1,13 | 1,97 | 1,91 | 2,46 | 2,68 | 1,50 | 1,53 | 4,13 | 3,24 |
| Total charcoal weight >2mm [g] | 549,92 | 40,38 | 15,80 | 5,32 | 2,63 | 4,90 | 20,18 | 16,98 | 50,14 | 2,16 | 41,09 | 25,37 | 20,57 | 5,01 | 104,10 | 19,14 | 19,65 | 95,50 | 61,00 |
| Analysed charcoal weight >2mm [g] | 84,30 | 7,29 | 4,91 | 3,09 | 0,77 | 3,74 | 3,51 | 0 | 5,49 | 1,58 | 4,14 | 0 | 3,12 | 1,99 | 16,05 | 4,74 | 6,69 | 11,87 | 5,32 |
| Analysed charcoal pieces | 1375 | 100 | 100 | 100 | 25 | 50 | 100 | 0 | 100 | 100 | 50 | 0 | 50 | 100 | 100 | 100 | 100 | 100 | 100 |
| Cereals | | | | | | | | | | | | | | | | | | | |
| Hordeum vulgare | 8 | · | · | · | · | · | · | 1 | · | 1 | · | · | 5 | · | 1 | · | · | · | · |
| Hordeum sp. | 2 | · | · | · | · | · | · | · | · | · | · | 2 | · | · | · | · | · | · | · |
| Cerealia indet. | 1 | · | · | · | · | · | · | · | · | · | · | 1 | · | · | · | · | · | · | · |
| Triticum spelta (spikelet fork) | 1 | · | · | · | · | · | · | · | · | · | · | · | 1 | · | · | · | · | · | · |
| Gathered plants | | | | | | | | | | | | | | | | | | | |
| Prunus spinosa | 1 | · | · | · | · | · | 1 | · | · | · | · | · | · | · | · | · | · | · | · |
| Rubus idaeus | 14 | 1 | 3 | 5 | 2 | · | 2 | 1 | · | · | · | · | · | · | · | · | · | · | · |
| Sambucus nigra | 4 | 1 | · | · | · | · | · | 2 | · | · | · | · | 1 | · | · | · | · | · | · |
| Sambucus sp. | 1 | · | · | · | · | · | 1 | · | · | · | · | · | · | · | · | · | · | · | · |
| Wild plants | | | | | | | | | | | | | | | | | | | |
| Avena sp. (awn fragment) | 1 | · | · | · | · | · | · | · | · | · | · | · | 1 | · | · | · | · | · | · |
| Chenopodium sp. | 3 | · | · | · | · | · | · | · | 3 | · | · | · | · | · | · | · | · | · | · |
| Fabaceae | 1 | 1 | · | · | · | · | · | · | · | · | · | · | · | · | · | · | · | · | · |
| Galium sp. | 1 | · | · | · | · | · | 1 | · | · | · | · | · | · | · | · | · | · | · | · |
| Polygonum aviculare | 1 | · | · | · | · | · | · | · | · | · | · | · | 1 | · | · | · | · | · | · |
| Silene sp. | 1 | 1 | · | · | · | · | · | · | · | · | · | · | · | · | · | · | · | · | · |
| Solanum nigrum | 1 | 1 | · | · | · | · | · | · | · | · | · | · | · | · | · | · | · | · | · |
| Spergularia sp. | 1 | · | · | · | · | · | · | · | · | · | · | · | 1 | · | · | · | · | · | · |
| Woodland | | | | | | | | | | | | | | | | | | | |
| Picea abies (needle fragment) | 1 | · | · | · | · | · | 1 | · | · | · | · | · | · | · | · | · | · | · | · |
| Others | | | | | | | | | | | | | | | | | | | |
| Charred plant fragments indet. | 148 | 13 | 25 | 26 | 26 | · | 23 | 9 | · | · | · | 1 | 3 | 15 | 7 | · | · | · | · |
| Osseous fragments | 38 | · | · | 5 | · | · | · | · | · | · | 2 | · | 30 | 1 | · | · | · | · | · |
| Scoraceous plant fragments indet. | 756 | 211 | 200 | 318 | · | · | · | · | · | 9 | 7 | · | 5 | 5 | · | · | · | · | 1 |
| Malachite fragments | 2 | · | · | · | · | · | · | · | · | · | · | · | · | · | · | · | · | · | 2 |

Figure 4. Summarizing table of all found charred plant remains from the excavation site at 'Bodaweg'

2006, 2008;Ludemann and Nelle 2002; Nelle 2002), was calculated for each sample using the following formula:

$$mD\ [cm] = [n_{(0\text{-}2cm)}*1+n_{(2\text{-}3cm)}*2.5+n_{(3\text{-}5cm)}*4+n_{(5\text{-}10cm)}*7.5+n_{(>10cm)}*15] / n_{total},$$

whereby n refers to the number of recorded pieces for the specific size classes (0-2 cm), (2-3 cm), etc. The mean diameter (mD) was calculated both for the total charcoal wood as well as subdivided for coniferous and deciduous wood per sample. Additionally the presence of reinforcements (no = 0, marginal = 1, intense = 2) in the tracheids of the coniferous wood as well the damages (no = 0, marginal = 1, intense = 2) caused by impact of fire were noted. In order to compare the different samples the weighted average values for reinforcement and damage (mean reinforcements = mRF; mean damages by Heat = mDH) are presented for each sample. Marginal damages are small formed cracks and deformations without influence to the identification of species. Intense damages are deformations, contaminations which complicated identification of the species as well as other parameters of the charcoal.

Finally for a rough estimate of the age of the burnt wood the outermost tree ring radius was compiled with the mean annual ring growth (maximum diameter of the charcoal piece in cross section (µm) shared through number of tree rings counted) of each single piece of charcoal. According to these estimations the wood age was ordered into the following age classes: years ≤10, 11-20, 21-40, 41-60, 61-80, 81-100, >100 and not specified (ns). The further numerical analysis (PCA) of the data was conducted with the software Canoco 4.5 (Ter Braak and Smilauer 2002).

## RESULTS

### Diaspors

The pits contained a few charred diaspors connected to farming activity and woodland uses (Figure 4). Among cereals barley (*Hordeum vulgare*) and a spikelet fork of spelt (*Triticum spelta*) were detected. Oat (*Avena* sp.) was documented with an awn fragment, but could not be determined to species level. Plants gathered for food and dietary supplement are represented by seeds of raspberry (*Rubus idaeus*) and elder (*Sambucus nigra*). Sloe (*Prunus spinosa*) was present as a single find of a drupe fragment. Identified weeds such as seeds of goosefoot (*Chenopodium* sp.), legumes (Fabaceae), bedstraw (*Galium* sp.), knotgrass (*Polygonum aviculare*), campion (*Silene* sp.), black nightshade (*Solanum nigrum*) and spurrey (*Spergularia* sp.) are plants connected with disturbed sites.

### Analysed Charcoal

Out of 16 samples a total of 507.57 g charcoal was isolated. 1375 randomly collected charcoal pieces were analysed with a total weight of 84.3 g (Figure 5). The species recorded split into two main groups: the first represents a mixed forest with dominance of spruce (*Picea/Larix* type) over fir (*Abies*) and beech (*Fagus sylvatica*), with admixed small amounts of maple (*Acer* sp.) and lime (*Tilia* sp.). The second group is composed by pioneer species predominated by hazel (*Corylus avellana*) and alder (*Alnus*) beneath small amounts of birch (*Betula* sp.), pomaceous trees (Maloideae) and pine (*Pinus sylvestris*).

A principle component analysis was conducted with the collected data (Figure 6). The composition of species per

| Barthoomäberg Bodaweg | Sample volume [l] | Sample number | Picea/Larix | cf. Abies | Fagus sylvatica | Acer sp. | Quercus sp. | Tilia sp. | Pinus sylvestris | Corylus avellana | Alnus sp. | Betula sp. | Betulaceae indet. | Maloideae indet. | cf. Clematis | cf. Euonymus | Pinaceae indet. | Deciduous indet. | indet. | Total charcoal weight >2mm [g] | Analysed charcoal weight >2mm [g] | Analysed charcoal pieces | Mean fragment weight [mg] | Sample N mD | Sample mD [cm] | Deciduous N mD | Deciduous mD [cm] | Coniferous N mD | Coniferous mD [cm] | Coniferous N mRF | Coniferous mRF | Sample N mHD | Sample mHD |
|---|---|---|---|---|---|---|---|---|---|---|---|---|---|---|---|---|---|---|---|---|---|---|---|---|---|---|---|---|---|---|---|---|---|
| MBA type 'b' | 10 | 1 | 78 | · | · | · | · | · | · | 20 | · | · | · | · | · | · | · | 2 | · | 20,6 | 3,1 | 50 | 62 | 50 | 5,96 | 11 | 5,77 | 39 | 6,01 | 39 | 0,41 | 50 | 1,62 |
| MBA type 'b' | 10 | 7 | 90 | · | · | · | · | · | · | 10 | · | · | · | · | · | · | · | · | · | 41,1 | 4,1 | 50 | 83 | 50 | 6,14 | 5 | 4,40 | 45 | 6,33 | 45 | 1,40 | 50 | 1,40 |
| MBA type 'b' | 10 | 11 | 74 | · | · | 1 | · | · | 2 | 14 | · | · | 3 | · | · | · | 2 | 4 | · | 5,0 | 2,0 | 100 | 20 | 92 | 4,32 | 19 | 4,32 | 73 | 4,32 | 78 | 1,14 | 96 | 1,26 |
| MBA type 'b' | 10 | 16 | 65 | · | · | · | · | · | 1 | 21 | 6 | 7 | · | · | · | · | · | · | · | 104,1 | 16,1 | 100 | 161 | 100 | 5,12 | 34 | 4,41 | 66 | 5,48 | 66 | 1,76 | 100 | 1,39 |
| MBA type 'b' | 10 | 18 | 39 | · | · | 3 | 2 | · | · | 39 | · | 8 | · | 9 | · | · | · | · | · | 19,1 | 4,7 | 100 | 47 | 100 | 5,94 | 61 | 5,84 | 39 | 6,08 | 38 | 1,11 | 100 | 1,28 |
| MBA type 'b' | 10 | 23 | 51 | · | · | · | · | · | · | 24 | 22 | · | · | 3 | · | · | · | · | · | 19,7 | 6,7 | 100 | 67 | 100 | 5,36 | 49 | 4,98 | 51 | 5,73 | 49 | 1,37 | 100 | 1,21 |
| MBA type 'b' | 10 | 24 | 50 | · | · | · | · | · | 4 | 20 | 25 | 1 | · | · | · | · | · | · | · | 95,5 | 11,9 | 100 | 119 | 100 | 7,50 | 46 | 7,30 | 54 | 7,66 | 53 | 1,51 | 100 | 1,05 |
| MBA type 'b' | 10 | 28 | 58 | · | · | · | · | · | 2 | 29 | 9 | · | · | · | · | · | · | 2 | · | 61,0 | 5,3 | 100 | 53 | 100 | 6,65 | 40 | 6,66 | 60 | 6,64 | 60 | 1,12 | 100 | 1,04 |
| MBA type 'a' | 10 | 2 | 84 | 3 | 9 | · | · | · | 2 | · | · | · | · | · | · | · | 1 | 1 | · | 20,2 | 3,5 | 100 | 35 | 100 | 5,48 | 10 | 4,95 | 90 | 5,53 | 89 | 1,06 | 100 | 1,61 |
| MBA type 'a' | 10 | 4 | 83 | · | · | · | · | · | · | · | 6 | 7 | 2 | · | · | · | · | 2 | · | 50,1 | 5,5 | 100 | 55 | 100 | 4,15 | 17 | 4,88 | 83 | 3,99 | 83 | 1,82 | 100 | 1,40 |
| MBA type 'a' | 10 | 12 | 59 | · | 3 | · | · | · | · | 10 | · | 3 | 1 | 9 | · | · | 5 | 10 | · | 2,2 | 1,6 | 100 | 16 | 94 | 2,91 | 33 | 2,80 | 61 | 2,98 | 57 | 1,39 | 99 | 1,28 |
| EBA type 'eba' | 10 | 8 | 89 | · | 2 | 4 | · | · | · | · | · | · | · | · | · | · | · | 2 | 3 | 40,4 | 7,3 | 100 | 73 | 93 | 6,35 | 6 | 4,33 | 87 | 6,49 | 89 | 1,04 | 98 | 1,37 |
| EBA type 'eba' | 10 | 9 | 81 | 2 | 4 | 4 | · | · | · | · | · | · | · | · | · | · | 1 | 8 | · | 15,8 | 4,9 | 100 | 49 | 93 | 5,91 | 12 | 5,54 | 81 | 5,97 | 83 | 1,19 | 100 | 1,32 |
| EBA type 'eba' | 10 | 10 | 59 | 10 | 6 | · | · | · | 1 | · | · | · | · | 3 | · | 1 | 19 | 1 | · | 5,3 | 3,1 | 100 | 31 | 95 | 5,16 | 9 | 5,11 | 86 | 5,16 | 69 | 0,91 | 96 | 1,00 |
| EBA type 'eba' | 10 | 13 | 56 | 8 | 28 | · | · | 8 | · | · | · | · | · | · | · | · | · | · | · | 2,6 | 0,8 | 25 | 31 | 25 | 5,38 | 9 | 5,83 | 16 | 5,13 | 16 | 1,50 | 25 | 1,68 |
| EBA type 'eba' | 5 | 39 | 46 | · | 34 | 2 | · | · | · | 4 | · | · | · | 2 | 4 | · | · | 8 | · | 4,9 | 3,7 | 50 | 75 | 50 | 5,86 | 27 | 5,46 | 23 | 6,33 | 23 | 1,30 | 50 | 1,96 |
| | | | | | | | | | | | | | | | | | | | [%] | | | | | | | | | | | | | | |
| Sum | 155 | all | | | | | | | | | | | | | | | | | | 507,6 | 84,3 | 1375 | 61 | 1342 | 5,48 | 388 | 5,37 | 954 | 5,52 | 937 | 1,30 | 1364 | 1,32 |
| | | | EBA Early Bronze Age<br>MBA Middle Bronze Age | | | | | | | | | | | | | | | | | N Number<br>mD mean Diameter | | | | mRF mean ReinForcements<br>mHD mean Heat Damages | | | | | | | | | |

Figure 5. Results of charcoal analysis. The pits are divided in order of their archaeological determination. The summarizing table shows the percentage values of identified species in order of their ecological distribution and some dendrological and taphonomical details. (mRF: mean Reinforcements and mHD: mean Heat Damages are calculated weighted averages and range between no = 0, marginal = 1, intense = 2)

| Axes | 1 | 2 | 3 | 4 | Total variance |
|---|---|---|---|---|---|
| Eigenvalues | 0.442 | 0.145 | 0.122 | 0.089 | 1.000 |
| Cumulative percentage variance of species data | 44.2 | 58.7 | 70.9 | 79.7 | |

Figure 6. Summary of the principle component analysis of the percentage values from the identified species from charcoal at 'Bodaweg'

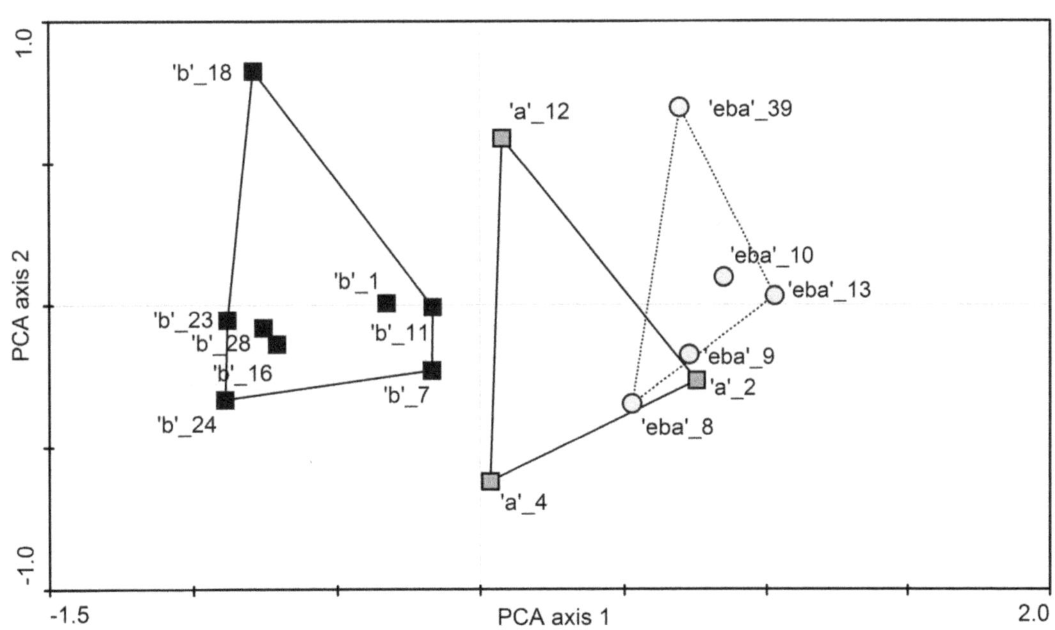


Figure 7. Labelling and ordering of the charcoal samples from 'Bodaweg' pits within the principle component analysis programme. The Early Bronze Age pits are labelled with 'eba' (circle) and the Middle Bronze Age pits are labelled with 'a' (grey box) or 'b' (dark box) depending on their archaeological determination

sample led to an ordering of the samples as shown in Figure 7. On PCA axis 1 the five samples from the Early Bronze Age cluster together, while the 11 samples from the Middle Bronze Age form two groups. The three samples of the Middle Bronze Age type 'a' are grouped beside the Early Bronze Age samples 'eba', while the eight samples of the Middle Bronze Age type 'b' are separated and close together. On PCA axis 2 the

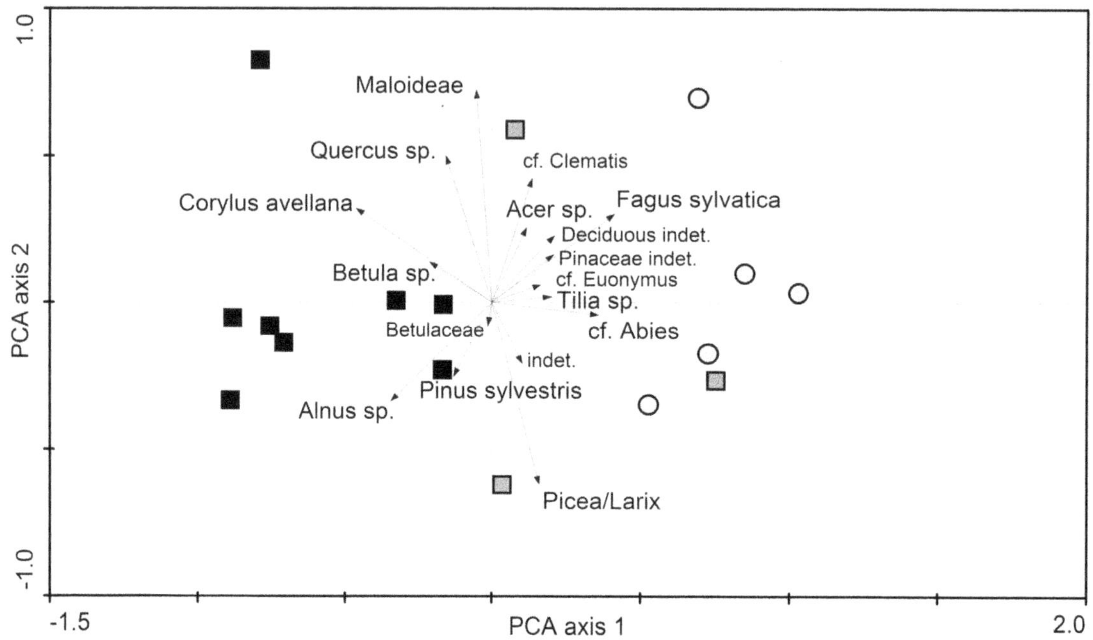


Figure 8. Distribution of species of the charcoal samples from 'Bodaweg' pits within the principle component analysis programme. The length and direction of the arrows show the weighting of species for arranging the samples at PCA axes. The Early Bronze Age pits are indrawn as a circle and the Middle Bronze Age pits type 'a' as a grey box and type 'b' as a dark box

differences between the samples in time and type of pits are obvious: the Early Bronze Age pits 'eba' and the Middle Bronze Age pits type 'a' show more differences within themselves than the Middle Bronze Age pits type 'b'. In summary the Early Bronze Age pits show within themselves more differences than all Middle Bronze Age pits. In fact, archaeologists determined that the Middle Bronze Age pits type 'a' had the same function as the Early Bronze Age pits 'eba' as usual settlement pits, while the Middle Bronze Age pits type 'b' are interpreted as unusual pits with unknown function so far. The most interesting fact is: charcoal from the Early Bronze Age pits reflects a mixed forest with few pioneer species whereas charcoal from the Middle Bronze Age pits reflects in high amounts pioneer species with few representatives of a mixed forest (Figure 8). The length and the direction of the arrows show the influence of the species for position of the sample in the diagram. Positive correlating arrows point to the same direction, while negative correlating species arrows show opposite directions.

Only single pieces were detected with a clear terminal ring, thus only a minimum diameter is reflected. The mean diameter of the charred wood is between 2.91 cm and 7.5 cm. No real differentiation is seen for the coniferous (mD = 2.98-7.66 cm) and the deciduous (mD = 2.80-7.30 cm) wood. When coniferous wood is exposed to tensile stress then the tracheids develop strengthening reinforcements. Such anatomical characteristics of the tracheids of the coniferous wood are recorded in all samples. The weighted average values range from 0.41 (less) to 1.82 (intense) while 87.5% of the samples reveal a value more than 1 (marginal) and the value of these reinforcements of all coniferous pieces is 1.30. This indicates a presence of strengthening reinforcements in the tracheids of timber of all samples.

All analysed charcoal was strongly fragmented. A good indicator for the status of the observed charcoal/sample is represented by the heat damages. The calculated total weighted average over all the samples is 1.32 and range from sample to sample between 1.00 (marginal) and 1.96 (intense). All the samples include charcoal with heat damages and indicate that the wood must have been burnt at high temperatures.

The results of the calculated minimum age of the burnt wood are divided into classes which are conformed to life time periods (Figure 9 and Figure 10). The assemblage at ages <60 years is obvious for the coniferous and the deciduous wood. The wood was cut when trees were in a profitable age for timber production.

## DISCUSSION

### Domesticates and subsistence

The recorded crop plants from the 'Bodaweg' alone are not sufficient to draw any fundamental conclusion but, in comparison with the results from the contemporary cultural layers of the nearby located fortified hilltop settlement 'Friaga' (Schmidl and Oeggl 2005), their representativness can be evaluated. The studies from the 'Friaga' cultural layers prove that barley (*Hordeum vulgare*) and spelt (*Triticum spelta*) were the main crops during the Bronze Age. Given that the most prominent plants have a higher probability to be recorded in cultural layers, the few findings of barley and spelt in the

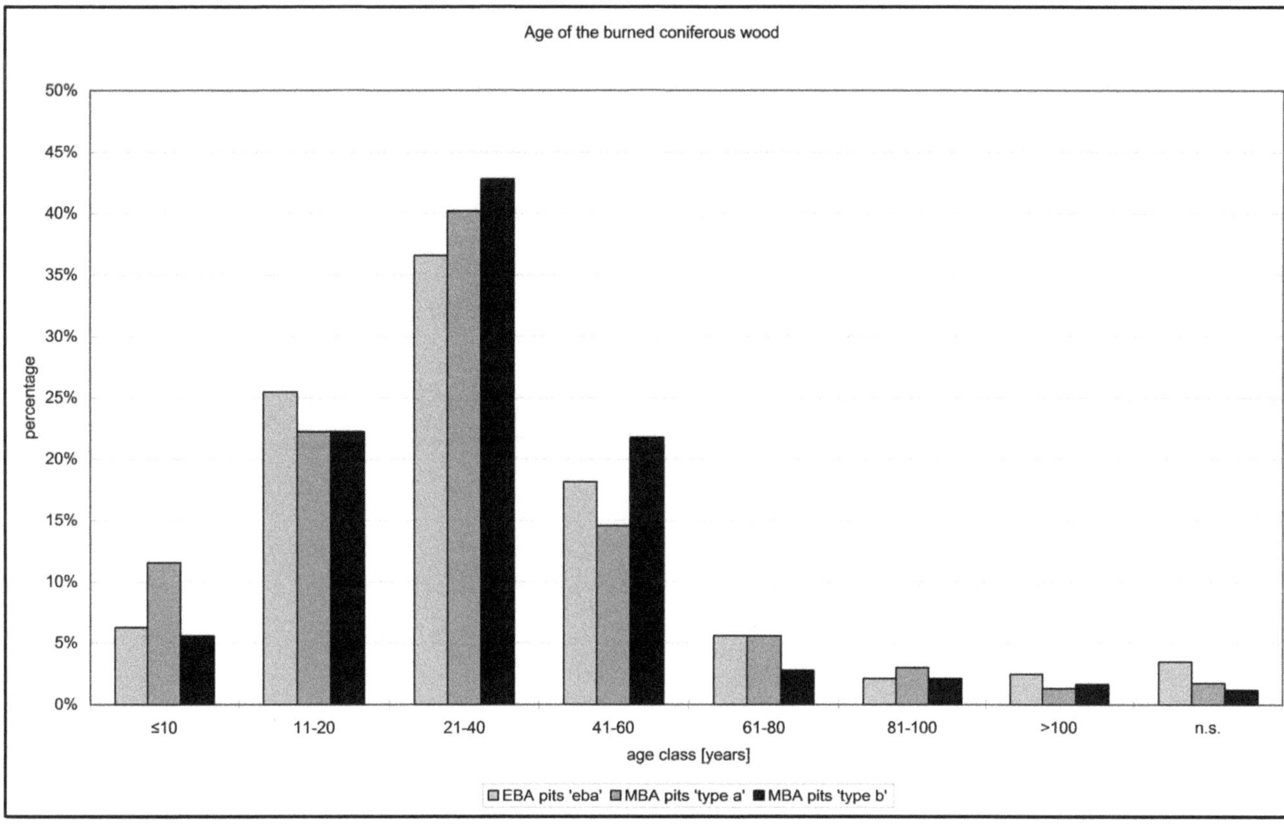


Figure 9. The calculated minimum age of the coniferous wood divided into classes (age classes in years, n.s. = not specified). The different type of the 'Bodaweg' pits are presented in order of their archaeological determination: Early Bronze Age 'eba'; Middle Bronze Age type 'a'; Middle Bronze Age type 'b'

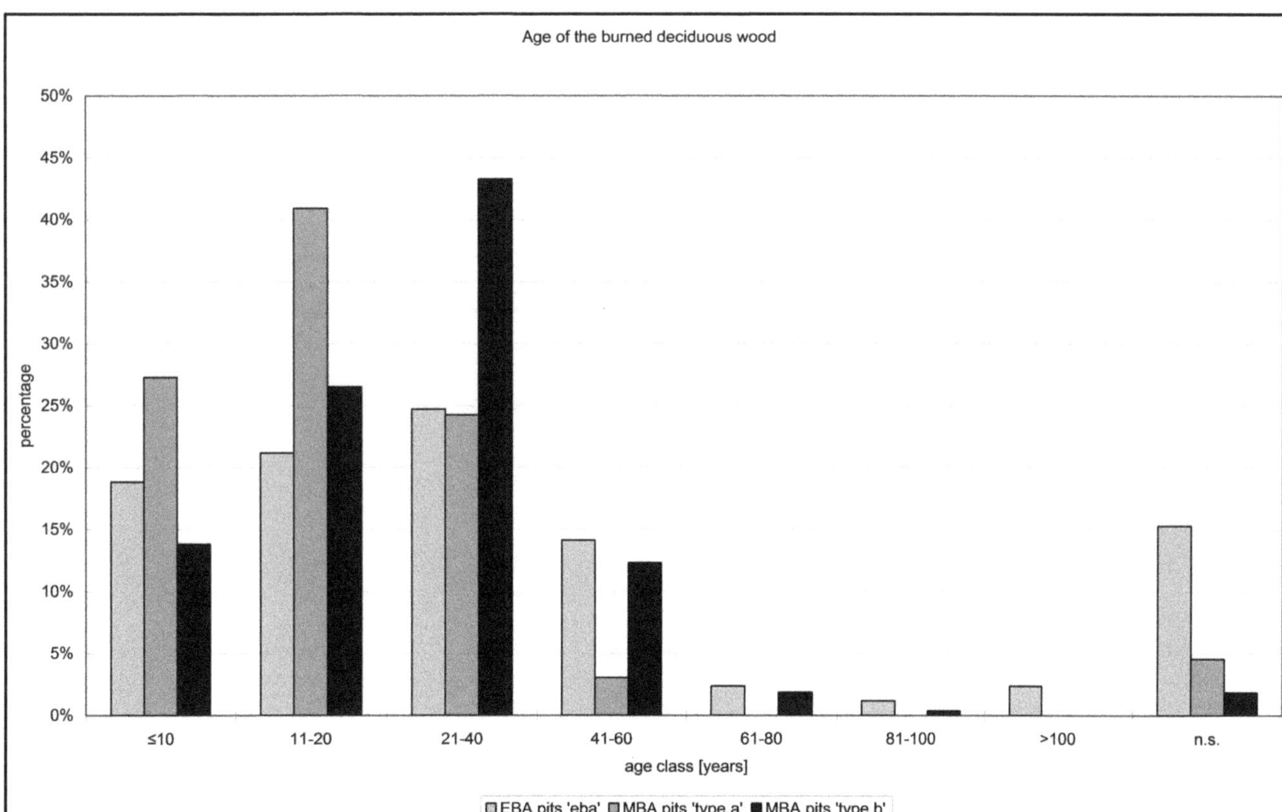


Figure 10. The calculated minimum age of the deciduous wood divided into classes (age classes in years, n.s. = not specified). The different type of the 'Bodaweg' pits are presented in order of their archaeological determination: Early Bronze Age 'eba'; Middle Bronze Age type 'a'; Middle Bronze Age type 'b'

Bodaweg layers mirror also the main crops of the Bronze Age.

Furthermore, similarities in the spectra of gathered plants are evidenced in both settlements, indicating the importance of wild fruits for the nutrition of the prehistoric settlers at that time. The common species raspberry (*Rubus idaeus*), elder (*Sambucus*) and sloe (*Prunus spinosa*) grow on cleared and nutrient rich areas. Open landscape is documented by the detected weeds: goosefoot (*Chenopodium* sp.), legumes (Fabaceae), bedstraw (*Galium* sp.), knotgrass (*Polygonum aviculare*), campion (*Silene* sp.), black nightshade (*Solanum nigrum*) and spurrey (*Spergularia* sp.) are all representatives of open, disturbed and nutrient rich habitats. Currently, these weeds grow in fields, woodland seams and opened woodland.

**Analysed charcoal and woodland use**

The analysed charcoal from the 'Bodaweg' pits enables the reconstruction of the local woodland for two time spans: the Early and the Middle Bronze Age. The charcoal spectra of the Early Bronze aged pits reflects an undisturbed mixed forest predominated by spruce (*Picea/Larix* type) with admixed beech (*Fagus sylvatica*) and fir (cf. *Abies*). The occasional occurrence of maple (*Acer*), lime (*Tilia*), hazel (*Corylus avellana*) and pomaceous woods (Maloideae) indicate a species rich semi-natural climax wood depending on humidity of the habitat for the Early Bronze Age.

On the contrary, the Middle Bronze aged pits show a different species composition in their charcoal spectra. Spruce (*Picea/Larix*) is still dominating but fir (*Abies*), beech (*Fagus sylvatica*) and maple (*Acer*) are distinctly reduced represented. Pioneer species like hazel (*Corylus avellana*), alder (*Alnus*) and birch (*Betula*) are more prominent, as well as pine (*Pinus mugo/sylvestris*). These species are light demanding and grow preferably in humid (alder) and dry habitats (birch, pine, hazel), e.g. shrubberies, hedges and woodland seams. The increase of these pioneer species indicates an extensive opening of the mixed woodland during the Middle Bronze Age.

These results are confirmed by pollen studies conducted on mires in the vicinity of the settlement (Oeggl *et al.* 2005), which reveal a spruce-fir dominated wood with beech since the Neolithic. Forests with lime (*Tilia*) and maple (*Acer*) were growing along rivulets. The first local human activities are documented at the transition of the Neolithic to the Bronze Age (2580-2400 BC). A second, more intensive period of wood clearance is recorded in the older part of the Early Bronze Age (2150-1969 BC), and a third local settlement period is reflected at the end of the Early Bronze Age (1780-1600 BC) by a decrease in pollen of climax trees, an increase of pollen from pioneer species (*Betula, Corylus avellana*) and pasture indicators, such as ribwort plantain (*Plantago lanceolata*) and dung associated fungi (*Sordaria, Sporormiella*) in connection with cultural indicators such as cereals.

**Use of wood**

The mean diameter of the coniferous wood with 2.98 to 7.66 cm indicates the burning of wood, which it is easily to handle. The same applies for the deciduous wood with a mean diameter range from 2.80-7.3 cm. Thereby a radial and tangential volumetric loss of 12-25% in wood tissue during carbonisation has to be considered (Schweingruber 1976). Nelle (2002; 2003) has defined that mD values of 4.3 cm are representative for a campfire sample, values of 8.3 cm for charcoal produced from coppiced wood, and values of 10.6 cm for charcoal from a kiln with timber diameters (*Fagus sylvatica*) between 10 and 30 cm. The 'Bodaweg' pits reveal mD of 2.80 to 7.66 cm and contained mainly charcoal from branches and twigs. This is corroborated by the observed reinforcements in the tracheids referring to branch wood. In addition all mean values of the samples range between marginal and intense. The diameters of the deciduous wood, especially the pioneer wood from the Middle Bronze aged pits, indicate the cutting of timber from a coppice or hedge. The diameters (mD up to 7.3 cm) are much higher than for campfire wood only. Therefore active cutting of timber is supposed. A rough age estimation can be given by the calculated minimum age of the used wood: for the coniferous (Figure 9), as well as for the deciduous (Figure 10) wood, the observed age ranges through all classes, but mainly in the class from 21 to 40 years. Less than 28% for the deciduous and less than 12% for the coniferous used wood was younger than 10 years while less than 6% for the deciduous and less than 11% for the coniferous used wood was older than 60 years. Wood was therefore adult and in a profitable age for timber production.

The calculated ages suppose the hypothesis of active woodland uses: indices for cyclic wood cutting are partly given from the deciduous and woodland opening from the coniferous wood. Burnt deciduous wood was mainly fuel wood from a coppice admixed with branches from adult trees such as beech or maple. Coniferous wood was mainly composed by branches and twigs from adult trees. Charcoal had in most cases marginal to intense damages caused by heat. The damages only occur if wood is not completely dry and the burning occurs rapidly at very high temperatures. The used wood was stored for short time periods only. This is seen for both periods: the Early and Middle Bronze Age.

## CONCLUSION

The charcoals studied from the 'Bodaweg' pits reflect the woodland composition and use during the Bronze Age. In the Early Bronze Age the mixed forest was used by cutting branches and small trees in the vicinity of the settlement. Over decades this led to a retraction of the climax species and therefore an opening of the forest vegetation is recorded. During the Middle Bronze Age a considerable increase of pioneer species is documented indicating the raised timber exploitation by man. Now two types of pits can be separated by differences in charcoal species composition. The function of these pit types is still unknown.

### Acknowledgments

This study was conducted within the special research programme HiMAT 'The History of Mining Activities in the Tyrol and Adjacent Areas – Impact on Environment & Human Societies' (www.uibk.ac.at/himat) funded by the Austrian National Science Fund FWF (grant nos: F3108, F3111), Land Vorarlberg, Stand Montafon and the communes Bartholomäberg and Silbertal. The authors acknowledge Dott.ssa Daniela Festi, University of Innsbruck, for valuable comments on the manuscript and Werner Kofler, University of Innsbruck, for software support and helpful comments on the data .

### References

BERGGREN, G. 1981. *Atlas of seeds and small fruits of Northwest-European plant species with morphological descriptions. Part 3*. Stockholm. The Swedish Museum of Natural History.

BERTSCH, K. 1941. Früchte und Samen, Ein Bestimmungsbuch zur Pflanzenkunde der vorgeschichtlichen Zeit. In H. Reinerth (ed.), *Handbücher der praktischen Vorgeschichtsforschung Band 1*. Stuttgart. Enke.

BRÄKER, O.U. 1979. *Angewandte Holzanalyse: Beitrag zur Rekonstruktion der Umwelt neolithischer Ufersiedlungen in Feldmeilen-Vorderfeld*. Academica Helvetica 3. Bern, Stuttgart. Haupt.

BROUWER, W. and STÄHLIN, A. 1955. *Handbuch für Samenkunde für Landwirtschaft, Gartenbau und Forstwirtschaft*. Frankfurt am Main. DLG-Verlags-GmbH.

CAPPERS, R.T.J., BEKKER, R.M. and JANS, J.E.A. 2006. *Digitale Zadenatlas van Nederland, Groningen Archaelogical Studies Volume 4*. Groningen University Library. Groningen.

EDK 2002. *Schweizerische Konferenz der kantonalen Erziehungsdirektoren (EDK):* Schweizer Weltatlas. Website URL: http://www.schweizerweltatlas.ch/images/swa/downloads/MEU3.jpg

GROSSER, D. 1977. *Die Hölzer Mitteleuropas. Ein mikrophotographischer Lehratlas*. Berlin Heidelberg New York. Springer.

HEISS, A.G. and OEGGL, K. 2008. Analysis of the fuel wood used in Late Bronze Age and Early Iron Age copper mining sites of the Schwaz and Brixlegg area (Tyrol, Austria). *Vegetation History and Archaeobotany* 17, 211-221.

JACOMET, S. and KREUZ, A. 1999. *Archäobotanik*. Stuttgart. Ulmer.

KRAUSE, R. 2001. Siedlungsarchäologie und Bergbauforschung: Ein interdisziplinäres Projekt zur Erforschung der inneralpinen Tallandschaft im Montafon/Vorarlberg (Österreich). *Jahrbuch des Vorarlberger Landesmuseums* 145, 43-61.

KRAUSE, R. 2005. Bronzezeitliche Burgen in den Alpen. Befestigte Siedlungen der frühen bis mittleren Bronzezeit. In B. Horejs and R. Jung and E. Kaiser and B. Teržan (eds.), *Interpretationsraum Bronzezeit. Bernhard Hänsel von seinen Schülern gewidmet. Universitätsforschungen zur prähistorischen Archäologie* 121, 389-413.

KRAUSE, R. 2006. Zur bronzezeitlichen Siedlungskammer im Montafon. Neue Ausgrabungen in Bartholomäberg. Bezirk Bludenz. *Jahrbuch des Vorarlberger Landesmuseums* 149, 47-60.

KRAUSE, R. 2009 in press. *Bronze Age copper production in the Alps. Organisation and social hierarchies in mining communities*. (in press 2009).

KRAUSE, R., OEGGL, K. and PERNICKA, E. 2004. Eine befestigte Burgsiedlung der Bronzezeit im Montafon, Vorarlberg. Interdisziplinäre Siedlungs-forschungen und Montanarchäologie in Bartholomäberg und im Silbertal. *Archäologie Österreichs* 15, 4-21.

LAND VORARLBERG 2010. VoGis. Website. URL: http://vogis.cnv.at/Ludemann, T. 1996. Die Wälder im Sulzbachtal (Südwest-Schwarzwald) und ihre Nutzung durch Bergbau und Köhlerei. *Mitteilungen Verein forstliche Standortskunde und Forstpflanzenzüchtung, Freiburg i. Br.* 38, 87-118.

LUDEMANN, T. 1999. Zur Brennstoffversorgung einer römischen Siedlung im Schwarzwald. – In: Brather, S., Bücker, Ch. & Hoeper, M. (eds.): Archäologie als Sozialgeschichte. Studien zu Siedlung, Wirtschaft und Gesellschaft im frühgeschichtlichen Mitteleuropa (Festschrift Steuer). *Internationale Archäologie* 9: Studia honoraria, 165-172. Rahden/Westf. (Leidorf).

LUDEMANN, T. 2006. Anthracological analysis of recent charcoal-burning in the Black Forest, SW Germany. – In: Dufraisse, A. (ed.): Charcoal Analysis: New Analytical Tools and Methods for Archaeology. Papers from the Table-Ronde held in Basel 2004. – *British Archaeological Reports International Series* 1483, 61-70. Oxford (BAR Publishing).

LUDEMANN, T. 2008. Experimental charcoal-burning with special regard to anthracological wood diameter analysis. – In: Fiorentino, G. & Magri, D. (eds.): Charcoals from the past: Cultural and palaeoenvironmental implications. – Proceedings of the Third International Meeting of Anthracology, Cavallino – Lecce (Italy) June 28th – July 1st 2004. – British *Archaeological Reports International Series* 1807, 147-157. Oxford (BAR Publishing).

LUDEMANN, T. and NELLE, O. 2002. Die Wälder am Schauinsland und ihre Nutzung durch Bergbau und Köhlerei. *Freiburger Forstliche Forschung* 15, 139 pp.

NELLE, O. 2002. Charcoal burning remains and forest stand structure examples from the Black Forest (south-west-Germany) and the Bavarian Forest (south-east-Germany). In S. Thiébault (ed.), Charcoal analysis. Methodological approaches, palaeoecological results and wood uses. Proceedings of the 2nd International Meeting of Anthracology, Paris, September 2000. *British Archaeological Reports. International Series* 1063, 201-207.

NELLE, O. 2003. Woodland history of the last 500 years revealed by anthracological studies of charcoal kiln sites in the Bavarian Forest, Germany. *Phytocoenologia* 33 (4), 667-682. Berlin Stuttgart.

OEGGL, K., KOFLER, W. and WAHLMÜLLER, N. 2005. Pollenanalytische Untersuchungen zu Vegetations- und Siedlungsgeschichte im Montafon. In J.M. Rollinger and R. Rollinger (eds.), *Montafon 1, Mensch-Geschichte-Naturraum*, 183-207. Dornbirn, Höfle.

SCHMIDL, A. and OEGGL, K. 2005. Subsistence strategies of two Bronze Age hill-top settlements in the eastern Alps—Friaga/Bartholomäberg (Vorarlberg, Austria) and Ganglegg/Schluderns (South Tyrol, Italy). *Vegetation History and Archaeobotany* 14, 303-312.

SCHOCH, W., HELLER, I., SCHWEINGRUBER, F.H. and KIENAST, F. 2004. *Wood anatomy of central European Species*. Online version: www.woodanatomy.ch.

SCHWEINGRUBER, F.H. 1976. *Prähistorisches Holz—Die Bedeutung von Holzfunden aus Mitteleuropa für die Lösung archäologischer und vegetations-kundlicher Probleme*. Academica helvetica 2. Bern. Haupt.

SCHWEINGRUBER, F.H. 1990. *Anatomie europäischer Hölzer*. Bern. Haupt.

TER BRAAK, C.J.F. and SMILAUER, P. 2002. Canoco reference manual and CanocoDraw for Windows User's guide: Software for Canonical Community Ordination (version 4.5). Microcomputer Power, 500 pp. New York. Ithaca.

VONBANK, E. 1966. Höhenfunde aus Vorarlberg und Liechtenstein. *Archaeologia Austriaca* 40, 80-92.

# CHARCOAL ANALYSIS OF LIME KILN REMAINS IN SOUTHERN FRANCE: AN ORIGINAL PROCESS OF MEDIAEVAL AND MODERN TRADITIONAL LIME BURNING

Christophe VASCHALDE, Aline DURAND, and Jacques THIRIOT
Laboratoire d'Archéologie Médiévale Méditerranéenne (UMR 6572) et Université d'Aix-Marseille I,
Maison Méditerranéenne des Sciences de l'Homme, 5, rue du Château de l'Horloge,
BP 647, 13 094 Aix-en-Provence cédex 1, France
vaschalde@mmsh.univ-aix.fr; adurand@mmsh.univ-aix.fr; thiriot@mmsh.univ-aix.fr

Isabel FIGUEIRAL
INRAP, DIR Méditerranée, 561 rue Etienne Lenoir, Km delta, 30900 Nîmes et Centre de Bioarchéologie et d'Ecologie (UMR 5059), Institut de Botanique, 163, rue Auguste Broussonnet, 34 090 Montpellier, France
figueir@univ-montp2.fr

**Abstract**: *In Europe, little is known of the traditional production of lime, in terms of fuel management and lay out of the fireplace. This is why charcoal analysis plays an important role when attempting to understand this activity. A new research program, at the basis of a PhD work, highlighted the weakness of data obtained previously, i.e. the lack of exhaustive charcoal sampling leading to partial information. As a result, a new sampling protocol, applied to the fireplaces used in lime kilns, has been established. Comprehensive results have been obtained for the first time, testifying to the potential existence of a specific Mediterranean technique using a high stokehole. Systematic measurement of the diameter of charcoal fragments testifies to the use of small branches and twigs of local, readily available plant species.*

**Key-words**: *limekiln with high stokehole, Southern France, sampling strategy, charring, craftsman's charcoal*

## INTRODUCTION

Limekilns have been used for a long time to transform limestone into lime, which had three main applications: mortar (in the construction of buildings), limewash (applied to outside walls of buildings) and compost (to improve soil fertility). We recall that the process of burning limestone releases quick lime (calcium oxide), which is chemically unstable in normal atmospheric conditions. Combined with water it produces slaked lime (calcium hydroxide).

Despite the extensive use of this process through time, little is known of the traditional production of lime as reflected by the scant archaeological publications, which focus only on theoretic information provided by scholars and engineers (Biringuccio 1572, Fourcroy de Ramecourt 1766, Biston 1836, Magnier 1881). The structures identified and excavated have been ranked by Adam and Varène (1985).

A review of charcoal data concerning this type of structure clearly shows that sampling was not carried out exhaustively partly due to the lack of time during field work, partly due to the absence of a keen interest in this type of activity. As a result, data obtained in the numerous structures excavated, in France and elsewhere in Europe, cannot be considered as totally representative of this activity.

According to published data, gallo-roman and medieval kilns from northern France appear to have been fed mainly with deciduous *Quercus* and/or *Fraxinus* (Solari 1990, Blin *et al.* 1999, Marguerie 2002).

Only once, were two phases of burning identified and sampled in a kiln (XIVth century), located at Colayrac-Saint-cirq (Ballarin *et al.* 2006, 136). Deciduous *Quercus*, *Corylus* and *Ulmus* were used during the first phase, while *Abies* replaced *Corylus* in the second phase.

Further south, more shrubby species, such as *Pistacia lentiscus* and *Cistaceae*, were also used (Devillers 1999, Chausserie-Laprèe 2000). The use of shrubby garrigue plants has been also identified in other kilns (Thinon in Brochier, Livache 2003 and 2004; Fabre 2001 and Fabre in Ginouvez 2001).

Elsewhere in Europe, and despite the reduced number of charcoal fragments available, a diverse plant spectrum of plants has been identified. In England, *Fagus*, *Populus*, *Quercus*, *Crataegus*, *Corylus* and *Acer* were used in the roman kilns at Weekley (Jackson *et al.* 1973), while in Germany (Rhine valley) kilns were fed with *Populus* and *Salix* (Sölter, 1970,19). In Switzerland, fuel wood used in the kilns of la Combe Tenon and Pâturage aux Boeufs comprised *Abies*, *Sorbus*, *Corylus*, *Populus*, *Fagus* and *Pinus sylvestris* (Gerber *et al.* 2002). The taxonomic diversity identified suggests that no species was particularly targeted for this activity. Local availability appears to have been the main factor.

Quantitative data concerning wood diameter is usually not included in the studies referred to above. In some cases, such as Nespouls (Figueiral, unpublished report) this was clearly impossible to carry out due to the absence of complete branches. We know however, that both branches of small calibre (Fabre, 2001; Brochier, Livache 2003, 12; Poirier in Ballarin *et al.* 2006, 137; Jackson *et al.* 1973) and bigger logs were used (Marguerie 2002).

Further information, concerning fuel used in kilns, is provided by ethnoarchaeology. In Morocco (at Moulay-Idris), the leaves of *Chamaerops humilis* ou *Hyphaene thebaica* were used to feed the fire (Herber 1923, 421). Dry herbaceous plants and branches of small steppe shrubs were used in Tunisia (at Kairouan) (Adam, Varène 1985, 95), along with branches of dry rosemary recovered from distilling factories (Chehaibi *et al.* 2008). In Grece (at Nauplie), and as recently as 1980, kilns were feed with olive oilcake and almond shells (Adam, Varène 1985, 93).

Figure 1. Traditional lime Kiln firing according to Fourcroy de Ramecourt

## A SPECIFIC TECHNIQUE OF CONVERTING LIMESTONE INTO LIME

Recent research carried out in Southern France, has allowed us to identify an unusual technique used for lime burning. It is generally accepted that the earliest limekilns were straightforward pit-like structures, becoming more important constructions from the Middle Ages onwards. These funnel-shaped constructions were usually built into a slope / hillside, natural depressions, so that the natural ramp allowed both the feeding of the fire and the transport of limestone, tipped down into the crucible. Inside, an arched sole made of limestone, is separated from the kiln pot and heating chamber (Figure 1). The opening of the kiln, located at the bottom, creates a direct access to the heating chamber. After the burning, the sole is also converted into lime, thus disappearing from the archaeological record.

Recent archaeological evidence has been obtained in 2007, while digging at Le vallon de l'Homme Mort (Peynier, Provence) (Vaschalde 2008a) and in 2008, at quartier de Bel Homme (Peynier, Provence) (Figure 2). The interest of the first limekiln results from the location of the fire place, 2 m under the small entrance (50 x 60 cm), with no other airing system. This unusual design raises problems concerning both the entrance of oxygen and the shape of the arch, dependant on the location of the entrance. Contrary to what is usual, a very thick layer (1,30 m) of charred remains was preserved beside the entrance. The fireplace is very well preserved, protected by a layer of lime.

The second limekiln, presents the same features, with the addition of a small "banquette" (Vaschalde 2008b).

The characteristics of these two kilns have been compared with those from other structures excavated in southern France. The kiln from the earthenware factory Favier (XVIIth century, Montpellier) presents the same stratigraphy, with an accumulation of charcoal fragments near the entrance. This kiln also presented two piers, which formed some kind of chimney between the entrance and the heating chamber (Ginouvez 2001). This device was described by Thiriot (2006) following the excavation of another of these structures, located at Bollène (Vaucluse). This kiln had two small heating chambers, each with a fireplace. This example, in association with ethnographic data from the Mediterranean region, mainly Tunisia (Adam, Varène 1985, Chehaibi *et al.* 2008), provides the basis to our reconstruction of the kiln from Peynier (Figure 3): two different arched areas on the top of the fire place establish the communication between the entrance and the small heating chamber.

In southern France, medieval and modern limekilns appear to run in a specific manner and have a characteristic lay out, with a deep pit and a kiln pot of small dimensions.

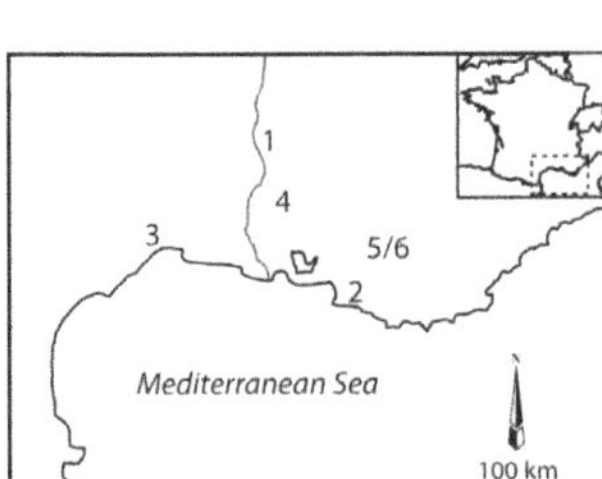


| Site | Period | Archaeologists | Anthracologist | References |
|---|---|---|---|---|
| 1 : Bollène - Saint-Blaise-de-Bauzon (Vaucluse) | XIII$^{th}$ | J. Thiriot | | Thiriot 2006 |
| 2 : Cassis - La Marcouline (Bouches-du-Rhône) | XIX$^{th}$ | Cl. Thomas, Chr. Vaschalde | Chr. Vaschalde | |
| 3 : Montpellier - Faïencerie Favier (Hérault) | XVII$^{th}$ | O. Ginouvez, J. Thiriot | L. Fabre | Ginouvez 2001, Fabre 2001 |
| 4 : Orgon - Le Défends (Bouches-du-Rhône) | XVII$^{th}$ | R. Gaday | I. Figueiral | Gaday 2007, Figueiral 2008 |
| 5 : Peynier - vallon de l'Homme Mort (Bouches-du-Rhône) | XVII$^{th}$ | Chr. Vaschalde | Chr. Vaschalde | Vaschalde 2008a |
| 6 : Peynier - Bel Homme (Bouches-du-Rhône) | ? | Chr. Vaschalde | | Vaschalde 2008b |

Figure 2. Location of sites referred to in the text

Figure 3. Hypothetical reconstruction of the lime kiln from Vallon de l'Homme Mort (Peynier)

This type of kiln is not often described in the specialized literature. The first description dates back to 1766 (Fourcroy de Ramecourt 1766). More recently, its use during the roman period is mentioned by Flach (1981), but no archaeological example is provided. And yet, this type of kiln appears to have been used in the whole of the Mediterranean world; well known examples include kilns from Rome (Sagui 1986), Kerkouane in Tunisia (Fantar 1984) and Moulay-Idris in Maroc (Herber 1923).

## METHODS

The results obtained at Vallon de l'Homme Mort (Peynier), in 2007, led us to set up a precise sampling protocol. In fact, each one of the samples studied had provided different data. The following year, at Cassis, a different strategy was followed: a quarter of the fireplace was sampled according to an artificial stratigraphy (each 5 cm). Furthermore, at Bel Homme (Peynier) a 50 cm grid was also used along with the same artificial stratigraphy. Results obtained indicate that sampling each 10 cm provides reliable data.

Water / dry sieving cannot always be used as fragile remains of charred leaves are often found. The sorting of charred material, with soft tweezers, is carried out directly during field work, as a result of exceptional conservation conditions. In fact, when the process of carbonisation reached its end, and the material was ready to be retrieved, a layer of lime was deposed on the top of the charcoal, thus sealing the kiln. Charcoal fragments are thus found in situ, without having suffered any mixture and can be sorted directly by hand. Only when sediment is also present, do we need to carry out water sieving.

Identification is carried out under a compound microscope. Pith to bark preservation in many of the fragments under study allowed the systematic measurement of diameters.

## RESULTS

The contents of three kilns were analysed (Figure 4): Peynier (Vallon de l'Homme Mort), Orgon and Cassis. The kiln of Bel Homme at Peynier is still under study. In the kiln of vallon de l'Homme Mort, three different samples were analysed, revealing the different distribution of plant species (Figure 5): high frequencies of *Quercus ilex-Q. coccifera* (c. 80%) are observed in the

| | Cassis* | | | | | | Orgon** | | | Peynier, vallon de l'Homme Mort* | | | | |
|---|---|---|---|---|---|---|---|---|---|---|---|---|---|---|
| | 0-5 cm | 10-15 cm | 20-25 cm | 30-35 cm | 40-45 cm | 45-50 cm | Sample 1 | Sample 5 | Sample 9 | US 33 | US 38 (Sample 1) | US 38 (Sample 2) | US 38 (Sample 3) | Total |
| *cf. Clematis* | | | | | | | | | 1 | | | | | 1 |
| *Cistaceae* | | | | | | | | 3 | | | | | | 3 |
| *Erica* | | 1 | | | | | | | | | | | | 1 |
| *Fabaceae* | | | | | | | 9 | 7 | 25 | | | | | 41 |
| *Juniperus sp.* | | | | | | | 4 | 13 | 3 | | | | 2 | 22 |
| *Lamiaceae* | | | | 1 | 1 | 2 | 7 | 15 | 11 | | | | | 37 |
| *Lamiaceae-Fabaceae* | | | | | | | | 3 | | 2 | | | 1 | 6 |
| *Pinus halepensis-P. pinea* | 72 | 1 | | | 1 | | | | | 28 | | 23 | 9 | 134 |
| *Pistacia sp.* | | | | | | | | | | | 2 | 1 | 1 | 4 |
| *cf. Quercus* | | | | | | | | | | 10 | | | | 10 |
| *Quercus* (deciduous) | | | | | 1 | | | | | 1 | 6 | 9 | 17 | 34 |
| *Quercus* (deciduous)-*Castanea* | | | | | | | | | | 6 | 3 | 22 | 3 | 34 |
| *Quercus* (evergreen) | | | | | | | 83 | 56 | 89 | | | | 1 | 229 |
| *Quercus ilex-Q. coccifera* | 12 | 77 | 70 | 70 | 111 | 80 | | | | 29 | 100 | 75 | 93 | 717 |
| *Rhamnus-Phillyrea* | | | 1 | 4 | | | | | | 1 | 8 | 1 | 2 | 17 |
| *Rosaceae maloidae* | | | | | | | | | | | | | 2 | 2 |
| *Rosaceae prunoidae* | | 2 | | | | | | | | | 3 | 4 | | 9 |
| *Rosmarinus officinalis* | 2 | 1 | 4 | | 4 | | 44 | 50 | 65 | | | 1 | | 171 |
| Bark | 3 | 1 | | | | | 2 | 1 | 6 | | | | | 13 |
| Indet. | 11 | 17 | 25 | 25 | 42 | 18 | 1 | 2 | | 23 | 4 | 14 | 9 | 191 |
| **Total** | **100** | **100** | **100** | **100** | **160** | **100** | **150** | **150** | **200** | **100** | **126** | **150** | **140** | **1676** |

* Chr. Vaschalde ; ** I. Figueiral.

Figure 4. Charcoal analysis table (Cassis, Orgon, Peynier)

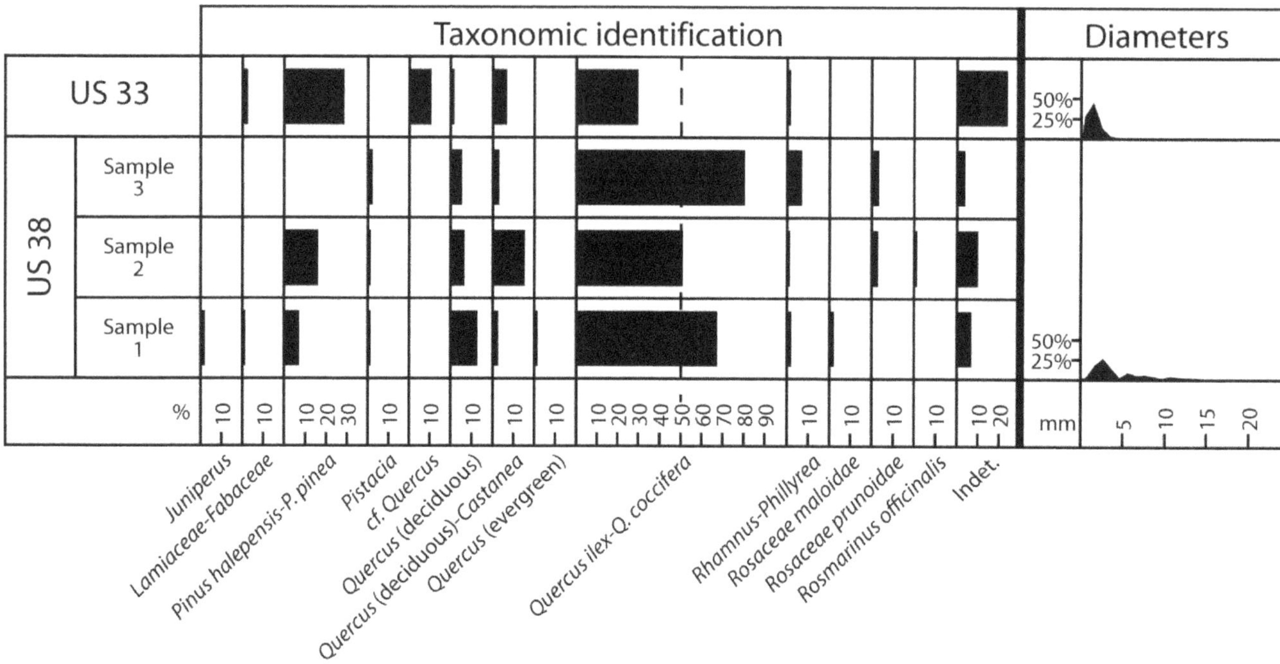


Figure 5. Charcoal analysis diagram (Peynier)

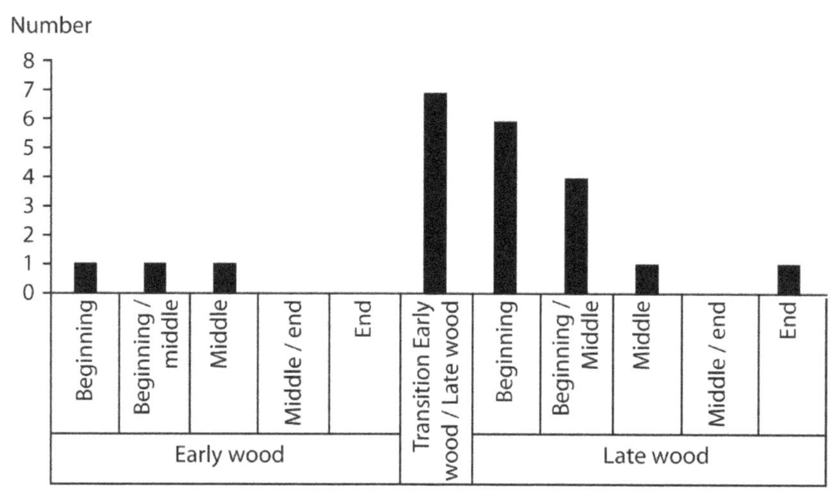


Figure 6. Cutting season identified in complete branches (Peynier)

first sample. Lower frequencies of this taxon are recorded in the second sample (50%) in association with *Quercus* (deciduous)-*Castanea* (c. 15%) and *Pinus halepensis-P. pinea* + *Pinus sp.* (15%). In the third sample, *Quercus ilex-Q. coccifera* reaches 66%, while *Quercus* deciduous and pine attain 12 and 6% respectively. The diameter of the charcoal fragments from these three samples varies between 1 and 14 mm, clearly showing that small twigs and branches were used as fuel wood. 15% of the specimens analysed allowed us to establish the cutting season (Figure 6): transition earlywood / latewood or beginning of latewood. It is therefore possible that the kiln was used during the beginning of the autumn.

At Orgon, taxa frequencies also fluctuate in the different samples studied. The analysis of the charcoal fragments (Figueiral, unpublished report) presents remarkable similarities with that of Peynier (Figure 7). Plant species used are characteristic of the surrounding 'garrigue' environment (evergreen *Quercus*, *Rosmarinus officinalis, Juniperus* and *Fabaceae*), with diameters varying between 2 and 13 mm.

Three phases of « construction » have been identified in the fireplace from Cassis (Figure 8). At the bottom, the first 10 cms comprise fragments of very small calibre (mean = 2 mm); *Quercus ilex-Q. coccifera* constitutes 70 – 80 % of charcoal fragments identified. Twigs and leaves of *Quercus ilex-Q. coccifera*, *Lavandula* and *Rosmarinus* are also abundant (Figure 9). In the upper layers, *Quercus* frequencies are still very important (70 %), but branch calibre increases (mean = 4 – 5 mm); small twigs and leaves are no longer recorded. The topmost level provides very different results. *Pinus halepensis-P. pinea* reaches 72%; Diameters were rarely mesured due to the high fragmentation rate. A branch of

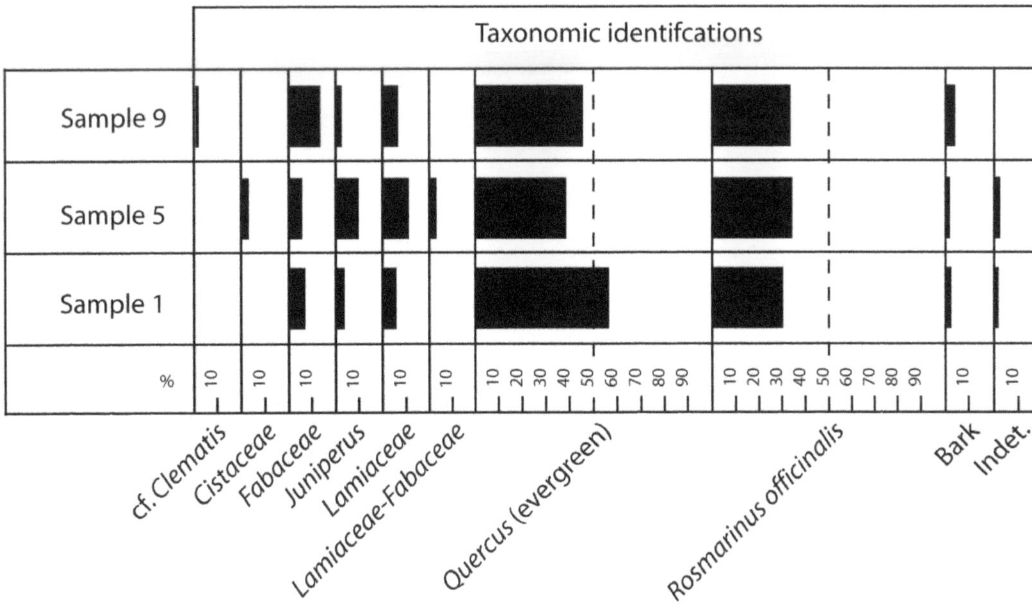


Figure 7. Charcoal analysis diagram (Orgon)

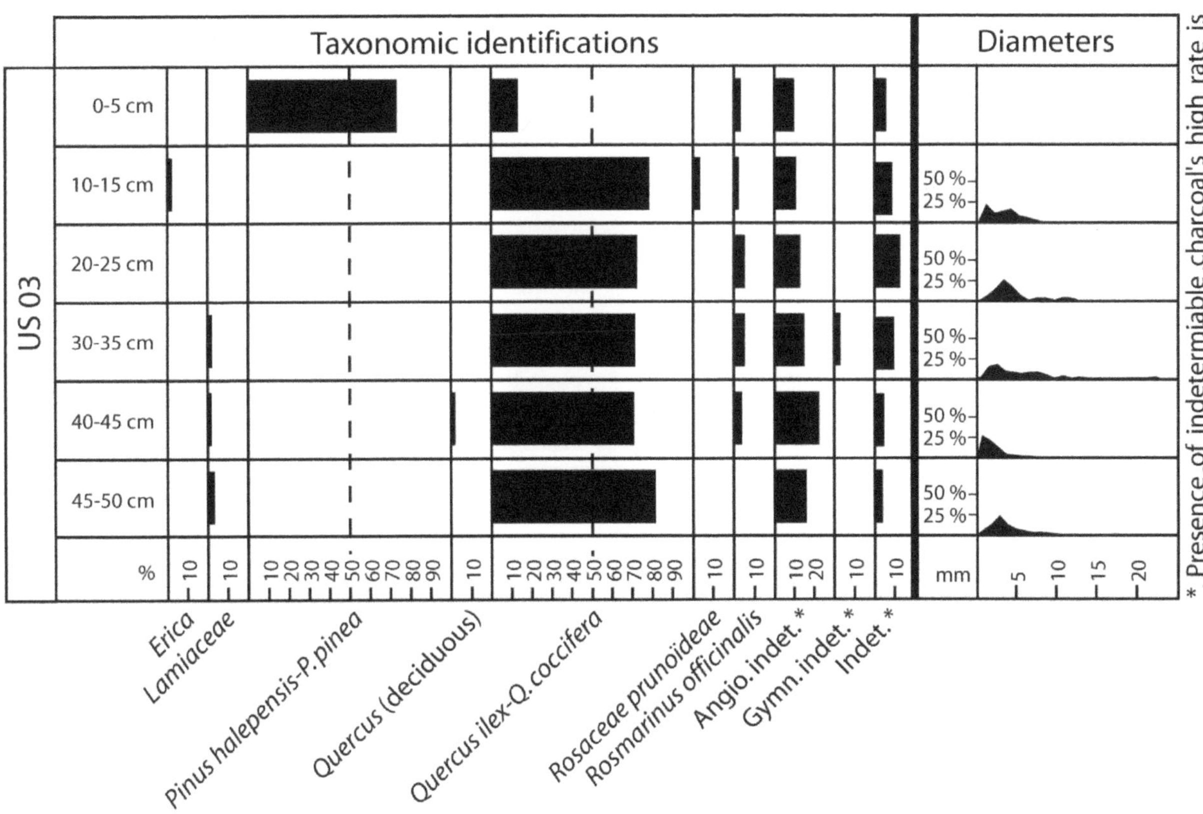


Figure 8. Charcoal analysis diagram (Cassis)

*Pinus halepensis-P. pinea* (diameter = 120 mm) was laid on the top of the charcoal fragments.

## DISCUSSION

### A specific fuel ?

Does this particular lime burning technique require special fuel management? The answer to this question requires a planned sampling strategy such as the one carried out at Le vallon de l'Homme Mort.

Charcoal was exhaustively sampled. The diameter clearly show that small twigs and branches branches were used as fuel wood. The measurements, carried out in 31% of the specimens, allowed the drawing of a calibration curve, which separates three different groups of specimens: those measuring c. 3 mm, c. 7 mm and c. 11 mm. It is difficult to explain the identification of these

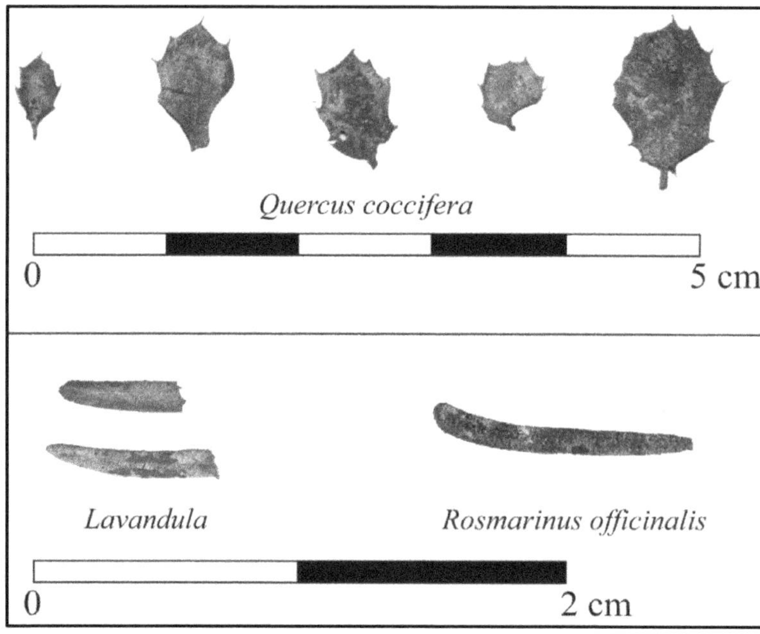


Figure 9. Charred leaves (Cassis)

three groups in the absence of the exact position of the samples in the kiln. It is likely, however, that this reflects only the natural morphology of the specimens thrown into the fire.

The study of this structure shows that lime burning in kilns with high firing chamber (high kiln pot) uses fuel well adapted to this activity / structure design. Twigs and small branches, probably in bundles, produce the flames required to the success of this activity (Fourcroy de Ramecourt 1766, Biston 1836). Furthermore, results obtained clearly show that different samples must be taken in the charcoal layer, as significant variations concerning species identified, their frequencies and calibre, have been recorded.

This new data, concerning the fuel used and its relation with the firing technique, shed new light on information from kilns excavated previously, such as the one from Orgon (Provence), excavated by R. Gaday (2007). This kiln, dating back to the XVIIIth century, was greatly damaged, making it impossible to recognize the existence of a high kiln pot. A lateral accumulation of charcoal identifies the fireplace area. The analysis of the charcoal fragments (Figueiral, unpublished report) presents remarkable similarities with that of Peynier (Figure 6).

The similarity of the charcoal data, the layout of the fireplace and the absence of a high stokehole suggest that the kiln from Orgon, also was a limestone kiln with high pot, despite the absence of structural remains.

**Towards a new sampling protocol**

The new sampling protocol has been established while excavating the kiln 'la Marcouline' at Cassis (Provence) aiming at a better understanding of the links between kiln construction, lay out and fire control. This kiln, apparently dating back to the 19th century, also presents a high kiln pot; and as at Peynier, the fireplace was protected by a layer of carbonated lime, preventing post-depositional disturbance. This certainly influenced the outcome of the dig in itself and of the charcoal analysis data. The important numbers of indeterminable specimens result from the reduced dimensions of the charcoal fragments.

This careful excavation and sampling was followed by micro-topography of the kiln, in order to identify the morphology / pattern of the fireplace. The precise location of the material studied was recorded. Three phases were recognized in its construction (Figure 8):

- The first 10 cm may include the remains of a taper, composed of twigs and leaves. When loading was complete, the high stokehole made it difficult to start the fire at the bottom. Kindling would have been made easier by the use of the taper. The fire spread upwards and the charge would slightly collapse over the taper, stopping air circulation and preserving the leaves.
- The intermediate layers (mean calibre = 4-5 cm), without leaves, correspond to the « height » of the process. Small calibre branches (majority *Quercus coccifera*) are added.
- At the top, changes in fuel used (*Pinus halepensis-P. pinea*) and in fragmentation, along with the appearance of ash testify to the end of the burning process, and the cooling of the lime. The presence of a large branch of *Pinus halepensis-P. pinea* (diameter = 120 mm) had no consequences in terms of fire efficiency..

## CONCLUSION

A survey of charcoal analysis work carried out in archaeological lime kilns emphasized the need of a sampling strategy applied to this type of structure. It is clear that data provided by a single sample only offer a partial image of the running of the fireplace. The new sampling protocol proposed allows us to recognize the different construction procedures and to understand how the fireplace was stoked. However, the success of this protocol depends on the good preservation of the fireplace, which only happens when this is sealed by a layer of carbonated lime.

Remains of a wick were found in the kiln from Cassis. This testifies to the importance of a thorough, exhaustive sampling and highlights how limited our knowledge is, concerning this activity. The presence of this wick provides evidence of an unusual way of starting the fire. At Peynier, the fire was located two meters under the kiln pot, and thus totally inaccessible.

In the long term, added improvements to our sampling strategy will hopefully provide precise three dimensional images of the fireplaces.

**Acknowledgements**

The authors thank M.P. Ruas for the identification of *Rosmarinus* and *Lavandula* leaves and P. Roiron for the identifications of *Quercus coccifera* leaves.

**References**

ADAM, J.-P. and VARÈNE, P. 1985. Fours à chaux artisanaux dans le bassin méditerranéen. *Histoire des techniques et sources documentaires, méthodes d'approche et expérimentation en région méditerranéenne, Cahier n°7, Actes du colloque du GIS, Aix-en-provence, 21-23 octobre 1982*. Aix-en-Provence, 87-100.

ARDISSON, S. and EXCOFFON, P. 2006. Nice, Cimiez. Zone 02, fouille du secteur 3, le four à chaux. *Bilan scientifique régional PACA 2005*, 88-89.

BALLARIN, C., CAILLAT, P., GUÉDON, Fr. and POIRIER, Ph. 2007. L'habitat de Naux à Colayrac-Saint-Cirq (Lot-et-Garonne): une maison forte de bourgeois? *Archéologie médiévale* 37, 111-145.

BENOÎT, F. 1977. *Cimiez, la ville antique (monuments, histoire)*. Paris, E. de Boccar.

BIRINGUCCIO, V. 1572. *La pyrotechnie, ou art du feu contenant dix livres ausquels est amplement traicté et diversité des minières, fusions et séparations des métaux, des formes et moules pour jetter artilleries, cloches et toutes autres figures; distillations, des mines...* Paris, C. Frémy.

BISTON, V. 1836. *Manuel théorique et pratique du chaufournier, contenant l'art de calciner la pierre à chaux et à plâtre; de composer toutes sortes de mortiers ordinaires et hydrauliques, cimens, pouzzolanes artificielles, bétons, mastics; briques crues, pierres et stucs ou marbres factices propres aux constructions*. Paris, Léonce Laget (réédition de 1981).

BLIN, O., BERNARD, V., COUTURIER, D., DIETRICH, A., GEBHARDT, A., MARGUERIE, D., MORIN, J.-M., LEROYER, C., ALLENET de RIBEMONT, G., MATTERNE, V., SEYRIESSOL, K. and ZWIERZINSKI, E. 1999. Impact anthropique et gestion du milieu durant l'Antiquité. Un exemple d'approche paléo-environnementale pluridisciplinaire sur le site d'agglomération secondaire antique de « La Ferme d'Ithe » à Jouars-Ponchartrain (Yvelines, France). *Les Nouvelles de l'Archéologie* 78, Paris, 45-56.

BROCHIER, J.-E. and LIVACHE, M. 2003. *Abri de la Fanfarline (Orgon, Bouches-du-Rhône). Campagne de fouille d'août-septembre 2003*. unpublished report.

BROCHIER, J.-E. and LIVACHE, M. 2004. Orgon, abri de la Fanfarline. *Bilan scientifique régional PACA 2003*. p. 148.

CHAUSSERIE-LAPRÉE, J. 2000. La romanisation. *Martigues, le temps retrouvé* (Biaggi, V., dir.). Equinoxe, 33-36.

CHEHAIBI, Z., HASSINE, Kh. and VASCHALDE, Chr. 2008. *Les fours à chaux de Tunisie. Rapport préliminaire d'enquête ethnoarchéologique (29 août-7 septembre 2008)*. unpublished.

COULTHARD, N. 1999. Les activités artisanales gallo-romaines à Touffréville (Calvados, France), et quelques réflexions sur leur importance dans le développement du site. *Artisanat et productions artisanales dans les provinces du nord-ouest de l'Empire Romain (Polfer M. dir.), Actes du colloque d'Erpeldange, mars 1999. Monographies Instrumentum* 9, 165-183.

CYWA, K. and TYSZLER, L. 2008. The use of wood by the population of the przeworsk and lusitanian cultures based on charcoal analysis form Konopnica (Central Poland). *Charcoal and microcharcoal. Continental and marine records. 4th International Meeting of Anthracology, Brussels, 8-13 september 2008 (Programme and abstracts)* (Damblon, Fr. and Court-Picon M. ed.), Royal Belgian Institute of Natural Sciences, Brussels, 54.

DEVILLERS, B. 1999. *Les formations sédimentaires et les indicateurs climato-anthropiques dans l'anse des Laurons*. Mémoire de maîtrise de Géographie dactylographié, Université de Provence.

FABRE, L. 2001. Approvisionnement en combustible de trois fours artisanaux de la première moitié du XVII$^{e}$ siècle à Montpellier (site archéologique de Pôle Echange Corum). *Montpellier (34). Tramway: Pôle d'Echanges-Corum. Fenêtre sur le Faubourg-du-Pila-Saint-Gély (XIV$^{e}$ – XX$^{e}$ siècles), faïencerie Favier* (Ginouvez (O.) dir.). D.F.S. de fouille préventive (13-24 décembre 1999 / 13 mars-30 avril 2000), AFAN-LAMM, SRA, DRAC Languedoc-Roussillon.

FANTAR, M.H. 1984. *Kerkouane, cité punique du Cap Bon (Tunisie)*. Tunis, Institut National d'Archéologie et d'Art.

FIGUEIRAL, I. 2008. *Orgon-le-Défends (Vaucluse). Etude anthracologique des fours à chaux et fosses associées.* Rapport d'analyse anthracologique, inédit.

FLACH, G.L. 1981. Les fours à chaux à l'époque gallo-romaine. *Revue archéologique, Sites* 12, 18-21.

FOURCROY de RAMECOURT, Ch.-R. 1766. *Art du chaufournier. Descriptions des arts et métiers faites et approuvées par MM. de l'Académie des sciences.* Paris.

GADAY, R. 2007. Orgon, le Défend. *Bulletin scientifique régional SRA PACA 2006*. p. 159.

GERBER, C., PORTMANN, M. and KÜNDIG, C. 2002. *Fours à chaux, four à fer et charbonnières dans le Jura bernois. Vestiges archéologiques médiévaux et modernes découverts entre Moutier et Roches sur le tracé de l'autoroute A16, 1995-1997.* Bern, Dok Verlag.

GIMPEL, J. 1975. *La révolution industrielle du Moyen Age.* Paris, Le Seuil.

GINOUVEZ, O. 2001. *Montpellier (34). Tramway: Pôle d'Echanges-Corum. Fenêtre sur le Faubourg-du-Pila-Saint-Gély (XIV^e^ – XX^e^ siècles), faïencerie Favier.* D.F.S. de fouille préventive (13-24 décembre 1999 / 13 mars-30 avril 2000), AFAN-LAMM, SRA, DRAC Languedoc-Roussillon.

HERBER, J. 1923. Un four à chaux à Moulay-Idris (Zerhoun). *Hespéris. Archives berbères et bulletin de l'Institut des hautes-études marocaines* 3, 421-424.

JACKSON, D.A., BIEK, L., and DIX, Br. 1973. A Roman Lime Kiln at Weekley, Northants. *Britannia* 4, 128-140.

MAGNIER, M.-D. 1881. *Nouveau manuel complet du chaufournier, du plâtrier, du carrier et du bitumier, traitant de l'exploitation des carrières, la fabrication de la chaux et du ciment, les propriétés des mortiers et des bétons, la fabrication du plâtre, des bitumes et des asphaltes.* Librairie encyclopédique de Roret, Paris.

MARGUERIE, D. 2002. Fuel from protohistorical and historical kilns in north-western France. *Charcoal Analysis. Methodological Approaches, Paleological Results and Wood Uses. Proceedings of the Second International Meeting of Anthracology, Paris, September 2000.* British Archaeological Reports, International Series 1063, 187-191.

SAGUI, L. 1986. Crypta Balbi (Roma): lo scavo nell'esedra del monumento romano. Seconda relazione preliminare. *Archeologia medievale* 13, 345-354.

SOLARI, M.-E. 1990. *Etude anthracologique des charbons de bois du site de Brétinoust.* unpublished report.

SÖLTER, W. 1970. *Römische kalkbrenner im Rheinland.* Rheinland-Verlag GMBH, Düsseldorf.

SUMÉRA, Fr. and VEYRAT, E. 1997. Les fours à chaux gallo-romains de « Brétinoust », commune de Sivry-Courtry (Seine-et-Marne). *Revue archéologique du centre de la France* 36, 99-130.

THIRIOT, J. 2006. Les fours à chaux à deux foyers superposés du XIII^e^ siècle de Saint-Blaise-de-Bauzon à Bollène (Vaucluse): une autre façon de cuire la pierre. *Archéologie du Midi Médiéval* 23-24, 247-263.

VASCHALDE, Chr. 2008a. Peynier, four à chaux du vallon de l'Homme Mort. *Bilan scientifique régional PACA 2007.* 157-159.

VASCHALDE, Chr. 2008b. *Le four à chaux PEY4 A1. Peynier, quartier de Bel Homme, Bouches-du-Rhône. Rapport de sondage programmé.* unpublished.

www.ingramcontent.com/pod-product-compliance
Ingram Content Group UK Ltd.
Pitfield, Milton Keynes, MK11 3LW, UK
UKHW061846041026
15591UKWH00015B/934

9 781407 311005